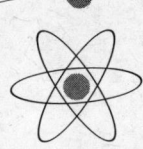

Elements of
Nuclear Theory

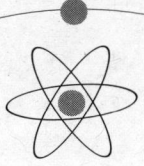

Elements of
Nuclear
Theory

SN Mukherjee PhD

Professor
Department of Physics
Banaras Hindu University
Varanasi, UP
India

CBS Publishers & Distributors Pvt Ltd

New Delhi • Bengaluru • Pune • Kochi • Chennai

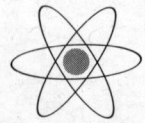

Elements of Nuclear Theory

ISBN: 978-81-239-1895-2

First Edition: 2010

Published by Satish Kumar Jain and produced by Vinod K. Jain for
CBS Publishers & Distributors Pvt Ltd
4819/XI Prahlad Street, 24 Ansari Road, Daryaganj
New Delhi 110 002, India.
Ph: 23289259, 23266861/67   Fax: +91-011-23243014

Website: www.cbspd.com
e-mail: delhi@cbspd.com;
cbspubs@vsnl.com;
cbspubs@airtelmail.in.

Branches

- Bengaluru: Seema House 2975, 17th Cross, K.R. Road, Banasankari 2nd Stage, Bengaluru 560 070, Karnataka
  Ph: +91-80-26771678/79   Fax: +91-80-26771680   e-mail: bangalore@cbspd.com

- Pune: Shaan Brahmha Complex, 631/632 Basement, Appa Balwant Chowk, Budhwar Peth, next to Ratan Talkies, Pune 411 002, Maharashtra
  Ph: +91-20-24464057/58   Fax: +91-20-24464059   e-mail: pune@cbspd.com

- Kochi: 36/14 Kalluvilakam, Lissie Hospital Road, Kochi 682 018, Kerala
  Ph: +91-484-4059061-65   Fax: +91-484-4059065   e-mail: cochin@cbspd.com

- Chennai: 20, West Park Road, Shenoy Nagar, Chennai 600 030, Tamilnadu
  Ph: +91-44-26260666, 26208620 Fax: +91-44-45530020   email: chennai@cbspd.com

Printed at India Binding House, Noida, (UP)

to
_____

the memory of
my uncle
Tarapada Bhattacharya

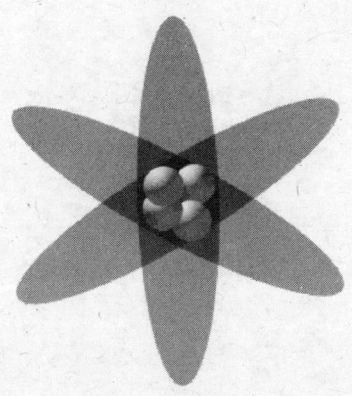

Preface

This textbook is written for honors and postgraduate students opting for a two-semester course in nuclear physics. The motivation for writing this book at an introductory level, is to acquaint the students with the basic concepts related to the developments in nuclear physics.

Over the years, recognition of the basic complexity of nuclear force, the evolution of nuclear theory in explaining data emerging from the new generation accelerators that are coupled with the high resolution detector facilities, which with the opportunities of using ideas taken from particle physics, provided the impetus for restructuring the course contents. Large-scale experiments and large international collaborations are quickly changing the face of nuclear physics. Some attempts are made in this book to give a snapshot view of the "new nuclear physics" emerging from the above changes.

The book is organized in three parts. First part describes the fundamentals and nucleon-nucleon interaction. Second part deals with nuclear structure theory covering liquid drop description, single particle and collective motions in nuclei. Third part presents nuclear reaction dynamics for light and heavy ion induced reactions at low, intermediate and relativistic energies. Finally, the last chapter gives some theoretical ideas which help to understand the weak interaction and neutrino physics. In the last two chapters we try to view the nucleus as a meeting ground between the nuclear physics, particle physics and astrophysics.

Topics which are relevant to nuclear physics are only discussed here. Simple theoretical descriptions with step-by-step derivation are given to understand the physics of nucleus. Experimental values are often quoted to ensure contact with reality. This book incorporates solved exercises as an integral part of each chapter. A list of some important references is given at the end of each chapter. Apologies are offered to the many authors of important papers whose works have not been quoted.

This book has been much improved by the comments of many physicists who have been kind enough to read the early drafts of some chapters and suggest corrections and improvements. In this context, I am particularly grateful to Professors George Rawitscher, E Henley, Manoj Pal and Peter Hodgson. I thank Professors TV Ramakrishnan, Kimshuk K Sinha and Dr VJ Menon for their generous help. I also thank Department of Science and Technology, Government of India, New Delhi, for financial support under its user's scheme, particularly Dr JBV Reddy.

I express my heartfelt thanks to Mr YN Arjuna of CBSP&D for his interest in the publication of this book, and to Mr Vidya Shankar Singh for typing and composing the manuscript. Finally my gratitude to Mr Jhilam Sadhukhan of Variable Energy Cyclotron Centre, Kolkata, for his help, and to my wife Reba for incredible support in my pursuit of all things good.

SN Mukherjee

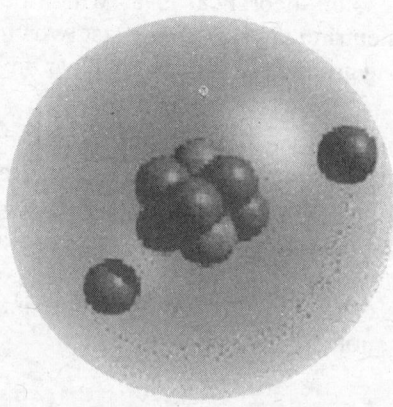

Contents

1

Understanding Physics of Nuclei

"Don't let yourself be talked into believing that the nucleus is not interesting.
It is so small and it has so few parts and still it shows a tremendous variety of phenomena. Its investigation requires the whole arsenal of presently available experimental techniques.
Its understanding makes use of almost all branches of theoretical physics. What a marvelous invention!
It is worth devoting a lifetime to it."

–V.F. Weisskopf[1]

1.1 HISTORICAL CHRONOLOGY

The discovery of nucleus by Rutherford[2] in 1911 through his celebrated alpha scattering experiment marks the beginning of nuclear physics. The first crude value for the nuclear radius obtained from his classical model for alpha scattering on gold was remarkably close to the value that we consider correct now a days. The departure from his famous $\mathrm{cosec}^4\theta/2$ law in the case of alpha scattering on helium was understood much later with the realization by Mott[3] that the identity of particles had quantum mechanical consequence for scattering experiment. Mott introduced the interference between two indistinguishable Coulomb amplitudes of colliding nuclei and this led to the famous Mott scattering formula.

The nuclear theory began humming after a long time with the discovery of neutron by Chadwick[4] in 1932. Soon after the discovery of neutron, Heisenberg[5] proposed a model for a nucleus in terms of neutron and proton constituents. He introduced an additional quantum number, the 'isotopic spin', to enable one to regard the neutron and proton as two different charge states

1

of one and the same particle – the nucleon. Soon it was found that all nuclei throughout the periodic table happen to have masses which are near multiple of the nucleon mass. Therefore, the description of nucleus in terms of the nucleon-degrees of freedom emerged. However, the understanding of the obscure nature of nuclear force needed imagination and physical intuition.

Before the discovery of neutron, the construction of theory of nucleus was beset with difficulties. Already quantum mechanics as a tool for understanding the atomic phenomena was developed. Its initial impetus came from Planck's quantum hypothesis for an explanation of the black-body radiation, Bohr's atomic theory for the description of hydrogen spectrum, De-Broglie's idea of matter waves and Heisenberg's uncertainty principle protecting the wave particle duality. The culmination of the theoretical framework to combine these concepts began with the introduction of the wave equation by Schrodinger:

> *"Erwin with his psi can do*
> *Calculations quite a few*
> *But one thing has not been seen*
> *Just what does psi really mean".*

- Erwin Schrodinger[6]

The validity of quantum mechanics for nuclear dynamics was doubted in the beginning. Though, as early as 1928, Gamow[7] made an important theoretical approach to explain alpha decay and its success provided a proof of quantum mechanics, yet it was the Fermi theory of beta decay[8] that ensured usefulness of quantum mechanics for developing nuclear theory. The idea of creating a particle (neutrino) first came from the beta decay theory and so it was the first example of the modern field theory.

Dirac[9], in an attempt to unify the theory of relativity and quantum theory made the formulation of the relativistic wave equation, which became an appropriate equation to account for properties and behavior of spin half particles, for example, the negatons. While the spins of neutron and proton gave indication that nucleons may obey the Dirac equation, the measured values of their magnetic moments (μ) showed that they could not be such simple particles.

We see from Table 1.1 that measured values of μ differ substantially from those predicted by the Dirac equation for spin half particles. This anomaly indicates that the nucleons are composite particles and the additional contributions to the magnetic moments are attributed to the 'meson clouds'.

The second evidence that protons can not be treated as point particles came from Hofstadter's pioneering experiment in 1957 on the scattering of high energy electrons by hydrogen. He found that the measured cross-section differs from the calculated cross-section for a point charge and he extracted the mean square radius of the charge and matter distributions of proton from his measurements[10].

Table 1.1: Nucleonic Observables (nm = nuclear magneton)

	Spin $(\hbar)$	μ_{Dirac} *(nm)*	$\mu_{expt.}$ *(nm)*	$<r^2>_{charge}$ *(fm²)*	*Mass* *(MeV/c²)*
Proton	$\frac{1}{2}$	1	+ 2.79	0.70	938.3
Neutron	$\frac{1}{2}$	0	−1.91	− 0.11	939.6

A. Symmetry Principles and Conservation Laws

Along with our changing conception of fundamental particle structure, our understanding of the fundamental interactions between particles evolved. Certain symmetry principles revealed the properties of the basic forces of nature.

An outcome of the Dirac equation is the complete symmetry between positive and negative electric charges. This associates every particle with an antiparticle. Therefore, nucleons obeying the Dirac equation must have antinucleons. After a hunt, lasting many years, the antiproton and antineutron were observed by Chamberlin et al.[11]. The complete symmetry between positive and negative charges is called the invariance under charge conjugation. The charge conjugation denoted by 'C', is one of the three reflection symmetries, which holds for the strong nuclear interaction. Charge conjugation interchanges particles and antiparticles and it is a non-space-time discrete operation. The other two discrete space-time symmetries are:

1. Invariance under time reversal.
2. Invariance under coordinate inversion.

Formally, the last symmetry is achieved by making equation of motion invariant under coordinate reflection. This is called the "Parity Operation". The parity operator denoted by P, sends $(t, \vec{r}) \rightarrow (t, -\vec{r})$ through the origin reversing the handedness of space. P is a unitary operator by definition:

$$P \, \vec{r} \, P^{-1} = -r \text{ and } P \, \vec{p} \, P^{-1} = -\vec{p} \,.$$

If the Hamiltonian H is invariant under space reflection, $[P,H] = 0$; the nuclear wave function can be chosen to be eigenfunctions of P and in the coordinate representation,

$$P \, \psi(r) = \psi(-r).$$

Operating the above equation again by P, we get

$$P^2\psi(r) = P\psi(-r) = \psi(r) \tag{1.1}$$

Therefore, the eigenvalue problem $P\psi(r) = \pi\,\psi(r)$ has a solution with $\pi = \pm 1$. In other words, if $\psi(-r) = +\psi(r)$, the wave function is said to have even parity and if $\psi(-r) = -\psi(r)$ then the wave function has odd parity. Parity is the property of non-relativistic quantum mechanics.

We categorize dynamical variables or operators as follows:

(a) Vectors such as $\vec{r}, \vec{v}, \dot{\vec{r}}, \vec{p}$ (and also $\vec{\nabla}_r$) and $\vec{E}$ (the electric field intensity), that are odd under space inversion e.g., $P\,\vec{\nabla}_r P^{-1} = -\vec{\nabla}_r$ etc., are called "polar vectors" operators.

(b) Vectors, such as $\vec{L} = \vec{r} \times \vec{p},\ \vec{S},\ \vec{H} = \vec{\nabla} \times \vec{A}$ (magnetic field), that are even under space inversion e.g., $P\,L\,P^{-1} = P\,\vec{r} \times \vec{p}\,P^{-1} = -\vec{r} \times (-\vec{p}) = +(\vec{r} \times \vec{p})$, are called axial vectors or pseudovector-operators.

(c) Scalars are the dot product of either two polar vectors or axial vectors.

(d) Pseudoscalars are dot products of one polar vector and one axial vector, and therefore odd under space inversion e.g., $\vec{r} \cdot \{\vec{r} \times \vec{p}\}, \vec{p} \cdot \vec{L}$ etc.

The predictions based on the conservation of parity have been verified to an accuracy of at least $1{:}10^{-8}$ in nuclear (strong) and electromagnetic interactions. Therefore, the energy operators in these cases do not contain any parity non-conserving term. These invariance requirements, most of which are connected with the conservation laws, provide a guiding principle in determining the form of nucleon-nucleon (N–N) force.

Time reversal denoted by T sends $(t,r) \to (-t,r)$. Since, classically momentum $(\vec{p} = d\vec{r}/dt)$ and angular momentum $(\vec{L} = \vec{r} \times \vec{p})$ change sign under T, therefore, a physical interpretation of T is motion reversal. The transformation properties of some common variables under the time reversal are:

$$r \to r, \quad p \to -p, \quad L \to -L, \quad \text{spin } \sigma \to -\sigma.$$

The effect of T on $\vec{E}$ or $\vec{H}$ follows from the Maxwell's equation:

$$\vec{E} \xrightarrow{T} \vec{E} \quad \text{and} \quad \vec{H} \xrightarrow{T} -\vec{H}.$$

In the beginning particles were classified by their masses: light particles (leptons), heavy particles (baryons) and particles in between (mesons). Baryons and mesons are subject to strong nuclear interaction and they are also called hadrons. Electrons, muons and neutrinos are leptons and are weakly interacting particles. In order to distinguish between various leptons and baryons, a lepton number L or a baryon number B is assigned to each particle/antiparticle. L and B remain conserved in a particular reaction.

In Table 1.2 we have shown Q, B and L values of some particles. One can see how the additive charges, baryon and lepton numbers, are conserved in particle decays.

Table 1.2: Charge conjugation and conservation of lepton number in some common particle decays

Particles	Q	B	L	Conservation of lepton number
Proton	$+e$	1	0	$n \rightarrow p + e^- + \bar{\nu}_e$
Anti-proton	$-e$	-1	0	$L : 0 \rightarrow 0 \quad 1 \quad -1$
Electron	$-e$	0	1	$\mu^+ \rightarrow e^+ + \nu_e + \nu_\mu$
Positron	$+e$	0	-1	$L : -1 \rightarrow -1 \quad 1 \quad -1$
Neutrino	0	0	1	$\pi^+ \rightarrow \pi^0 + e^+ + \nu_e$
Anti-neutrino	0	0	-1	$L : 0 \rightarrow 0 \quad -1 \quad 1$

The charge conjugation consists of changing a particle to antiparticle in any process while maintaining the dynamical variables, momentum, spin etc. unchanged. If we specify a state by $|A, p, \sigma...\rangle$, where A stands for the additive quantum numbers usually called the number operators such as Q, B, L, etc. and p and σ represent dynamical variables, then the charge conjugation operation C changes the sign of the additive quantum number A and leaves p, σ etc. unchanged. That is,

$$C \, | A, p, \sigma \cdots \rangle \ = \ | \tilde{A}, p, \sigma \cdots \rangle \tag{1.2}$$

We use the notation $\tilde{A}$ to signify change of particle $\rightarrow$ antiparticle.

$$C^2 \, | A, p, \sigma \rangle \ = \ C \, | \tilde{A}, p, \sigma \rangle = 1 \, | A, p, \sigma \rangle.$$

Example: $\quad C \, | \pi^+ \rangle = | \pi^- \rangle \neq \pm | \pi^+ \rangle.$

In the case of neutral pion π^0, the particle transforms into itself under charge conjugation so that, $C \, | \pi^0 \rangle = \pm 1 \, | \pi^0 \rangle$.

Symmetry Breaking Effects: Certainly one of the most exciting discoveries in the mid 1950's is the breakdown of the law of parity conservation. Lee and Yang[12], as an explanation of an anomaly in the K^- meson decay (for details see Ch. 7), suggested that parity is not conserved in the above process and this led to another form of interaction – the weak interaction besides electromagnetic, strong and gravitational interactions.

If a particle is assumed to possess electric dipole moment μ_e due to the center of charge not coinciding with the center of mass and if its magnetic dipole moment is μ_m, then the electromagnetic interaction Hamiltonian

$$H_{\text{int}} = \mu_m \vec{\sigma} \cdot \vec{H} + \mu_e \vec{\sigma} \cdot \vec{E}, \tag{1.3}$$

where σ is the particle spin. Under the time reversal we have

$$\vec{\sigma} \cdot \vec{H} \xrightarrow{\quad T \quad} \vec{\sigma} \cdot \vec{H} \quad \text{and} \quad \vec{\sigma} \cdot \vec{E} \xrightarrow{\quad T \quad} -\vec{\sigma} \cdot \vec{E} \qquad (1.4)$$

and under the parity operation

$$\vec{\sigma} \cdot \vec{H} \xrightarrow{\quad P \quad} \vec{\sigma} \cdot \vec{H} \quad \text{and} \quad \vec{\sigma} \cdot \vec{E} \xrightarrow{\quad P \quad} -\vec{\sigma} \cdot \vec{E} \qquad (1.5)$$

The existence of a particle with an electric-dipole moment would then be evidence for the violation of time reversal invariance and parity conservation in the electromagnetic interaction. The most sensitive tests of P and T violations are measurements of the electric-dipole moment of neutron and the current experimental limit[13] is $\mu_e \sim (-4 \pm 6) \times 10^{-23} e \, cm$.

The CPT theorem states that any Hamiltonian, which is invariant under the proper Lorentz transformation, is also invariant under the combined operation of Charge conjugation, Parity and Time reversal (CPT), whether or not it is invariant under C, P or T separately. The consequences of the CPT theorem are that masses and lifetimes of particle and antiparticle should be the same. No exception to the prediction has been found. Another consequence of the theorem is that if any individual or a pair of symmetries is violated there must be a compensating violation in other operations or operation so that the CPT symmetry is maintained (Table 1.3).

Table 1.3: Properties of the fundamental interactions

Interaction	Coupling constant	Charge conjugation C	Parity P	Time reversal T	CPT
Strong	100α	Yes	Yes	Yes	Yes
Electromagnetic	$\alpha \approx 1/137$	Yes	Yes	Yes	Yes
Weak	4α	No	No	No	Yes

B. Isospin Formalism

Recognizing that certain symmetries apply between interacting particles helped us in understanding forces acting between them. Probing how these symmetries are broken also reveals important details about particle interactions. Isospin was the first internal symmetry postulated for the *N-N* interaction.

The nucleon isospin is denoted by $|\vec{t}| = 1/2$, accordingly in the particle

physics sign convention $t_3 = +\dfrac{1}{2}\left(-\dfrac{1}{2}\right)$ for proton (neutron). $\qquad (1.6)$

The proton and neutron states, thus, can be expressed as $p \equiv \left| \frac{1}{2} \, \frac{1}{2} \right\rangle$ and $n \equiv \left| \frac{1}{2} -\frac{1}{2} \right\rangle$. The charge of the particle is given by

$$q = e\left(t_3 + \frac{1}{2}\right),$$ (1.7)

with the values of third component t_3 given by Eq. (1.6), the proton has a charge e and neutron has no charge. The isospin vector is $\vec{t} = \hat{i}t_1 + \hat{j}t_2 + \hat{k}t_3$, where $\hat{i}$, $\hat{j}$ and $\hat{k}$ are respectively unit vectors along X, Y, Z (or 1, 2, 3) directions.

The total isospin quantum number T for a nucleus having N neutrons and Z protons, is obtained by the vector addition of individual nucleon isospin:

$$\vec{T} = \sum_{i=1}^{A} \vec{t}_i,$$ (1.8)

where A denotes the mass number and T is integral or half integral depending on whether A is even or odd. The total isospin projection quantum number T_3 likewise is obtained by the algebraic addition of the individual nucleon isospin projection quantum number t_3, leading to the expression

$$T_3 = -(N - Z)/2,$$ (1.9)

where $-2T_3$ gives the neutron excess. In view of the vector character of $\vec{T}$, the assignment of the total isospin quantum number T is not so simple. There are A isospin vectors with $t = 1/2$, which add vectorially, thus, the maximum value of T is $A/2$ and minimum value of $|T| = |(N - Z)/2|$, because a vector cannot be smaller than one of its components. Therefore, $(N - Z)/2 \leq T \leq A/2$.

The total charge, $q = e\sum_{i=1}^{A}\left[t_{i3} + \frac{1}{2}\right] = e\left[T_3 + \frac{A}{2}\right] = eZ$. (1.10)

Further,

$$T_+ = T_1 + iT_2 \text{ with } T_+ = \sum_i t_+(i)$$ (1.11)

and $$T_- = T_1 - iT_2 \text{ with } T_- = \sum_i t_-(i)$$ (1.12)

$t_+(t_-)$ changes neutron $\rightarrow$ proton (proton $\rightarrow$ neutron).

Along with space-spin symmetry there is now another fundamental symmetry connected with spin and isospin exchange of identical particles. For two identical fermions the wave function may be written as

$$\psi = \psi_{space}\,\psi_{spin}\,\psi_{isospin},$$ (1.13)

which must be antisymmetric with respect to changes of two nucleons. This requirement is called the 'extended Pauli principle'.

Deuteron is a $L = 0$ bound state and has $S = 1$. Thus, it has $T = 0$ in order that total wave function of two nucleon system is antisymmetric. Hence, deuteron should have no isospin partner. This is also experimentally verified – no bound di-neutron or di-proton exists.

It was also found from the analysis of proton-proton scattering that the nucleon-nucleon force is charge independent. Let us examine the concept of charge independence of the nucleon-nucleon force in the light of the isospin formalism stated above. For two identical fermions (nucleons), with $t = 1/2$, the total isospin T can be 1 or 0. The $T = 1$ triplet is

$$|1,1\rangle = \left|\tfrac{1}{2}, \tfrac{1}{2}\right\rangle \left|\tfrac{1}{2}, \tfrac{1}{2}\right\rangle \equiv |p\rangle |p\rangle, \quad |1,-1\rangle = \left|\tfrac{1}{2}, -\tfrac{1}{2}\right\rangle \left|\tfrac{1}{2}, -\tfrac{1}{2}\right\rangle \equiv |n\rangle |n\rangle,$$

$$|1,0\rangle = \frac{1}{\sqrt{2}}\left[\left|\tfrac{1}{2}, \tfrac{1}{2}\right\rangle \left|\tfrac{1}{2}, \tfrac{1}{2}\right\rangle + \left|\tfrac{1}{2}, -\tfrac{1}{2}\right\rangle \left|\tfrac{1}{2}, \tfrac{1}{2}\right\rangle\right] \equiv \frac{1}{\sqrt{2}}\left[|p\rangle|n\rangle + |n\rangle |p\rangle\right],$$

and the $T = 0$ singlet is

$$|0,0\rangle = \frac{1}{\sqrt{2}}\left[\left|\tfrac{1}{2}, \tfrac{1}{2}\right\rangle\left|\tfrac{1}{2} - \tfrac{1}{2}\right\rangle - \left|\tfrac{1}{2} - \tfrac{1}{2}\right\rangle\left|\tfrac{1}{2}\ \tfrac{1}{2}\right\rangle\right] \equiv \frac{1}{\sqrt{2}}\left[|p\rangle|n\rangle - |n\rangle|p\rangle\right].$$

The factor $1/\sqrt{2}$ is the appropriate Clebsch-Gordon ('C') coefficient (Appendix D2). It is evident that the triplet states are symmetric with respect to interchange of the two nucleons, while the singlet state is antisymmetric. There is no a priori reason why the force between two nucleons should be the same in $T = 1$ states and $T = 0$ state. Indeed, the antisymmetry of the wave function required by the extended Pauli principle, implies that $\Psi_{space} \Psi_{spin}$ is symmetric for $T = 0$ with antisymmetric $\Psi_{isospin}$. Conversely, $\Psi_{space} \Psi_{spin}$ is antisymmetric for $T = 1$ with symmetric $\Psi_{isospin}$. Because of the different space-spin wave functions one would expect that the force between two nucleons in the $T = 1$ state would differ from that in $T = 0$ state. Charge independence of the nucleon-nucleon force then applies only to $T = 1$ state and it states that the pp, np and nn force depends on the total isospin T and not on T_3.

If the energy level diagrams of several nuclei of the same A but different (N, Z) are compared, some marked similarities become apparent. Let us consider a nucleus formed from A nucleons, Z protons and N neutrons. T is a good quantum number if the electromagnetic interaction is switched off and neutron-proton mass difference neglected. The electromagnetic force produces two effects. A repulsion between the protons in the nucleus and a mass difference between neutrons and protons. The energy difference between members of an isospin multiplet in isobars $(A, Z + 1)$ and (A, Z) is

$$\Delta E = E(A, Z+1) - E(A, Z) = \Delta E_{coul} - (m_n - m_p) c^2, \tag{1.14}$$

It is observed that energy spectra of mirror nuclei obtained experimentally are very much alike. Mirror nuclei are nuclei which transform into each other by an interchange of protons and neutrons. An example is given in Fig. 1.1 for odd mass mirror pair $Mg_{13} - Al_{12}$. The differences of excitation energies are very small. We see from the level scheme that apart from a Coulomb contribution and n-p mass difference a substitution of all protons by neutrons and vice versa does not affect the binding energies and level schemes. It can be concluded that the n-n force is the same as the p-p force. This property of nuclear forces is called the charge symmetry.

Isobaric Analog State (IAS) is a highly excited state of a nucleus with the same mass number A but with one higher atomic number Z, i.e. a state with same A, same T but with T_3 increased by 1.

Example: For $^{81}_{35}Br_{46}$, $|T, T_3\rangle \equiv \left|\frac{11}{2}, -\frac{11}{2}\right\rangle$, the corresponding IAS in $^{81}_{36}Kr_{45}$ is

$|T, T_3\rangle \equiv \left|\frac{11}{2}, -\frac{9}{2}\right\rangle$. Also $I^{\pi}(^{81}Kr(IAS)) = I^{\pi}(^{81}Br)$.

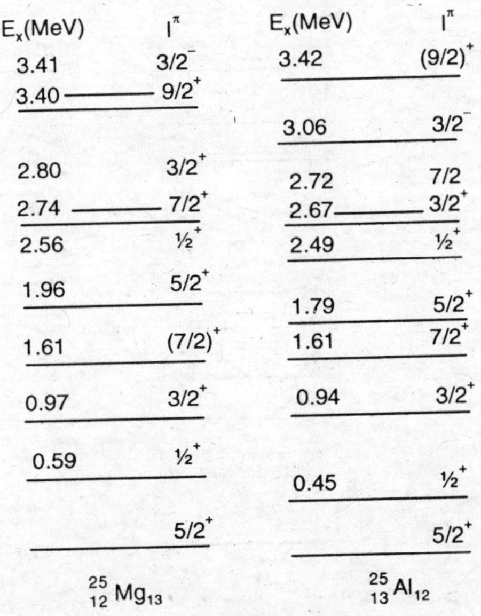

E_x(MeV)	I^{π}	E_x(MeV)	I^{π}
3.41	$3/2^-$	3.42	$(9/2)^+$
3.40	$9/2^+$		
		3.06	$3/2^-$
2.80	$3/2^+$	2.72	$7/2$
2.74	$7/2^+$	2.67	$3/2^+$
2.56	$1/2^+$	2.49	$1/2^+$
1.96	$5/2^+$	1.79	$5/2^+$
1.61	$(7/2)^+$	1.61	$7/2^+$
0.97	$3/2^+$	0.94	$3/2^+$
0.59	$1/2^+$	0.45	$1/2^+$
	$5/2^+$		$5/2^+$
$^{25}_{12}Mg_{13}$		$^{25}_{13}Al_{12}$	

Fig. 1.1: Similarity between the level spacings of $A = 25$ mirror nuclei (Isospin doublet, $T = 1/2$, $T_3 = \pm 1/2$)

Next let us consider isobaric mass triplet such as given in Fig. 1.2 for $^{30}_{14}Si_{16}$, $^{30}_{15}P_{15}$ and $^{30}_{16}S_{14}$. From the similarity of the ^{30}Si and ^{30}S spectra only charge symmetry follows as was the case for $^{25}_{12}Mg_{13} - ^{25}_{13}Al_{12}$. From a comparison of the spectrum of ^{30}P with those of ^{30}Si and ^{30}S another property of the nuclear forces emerges. Consider $I^{\pi} = 0^+$ ground state of ^{30}Si and ^{30}S.

A $I^\pi = 0^+$ level occurs also in ^{30}P as the first excited state. The absolute binding energies of these corresponding states are identical once they are corrected for the Coulomb energies and *n-p* mass difference. The analog of the first $I^\pi = 2^+$ state in ^{30}Si and ^{30}S is also observed in $^{30}P_0$ at an energy above the $I^\pi = 0^+$ first excited state that agrees well with the excitation energy of the lowest $I^\pi = 2^+$ state in ^{30}Si and ^{30}S. Also for higher excited states in ^{30}Si and ^{30}S, corresponding states can be found in ^{30}P. The states in ^{30}P, corresponding to ^{30}Si and ^{30}S states are called isobaric analogue states. One may conclude that when one neutron in $^{30}_{14}Si_{16}$ is replaced by a proton to give $^{30}_{15}P_{15}$, the total interaction energy of the many nucleon system is not changed, apart from Coulomb energy. In ^{30}Si this single neutron interacts with all other nucleons by *n-n* and *n-p* forces, while in ^{30}P these interactions are replaced by *p-n* and *p-p* forces, respectively. This demonstrates the charge independence of nuclear forces, i.e. equality of *n-n*, *n-p* and *p-p* forces.

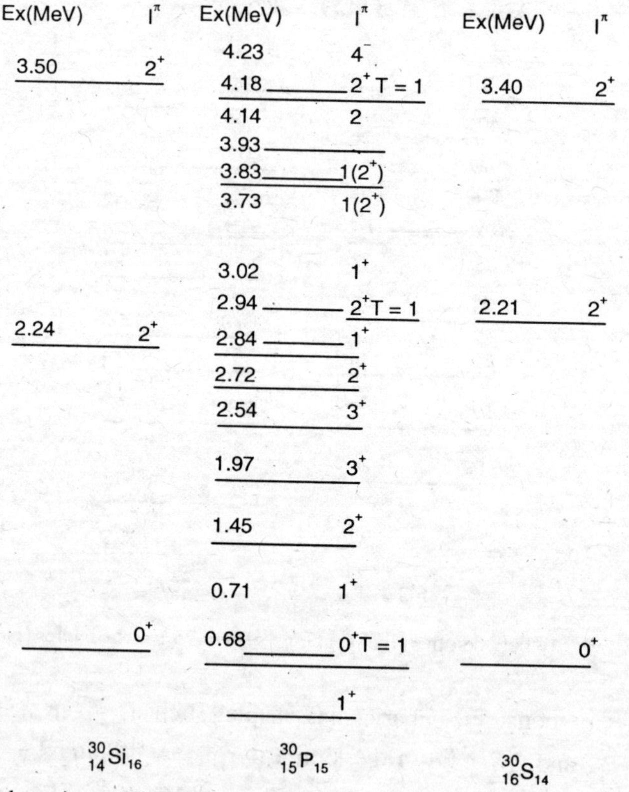

Ex(MeV)	I^π	Ex(MeV)	I^π	Ex(MeV)	I^π
		4.23	4^-		
3.50	2^+	4.18	2^+ T = 1	3.40	2^+
		4.14	2		
		3.93			
		3.83	$1(2^+)$		
		3.73	$1(2^+)$		
		3.02	1^+		
		2.94	2^+ T = 1	2.21	2^+
2.24	2^+	2.84	1^+		
		2.72	2^+		
		2.54	3^+		
		1.97	3^+		
		1.45	2^+		
		0.71	1^+		
	0^+	0.68	0^+ T = 1		0^+
			1^+		
$^{30}_{14}Si_{16}$		$^{30}_{15}P_{15}$		$^{30}_{16}S_{14}$	

Fig. 1.2: Level spacings of isospin triplets ($T = 1$, $T_3 = \pm 1$, 0) for $A = 30$ isobars. For $^{30}_{15}P_{15}$ levels, below about 2.84 MeV excitation all belong to the $T = 0$ configuration while those above belong to $T = 1$ configuration

It is also seen in Fig. 1.2 that ^{30}P has also levels not found in ^{30}Si and ^{30}S. This feature can be qualitatively understood as follows: A neutron and proton can occupy the same state without violating the Pauli principle. This is not possible for identical nucleons when both nucleons are protons or neutrons. Nucleons in $^{30}_{14}Si_{16}$ and $^{30}_{16}S_{14}$ each one has only 14 pairs together with two identical nucleons. Thus, last pair in $^{30}_{15}P_{15}$ can produce states not allowed by the Pauli principle in ^{30}S and ^{30}Si.

"Charge symmetry" is a less restrictive form of isospin invariance. It requires that the strong force be invariant for 180° rotations about any axis perpendicular to the charge axis (3-axis). The charge symmetry operator and its action on a state with isospin quantum members given by T and T_3 can be written as

$$P_{CS} = e^{-i\pi T_2}$$

and

$$P_{CS}|T, T_3\rangle = (-)^{T+T_3}|T, -T_3\rangle .$$

That is, it effectively transforms the quantum number T_3 *into* $-T_3$. *For two nucleon system this implies that n-n and p-p forces are* identical, but does not relate them to *n-p* force.

In analogy to the Pauli spin matrices $\vec{\sigma}$ and its relationship to spin operator $\vec{S} = (1/2)\vec{\sigma}$, we introduce isospin operator $\vec{t} = (1/2)\vec{\tau}$, where the τ-matrices are isospin version of the Pauli matrices, given by:

$$\tau_1 = \begin{pmatrix} 0 & 1 \\ 1 & 0 \end{pmatrix}, \ \tau_2 = \begin{pmatrix} 0 & -i \\ i & 0 \end{pmatrix}, \ \tau_3 = \begin{pmatrix} 1 & 0 \\ 0 & -1 \end{pmatrix} \qquad (1.15a)$$

τ-matrices act on proton and neutron states represented by

$$|p\rangle = \begin{pmatrix} 1 \\ 0 \end{pmatrix}, \quad |n\rangle = \begin{pmatrix} 0 \\ 1 \end{pmatrix}.$$

Now $\qquad q = e\left[t_3 + \dfrac{1}{2}I\right] = e\begin{pmatrix} 1 & 0 \\ 0 & 0 \end{pmatrix}$ where $I = \begin{pmatrix} 1 & 0 \\ 0 & 1 \end{pmatrix}.$

The operators τ_+ and τ_- which respectively increase or decrease charge of a nucleon by one unit are: $\tau_+ = \tau_1 + i\tau_2$ and $\tau_- = \tau_1 - i\tau_2$.

Hence, $\tau_+|n\rangle = \sqrt{2}|p\rangle$ and $\tau_-|p\rangle = \sqrt{2}|n\rangle$

$$[\tau_i, \tau_j] = 2i \in_{ijk} \tau_k, \qquad (1.15b)$$

where $\in_{ijk}$ (= 1 or −1 for an even or odd number of permutations) are called structure constants.

Exercise 1.1. If $\hat{B}$ is a baryon number operator, which counts the number of baryons, show that $C\hat{B} \neq \hat{B}C$.

Now,

$$\hat{B}|B=1\rangle = +1|B=1\rangle$$

$$C\hat{B}|B=1\rangle = +C|B=1\rangle = 1|B=-1\rangle.$$

On the other hand,

$$\hat{B}C|B=1\rangle = \hat{B}|B=-1\rangle = -1|B=-1\rangle.$$

Hence,

$$C\hat{B} \neq \hat{B}C.$$

Exercise 1.2. Derive a mass formula for nuclei of isospin multiplet.

The mass differences between the members of isospin multiplet of a nucleus (mass number A, atomic number Z, neutron number N and radius r_0) can be derived by the Coulomb energy and neutron-proton mass difference. Therefore,

$$M = \frac{3}{5}\frac{e^2 Z^2}{r_0} + m_n N + m_p Z$$

$$= B\left(\frac{A}{2} + T_3\right)^2 + m_n\left(\frac{A}{2} - T_3\right) + m_p\left(\frac{A}{2} + T_3\right)$$

$$= B\frac{A^2}{4} + ABT_3 + BT_3^2 + (m_p - m_n)T_3 + \frac{m_p + m_n}{2}A$$

$$= a + bT_3 + cT_3^2,$$

where

$$a = \frac{BA^2}{4} + M_0, \quad b = BA + (m_p - m_n), \quad C = B,$$

$$B = \frac{3}{5}\frac{e^2}{r_0} \text{ and } M_0 = \frac{m_p + m_n}{2}A.$$

There are three constants, a, b and c in the formula, so three independent linear equations are needed for their determination. Therefore, we need at least $T = 1$.

Exercise 1.3. Which of the following reactions is possible as strong interaction process:

$$^{10}B + p \rightarrow {}^{10}B^* + p' \tag{I}$$

$$^{10}B + d \to {}^{10}B + d' \qquad\qquad (II)$$

The reaction (I) : $\qquad\qquad {}^{10}B + p \to {}^{10}B^* + p'$

$(T = 0\ T = 1/2) \to (T = 1\ T = 1/2)$ is allowed because T is conserved.

But in the reaction (II) : $\ {}^{10}B + d \to {}^{10}B^* + d'$

$(T = 0\ T = 0) \to (T = 1\ T = 0)$ is disallowed because T is not conserved.

Exercise 1.4. Consider the following reactions: $p + d \to \pi^+ + H^3$ and $p + d$ $\to \pi^0 + He^3$. Show that the ratio of their cross-sections $= 1/2$.

Now H^3 and He^3 form a $T = 1/2$ doublet, $\begin{pmatrix} He^3 \\ H^3 \end{pmatrix}$.

The initial state corresponds to $T = \dfrac{1}{2}, T_3 = \dfrac{1}{2}$. With appropriate 'C' coefficients, the wave function

$$|1/2,\ 1/2\rangle = \sqrt{\frac{1}{3}} |\pi^+ \ H^3\rangle - \sqrt{\frac{2}{3}} \ |\pi^0 \ He^3\rangle .$$

Employing the above wave function, we get

$$\frac{\sigma(p + d \to \pi^+ + H^3)}{\sigma(p + d \to \pi^0 + He^3)} = \left(\sqrt{\frac{1}{3}}\right)^2 \div \left(\sqrt{\frac{2}{3}}\right)^2 = \frac{1}{2} .$$

Exercise 1.5a. Confirm $T = 1$ assignment to pion from the following reactions:

$$n + p \to \pi^0 + d \text{ and } p + p \to \pi^+ + d$$

In the initial state for the reaction, the *np* system is in $T = 1$ only half of the time while the *pp* system is always $T = 1$. Hence, we can expect

$$\frac{\sigma(n + p \to \pi^0 + d)}{\sigma(p + p \to \pi^+ + d)} = \frac{1}{2}, \text{ which is confirmed experimentally.}$$

(b) Show that the reaction $d + d \to {}^4 He + \pi^0$ is forbidden in a strong interaction.

	d	+	d	$\to$	$^4 He$	+	π^0
T	0		0		0		1
T_3	0		0		0		0

Since the isospin is not conserved, the above reaction is forbidden in strong interaction.

Exercise 1.6. Find an isobaric analog state for the ground state of $_{82}^{208}Pb_{126}$. What is the energy difference of these states?

We employ Eq. (1.11) and get $IAS\left(_{83}^{208}Bi_{125}\right) = T_{+}\left|_{82}^{208}Pb_{126}\right\rangle_{gs}$.

In $_{82}^{208}Pb_{126}$, $I^{\pi} = 0^{+}$ state has $T = 22$, $T_3 = -22$ and in $_{83}^{208}Bi_{125}$, $I^{\pi} = 0^{+}$ state has $T = 22$, $T_3 = -21$, which is the isobaric analog state of $_{82}^{208}Pb_{126}$.

The energy difference between two isobaric analog states is (*see* Eq. 1.14)

$$\Delta E = \Delta E_{Coul} - (m_n - m_p)c^2 = \frac{6}{5}\frac{Ze^2}{r_0} - (m_n - m_p)c^2 = 18.65 \text{ MeV}.$$

1.2 FIRST GLANCE AT THE GLOBAL PROPERTIES

The attractive nuclear force between neutrons or between a neutron and a proton gives rise to an attractive (negative) potential energy, whereas the Coulomb repulsion between protons results in a repulsion (positive) potential energy. Since the Coulomb force is very much weaker than the nuclear force, there is a net attractive nuclear potential energy (V). The motion of nucleons inside the nucleus also gives a positive kinetic energy (T). However, for the nucleus to be bound, the total energy E ($=T + V$) is necessarily negative. The magnitude $|E|$ of the attractive total energy E is called the Binding Energy (*BE*), which is related to the mass of the nucleus via mass energy equivalence of the special theory of relativity as follows:

The mass of a nucleus is always less than the masses of the protons and neutrons comprising the nucleus. The energy of the nucleus in the ground state

$$E(N,Z) = \left\{M(A,Z) - Zm_p - (A - Z)m_n\right\}c^2, \tag{1.16a}$$

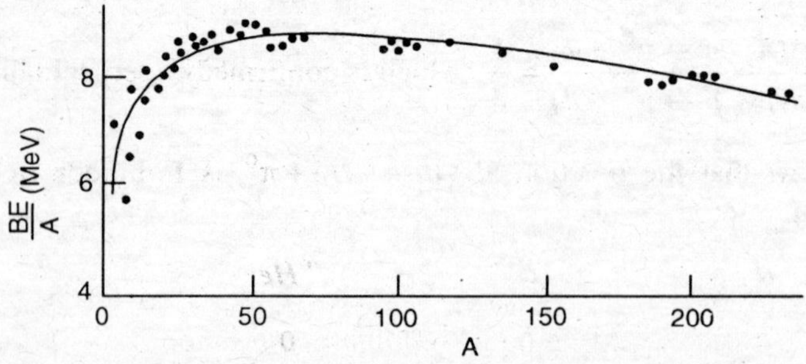

Fig. 1.3: Variation of binding energy per nucleon *BE/A* versus *A*

where A and Z are respectively the mass number and the atomic number of the nucleus, $M(A, Z)$ is the mass of the nucleus and $m_p(m_n)$ the mass of proton (neutron).

The experimental value of $E(N,Z)$ may be obtained from the precise measurement of atomic masses using a mass spectrometer. Therefore, one considers the mass of the whole atom and not just that of the nucleus alone. Accordingly, Eq. (1.16a) is written as

$$E(N,Z) = \left[M(A,Z) - Zm_H + (A-Z)m_n \right] c^2, \qquad (1.16b)$$

where $m_H (= m_p + m_e)$ is the mass of the hydrogen atom, m_e is the electron mass and $M(A,Z)$ is the atomic mass. The binding energy of the hydrogen atom (= 13.6 eV) is considered negligible. Consideration of the atomic mass automatically takes into account the rest mass of the electron created in the decay. Figure 1.3 shows schematically the results of binding energies per nucleon (BE/A) measured for stable nuclei. Apart from the lighter elements BE/A for most stable nuclei is nearly constant, around 7-8 MeV.

A. Shape and Size of Nuclei

Historically, the shape of the nucleus was first investigated by the determination of nuclear quadrupole moment from hyperfine structure of atomic spectra. The hyperfine atomic transitions are produced by the interaction energy of the nuclear charge distribution with the field in the vicinity of the nucleus generated by the atomic electrons. We calculate this interaction energy by a semi-classical treatment.

We assume the nucleus to be situated in a known electrostatic potential Φ (r) which is the result of non-nuclear charges (or the external applied field) and denote the localized charge distribution of the nucleus by $\rho(r)$. Then the electrostatic interaction energy

$$W = \int \rho(r) \Phi(r) d^3r. \qquad (1.17)$$

Of interest is the level shift of initial and final atomic states due to the perturbation of the form W above. Expanding the potential $\Phi(r)$ in Taylor's series around the centre of nuclear charge taken to be the origin, we have

$$\Phi(r) = \Phi(0) + \vec{r} \cdot \vec{\nabla} \, \Phi(0) + \frac{1}{2} \sum_{i,j} x_i x_j \left(\frac{\partial^2 \Phi}{\partial x_i \, \partial x_j} \right)_0 + \cdots$$

$$= \Phi(0) + \vec{r} \cdot \vec{\nabla} \, \Phi(0) + \frac{1}{6} \sum_{i,j} (r^2 \delta_{ij} + q_{ij}) \left(\frac{\partial^2 \Phi}{\partial x_i \, \partial x_j} \right)_0 + \cdots \qquad (1.18)$$

where $q_{ij} = 3x_i x_j - r^2 \delta_{ij}$. In going from Eq. (1.17) to Eq. (1.18), we have

added and subtracted $\dfrac{1}{6} r^2 \delta_{ij} \left(\dfrac{\partial^2 \Phi}{\partial x_i \, \partial x_j} \right)_0$. Substituting $\Phi(r)$ in Eq. (1.17), we

can rewrite (1.17) in the form

$$W = W_a + W_e, \qquad (1.19)$$

where

$$W_a = \frac{1}{6} (\nabla^2 \Phi)_0 \int \rho(r) \, r^2 d^3 r \qquad (a)$$

and

$$W_e = q \, \Phi(0) - \vec{D} \cdot \vec{E}(0) - \frac{1}{6} \sum_{i,j} Q_{ij} \left(\frac{\partial E_j}{\partial x_i} \right)_0 - \cdots \qquad (b)$$

with

$$q = \int \rho(r) d^3 r = Ze, \quad \vec{E} = -\vec{\nabla} \Phi, \quad \vec{D} = \int r \rho(r) d^3 r \qquad (c)$$

and

$$Q_{ij} = \int \rho(r) \left(3x_i x_j - \delta_{ij} r^2 \right) d^3 r$$

The symbols q, D and Q_{ij} stand for the total nuclear charge, electric dipole moment and electric quadrupole tensor respectively. Energy W_a is associated with atomic spectra. If the potential $\Phi(r)$ is due to an external electric field, then according to the Laplace equation $\nabla^2 \Phi = 0$, we have $W = W_e$. The electric quadruple tensor has the following properties: It is symmetric, $Q_{ij} = Q_{ji}$ and its trace (the sum of diagonal elements) vanishes, i.e. $Q_{xx} + Q_{yy} + Q_{zz} = 3(x^2 + y^2 + z^2) - 3r^2 = 0$. It is customary to call the quantity Q_{zz} component as quadrupole moment

$$eQ = \int 3(z^2 - r^2) \, \rho(r) \, d^3 r = \int \sqrt{\frac{16\pi}{5}} \, r^2 \, Y_2^0(\theta) \rho(r) d^3 r. \qquad (1.20)$$

The quantum mechanical expectation values for the operators corresponding to the total electric dipole moment and electric quadrupole moment are obtained by writing the effective charge density

$$\rho_0(r) = e \sum_{k=1}^{Z} P_k(r),$$

where $P_k(r)$ is the probability of the kth proton being at r. $P_k(r)$ can be written in terms of ground state nuclear wave function as

$$P_k(r) = \int \left| \psi_0(r_1 \ldots\ldots r_A) \right|^2 d\tau_k,$$

where $d\tau_k = dr_1 \ldots\ldots dr_A$ excluding dr_k. The integration is carried over the coordinates of all the nucleons except k^{th}. Hence, the electric dipole moment

$$\vec{D} = e \sum_{k=1}^{Z} \int \vec{r}_k \left|\psi(r_1 \cdots\cdots r_A)\right|^2 d\tau, \qquad (1.21a)$$

where $d\tau$ is the volume element for coordinates of all the nucleons including the kth $d\tau = dr_1 \cdots\cdots dr_k \cdots\cdots dr_A$.

Similarly,

$$eQ = \sum_{k=1}^{Z} e\sqrt{\frac{16\pi}{5}} \int r_k^2 \, Y_2^0(\theta_k) \left|\psi_0(r_1 \cdots\cdots r_A)\right|^2 d\tau = \sum_{k=1}^{Z} Q_k, \qquad (1.21b)$$

where $\qquad Q_k = \left\langle \psi_0 \left| \sqrt{\frac{16\pi}{5}} \, r^2 Y_2^0(\theta) \right| \psi_0 \right\rangle_k. \qquad (1.21c)$

The quadrupole moment operator has the form:

$$\hat{Q} = \sum_{k=1}^{Z} \sqrt{\frac{16\pi}{5}} \, r_k^2 Y_2^0(\theta_k),$$

The observed quadrupole moment is the expectation value of $\hat{Q}$ in a state $\psi_{I,M_I=I}$ (angular momentum I has its maximum projection $m_I = I$). The quadrupole moment expressed this way has a unit of cm^2. In general,

$$Q_{\ell m} = \int r^{\ell} Y_{\ell}^{m}(\theta, \phi) \left|\psi(r_1 \cdots\cdots r_A)\right|^2 d\tau. \qquad (1.21d)$$

High Energy Electron Probe: When we want to resolve the molecular structure of a material we may examine it by using an electron microscope. When we use a very high powered electron microscope, i.e. an electron accelerator, we find successively evidence of atom, nucleus, nucleon and ultimately nucleon substructure as the resolving power of the machine is increased.

Let us consider the simple case of elastic scattering of a Dirac electron of energy E on a spin zero target having point charge Ze, and mass M_T. This process can be described quite satisfactorily with Quantum Electro Dynamics (QED) calculations on the assumption that they occur through the mediation of a single virtual photon (Fig. 1.4).

We employ the Born approximation (see Quantum Mechanics by Schiff)[18] to express the elastic scattering amplitude

$$f(\vec{p}, \vec{p}') = -\frac{m}{2\pi} \int V(x) \, e^{i\vec{q}\cdot\vec{x}} \, d^3x \,,$$

where
$$\vec{q} = \vec{p}' - \vec{p} \ , \ q = |\vec{p}' - \vec{p}|$$

and the Coulomb interaction $V(x) = \int \dfrac{\rho(\vec{r}) \, Ze}{|\vec{x} - \vec{r}|} \, d^3r,$

where $\vec{x}$ is the location vector of the electron with the nucleus at the origin.

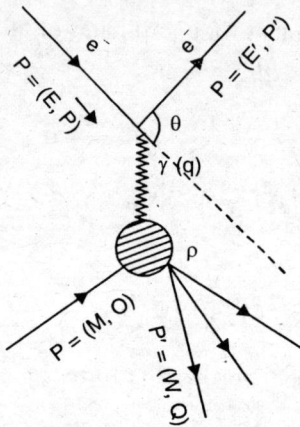

Fig. 1.4: Electron-nucleus elastic scattering in one photon exchange approximation

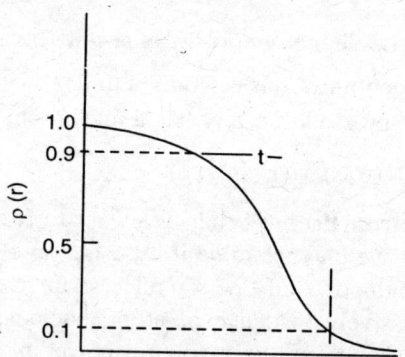

Fig. 1.5: Fermi distribution of the charge density

Now, $f(\vec{p}, \vec{p}') = -\dfrac{m}{2\pi} \int d^3r \, \rho(r) \, e^{i\vec{q}\cdot\vec{r}} \int \dfrac{e^{i\vec{q}\cdot(\vec{x}-\vec{r})}}{|\vec{x} - \vec{r}|} \, d^3x$

$$= -\dfrac{m}{2\pi} \int d^3r \ \rho(r) \, e^{i\vec{q}\cdot\vec{r}} \times \int \dfrac{e^{i\vec{q}\cdot\vec{x}'} \, Ze}{x'} \, d^3x',$$

where $\vec{x} - \vec{r} = \vec{x}'$, $d^3x = d^3x'$ and for a point nucleus $\rho(\vec{r}) = Ze \, \delta(\vec{x} - \vec{r})$;

$$V(x) = \int \frac{Ze \, \delta(\vec{x} - \vec{r}) \, d^3r}{|\vec{x} - \vec{r}|} = \frac{Ze}{x}.$$

Dropping the prime on x', we can write

$$f(\vec{p}, \vec{p}') = f_{\text{point}}(\vec{p}, \vec{p}') \int d^3r \, \rho(r) \, e^{i\vec{q}\cdot\vec{r}},$$

where $f_{\text{point}}(\vec{p}, \vec{p}') = -\frac{m}{2\pi} \int e^{i\vec{q}\cdot\vec{x}} \frac{Ze}{x} d^3x.$

Hence $\qquad \left(\frac{d\sigma}{d\Omega}\right) = \left(\frac{d\sigma}{d\Omega}\right)_{\text{point}} \left|F(q^2)\right|^2,$

where $\qquad F(q^2) = \int d^3r \, \rho(\vec{r}) \, e^{i\vec{q}\cdot\vec{r}}.$ $\hfill$ (1.22a)

Upto now we have neglected the spins of the electron and of the target. At the relativistic energies Mott's cross-section given below includes effect due to the electron spin.

$$\left(\frac{d\sigma}{d\Omega}\right)_{Mott} = \left(\frac{Ze^2}{2E}\right)^2 \frac{1}{\sin^4 \theta/2} \left[\frac{\cos^2(\theta/2)}{1 + \frac{2E}{M_T} \sin^2 \theta/2}\right], \qquad (1.22b)$$

where θ and Ω are measured in the Lab system.

The first factor in (1.22b) is the usual Rutherford scattering and the second factor takes into account electron spin and target recoil. The expression shows that at relativistic energies, the Mott cross section drops off more rapidly at large scattering angles than does the Rutherford cross section. In order to consider scattering from finite size nucleus, we express

$$\left(\frac{d\sigma}{d\Omega}\right)_{Expt.} = \left(\frac{d\sigma}{d\Omega}\right)_{Mott} \left|F(q^2)\right|^2, \qquad (1.23)$$

where $q^2 = (\vec{p} - \vec{p}')^2$ is the momentum transfer squared and $F(q^2)$ is the form factor, related to the charge distribution.

Experimentally, the magnitude of the form factor is determined by the ratio of the measured cross section to the Mott cross section. One, therefore, measures the cross section for the fixed beam energy at various angles and divides it by the calculated Mott cross section. We shall now discuss form factors of spherically symmetric systems which have no preferred orientation in space. In this case the form factor only depends on the momentum transfer q. We symbolize this fact by writing the form factor as $F(q^2)$.

In the Born approximation with negligible nuclear recoil, the form factor is the Fourier transform of the charge distribution $\rho\,(\vec{r})$:

$$F(q^2) = \int e^{i\vec{q}\cdot\vec{r}} \rho\,(\vec{r})\, d^3r = \iiint e^{iqr\cos\theta} \rho(r) r^2 dr\, \sin\theta\, d\theta\, d\phi,$$

where $\qquad\qquad\qquad \rho\,(\vec{r}) = \rho\,(r)$ and $\theta < (q, r)$.

Substituting $\cos\theta = t$, we carry the integration and get

$$F(q^2) = 4\pi \int_0^\infty \rho(r) \frac{\sin qr}{qr} r^2 dr \text{ with } 4\pi \int_0^\infty \rho(r)\, r^2 dr = 1. \qquad (1.24)$$

The charge density defined this way is related to the effective charge density $\rho_0(r)$ in the ground state of the nucleus of charge Ze as

$$\rho_0(r) = Ze|\psi_0(r)|^2, \text{ where}$$

$$\rho\,(r) = \rho_0(r)/Ze \text{ and } \int |\psi_0(r)|^2\, d^3r = 1.$$

In the above, r is the position vector of a charged volume element in the nucleus measured from its c.m. In principle, the radial charge distribution could be determined by the inverse Fourier transform:

$$\rho\,(r) = \frac{1}{(2\pi)^3} \int F(q^2)\, e^{-i\vec{q}\cdot\vec{r}} d^3q, \qquad (1.25)$$

A charged distribution, which drops off gently, corresponds to a smooth form factor. The more extended the charge distribution, the stronger the fall off of the form factor with q^2. On the other hand, if the target is small, the form factor falls off slowly. For $\rho(r) = 4\pi\delta(r)$ for a point like target,

$$F(q^2) = 4\pi \int_0^\infty e^{-\vec{q}\cdot\vec{r}}\, \delta(r)\, d^3r = 1.$$

Information about the nuclear radius can be obtained from the behaviour of the form factor for $q^2 \to 0$. If q is very small compared to nuclear radius r_0 then $qr_0 \ll 1$ and the form factor $F(q^2)$ can be expanded by expression of $\sin qr$ in the power of qr as

$$F(q^2) = 4\pi \int_0^\infty \rho(r) \left[1 - \frac{1}{6} r^2 q^2 + \cdots \right] r^2 dr = 1 - \frac{1}{6} q^2 \langle r^2 \rangle + \cdots, (1.26)$$

where $\qquad \langle r^2 \rangle = \int_0^\infty \left[r^2 \rho\,(r) \right] r^2\, dr.$

Hence, it is necessary to measure $F(q^2)$ down to very small value in order to determine $\langle r^2 \rangle$. For small q^2 then

$$\langle r^2 \rangle = -6 \left. \frac{\partial F(q^2)}{\partial q^2} \right|_{q^2=0} \qquad (1.27)$$

is the mean square radius of the charge distribution of the target, which is obtained from the slope of $F(q^2)$ versus q^2 plot.

One of the principal results of electron scattering experiments is the conclusion that nuclei do not have sharp edges but rather there is an appreciable region over which the nucleon density falls away. One form of density dependence is the Fermi density function with two parameters:

$$\rho_F(r) = \frac{\rho_0}{1 + \exp[(r-c)/a]} , \text{ with } \rho_0 = \frac{3Ze}{4\pi c^3} \left(1 + \frac{\pi^2 a^2}{c^2} \right)^{-1} \qquad (1.28a)$$

corresponding to the normalization $4\pi \int\limits_0^\infty \rho(r) \, r^2 dr = Ze$.

In (1.28a) c is the radius at which the density $\rho_F(r)$ falls to half of its maximum value and a is related to another parameter t, which denotes the distance in the surface layer over which the density decreases from 90% to 10% of its maximum, such that $t = 4.4a$ (Fig. 1.5). Empirically, for larger nuclei c and a are: $c = 1.07 \, A^{1/3}$ fm and $a = 0.54$ fm.

All experiments confirm that for heavy nuclei, nuclear density is constant upto a certain distance from the centre and that it decreases to zero in a further distance which is small compared to the former. Due to large incompressibility of nuclear matter, for such nuclei we can possibly assign an average nuclear radius, $r_0 = a_0 A^{1/3}$ in which the constant a_0 may well be different for different measuring methods. Usually one takes an average value of $a_0 = 1.2$ fm. Therefore, nucleon density in a nucleus in the ground state

$$\rho_0 = A \big/ 4\pi r_0^3 = \frac{3}{4\pi} \frac{1}{(1.2)^3 \, fm^3} = 0.14 \, fm^{-3} \qquad (1.28b)$$

B. Visualizing the Nucleus

"Utmost any theory can do [is] to be instrumental in suggesting and guiding new developments beyond its original scope".

–Niels Bohr[14]

The saturation characteristic of nuclear force led Bohr[14] to treat a nucleus analogous to an incompressible classical charged liquid drop having the known properties: constant density, surface tension, short range forces, saturation and deformability.

The quantum nature of the nucleus was then incorporated by Bethe and Bacher[15] by considering the nucleus as degenerate Fermi gas of nucleons. A systematic formulation of nuclear binding energy based on the above physically distinguishable ideas was carried out by Weizsäcker and this led finally to the well known Bethe-Weizsäcker mass formula[16].

Volume energy: The experimental BE/A vs A curve (Fig. 1.3) revealing saturation characteristic of nuclear forces, implies $BE \propto A$. Therefore, the main contribution to the nuclear binding energy comes from the total number of nucleons in the entire volume of the nucleus. So that the volume energy,

$$(E)_{\text{vol}} = -a_v A , \tag{1.29}$$

where a_v is a constant. This equation implies that not all pairs of nucleons interact, since the number of pairs of nucleons is $A(A-1)/2$. This is the leading term for the expression of binding energy and it provides predominantly attractive contribution to the total energy $E(N,Z)$.

Surface energy: The nucleons that find themselves at the surface of the nuclear liquid drop are not bound as tightly as others. The volume term (1.29) must be reduced due to so called nuclear 'surface tension effect' by an amount

$$(E)_{surface} = a_s A^{2/3} , \tag{1.30}$$

since, $(E)_{surface} = \gamma(4\pi r_0^2) = \gamma(4\pi) a_0^2 A^{2/3}$, where γ is the nuclear surface tension constant, expressed in units of MeV/fm^2. $a_s = 4\pi\gamma a_0^2$ and $r_0 = a_0 A^{1/3}$.

Coulomb energy: The surface tension force acts as a restoring force and counteracts the force of Coulomb repulsion inside the charged liquid drop. The self energy of the charged nuclear sphere can be calculated from the formula of electrostatics. We assume that the protons are distributed over a sphere of radius r_0 with the last proton smeared out over it such that the charge density $\rho = \left(3Ze/4\pi r_0^3\right)$. The self Coulomb energy of the charged nuclear liquid drop

$$E_c = \frac{1}{2}\int_0^{r_0} V(r)\,\rho\,dv = \frac{1}{2}\int_0^{r_0} \frac{Ze}{2r_0}\left(3 - \frac{r^2}{r_0^2}\right)\frac{3Ze}{4\pi r_0^3}4\pi r^2\,dr = \frac{3}{5}\frac{Z^2 e^2}{r_0} .$$

The factor 1/2 occurs because the elemental charges must not be counted twice when summed to give total E_c. The self Coulomb energy of the charged nucleus overestimates the electrical repulsion. So Z^2 should be replaced by $Z(Z-1)$. Usually E_c is expressed as

$$E_c = a_c \, Z(Z-1)/A^{1/3} \text{ where } a_c = \frac{3}{5} \frac{e^2}{a_0} = 0.72 \text{ MeV}. \qquad (1.31)$$

Symmetry energy: As long as mass numbers are small, nuclei tend to have same number of protons and neutrons. Heavier nuclei accumulate more and more neutrons to partly compensate for the increasing Coulomb repulsion by increasing the nuclear force. This creates an asymmetry in the number of neutrons and protons. The contribution to the binding energy from this asymmetry is obtained by calculating the kinetic energy of N neutrons and Z-protons obeying the Pauli principle as follows:

Nucleus is a quantum mechanical system consisting of fermions. We can regard the ground state of the nucleus as "degenerate Fermi gas of nucleons". We assume that each nucleon moves independently in an average uniform potential well with a constraint to move in a spherical cavity. Therefore, there is no force on a nucleon except when it is at the edge of the well. Interaction between pairs of nucleons is neglected.

Further, we assume that all the lowest available energy states of the system are occupied. Hence, transfer of momentum and energy between particles which would be normal consequence of the strong forces acting between them is prevented by the Pauli principle, as there will be no empty states where they can scatter to. In other words, the mean free path of a particle in a degenerate Fermi gas is long compared to the volume dimension. Therefore, in the above picture of nucleus, the number of possible states for a particle within the momentum range p and $p + dp$ is

$$dn = 2 \cdot \frac{4\pi}{3} r_0^3 \frac{4\pi p^2 \, dp}{(2\pi\hbar)^3} = \frac{4}{3\pi\hbar^3} a_0^3 \, A p^2 \, dp \,, \qquad (1.32)$$

where $(2\pi\hbar)^3$ is the volume of a phase cell and the factor 2 appears because of the two possible spin projections. First, we consider a proton gas. If p_0 is the maximum proton momentum and if each of protons occupies a quantum state, then

$$Z = n = \int_0^{p_0} \frac{4 a_0^3 A}{3\pi\hbar^3} p^2 \, dp \text{ or } p_0 = \left(\frac{9\pi}{4}\right)^{1/3} (Z/A)^{1/3} \frac{\hbar}{a_0}.$$

The total kinetic energy (KE) of all protons (m = mass of proton),

$$(KE)_{\text{Proton}} = \int_0^{p_0} \frac{p^2}{2m} \frac{4A}{3\pi\hbar^3} a_0^3 p^2 \, dp = \frac{2}{3\pi} \frac{A}{m} \frac{a_0^3}{\hbar^3} \frac{p_0^5}{5} = CZ \left(\frac{Z}{A}\right)^{2/3}. \qquad (1.33)$$

Similarly, $\qquad\qquad\qquad (KE)_{\text{neutron}} = CN(N/A)^{2/3},$

where
$$C = \frac{3}{10}\left(\frac{9\pi}{4}\right)^{2/3}\frac{\hbar^2}{ma_0^2} = 32.6 \text{ MeV}.$$

We have also assumed here that the radius r_0 of the proton distribution is the same as that of the neutron distribution, i.e. the charge and the matter radii in a nucleus are nearly the same. Therefore, for the whole nucleus ($A = Z + N$), the total kinetic energy

$$T = C\left[Z\cdot\left(\frac{Z}{A}\right)^{2/3} + N\cdot\left(\frac{N}{A}\right)^{2/3}\right]$$

$$= \frac{C}{2}\left(\frac{1}{2}\right)^{2/3}\left[(A-\Delta)\left(1-\frac{\Delta}{A}\right)^{2/3} + (A+\Delta)\left(1+\frac{\Delta}{A}\right)^{2/3}\right], \text{ where } \Delta = N - Z$$

Expanding in power series and retaining terms upto Δ^2/A we get

$$T \approx \frac{C}{2}\left(\frac{1}{2}\right)^{2/3}\left[2A + \frac{10}{9}\frac{\Delta^2}{A}\right] = bA + a_a\frac{(N-Z)^2}{A}, \qquad (1.34)$$

where $b = C\left(\frac{1}{4}\right)^{1/3}$ and $a_a = \frac{5}{9}b$. Here $b = 20.5$ MeV and $a_a = 11.4$ MeV.

The kinetic energy, thus, contributes 11.4 MeV to the symmetry energy coefficient. The first term in (1.34) contributes to the volume energy.

Momentum distribution of nucleons in Fermi gas: An important property of the Fermi gas is that the particle density is fixed, and for a given number of particles, the system has completely defined volume Ω. The kinetic energy of a particle with the well defined momentum is $\left(\hbar^2 k^2/2m\right)$. By virtue of the Pauli principle no more than four nucleons, differing in the values of their components of spin and isospin, can be found in each of the states with given value of the energy. Hence, Eq. (1.32) can be rewritten in the form

$$dn_k = 4\Omega\cdot\frac{4\pi k^2 dk}{(2\pi)^3}.$$

Since, in the ground state all levels with $k < k_F$ are occupied and also the number of nucleons is equal to A, the limiting Fermi momentum k_F is determined as follows:

$$A = 4\Omega\int_0^{k_F}\frac{4\pi}{(2\pi)^3}k^2 dk,$$

or
$$k_F = \left(\frac{3\pi^2}{2}\right)^{1/3} \left(\frac{A}{\Omega}\right)^{1/3} = \left(\frac{3\pi^2}{2}\right)^{1/3} \rho^{1/3} = \left(\frac{9\pi}{8}\right)^{1/3} \left(\frac{1}{a_0}\right), \quad (1.35a)$$

where the nuclear density $\rho = \dfrac{A}{\Omega}$, $\Omega = \dfrac{4\pi}{3} r_0^3$ and $r_0 = a_0 A^{1/3}$.

For $a_0 = 1.1$ fm, the limiting value of the Fermi momentum $k_F = 1.38$ fm^{-1} and, hence,

$$T_F = \frac{\hbar^2 k_F^2}{2m} = 39 \text{ MeV}. \quad (1.35b)$$

The total kinetic energy of the system in the ground state,

$$T = 4\Omega \int_0^{k_F} \frac{d^3k}{(2\pi)^3} \frac{k^2\hbar^2}{2m} = \frac{3}{5} \frac{\hbar^2}{2m} \left(\frac{3\pi^2}{2}\right)^{2/3} \rho^{2/3} A.$$

The mean kinetic energy per nucleon $\bar{T} = \text{constant} \cdot \rho^{2/3} = \dfrac{3}{5} T_F.$ (1.36a)

The mean nucleon momentum $\left\langle p^2 \right\rangle = \dfrac{3}{5} p_F^2.$ (1.36b)

The minimum energy corresponds to the ground state of the system in which all single particle levels with $k < k_F$ are occupied and higher levels are vacant. The momentum distribution of nucleons in an ideal Fermi gas is shown in Fig. 1.6.

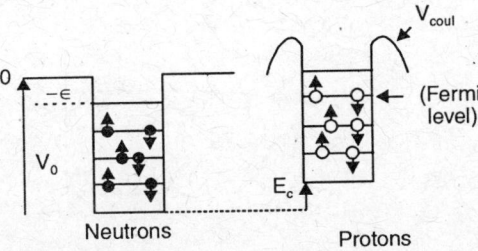

Fig. 1.6: Momentum distribution of nucleons in an ideal Fermi gas

Pairing energy: Two identical nucleons in the degenerate Fermi gas model have tendency to align their spins in opposite directions. A state that contains such a saturated pair lies lower in energy than a state that has two individual nucleons. As a result of this, even Z and even N nuclei are most strongly bound. Odd Z, even N or even Z, odd N nuclei have medium binding and odd Z; odd N nuclei are the least bound. This odd-even effect arising due to the pairing correlation is also evident from the fact that there are more number of stable isotopes for even-even nuclei compared to odd-even or odd-odd nuclei.

The additional pairing energy depends on the mass number and it is, therefore, expressed by an empirical relation by inserting an extra attractive and repulsive contribution in the even-even and odd-odd nuclei respectively, whereas no such extra contribution is made for odd mass nuclei. Accordingly,

$$(\delta E)_{pairing} = \left[(-1)^N + (-1)^Z \right] \delta(A) = \lambda \delta(A), \qquad (1.37)$$

where $\lambda = 0$ for odd A.

We find from the measured data that $\delta(A) \approx a_\delta A^{-1/2}$, where a_δ is a constant. The constant a_δ is found nearly the same for the neutron pairs and proton pairs. The origin of this pairing effect lies in the special property of N-N force (*see* Chapter 3). The pairing energy term, $\delta(A)$ in a way compensates for the lack of spin consideration in the mass formula. Adding up all terms we finally get

$$E(N,Z) = -(a_v - b)A + a_s A^{2/3} + a_c \frac{Z^2}{A^{1/3}} + a_a \frac{(N-Z)^2}{A} + a_\delta A^{-1/2}. \quad (1.38a)$$

This is also called the Bethe Weizsäcker (BW) formula[16] for the nuclear binding energy, $BE(N,Z) = -E(N,Z)$. The corresponding mass formula is

$$M(N,Z) = Zm_p + Nm_n - \frac{BE(N,Z)}{c^2}. \qquad (1.38b)$$

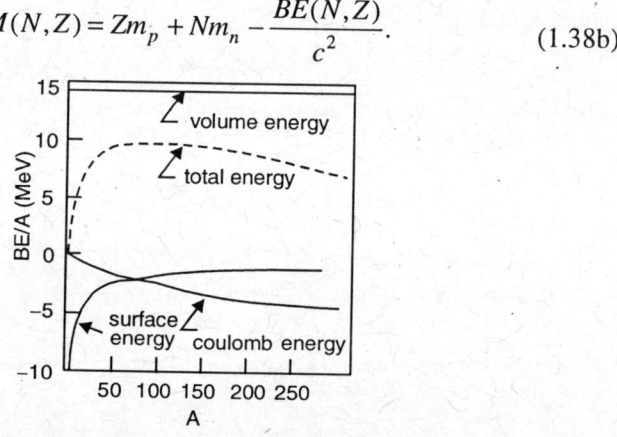

Fig. 1.7: The contribution of each term to the BW formula

The contribution of each of the terms of the BW formula is shown in Fig. 1.7. We notice that the deviations from the BW formula slightly occur for $A < 10$. Except the parameter a_v, all other parameters of the mass formula (a_s, a_c, a_a and a_δ) can be calculated. Calculation of a_v is one of the main objectives of the nuclear matter theory. Sometimes the BW formula is written in the form

$$E(N,Z) = -a_1 A + a_2 A^{2/3} + a_3 \frac{Z^2}{A^{1/3}} + a_4 \frac{(N-Z)^2}{A} + a_5 A^{-1/2} \qquad (1.39a)$$

where the constants a_1 ...... a_5 are determined by fitting the experimental values of BE. The exact values of these constants depend on the range of masses for which they are optimized. One of the best fit parameters expressed in MeV are given below:

$$a_1 = 15.6, \ a_2 = 17.23, \ a_3 = 0.714, \ a_4 = 23.28 \text{ and } a_5 = 11.2. \qquad (1.39b)$$

For the atomic mass

$$M(N,Z) = Z m_H + (A - Z) m_n$$

$$- \frac{1}{c^2} \left[a_1 A - a_2 A^{2/3} - a_3 \frac{Z^2}{A^{1/3}} - a_4 \frac{(N-Z)^2}{A} - a_5 A^{-1/2} \right], \qquad (1.40a)$$

where m_H is the mass of hydrogen atom. The equation for the most stable isobar is found from the condition

$$\left[\frac{\partial M(N,Z)}{\partial Z} \right]_{A = \text{constant}} = 0. \qquad (1.40b)$$

This gives the minimum value of $Z = Z_m$ as follows:

$$\frac{\partial M}{\partial Z} = m_p - m_e - m_n - \frac{1}{c^2} \left[-2a_3 \frac{Z}{A^{1/3}} + \frac{4a_4}{A}(A - 2Z) \right]_{Z = Z_m} = 0$$

or $\qquad Z_m = \frac{A}{2} \left\{ \left[\frac{(m_n - m_p - m_e)c^2}{4a_4} + 1 \right] \middle/ \left[\left(\frac{a_3}{4a_4} A^{2/3} \right) + 1 \right] \right\} .$

Since a_4 and a_3 are fractional constants, Z_m is usually a constant. The value of Z_m is chosen to be the one nearest to the integer. For the neutron excess in most stable isobar we then obtain

$$A - 2Z_m = \frac{(a_3/4a_4)A^{2/3} - (m_n - m_p - m_e)c^2/4a_4}{1 + (a_3/4a_4)A^{2/3}} A . \qquad (1.41)$$

We rewrite (1.40a) for fixed A and express $M(N,Z)$ as a function of Z:

$$M(N,Z) = Z m_H + (A - Z) m_n$$

$$- \frac{1}{c^2} \left[-a_1 A - a_2 A^{2/3} - a_3 \frac{Z^2}{A^{1/3}} - a_4 \frac{(A - 2Z)^2}{A} - a_5 A^{-1/2} \right]$$

$$= \left[\alpha A + \beta Z + \gamma Z^2 + \lambda \delta(A) \right] / c^2 , \qquad (1.42)$$

where $\quad \alpha = m_n\, c^2 - \left(a_1 - \dfrac{a_2}{A^{1/3}} - a_4 \right),$

$$\beta = -4a_4 - (m_n - m_p - m_e)\, c^2 \quad \text{and} \quad \gamma = \frac{4a_4}{A} + \frac{a_3}{A^{1/3}}. \tag{1.43}$$

Clearly (1.42), which equates $M(N, Z)$ with a quadratic function of Z for constant A, is an equation of a parabola. For odd A, the pairing term $\lambda\,\delta(A) = 0$ and we get a single parabola while for even A; $\delta(A) = \pm a_s A^{-\frac{1}{2}}$ corresponding to $\lambda = \pm 1$, and we get two parabolas depending on whether both N and Z are even or odd.

Separation energy: We define the neutron separation energy $S_n(N,Z)$ in the nucleus (N, Z) as the energy which must be supplied to separate (N,Z) into $(N - 1, Z)$ plus a neutron at rest. Therefore,

$$E(N,Z) = E(N - 1, Z) + E_{int}\left[n \cdot (N - 1, Z) \right], \tag{1.44}$$

where the last term is the interaction of neutron with core. Therefore,

$$E_{int}(n \cdot (N - 1, Z)) = E(N,Z) - E(N - 1, Z) = -S_n(N,Z)$$

or $\qquad\qquad S_n(N,Z) = B(N,Z) - B(N - 1, Z). \tag{1.45a}$

Similarly, the proton separation energy,

$$S_p(N,Z) = E(N, Z - 1) - E(N,Z) = B(N,Z) - B(N, Z - 1). \tag{1.45b}$$

Exercise 1.7. Show that the permanent electric dipole moment of a nucleus is zero.

We apply the parity operator to Eq. (1.19c), i.e. make the transformation r_k to $-r_k$ and use the fact that the nucleus has definite parity, i.e.

$$\left| \psi_0(r_1, \cdots r_A) \right|^2 = \left| \psi_0(-r_1, \cdots r_A) \right|^2.$$

Hence, $\qquad\qquad\qquad \vec{D} = -\vec{D} \ \text{ or } \ \vec{D} = 0.$

Exercise 1.8. Show that for an exponential charge distribution,

$\rho(r) = \dfrac{a^3}{8\pi} e^{-ar}$, the form factor $F(q^2) = (1 + q^2/a^2)^{-2}$.

Substituting $\rho(r)$ in (1.24) and $qr = x$, we get

$$F(q^2) = \int\limits_0^\infty \frac{a^3}{8\pi} e^{-ar} \frac{\sin qr}{qr} r^2\, dr\, d\Omega = \int\limits_0^\infty \frac{a^3}{2q^3} e^{-\frac{a}{q}x} x \sin x\, dx.$$

We use the standard integral

$$\int\limits_0^\infty e^{-mx} x \sin px\, dx = \frac{2mp}{\left(m^2 + p^2 \right)^2} \quad \text{for } m > 0$$

We get

$$F(q^2) = (1 + q^2/a^2)^{-2}.$$

This is known as the "dipole form factor". Thus, the form factor can be calculated analytically for certain charge distributions as mentioned above.

For example, for Gaussian $\rho(r) = \left(\dfrac{a^2}{2\pi}\right)^{3/2} e^{-a^2 r^2/2}$: $F(q^2) = e^{-q^2/2a^2}$.

Exercise 1.9. Calculate the minimum excitation energy for the volume oscillation of a nucleus employing the Fermi gas model with equal number of protons and neutrons and assuming harmonic oscillation.

The elastic force for volume oscillation or the incompressibility is defined as the second derivative of the total energy E at equilibrium:

$$E = A\left[\frac{3}{5}T_F + \frac{1}{2}m\omega^2 r_0^2\right].$$

Therefore, $k = \dfrac{\partial^2}{\partial r_0^2}\left[\dfrac{3}{5}AT_F\right]_0$ neglecting $\dfrac{\partial}{\partial r_0^2}\left(\dfrac{A}{2}m\omega^2 r_0^2\right) = Am\omega^2$

Employing (1.35) we get

$$k = \frac{18}{5}T_F^2 A^{1/3}\left(\frac{8}{9\pi}\right)^{2/3}\left(\frac{2m}{\hbar^2}\right).$$

For harmonic oscillator, the minimum excitation energy

$$\hbar\omega = \left[\frac{\hbar^2 k}{mA}\right]^{1/2}$$

$$= \left[\frac{36}{5}\left(\frac{8}{9\pi}\right)^{2/3}\right]^{1/2}\frac{T_F}{A^{1/3}} = \frac{1.75\,T_F}{A^{1/3}} \approx \frac{70}{A^{1/3}}\ \text{MeV}.$$

Exercise 1.10. Show that for isobaric stable nuclei the difference of energies required to remove a proton and a neutron from a nucleus, (A,Z) is proportional to $A^{4/3}$.

$$\left|S_p - S_n\right| = BE(Z, A-1) - BE(Z-1, A-1)$$

$$= a_3(A-1)^{-1/3}\left[(1-2Z) + 4a_4(A-1)^{-1}(A-2Z)\right]$$

For stable isobaric nuclei, $\dfrac{\partial BE(A,Z)}{\partial Z} = 0$, which yields $Z \cong \left(1 - \dfrac{a_3}{4a_4}A^{2/3}\right)$.

Therefore, $\left|S_p - S_n\right| = \dfrac{a_3}{(A-1)}\left[A^{5/3} - (A-1)^{5/3} + (A-1)^{2/3} A^{5/3} \dfrac{a_3}{4a_4}\right].$

For heavy nuclei $A \gg 1$, $\left|S_p - S_n\right| = \dfrac{a_3^2}{4a_4}\, A^{4/3} = 5.5\, A^{4/3}.$

1.3 ALPHA CLUSTERING AND DECAY

There are numerous phenomena of nuclear behaviour that suggest the clustering of nucleons within a nucleus. Heavy nuclei that spontaneously decay by α-emission have decay rates suggesting at least a tendency for preformation of α-particle clusters in nuclei. Alpha-particle is a tightly bound structure due to its spin and isospin symmetry and so for many purposes it behaves as a single entity. Many features of nuclear structure and nuclear reactions can, therefore, be treated by considering alpha-particle this way, without explicit reference to its constituent neutrons and protons.

A. Alpha Clusters

The evidence that nuclei tend to form alpha particle clusters is found in the binding energy data. The binding energies of some even-even ($N = Z$) nuclei can be largely accounted for as multiples of the alpha particle binding energy, $\epsilon_\alpha = 28.4$ MeV. Viewing such nuclei as consisting of n weakly bound alpha particle clusters, we can express their binding energies E_b as

$$E_b = n\,\epsilon_\alpha + mC, \qquad (1.46)$$

where C is the inter-cluster bond energy and m is number of bonds for tightly packed alpha clusters. In Table 1.4 we compare the nuclear binding energies predicted by the above Cluster Model (CM) with the experimental values.

Table 1.4: Alpha cluster model predictions of binding energies

Nucleus	n	m	C (MeV)	$n(BE)_\alpha$ MeV	$(BE)_{CM}$ MeV (Eq. 1.46)	$(BE)_{Expt.}$ MeV
C^{12}	3	3	2.46	85.2	92.78	92.5
O^{16}	4	6	2.41	113.6	128.06	127.0

In the self-conjugate $4n$ nuclei the study of structure change is schematically summarized by the so called Ikeda diagram[17] (Fig. 1.8), where the unlabelled smaller circles represent α-clusters. If we consider, for example ^{16}O, we find the binding energy to be 127 MeV. The sum of binding energies of four α-particles is 4 times 28.4 MeV. This is about 113.6 MeV. The α-particle cluster model thus predicts 90 per cent of the binding energy of ^{16}O. The remaining difference is the inter α binding energy. The inter-alpha binding energy per bond is nearly constant from ^{12}C to ^{32}S (around 2.4 MeV). It would be interesting to find out whether the inter nucleon forces give us physical reasoning for preferring a cluster substructure.

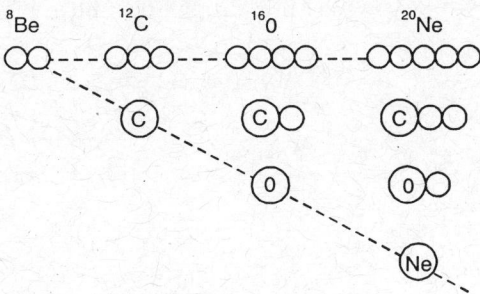

Fig. 1.8: Ikeda Diagram

Let us consider ^6Li nucleus. In the ground and low lying states we should expect four of the nucleons to correlate as an α cluster. However, because the Pauli principle prevents more than four nucleons from entering into the mutual short ranged interaction region, we expect the remaining two nucleons to correlate most strongly with each other, forming a deuteron cluster. Therefore, ^6Li can be considered as a cluster of $\alpha + d$.

B. Leakage of Alpha Particles

Most of heavy nuclei ($Z \geq 84$, $A \geq 208$) decay by α-emission according to the relation

$$_Z^A P \rightarrow {}_{Z-2}^{A-4}D + {}_2^4 He,\qquad (1.47)$$

with Q value of the process $=\left[m_P - (m_D + m_\alpha)\right]c^2$, which is positive, thus releasing energy in α emission.

By the analysis of a large number of experimental values of alpha decay constant (λ) and alpha energy (E), Geiger and Nuttall obtained an empirical relationship between them (Geiger-Nuttall law):

$$\log \lambda = aE^{-1/2} + b.\qquad (1.48)$$

It is also observed that α-particles are emitted with energies less than that necessary to penetrate the nuclear Coulomb barrier. For example, in the case of $_{92}^{238}U$ the height of the Coulomb potential barrier is about 30 MeV, whereas the kinetic energy of emitted α-particle is only 4.2 MeV. Gamow[7] was the first to explain α-emission by employing the concept of quantum mechanical tunneling, *i.e.* leakage of α-particles through potential barrier. He also derived the Geiger Nuttall law[7] on a sound theoretical basis.

In order to understand the mechanism of tunneling of alpha particle through a potential barrier, let us consider a simple potential barrier (Fig. 1.9) which is spherically symmetric. Therefore, we consider the positive energy ($E > 0$) solution of radial Schrodinger equation (S. Eq.) representing the motion of a spinless particle with $\ell = 0$, passing through the above barrier:

$$\frac{d^2u}{dr^2} + \frac{2\mu}{\hbar^2}\left[E - V_{eff}(r)\right]u = 0, \tag{1.49}$$

$$V_{eff}(r) = V_N(r) + \frac{C}{r} + \frac{\hbar^2 \ell(\ell+1)}{2\mu r^2} = V_N(r) + V_C(r) + V_\ell(r), \tag{1.50}$$

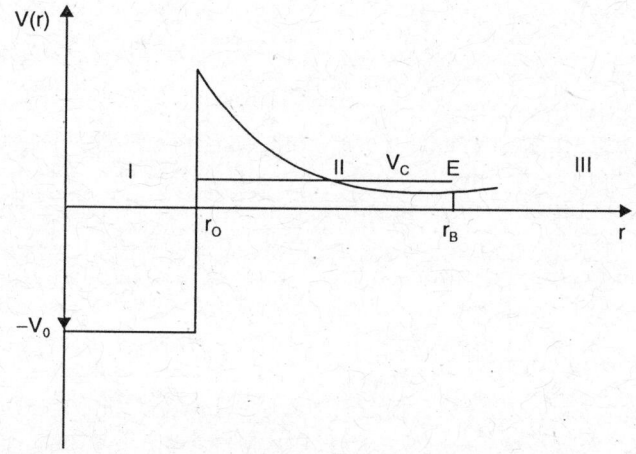

Fig. 1.9: Shape of the Potential Barrier (*see* Text)

Since Gamow[7] gave the theory of alpha decay, a large number of workers have treated this problem either refining the mathematical approximations or introducing certain physical effects which were first ignored. However, we present here an elementary theory of α-decay by solving S. Eq. in question, using WKB approximation (see Quantum Mechanics by L.I. Schiff (Ref. 18)). In the WKB method we start with a three dimensional S. Eq. which on the assumption of spherical symmetry reduces to one dimension in radial coordinates (Appendix B1) as Eq. (1.49). In Eq. (1.50), $V_{eff}(r)$ is the sum of nuclear, Coulomb and centrifugal potentials, ℓ the angular momentum,

$C = 2(Z-2)e^2$ and $\mu\ \left\{\ = m_\alpha m_D \big/ (m_\alpha + m_D)\right\}$ is the reduced mass for the α-nucleus relative motion. For, $\ell > 0$ the centrifugal barrier increases the height of the barrier that must be traversed. However, for large Z, the influence of the centrifugal term is small compared to the Coulomb term and therefore, the decay probability does not depend strongly on ℓ (Exercise 1.11). The WKB method is valid when $V_{eff}(r)$ is smoothly varying with coordinate r. For $E < V_{eff}$, there is finite probability for α-particle to leak through the potential barrier, which is being calculated by the WKB method. The method consists in introducing an expansion of the wave function u in powers of $\hbar$ and neglecting terms of higher order than $\hbar^2$. One thus replaces S. Eq. at least in some regions of space where classical interpretation is meaningless (regions $E < V_{eff}$, which are inaccessible to classical particles). The transmission coefficient in WKB method[18]

$$T_\ell^{WKB} = e^{-2S},\qquad(1.51)$$

where $$S = \int_{r_i}^{r_B} k(r)\, dr \text{ and } k(r) = \left[\frac{2\mu}{\hbar^2}\left(V_{eff}(r) - E\right)\right]^{1/2}.\qquad(1.52)$$

Here r_i and r_B are the classical turning points, r_i on inside the well and $r_B = C/E$ on the outside. To make quantitative estimate of T_ℓ^{WKB}, we begin with $\ell = 0$. The integral in Eq. (1.52) becomes

$$S = \sqrt{\frac{2\mu}{\hbar^2}} \int_{r_i}^{r_B}\left[\frac{C}{r} - E\right]^{1/2} dr = \sqrt{\frac{2\mu E}{\hbar^2}} \int_{r_0}^{r_B}\left(\frac{r_B}{r} - 1\right)^{1/2} dr,\qquad(1.53)$$

for a square well potential for $V_N(r)$. The turning points are then the nuclear radius $r_0 (= r_i)$ on the inside, and $r_B = C/E$ on the outside. To integrate we substitute $r = r_B \cos^2\theta$ and get

$$S = \sqrt{\frac{2\mu C r_B}{\hbar^2}}\left[\arccos\left(\frac{r_0}{r_B}\right)^{\frac{1}{2}} - \left(\frac{r_0}{r_B} - \frac{r_0^3}{r_B^3}\right)^{\frac{1}{2}}\right]$$

$$= \sqrt{\frac{2\mu C r_B}{\hbar^2}}\left[\frac{\pi}{2} - \left(\frac{r_0}{r_B}\right)^{\frac{1}{2}}\right] \qquad \text{for } \frac{r_0}{r_B} \to 0 \qquad(1.54)$$

or $$S = \sqrt{2\mu c^2}\left(\frac{e^2}{\hbar c}\right)2(Z - 2)\left[\frac{\pi}{2}\right]E^{-1/2} \qquad \text{for } r_0 \ll r_B.\qquad(1.55)$$

In order to calculate α-decay probability we further assume that α particles are preformed inside the nucleus. The velocity of the α-particle inside the well can be calculated for a square well potential as

$$v = \frac{\kappa \hbar}{\mu} \qquad \text{where } \kappa = \left[\frac{2\mu}{\hbar^2} \left(E + V_0 - \frac{C}{r_0} \right) \right]^{1/2} \tag{1.56}$$

The α-particle collides v/r_0 times per unit time on the walls of the nucleus. Therefore, the probability of decay per unit time (or equivalently $t_{1/2}$) is

$$\lambda = \frac{v}{r_0} T_{\ell=0}^{WKB} \tag{1.57}$$

$$\log \lambda = -2S + \log \frac{v}{r_0}$$

$$= -\sqrt{2\mu c^2} \left(\frac{e^2}{\hbar c} \right) 2(Z-2)\pi \, E^{-1/2} + \log \frac{v}{r_0} = a \, E^{-1/2} + b, \tag{1.58}$$

where a and b are constants. Eq. (1.58) again is the well known Geiger-Nuttal law. Please note that the energy E appears as a logarithmic quantity in constant b and therefore, will not disturb $E^{-1/2}$ dependence. Although Eq. (1.58) has been derived by using strongly simplifying assumptions (such as the use of square well potential and pre-existence of α-particle in the nucleus), the expression represents the experimental values correctly to within a factor of 50. This degree of agreement is good if we consider that half lives of naturally occurring α-radioactive nuclei range between 10^{-7} sec to 10^{15} years. The barrier height enters in the exponent, and is therefore decisive for T. Its contributions from the Coulomb and the centrifugal potentials are roughly given by their values slightly outside the nuclear radius ($r_B \approx 1.4 \, A^{1/3}$ fm).

Exercise 1.11. For alpha decay of a nucleus ($A = 234$, $Z = 92$), show that the ratio

$$R = \frac{\text{Centrifugal barrier}}{\text{Coulomb barrier}} = \frac{\ell(\ell+1)}{500}.$$

$$R = \frac{\hbar^2 c^2}{2\mu c^2} \frac{\ell(\ell+1)}{2(Z-2)e^2} \cdot \frac{1}{r}$$

Substituting $r = 1.1(4^{1/3} + A^{1/3})$ fm, $e^2 = 1.44$ MeV fm,

$\hbar c = 197$ MeV fm and $\mu = \dfrac{4 \times 230}{234} \cdot \dfrac{938 \text{ MeV}}{c^2}$, we get

$$R = \frac{\ell(\ell+1)}{500}.$$

1.4 INTERACTION OF ELECTROMAGNETIC RADIATION WITH NUCLEI

"Only through its light do all these others shine. And by its shining all is seen".[19]

Interaction of electromagnetic fields with nuclear systems has helped us in determining the nature of nuclear force and many nuclear properties. It is possible to formulate the electromagnetic interaction with great accuracy. Due to the relative weak nature of the electromagnetic interaction on one hand and the strong nature of the nuclear interaction on the other hand, at typical inter nucleon distance (~ of the order of 1 fm), the structure of the nucleus is by and large dictated by the strong interaction. At the same time comparatively, the weak nature of electromagnetic interaction allows it to be used to probe the structure without altering the later.

The electromagnetic field in a uniform isotropic medium in free space is given by the Maxwell equations, which when expressed in mixed Gaussian units are:

$$\vec{\nabla} \cdot \vec{H} = 0 \qquad\qquad \vec{\nabla} \cdot \vec{H} = 0 \qquad\qquad (1.59)$$

$$\vec{\nabla} \times \vec{H} = \frac{1}{c} \frac{\partial \vec{E}}{\partial t}, \qquad \vec{\nabla} \times \vec{E} = -\frac{1}{c} \frac{\partial \vec{H}}{\partial t}. \qquad (1.60)$$

The Ampere law $\vec{\nabla} \times \vec{H} - \dfrac{1}{c} \dfrac{\partial \vec{E}}{\partial t} = \vec{j}$, and Gauss law $\vec{\nabla} \cdot \vec{E} = \rho$ $\qquad (1.61)$

follow from the Maxwell equations. Here ρ and $\vec{j}$ are charge and current respectively. The electric (magnetic) fields $\vec{E}(\vec{r},t)(\vec{H}(\vec{r},t))$ are continuous vector functions and c is the velocity of light. The vector fields $\vec{E}(\vec{r})$ and $\vec{H}(\vec{r})$ appearing in (1.60) obey

$$\vec{\nabla} \times \vec{H}(\vec{r}) = -ik\vec{E}(\vec{r}) \qquad\qquad (1.62)$$

and $\qquad \vec{\nabla} \times \vec{E}(\vec{r}) = -ik\vec{H}(\vec{r})$, where $\qquad k = \dfrac{\omega}{c}$.

These equations again reduce to the following wave equations:

$$(\nabla^2 + k^2)\vec{H}(\vec{r}) = 0 \text{ and } (\nabla^2 + k^2)\vec{E}(\vec{r}) = 0 \cdot \qquad (1.63)$$

The Hamiltonian operator for the radiation field

$$H_{rad} = \frac{1}{8\pi}\left[\int (E^2(r,t) + H^2(r,t))\, d^3r \right]. \qquad (1.64)$$

We can characterize also the electromagnetic field using only a single divergence-less vector field $\vec{A}(\vec{r}, t)$ in terms of which $\vec{E}(\vec{r}, t)$ and $\vec{H}(\vec{r}, t)$ are defined as

$$\vec{E}(\vec{r},t) = -\frac{1}{c}\frac{\partial \vec{A}(\vec{r},t)}{\partial t} \ , \ \vec{H}(\vec{r},t) = \vec{\nabla} \times \vec{A}(\vec{r},t) \text{ and } \vec{\nabla} \cdot \vec{A}(\vec{r},t) = 0. \tag{1.65}$$

The wave equation for the vector potential field is obtained by differentiating with respect to 't'. Substituting $\vec{H}$ from (1.65), and using the vector relation $\vec{\nabla} \times \vec{\nabla} \times \vec{A} = \vec{\nabla}(\vec{\nabla} \cdot \vec{A}) - \nabla^2 \vec{A}$ and the condition $\vec{\nabla} \cdot \vec{A} = 0$ we get

$$\nabla^2 \vec{A}(\vec{r},t) - \frac{1}{c^2}\frac{\partial^2 \vec{A}(\vec{r},t)}{\partial t^2} = 0. \tag{1.66}$$

Again we consider single frequency harmonic time dependence

$$\vec{A}(\vec{r},t) = \vec{A}(\vec{r}) \ e^{-i\omega t} + \vec{A}^*(\vec{r}) \ e^{i\omega t} = \vec{A}^*(\vec{r},t),$$

so that $\qquad \vec{E}(\vec{r}) = -ik\vec{A}(\vec{r}) \text{ and } \vec{H}(\vec{r}) = \vec{\nabla} \times \vec{A}(r).$ (1.67a)

Using (1.66) and (1.67a) we get the Helmholtz equation

$$(\nabla^2 + k^2)\vec{A}(\vec{r}) = 0. \tag{1.67b}$$

We postulate that the system is enclosed in a large spherical box of radius r_0 and we adopt a boundary condition $\vec{A} \times \vec{r} = 0$ and $r = r_0$. We attempt to represent $\vec{A}$ in terms of angular momentum operator and eigenfunctions. If we find an angular momentum operator that commutes with $(-k^2\vec{\nabla} \times \vec{\nabla}\times)$, then it has within a constant factor the same eigen functions as $\vec{A}$.

J^2 and J_z commute with the operator $\vec{\nabla} \times \vec{\nabla}\times$:

$$J_z\vec{\nabla} \times \vec{\nabla}\times = \vec{\nabla}J_z\vec{\nabla}\times = \vec{\nabla} \times \vec{\nabla} \times J_z \text{ and}$$

$$J^2\vec{\nabla} \times \vec{\nabla}\times = \vec{\nabla} \times J^2\nabla\times = \vec{\nabla}\times \ \vec{\nabla}\times J^2.$$

Hence, J_z, J^2 commute with the vector $\nabla \times \nabla \times$ and also with $(-k^2\vec{\nabla} \times \vec{\nabla}\times)$.

Therefore the required eigen functions of (1.63) can be expressed in terms of eigen functions of J^2, J_z. So $\vec{A}(\vec{r}, t)$ can be expressed in terms of vector eigen functions, called 'Vector Spherical Harmonics' (*see* Appendix D) and some radial functions. In contrast the scalar Helmholtz equation has solution $j_\lambda(kr) \ Y_\lambda^\mu(\theta, \phi)$ where $j_\lambda(kr)$ and $Y_\lambda^\mu(\theta, \phi)$ are respectively spherical Bessel function and spherical harmonics.

A. Gamma Transitions and Selection Rules

The angular dependence of the vector potential $\vec{A}(\vec{r})$ (and thus of fields $\vec{E}$ and $\vec{H}$) is now expressed by y_{L1J}^M. Since y_{L1J}^M form a complete set of vector functions, $\vec{A}, \vec{E}$ and $\vec{H}$ can be expanded in terms of them after multiplying by suitable radial functions. In particular, we note that for $L = J$, we have

$$y_{L1J}^M(\theta, \phi) = \frac{\vec{L}\, Y_J^M(\theta, \phi)}{\sqrt{J(J+1)}} = X_{JM}(\theta, \phi) \text{ where } \vec{L} = -i\vec{r} \times \vec{\nabla}.$$

It is useful to use the standard multipole classification of electromagnetic fields. In order to identify separately the nuclear total angular momentum with angular momentum carried by gamma photon, we rewrite JM as $\lambda\mu$. The multipole order is referred to as 2^λ pole: $\lambda = 1$ (dipole), $\lambda = 2$ (quadrupole) and $\lambda = 3$ (octupole) etc. The multipole is further designated as an electric multipole (E-mode) and magnetic multipole (M-mode) on the basis of spatial parity of A as follows:

Multipole nature	Parity	Multipole nature	Parity
2^λ (E-mode)	$(-)^\lambda$	2^λ (M-mode)	$(-)^{\lambda+1}$

We single out the emission process involving the E-mode and express:

$$\vec{H}_{\lambda\mu}^E(\vec{r}) = N_{\lambda\mu}^E\, j_\lambda(kr) X_{\lambda\mu}(\theta, \phi) \text{ and } \vec{E}_{\lambda\mu}^E(\vec{r}) = \frac{i}{k} \vec{\nabla} \times \vec{H}_{\lambda\mu}^E(\vec{r}).$$

Similarly, for M-mode, the multipole fields are

$$\vec{E}_{\lambda\mu} = N_{\lambda\mu}^M\, j_\lambda(kr)\, X_{\lambda\mu}(\theta, \phi),$$

$$\vec{H}_{\lambda\mu}^M = -\frac{i}{k} \vec{\nabla} \times \vec{E}_{\lambda\mu}^M$$

and $\vec{A}_{\lambda\mu}^E(\vec{r}) = \dfrac{1}{k^2} \vec{\nabla} \times \vec{H}_{\lambda\mu}^E(\vec{r})$, where $X_{\lambda\mu}(\theta, \phi)$ are the vector spherical harmonics.

The normalization of the vector potential is chosen to give a single photon of energy $\hbar\omega$ of multipole order $\lambda\mu$, confined to a sphere of radius r_0. The normalization constant $N_{\lambda\mu}$ is determined by using properties of vector spherical harmonics and asymptotic form of the fields ($kr \gg 1$), with r_0 taken large enough. Since for $kr \gg 1$, $E^*E \approx H^*H$.

$$\hbar\omega = \left(\frac{N_{\lambda\mu}^E}{2\pi}\right)^2 \int_0^{r_0} j_\lambda^2(kr) r^2 dr \int X_{\lambda\mu}^*(\theta, \phi)\, X_{\lambda\mu}(\theta, \phi) d\Omega.$$

The radial integral, $\int_0^{r_0} j_\lambda^2(kr)r^2 dr = r_0/2k^2$ and the angular integral is just unity.

Hence, $N_{\lambda\mu}^E = \left(\dfrac{4\pi\hbar\, k^3 c}{r_0}\right)^{1/2}$. Hence, one of the solutions of 1.67b (E-mode) is

$$\vec{A}_{\lambda\mu}^E(\vec{r}) = \frac{1}{k}\left(\frac{4\pi\hbar\, kc}{\lambda(\lambda+1)r_0}\right)^{1/2} \vec{\nabla}\times \vec{L}u_{\lambda\mu}(r,\theta,\phi), \qquad (1.68a)$$

where $u_{\lambda\mu}(r,\theta,\phi) = j_\lambda(kr)\, Y_\lambda^\mu(\theta,\phi)$, $j_\lambda(kr)$ are the spherical Bessel functions and $Y_\lambda^\mu(\theta,\phi)$ are the usual spherical harmonics. The other unnormalized solution (M-mode) is expressed as

$$\vec{A}_{\lambda\mu}^M(\vec{r}) = -\frac{1}{ik}\,\vec{E}_{\lambda\mu}^M = \frac{i}{k}\, j_\lambda(kr)\, X_{\lambda\mu}(\theta,\phi). \qquad (1.68b)$$

To simplify (1.68a), we use the vector operator identity

$$\vec{\nabla}\times\vec{L} = i\left[\vec{\nabla}\left(1 + r\,\frac{\partial}{\partial r}\right) - \vec{r}\,\vec{\nabla}^2\right] \qquad (1.68c)$$

and the recurrence relation for the spherical Bessel functions

$$\left(1 + r\,\frac{\partial}{\partial r}\right)j_\lambda(kr) = (\lambda+1)\, j_\lambda(kr) - kr\, j_{\lambda+1}(kr). \qquad (1.68d)$$

Using (1.68c) and (1.68d), we get

$$\vec{A}_{\lambda\mu}^E(\vec{r}) = \frac{i}{k}\left(\frac{4\pi\hbar\, kc}{\lambda(\lambda+1)r_0}\right)^{1/2}\left\{\vec{\nabla}\left[(\lambda+1)u_{\lambda\mu} - kr j_{\lambda+1}(kr)\, Y_\lambda^\mu\right] + \vec{r}k^2 u_{\lambda\mu}\right\},$$

where we have also used $\nabla^2 u_{\lambda\mu} = -k^2 u_{\lambda\mu}$. Retaining only the first term, we get

$$\vec{A}_{\lambda\mu}^E(\vec{r}) = \frac{i}{k}\left(\frac{4\pi\hbar\, kc}{r_0}\frac{\lambda+1}{\lambda}\right)^{1/2} \vec{\nabla}u_{\lambda\mu}. \qquad (1.68e)$$

We find from (1.68a) that $\vec{L}$ applied to Y_0^0 gives zero as expected, because photon must carry at least unit angular momentum. In the special case

$kr \ll 1$, (long wavelength approximation, i.e. the radius of the nucleus is very much smaller than the wave length of radiation), the Bessel functions satisfy

$$j_\lambda(kr) = \frac{(kr)^\lambda}{(2\lambda+1)!!}, \quad kr \ll 1 \qquad (1.69)$$

where $(2\lambda+1)!! = 1.3.5 \ldots (2\lambda+1)$. For typical values of $r \approx 5$ fm and gamma ray energy of 3 MeV, $kr \approx 1/10$. This is also referred as the non-retarded approximation.

The fields $\vec{E}$ and $\vec{H}$ are related to $\dot{\vec{A}}_{\lambda\mu}^M(r)$. We use (1.68b) for $\vec{A}_{\lambda\mu}^M(r)$, (1.69) for $j_\lambda(kr)$ and retain only the lowest power of (kr) so that for the magnetic radiation $\vec{A}_{\lambda\mu}^M(r)$, we have $|\vec{E}| \sim (kr)^\lambda$ and $|\vec{H}| \sim (kr)^{\lambda-1}$. Hence, for magnetic radiation $\vec{A}_{\lambda\mu}^M(r)$, $|\vec{H}| \gg |\vec{E}|$. Again we use (1.68e) for $\vec{A}_{\lambda\mu}^E(r)$, the relations (1.65) for $\vec{E}$, $\vec{H}$ and $\vec{A}$, and the expansion (1.69) to get $|\vec{E}| \sim (kr)^{\lambda-1}$ and $|\vec{H}| \sim (kr)^{\lambda-2}$. Hence, for the electric radiation $\vec{A}_{\lambda\mu}^E(r)$, $|\vec{E}| \gg |\vec{H}|$. The above relations show the significance of the labels electric (E) and magnetic (M). The vector field $\vec{A}(\vec{r})$ can be expressed as a combination of $\vec{A}_{\lambda\mu}^M(\vec{r})$ and $\vec{A}_{\lambda\mu}^E(\vec{r})$, which are the two independent solutions of (1.67b)

$$\vec{A}(\vec{r}) \sim \vec{A}_{\lambda\mu}^M(\vec{r}) + \vec{A}_{\lambda\mu}^E(\vec{r}).$$

We shall use a simple time dependent perturbation theory to calculate the spontaneous decay rate of an excited state of a nucleus. We imagine the nucleus in question and the radiation field to be confined to the interior of a large spherical box of radius r_0 with perfectly conducting wall. The system Hamiltonian consists of three parts:

$$H = H_n + H_r + H_{\text{int}}, \qquad (1.70)$$

where H_n is the Hamiltonian of the nucleus, H_r the Hamiltonian of the radiation field and H_{int} the interaction Hamiltonian of the nucleus and the radiation field. A nucleus is a system of nucleons bound by a two-body force v_{ij} in the presence of an electromagnetic self-field generated by the nucleon convection currents and spin magnetic currents. The electromagnetic field is specified by the vector potential $\vec{A}$. Therefore,

$$H_n + H_{\text{int}} = \sum_{j=1}^{A} \frac{1}{2m_j}(\vec{p}_j - q_j\vec{A}/c)^2 - \sum_j \frac{e\hbar}{2m_jc}\vec{\mu}_j \cdot \vec{H} + \frac{1}{2}\sum_{\substack{j,j' \\ j \neq j'}} v_{jj'},$$

where j (and j') refer to nucleons in the nucleus, individually having respectively mass, charge and magnetic moment (in nuclear magnetons), m_j, q_j and and μ_j. We can take mass of neutron and proton to be the same and simply write $m_j = M$; the nucleon charge $q_j = +e$ for proton and zero for neutron and $\mu_j = \mu_p \vec{\sigma}_p \left(= 2.79 \, \vec{\sigma}_p\right)$ for proton and $\mu_j = \mu_n \vec{\sigma}_n \left(= -1.91 \, \vec{\sigma}_n\right)$ for neutron, where $\vec{\sigma}$ is the Pauli spin operator. We can expand the first term to give

$$\frac{1}{2M}\left(p_j - \frac{q_j}{c}\vec{A}\right)^2 = (p_j^2/2M) + \frac{q_j}{2Me}(\vec{p}_i \cdot \vec{A} + \vec{A} \cdot \vec{p}_j) + \frac{q_j^2}{2Mc^2}A^2.$$

Since, we shall be using a weak field perturbing calculation, the quadratic term in vector potential is small. Further, we use the Coulomb gauge $\vec{\nabla} \cdot \vec{A} = 0$ and note that $\vec{p}$ and $\vec{A}$ commute, so that $\vec{p}_j \cdot \vec{A} + \vec{A} \cdot \vec{p}_j = 2\vec{p}_j \cdot \vec{A}$. Replacing $\vec{H}$ by $\vec{\nabla} \times \vec{A}$ we get

$$H_n + H_{int} = \sum_j \left(p_j^2/2M\right) + \frac{1}{2}\sum_{\substack{j,j' \\ j \neq j'}} v_{jj'}$$

$$-\sum_j \left[\frac{q_j}{Mc}\vec{p}_j \cdot \vec{A} + \frac{e\hbar}{2Mc}\vec{\mu}_j \cdot (\vec{\nabla} \times \vec{A})\right].$$

We identify first two terms with H_n and the last two terms with H_{int} so that

$$H_n = \sum_j \left(p_j^2/2M\right) + \frac{1}{2}\sum_{\substack{jj' \\ j \neq j'}} v_{jj'} \qquad (1.71a)$$

$$H_{int} = -\sum_j \frac{q_j}{Mc}\vec{p}_j \cdot \vec{A} + \frac{e\hbar}{2Mc}\vec{\mu}_j \cdot (\vec{\nabla} \times \vec{A}), \qquad (1.71b)$$

where $\vec{\mu}_j = \left(1/2\right) g_j^s \, \vec{\sigma}_j$ and g_j^s denotes the gyromagnetic ratio.

The potential energy term $v_{jj'}$ is presumed to include the static electromagnetic interactions of the nucleons with each other via the Coulomb and the magnetic interactions. It is assumed that the unperturbed eigen states of H_n are known. The current density $\vec{j}(\vec{r}) = \sum_j (1/c) q_j (\vec{p}_j/M) = \sum_j (1/c) \, q_j \vec{v}_j$ for the point charges and $\vec{v}_j$ is the velocity of the jth nucleon. It should be noted that (1.71b) is the quantum mechanical operator associated with the classical interaction

$$\Delta E(classical) = -\left[\frac{1}{c}\vec{A}(\vec{r}) \cdot \vec{j}(\vec{r}) + \vec{\mu}(\vec{r}) \cdot \vec{H}(\vec{r})\right], \qquad (1.72a)$$

where $\vec{j}(\vec{r})$ is the current density and $\vec{\mu}(\vec{r})$ is the magnetic dipole moment per unit volume. For electric transition we use only the first term of (1.72a) and write

$$H_{int} = -\frac{1}{c}\vec{A}(\vec{r}) \cdot \vec{j}(\vec{r}). \qquad (1.72b)$$

The interaction operator $\vec{A}(\vec{r}) \cdot \vec{j}(\vec{r})$ is impossible to treat exactly because we can know the current $\vec{j}$ only if we know the total Hamiltonian (1.70). The Siegert theorem[20] gives us an exact expression for the interaction operators of electric multipoles in the limit of long wavelengths, $(kr) \ll 1$. The theorem states that the current operator can be expressed in terms of a charge operator (*see* Exercise 1.13). The transition probability (decay rate for spontaneous γ emission) is given by the Golden Rule:

$$T_{fi} = \frac{2\pi}{\hbar}\left|\langle f|H_{int}|i\rangle\right|^2 \rho(E_f)\delta(E_f - E_i + \hbar\omega)$$

$$= \left(\frac{2\pi}{\hbar c^2}\right)\left|\langle f|\vec{A}(\vec{r}) \cdot \vec{j}(\vec{r})|i\rangle\right|^2 \rho(E_f) \qquad (1.73)$$

The delta function ensures conservation of energy in the transition. The density of the final states $\rho(E_f) = \dfrac{dn}{dE_f}$ where, $dE_f = d\hbar\omega = \hbar c dk$.

Using the relation for perfectly conducting walls $j_\lambda(kr_0) = 0$ or $(kr_0 - \frac{1}{2}\pi\lambda) = n\pi$, we get $r_0\,dk = \pi\,dn$, where n is an integer treated here approximately as a continuous variable with n large. Hence,

$$\rho(E_f) = r_0 / \pi\hbar c. \qquad (1.74)$$

Employing (1.72-1.74) and the approximation (1.69), the transition probability for electric 2^λ pole transition is obtained as

$$T(E,\lambda\mu) = \frac{2\pi}{\hbar}\frac{4\pi\hbar kc}{r_0}\frac{\lambda+1}{\lambda}\frac{k^{2\lambda}}{[(2\lambda+1)!!]^2}\left|Q_{\lambda\mu}\right|^2\frac{r_0}{\pi\hbar c}.$$

$$= \frac{8\pi}{\hbar}\frac{(\lambda+1)}{\lambda[(2\lambda+1)!!]^2}k^{2\lambda+1}\left|Q_{\lambda\mu}\right|^2,$$

where $Q_{\lambda\mu} = \int \psi_f^* \, e\sum_{j=1}^{z} r_j^\lambda Y_\lambda^{\mu^*}(\theta_j,\phi_j)\,\psi_i d\tau.$ $\qquad (1.75a)$

The reduced 2^λ pole electric transition probability

$$B(E\lambda) = |Q_{\lambda\mu}|^2.\tag{1.75b}$$

The magnetic multipole transition rates are computed in a manner quite similar to that for the electric multipole transition. We simply cite the results

$$T(M,\lambda\mu) = \frac{8\pi}{\hbar}\frac{(\lambda+1)}{\lambda[(2\lambda+1)!!]^2}k^{2\lambda+1}|M_{\lambda\mu}|^2$$

where $M_{\lambda\mu} = \int \psi_f^* \mu_0 \left\{ \sum_j (g_j^s \, \bar{s}_j + \frac{2}{\lambda+1}\, g_j^\ell \, \bar{\ell}_j) \cdot \vec{\nabla}(r^\lambda \, Y_\lambda^\mu)_j \right\} \psi_i \, d\tau.$

$$\tag{1.76a}$$

The reduced 2^λ pole magnetic transition probability

$$B(M\lambda) = |M_{\lambda\mu}|^2.\tag{1.76b}$$

Selection Rules

In the gamma emission process the selection rules are important. In an electric or magnetic 2^λ multipole radiation, the γ-quantum carries away $\lambda\hbar$ angular momentum. The parity (π) of wave function of a nuclear state may be +ve or −ve. Furthermore, the parity is conserved in close systems in strong and electromagnetic interactions. The parity conservation law can be formulated as follows: $\pi_i \, \pi_f \, \pi_r = +1$, where π_i and π_f mean the parity of initial and final states and π_r is the parity of electric or magnetic radiation.

The parity of $Q_{\lambda\mu}$ is that of Y_λ^μ, namely $(-)^\lambda$, but $M_{\lambda\mu}$ has parity $(-)(-)^\lambda$, since in addition to Y_λ^μ it contains a vector in the gradient operator and pseudo vectors $\bar{s}$ and $\bar{\ell}$. For given states $|i\rangle$ and $|f\rangle$ characterized by $|J_i M_i\rangle$ and $|J_f M_f\rangle$, now vanishing transition rates determined from (1.73) occur only for those multipole operators that obey certain selection rules: From the dominant term in Eq. (1.75) for the electric transitions the operator $r_j^\lambda \, Y_\lambda^{\mu*}$ only gives non-vanishing matrix elements if

$$|J_f - J_i| \le \lambda \le J_f + J_i, \; M_i + \mu = M_f \text{ and } \pi_i\pi_f = (-)^\lambda.$$

When $\lambda = |J_f - J_i|$ is permitted by the parity condition, thus, allowing electric radiation of this multipolarity, one speaks of transition as being parity favoured. If this lowest value of λ, $\lambda = |J_f - J_i|$ satisfies instead the parity condition $\pi_i\pi_f = (-)^{\lambda+1}$, leading to magnetic transition of this polarity; the transition is described as parity favoured.

When the gamma transitions are parity favoured, they are almost always pure electric transitions of multipolarity $\lambda = |J_f - J_i|$. However, when the transitions are parity unfavoured, mixtures of magnetic radiation of multipolarity $\lambda = |J_f - J_i|$ and electric radiation of multipolarity $\lambda + 1$ are commonly encountered. A common example is the mixing of ($M1$) and ($E2$) radiations.

B. Giant Resonances and Sum Rules

The absorption of γ-ray by a nucleus leads either to the formation of the compound nucleus that eventually decays or to photodisintegration. Photodisintegration cross sections exhibit two very interesting characteristics. Firstly, the absorption cross section is resonant like in 10-30 MeV region; secondly, the integrated cross-section far exceeds to that attributed to the motion of one or two nucleons. Goldhaber and Teller[21] first called attention to the existence of "giant resonance" and provided a basically correct picture to describe it. They assumed that the photon excites a motion in the nucleus in which the bulk of the protons move in one direction while the neutrons move in the opposite direction. They called this motion "dipole vibration" (Fig. 1.10*b*).

Nuclei possess no permanent electric dipole moment (Exercise 1.7). However, if a nucleon moves relative to the center of mass of the remaining nucleons then there appears an instantaneous electric dipole moment $\vec{D}$ that can interact with electric field $\vec{E}$. The interaction energy is given by $H_{el} = \vec{D} \cdot \vec{E}$. If we regard the nucleus of nucleon number A as instantaneously divided into $(A-1)$ nucleons with charge e_{A-1} and a single nucleon of charge e_1, then it has a dipole moment of magnitude (Fig. 1.10*a*)

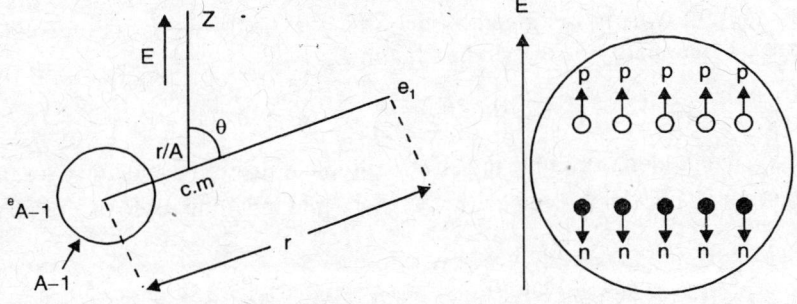

Fig. 1.10: (*a*) Instantaneous dipole moment, (*b*) E pushes protons up and neutrons go down to conserve centre of mass

$$|\vec{D}| = e_1 \frac{A-1}{A} r - e_{A-1} \frac{r}{A} = r\left[\frac{A-1}{A} e_1 - \frac{1}{A} e_{A-1}\right]. \qquad (1.77)$$

If the single nucleon is a proton then $e_1 = e$ and $e_{A-1} = (Z-1)e$, so that

$$|\vec{D}|_{proton} = r\left[\frac{A-1}{A} e - \frac{Z-1}{A} e\right] = \frac{N}{A} er.$$

If we are dealing with a single neutron then $e_1 = 0$ and $e_{A-1} = Ze$, so that

$$|\vec{D}|_{neutron} = r\left[-\frac{Z}{A} e\right] = -\frac{Z}{A} er.$$

The number (eN/A) is called the effective charge of proton and $(-Ze/A)$ the effective charge of neutron. The quantity r is the distance of the nucleon in question from the c.m. of the other $A - 1$ nucleons in the nucleus. Therefore, we write the Z component of the dipole operator independent of c.m:

$$eD_z = e\left[\frac{N}{A} \sum_{proton} z_i - \frac{Z}{A} \sum_{neutron} z_i\right].$$

If the Z-direction is chosen as direction of polarization of photon as shown in Fig. 1.10 then the perturbation term

$$H_{int} = -e\, D_z\, E_z = -f\, e\, E_z\, z,$$

where $f = N/A$ for proton and $f = -Z/A$ for neutron. The time dependence of electric field vector is $E_z(e^{-i\omega t} + e^{i\omega t}) = E_z(2\cos\omega t)$. It follows from the electrodynamics of plane waves that the average of the Poynting vector $\vec{P}$ is

$$|\vec{P}| = \frac{c}{4\pi}|\vec{E} \times \vec{H}| = \frac{c}{4\pi}E_z^2\, \overline{4\cos^2\omega t} = \frac{c}{2\pi} E_z^2.$$

The current density j is given by the time average of the Poynting vector divided by the energy of the single photon E_γ:

$$|\vec{j}| = |\vec{P}|/E_\gamma = \frac{c}{2\pi E_\gamma} E_z^2.$$

Again the golden rules give the electric dipole transition probability for the absorption of photon by nuclear $gs \equiv |\psi_0\rangle$, making transition to the state $|\psi_n\rangle$ as

$$T(E1) = \frac{2\pi}{\hbar}\, |\langle\psi_n| - f e E_z z |\psi_0\rangle|^2\, \rho\,(E_n),$$

where $\rho(E_n)$ is the number of states per energy interval at a final energy E_n of the whole system. The electric dipole absorption cross section

$$\sigma(E_\gamma) = \frac{T(E1)}{|\vec{j}|} = \frac{2\pi}{\hbar} f^2 e^2 E_z^2 \frac{2\pi E_\gamma}{cE_z^2} |\langle \psi_n |z| \psi_0 \rangle|^2 \rho(E_n)$$

$$= 4\pi^2 \left(\frac{e^2}{\hbar c}\right) E_\gamma \left|\int \psi_n^* \, D_z \, \psi_0 \, d\tau\right|^2 \rho(E_n) \text{, where } E_\gamma = E_n - E_0 \,. \quad (1.78)$$

The relation (1.78) represents a general connection between the cross section for the absorption of electric dipole radiation and the properties of the nucleus. Two methods are generally used for calculating photon absorption cross section of nuclei. The first, by using perturbation theory in which the dipole matrix element can be calculated by making explicit assumption of the ground state and excited state wave functions based on some nuclear models.

The second method makes use of the sum-rules and involves the knowledge of the nuclear wave function of the ground state. The sum rule calculations not only present a convenient way to compare all the experimental data at once but also lead to a simplification of the theoretical calculations as they do not involve the excited states. With the employment of the ground state wave function and N-N interaction, one can calculate the first few moments

of the photonuclear sum rules, where moments $\sigma_n = \int\limits_0^\infty \sigma(E_\gamma) E_\gamma^n \, dE_\gamma$, with E_γ

being the photon energy. σ_0 is called the integrated cross section (σ_{int}) and σ_{-1} is called the weighted cross section (σ_b). Even though the information about the excited states is not required in the sum-rule calculations, it indirectly affects the sum rule via exchange forces.

When a photon is absorbed by the target nucleus in the ground state, and produces a discrete non-degenerate excited state, then by integration over the energy range of this resonance, we have (σ_b). Integrating both sides of (1.78) with only this one state populated and assuming that all other quantities in (1.78) do not depend on E_f, we have for $n = 0$,

$$\sigma_{int} = \int \sigma(E_\gamma) \, dE_\gamma = 4\pi^2 \left(\frac{e^2}{\hbar c}\right) E_\gamma \left|\int \psi_0^* \, D_z \, \psi_0 \, d\tau\right|^2 . \quad (1.79a)$$

For a discrete spectrum of states which can be reached from the ground state, we have for $E_0 = 0$ and $E_\gamma \equiv E_n$:

$$\sigma_{int} = 4\pi^2 \left(\frac{e^2}{\hbar c}\right) \sum_n E_n \left|\int \psi_n^* \, D_z \, \psi_0 \, d\tau\right|^2 . \quad (1.79b)$$

Two most important properties of giant dipole resonances are its position and photon energy integrated cross section

$$\sigma_{int} = 4\pi^2 \left(\frac{e^2}{\hbar c} \right) \sum_n E_n |\langle n| D_z |0\rangle|^2 . \tag{1.80}$$

The summation is over all final states that can be reached by a dipole transition from the ground state $|0\rangle$. The right hand side of (1.80) can be expressed by the ground state expectation value of the double commutator of the Hamiltonian with respect to the dipole operator:

$$\sum_n |\langle n|D_z|0\rangle|^2 = \frac{1}{2}\langle 0|[D,[H,D_z]]|0\rangle.$$

Now $H|0\rangle = 0$ and $H|n\rangle = E_n|n\rangle$, so that

$$\sum_n E_n |\langle n|D_z|0\rangle|^2 = \sum_n \langle 0|D_z E_z|n\rangle \langle n|D_z|0\rangle$$

$$= \frac{1}{2}\langle 0| D_z H D_z - D_z^2 H - H D_z^2 + D_z H D_z|0\rangle \frac{1}{2} \langle 0|[D_z,[H,D_z]]|0\rangle.$$

Therefore, $\sigma_{int} = 2\pi^2 \dfrac{e^2}{\hbar c}\langle 0|[D_z,[H,D_z]]|0\rangle.$ \hfill (1.81a)

The above relation is called the TRK (Thomas-Reiche-Kuhn) sum rule, connecting the integrated cross section and matrix element through the use of algebraic relation among operators. The advantage of sum rule is that the measureable quantity σ_{int} can be easily calculated without the knowledge of the excited states. To evaluate the double commutator in (1.81a), we write $H = K + V$, where $K = (1/2M)\sum p_i^2$ is the kinetic energy and V = two-body N-N interaction. Since, $[V, D_z] = 0$ then

$$\langle O|[D_z,[H,D_z]]\langle O| = \frac{\hbar^2}{M} \frac{NZ}{A}$$

Hence, $\sigma_{int} = \dfrac{2\pi^2 e^2 \hbar}{Mc} \dfrac{NZ}{A} = 60 \dfrac{NZ}{A}$ MeV-mb . \hfill (1.81b)

or $\quad \sigma_{int} = 15$ A MeV-mb for $N = Z$.

Those parts of the potential energy term that do not commute with the dipole operator make additional contribution to the integrated cross section. We define a parameter k called the enhancement factor as

$$k = \frac{A}{NZ} \frac{M}{\hbar^2} \langle 0| \left[D_z, [V, D_z] \right] |0 \rangle, \text{ so that}$$

$$\sigma_{\text{int}} = \frac{2\pi^2 e^2 \hbar}{Mc} \frac{NZ}{A} [1 + k]. \tag{1.81c}$$

Levinger and Bethe[22] evaluated the increase in the dipole sum for an equal mixture of ordinary and exchange forces as follows:

$$V = V(r_{ij}) P_{ij}^M, \text{ where } P_{ij}^M \text{ is the Majorana exchange operator.}$$

Now,

$$[V, z_i] = V(r_{ij}) P_{ij}^M z_i - z_i V(r_{ij}) P_{ij}^M = (z_j - z_i) V(r_{ij}) P_{ij}^M$$

or

$$\langle 0| \left[z_i, [V, z_i] \right] |0 \rangle = - \langle 0| (z_i - z_j)^2 V(r_{ij}) P_{ij}^M |0 \rangle$$

$$= -\frac{1}{3} \langle 0| r_{ij}^2 V(r_{ij}) P_{ij}^M |0 \rangle \text{ for the spherical nucleus.}$$

Hence, the integrated cross section for photonuclear reactions obtained by Levinger and Bethe may be written as

$$\sigma_{\text{int}} = \frac{2\pi^2 e^2 \hbar}{Mc} \left[\frac{NZ}{A} - \frac{M}{3\hbar^2} \int \psi_0^* \left\{ \sum_i \sum_j r_{ij}^2 V(r_{ij}) P_{ij}^M \right\} \psi_0 \, d\tau \right], \tag{1.82}$$

where ψ_0 is the complete wave function of the nuclear ground state. Levinger and Bethe made a definite calculation of the increase attributable to exchange forces. If the exchange forces are a fraction 'x' of two-body forces then

$$\sigma_{\text{int}} = 15A (1 + 0.8x) \text{ MeV-mb.}$$

Bethe also calculated energy weighted cross section namely,

$$\sigma_b = \int \left[\sigma(E1) / E_\gamma \right] dE_\gamma = \frac{2\pi^2 e^2 \hbar}{Mc} \frac{2M}{\hbar^2} \sum_n \left| \left(\frac{N}{A} \right) (z_i)_{on} - \left(\frac{Z}{A} \right) (z_j)_{on} \right|^2$$

$$= 4\pi^2 \left(\frac{e^2}{\hbar c} \right) \frac{NZ}{A} \langle z^2 \rangle_{00}, \tag{1.83}$$

where we have used relation (1.78), closure property and $z = |z_i - z_j|$.

For a spherical heavy nucleus, $\sigma_b = \frac{4\pi^2}{3} \left(\frac{e^2}{\hbar c} \right) \frac{NZ}{A} \langle r^2 \rangle_{00}$ and for a light nucleus[23],

$$\sigma_b = \frac{4\pi^2}{3}\left(\frac{e^2}{\hbar c}\right)\frac{NZ}{A-1}\langle r^2 \rangle_{00},$$

where $\langle r^2 \rangle_{00}$ = mean square radius of the charge distribution.

Exercise 1.12. Show that $\left[D_z,\left[K,D_z\right]\right]=\dfrac{\hbar^2}{M}\dfrac{NZ}{A}.$

$$[K,D_z]=\left[\frac{1}{2M}\sum p_z^2, D_z\right]=\frac{i\hbar}{M}\left[\frac{N}{A}\sum_{proton} p_z - \frac{Z}{A}\sum_{neutron} p_z\right].$$

In the above we have used

$$[p_z,z]=-i\hbar \text{ and } [p_z^2,z]=p_z[p_z,z]+[p_z,z]p_z=-2i\hbar p_z$$

Hence, $\left[D_z,\left[K,D_z\right]\right]=\dfrac{\hbar^2}{M}\dfrac{NZ}{A}.$

Exercise 1.13. Prove Siegert's theorem for the electric multipole transition in $kr \ll 1$ approximation.

The fact that $A_{\lambda\mu}^E(r)$ in (1.68) is gradient of a scalar for $kr \ll 1$, we can write

$$\vec{\nabla} \equiv \vec{\nabla}_r \text{ and } A_{\lambda\mu}^E(r) = \nabla\Phi,$$

where
$$\Phi = \frac{i}{k}\left(\frac{4\pi\hbar kc}{r_0}\frac{\lambda+1}{\lambda}\right)^{\frac{1}{2}}\frac{1}{(2\lambda+1)!!}(kr)^\lambda Y_\lambda^\mu.$$

The transition matrix element of (1.73) is

$$\int \vec{\nabla}\Phi \cdot \langle f|\vec{j}(r)|i\rangle \, d^3r = \int \vec{\nabla}\Phi \cdot \left[\psi_f^* \, \vec{j}(r) \, \psi_i\right] d^3r_1 \cdots d^3r_A$$

We use the general identity, $\vec{\nabla}\cdot(u\vec{v})=\vec{v}\cdot\vec{\nabla}u+u\vec{\nabla}\cdot\vec{v}$ and get

$$\langle \psi_f|\vec{j}(r)||\psi_i\rangle\cdot\vec{\nabla}\Phi = \int \vec{\nabla}\Phi\cdot\psi_f^*\vec{j}(r)\psi_i d\tau - \int \Phi\{\vec{\nabla}\cdot\langle\psi_f|\vec{j}(r)|\psi_i\rangle\} \, d\tau.$$

Using Green's theorem, the first term in the right hand side of the above equation can be converted into a surface integral which vanishes if the surface is selected outside the nuclear volume but still satisfying $kr \ll 1$. Therefore,

$$\langle\psi_f|\vec{A}_{\lambda\mu}^E(r)\cdot\vec{j}(r)|\psi_i\rangle = -\int\Phi\{\langle\psi_f|\ \vec{\nabla}\cdot\vec{j}(r)\,|\psi_i\rangle \, d\tau\}$$

$$= \int \Phi \left\{ \left\langle \psi_f \left| \frac{\partial \rho(r)}{\partial t} \right| \psi_i \right\rangle \right\} d\tau,$$

from equation of continuity for charge conservation (Appendix B1). For point like charges we can write $\rho(r) = \sum_j q_j \delta(r - r_j)$. Therefore,

$$\left\langle \psi_f \left| \vec{A}_{\lambda\mu}^E(r) \cdot \vec{j}(r) \right| \psi_i \right\rangle = \int \Phi \left\{ \frac{\partial}{\partial t} \left\langle \psi_f \left| \rho(r) \right| \psi_i \right\rangle \right\} d\tau$$

$$= \frac{i(E_i - E_f)}{\hbar} \int \Phi \left\langle \psi_f \left| \rho(r) \right| \psi_i \right\rangle d\tau,$$

which proves the statement of the Siegert theorem. Here $(E_i - E_f)/\hbar = ck$.

1.5 CLASSIFICATION OF ELEMENTARY PARTICLES

Arrangement of elements in order of their increasing atomic weights by Mendeleef in 1869 led to the discovery that elements having similar properties recur after regular intervals (Periodic Law). It was the first example of great synthesis achieved semi empirically when theoretical basis was still lacking. There have been many attempts to develop a classification scheme for particles on the assumption that some are more elementary than others. In an attempt to make broad categories of particles, the particle directory had to be expanded.

Hyperon: Neutron and proton are not only $A = 1$ hadrons. Other hadrons with mass number $A = 1$ exist; they are called hyperons. Historically, lamda particle (Λ), the first hyperon having mass of 1116 MeV and charge zero was observed in $\pi^- + p$ collisions through a hydrogen bubble chamber. The negative pion first disappears and further down stream appears as two V-like events. The first particle is neutral kaon; a meson and the second particle is called hyperon or lamda $\left(\Lambda^0 \right)$ (the name of course, refers to the characteristic appearance of the tracks of proton and pion). Λ^0 is not the only hyperon, a number of other hadronically stable particles of similar character have been found. For example, Σ of mass 1189.4 MeV and charge $+e$ and Ω of mass 1672.5 MeV.

It was realized soon after the discovery of these new particles that they are produced in strong interaction, their mean lives of decay are very long $\left(10^{-8} - 10^{-10} \text{ s} \right)$ on nuclear time scale $\left(10^{-23} \text{ s} \right)$. Further, they are never produced singly but pairs of them are produced in association with each other (associated production). These characteristics of the new particles are described in terms of a new property, called "strangeness", a concept mooted by Gell-Mann and Nishijima[24]. Hence, these particles are known as strange particles.

Resonances: Apart from the stable and semi-stable particles discussed so far, many other particles have been discovered having life time comparable to the time taken by a pion to travel past a proton ($\sim 4 \times 10^{-24}$ sec). These are known as particle resonances. Mesons, on interacting with the nucleons excite them to their resonant states. For an energetic pion beam incident on a proton, the nucleon resonances show up as distinct peaks in the invariant mass spectrum of the $\pi - p$ system. The width of these peaks gives mean lives through the uncertainty relation $\tau = \hbar/\Delta E$, while their position gives masses: The lightest and most dominant of these resonances is the Δ-resonance (*see* Fig. 1.11). It occurs at excitation energy of about 300 MeV and is an extremely short lived resonance. Its life time is of the order of $\tau = \hbar/\Delta E$ sec, which corresponds to a width of about 120 MeV. It has spin and isospin, both equal to 3/2 and hence, appear into four charge states e.g., Δ^{++}, Δ^{+}, Δ^{0} and Δ^{-}. The measured cross section for the $(\pi^{+}p)$ scattering at the resonance is 200 mb at the pion energy of 190 MeV. We can assign spin and isospin values of the resonance state as follows: $(\sigma_{max}^{sc})_{expt.} \approx 200$ mb.

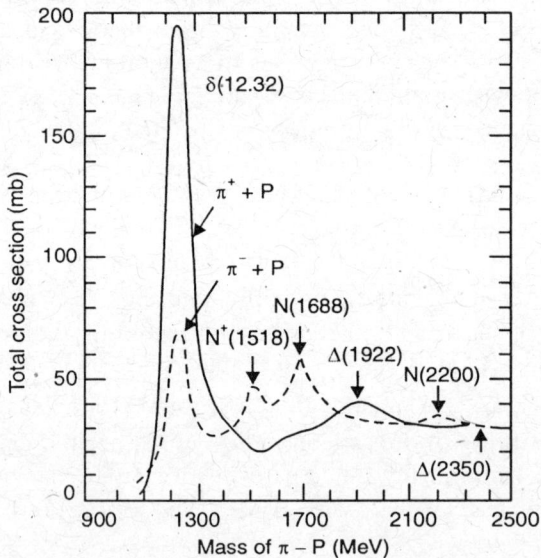

Fig. 1.11: Particle resonances

Employing elastic resonance scattering total cross section (*see* Eq. 5.42), we get

$$(\sigma_{max}^{sc})_{theo.} = 4\pi\lambda^{2}g, \text{ where } g = \frac{(2I+1)}{(2S_{\pi}+1)\,(2S_{p}+1)} = \frac{(2I+1)}{2}.$$

or $$(\sigma_{max}^{sc})_{theo.} = 2\pi\lambda^2(2I+1).$$

where I is the angular momentum of the compound state, s_π and s_p are respectively spin of pion and proton and λ is the wavelength of incident pion ($= 0.833$ fm for $E_\lambda = 190$ MeV). Comparing $(\sigma_{max}^{sc})_{expt.}$ with $(\sigma_{max}^{sc})_{theo.}$ we can assign $I = 3/2$ to the compound (resonance) state C^*. Now isospin $T_\pi = 1$ and $T_N = 1/2$ respectively for pion and nucleon. The isospin of the compound state, $\vec{T}_{C^*} = \vec{T}_\pi + \vec{T}_N$ and $T_3^{C^*} = T_3^\pi + T_3^N$. Hence, for the reaction

$$\pi^+ + p \to C^* \ i.e., (T = 1 + 1/2 \to 3/2) \ \text{and} \ (T_3 = 1 + 1/2 \to 3/2),$$

we assign $T^{C^*} = 3/2$ and $T_3^{C^*} = +3/2$ to the resonance state.

A. Strangeness and Hypercharge

The whole subject of production and decay of different particles was systematized by introducing the new quantum number called strangeness 'S' and by postulating further that in strong interaction strangeness must be conserved. A semi-empirical relationship between particle charge Q (in units of proton charge e), the generalized isospin T_3, the hypercharge Y, the baryon number B and strangeness S was also given by Gell-Mann and Nishijima[24] as

$$Q = \left(T_3 + \frac{1}{2}Y\right) \ \text{where} \ Y = B + S. \tag{1.84}$$

For non-strange particles like nucleons and pions $S = 0$. For nucleons $B = 1$, $Y = 1$. The charge operator

$$Q = T_3 + \frac{B+S}{2}.$$

SYMMETRY CLASSIFICATION

After the discovery of strange particle, Sakata[25] considered the neutron, proton and Λ as fundamental particles and tried to build a model in which all particles are build by these three and their antiparticles. The three particles p, n and Λ were called Sakatons. Sakata's model was not very successful because it indicated certain lack of symmetry in combined system and it predicted particles that were not observed. Gell-Mann and Neeman[26] suggested that it is possible to collect all presently known particles and resonances in group containing 3 or 8 or 10 or 27 members. This classification known as "The Eightfold Way" comes from the symmetry properties of group SU(3) (*see* Appendix D). The group containing three members (basic triplet) consists of so called up (u), down (d) and strange (s) quarks, whose properties are given in Table 1.5. These quarks are the building blocks of various hadrons (mesons and baryons).

B. The Quark Model of Hadrons

In the quark model, mesons are quark-antiquark $(q\bar{q})$ pairs bound by their interactions. The states with the lowest energy or mass are expected to be spatial s-states (with relative angular momentum $\ell = 0$). The total spin can be $S = 0$ or 1. The quantum number T_3, S, Y of antiquark are negative of those of quarks.

Table 1.5: Quark properties

Quark flavour	Spin	Q	B	T_3	S	Y	Mass (MeV/c^2)
u	$\dfrac{1}{2}$	$\dfrac{2}{3}$	$\dfrac{1}{3}$	$\dfrac{1}{2}$	0	$\dfrac{1}{3}$	390
d	$\dfrac{1}{2}$	$-\dfrac{1}{3}$	$\dfrac{1}{3}$	$-\dfrac{1}{2}$	0	$\dfrac{1}{3}$	390
s	$\dfrac{1}{2}$	$-\dfrac{1}{3}$	$\dfrac{1}{3}$	0	-1	$-\dfrac{2}{3}$	510

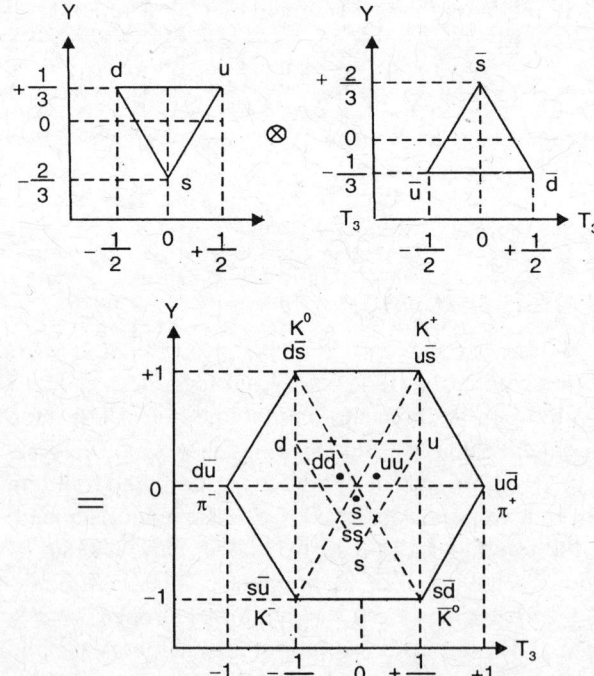

Fig. 1.12: (left): Flavour structure of $J^\pi = 0^-$ mesons. (right): Weight diagram

This scheme of classification on the basis of octet symmetry is best demonstrated by making a graphical plot (weight diagram) of the mesons in the $T_3 - Y$ plane shown in Fig. 1.12. The members of the octet (super-multiplet) form a symmetric hexagon with one at each corner and two at the centre of the hexagon. As can be seen from Fig.1.12.

We can also display the flavour structure of the $J^\pi = 0^-$ mesons into 3×3 matrix:

$$\begin{pmatrix} u \\ d \\ s \end{pmatrix} \otimes (\bar{u} \; \bar{d} \; \bar{s}) = \begin{pmatrix} u\bar{u} & u\bar{d} & u\bar{s} \\ d\bar{u} & d\bar{d} & d\bar{s} \\ s\bar{u} & s\bar{d} & s\bar{s} \end{pmatrix} = \frac{1}{3} \begin{pmatrix} tr & 0 & 0 \\ 0 & tr & 0 \\ 0 & 0 & tr \end{pmatrix} + \frac{1}{3} \begin{pmatrix} A & u\bar{d} & u\bar{s} \\ d\bar{u} & B & d\bar{s} \\ s\bar{u} & s\bar{d} & C \end{pmatrix},$$

where
$$tr = u\bar{u} + d\bar{d} + s\bar{s} \text{ and } A = (2u\bar{u} - d\bar{d} - s\bar{s}),$$
$$B - (2d\bar{d} - \bar{u}u - \bar{s}s) \text{ and } C = (2\bar{s}s - \bar{u}u - \bar{d}d).$$

The set $j^\pi = 0^-$ mesons has the following members π^+, π^-, π^0, η^0, K^+, K^-, K^0, $\bar{K}^0$. Properties of these mesons are given in Table 1.6. Similarly, the triplet state of the quark-antiquark system with orbital angular

Table 1.6: Properties of $J^\pi = 0^-$ meson nonet

Meson	Principal decay mode	$q\bar{q}$ System	T_3	S	Y	Mass (MeV/c^2)	t (sec)
K^+	$\mu^+\nu$, $\pi^+\pi^0$	$u\bar{s}$	1	1	1	493.6	1.24×10^{-8}
K^-	$\mu^-\nu$, $\pi^-\pi^0$	$s\bar{u}$	-1	-1	-1	493.6	,,
K^0	$\pi^+, \pi^-, \pi^0\pi^0$	$d\bar{s}$	1	1	1	497.7	0.89×10^{-10}
$\bar{K}^0$		$s\bar{d}$	-1	-1	-1	497.67	
π^+	$\mu^+\nu$	$u\bar{d}$	0	0	0	139.6	2.60×10^{-8}
π^-	$\mu^-\nu$	$d\bar{u}$	0	0	0	139.6	
π^0	$\gamma\gamma$	$\frac{1}{\sqrt{2}}(u\bar{u} - d\bar{d})$	0	0	0	134.9	0.83×10^{-16}
η^0	$\gamma\gamma$, $3\pi^0$	$\frac{1}{\sqrt{2}}(u\bar{u} - d\bar{d})$	0	0	0	548.8	$\sim 0.5 \times 10^{-18}$
η^{0*} singlet		$\frac{1}{\sqrt{3}}(s\bar{s} + d\bar{d} + u\bar{u})s\bar{s}$	0	0	0	957.5	

momentum zero gives the observed $J^\pi = 1^-$ vector mesons which again can be assigned to an $SU(3)$ octet and a singlet with masses given in the traditional notation ρ (770), ω (782), ϕ (1020), δ (980). The simplest baryons of integral charges and baryon number $B = 1$ can be constructed from three quarks (qqq).

The states of lowest energies are again expected to be spatial S-state ($\ell = 0$). The intrinsic spin can be $S = 3/2$ or $1/2$. Baryons tend to be much more massive than mesons, so that the mass difference between s and u, d quarks are not as significant. The flavour $SU(3)$ decomposition of the 27 possible qqq combinations, resulting from the choice of 3 spin $1/2$ quarks from any three flavours is done by some symmetry principle peculiar to members of a particular multiplet. The above combinations are written symbolically as

$$3 \otimes 3 \otimes 3 = 10_S \oplus 8_{MS} \oplus 8_{MS} \oplus 1_A,$$

where the subscripts indicate symmetric, mixed symmetry and antisymmetry states under interchange of any two quarks. We first combine two of the quarks so that nine qq combinations arrange themselves into two $SU(3)$ multiplets, $\{3\} \otimes \{3\} = \{6\} \oplus \{3^*\}$, i.e. a triplet of antisymmetric combinations:

$$\{3^*\} = [ud], [us], [ds],$$

where
$$[q_1 q_2] = \frac{1}{\sqrt{2}} (q_1 q_2 - q_2 q_1)$$

and a sextet of symmetric combinations

$$\{6\} = uu, dd, ss, \{ud\}, \{us\}, \{ds\},$$

where $\{q_2 q_2\} = (q_1 q_2 + q_2 q_1)/\sqrt{2}$. For baryons $Y = B+S = 1+S$. The two nucleons ($S = 0$) with $T_3 = \pm(1/2)$ fall on the line $Y = B + S = 1$; the three Σ hyperons ($S = -1$) with $T_3 = \pm 1, 0$ fall on the line $Y = 0$ and two X hyperons ($S = -2$) with $T_3 = \pm 1, 0$ fall on the line $Y = -1$. Finally, the single L hyperon ($S = -1$) with $T_3 = 0$ is at the centre of the figure (Fig. 1.13).

To figure out the T_3, Y content of the baryons we look at the corresponding weight diagram Fig. (1.13). $J^\pi = \frac{1}{2}^+$ baryons have the members:

$$pn\Lambda \sum{}^+ \sum{}^0 \sum{}^- \Xi^0 \Xi^-. \text{ (Table 1.7)}.$$

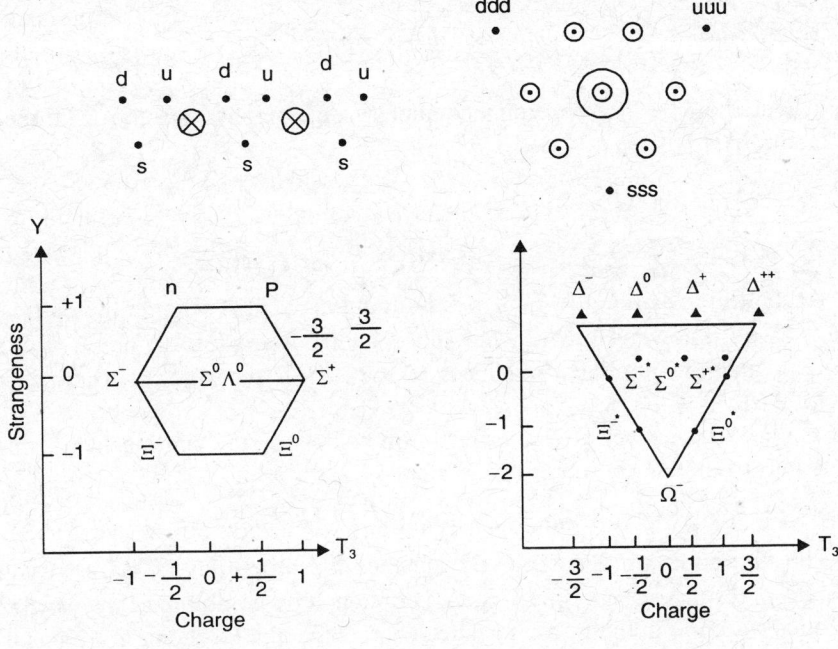

Fig.1.13: (*a*) $J^\pi = \frac{1}{2}+$ baryon octet   **Fig. 1.13:** (*b*) $J^\pi = \frac{3}{2}+$ baryon decuplet

Table 1.7: Properties of members of $J^\pi = \frac{1}{2}^+$ octet baryons

Baryons	Principal decay mode	qqq system	S	Y	T₃·	Mass (MeV/c²)	τ (sec)
p	Stable	uud	0	1	+1/2	938.3	∞
n	$pe^-\bar{\nu}$	udd	0	0	−1/2	939.6	898
Σ^-	$n\pi^-$	sdd	−1	0	−1	1197.3	1.5×10^{-10}
Σ^+	$p\pi^0, n\pi^+$	suu	−1	0	1	1189.4	0.8×10^{-10}
π^+	$\Lambda\pi^-$	ssd	−2	−1	0	1321.3	1.6×10^{-10}
Ξ^0	$\Lambda\pi^0$	ssu	−2	−1	0	1314.9	2.9×10^{-10}
Σ^0	$\Lambda\gamma$	s{du}	−1	0	0	1192.5	5.8×10^{-10}
Λ^0	$p\pi^-, n\pi^0$	s[du]	−1	0	0	1115.6	2.6×10^{-10}

Here, $\{du\} = \dfrac{1}{\sqrt{2}}\,(du + ud)$ and $[du] = \dfrac{1}{\sqrt{2}}\,(du - ud)$.

Again the $J^{\pi} = 3^+/2$ baryon decuplet (group of ten) has the following members:

$\Delta^-(1232),\ \Delta^0, \Delta^+, \Delta^{++}$

$\sum{}^{*-}(1285),\ \sum{}^{*0},\ \sum{}^{*+},\ \Xi^{*-}\ (1430),\ \Xi^{*0}$ and $\Omega^-(1680)$.

At the time when Gell-Mann was building up SU (3) scheme the particle Ω^- was not known. However, the subsequent observation of the Ω^- with correct quantum numbers and mass = 1672 MeV measured a triumph of flavor $SU(3)$.

The Color Degrees of Freedom: In implementing the quark scheme, one runs into trouble at the next logical step:

$$p = uud \qquad n = udd \qquad \Delta^{++} = uuu$$

The *uuu* configuration correctly matches the properties of Δ^{++} baryon (the $\pi^+ p$ resonance). Its spin $J = 3/2$ is obtained by combining three $J = 1/2$ identical fermion in their ground state. That is, the quark scheme forces us to combine identical fermions u in a completely symmetric ground state uuu. This is forbidden by the Pauli principle. In addition to the Pauli principle fiasco, the constituent quark model allows particles like uu, $\bar{u}\,\bar{u}$ etc. No particles with charge 4/3 have ever been observed. These problems can be resolved by introducing a new property or quantum number for quarks: "Color". It is proposed that quarks come in three colors: red, green and blue, denoted symbolically by R, G, B. These colors have of course no relation to real colors of every day life. We then rewrite the quark wave function for Δ^{++} as $u_R u_G u_B$. By this choice we clearly implement the Pauli principle correctly.

The three quarks that make up the Δ-state are distinguishable by their color quantum number. Though by introducing the color quantum number we have overcome statistics problem, yet there is another problem. If $u_R u_G u_B$ is Fermi's Δ^{++}, then we end up with many states for proton : $u_R u_G d_B$, $u_R u_G d_G$, $u_R u_G d_G$ and so on, yet only one state exists. This problem is resolved by demanding that all particles in nature are colorless or white or color singlet (or to be more precise unchanged by an exchange in R, G, B space). The antiquarks are the complementary colors : cyan ($\bar{R}$), magenta ($\bar{G}$) and yellow ($\bar{B}$).

This artificial looking color hypothesis was confirmed in the $e^+ e^-$ annihilation experiments. The $e^+ e^-$ annihilation takes place as multistep process :

$$e^+ e^- \rightarrow \gamma \rightarrow \sum q\bar{q} \rightarrow \text{hadrons} \qquad\qquad (I)$$

and also $$e^+e^- \to \gamma \to \mu^+\mu^-.$$ (II)

The process (I) mediates the change of an electron pair into a quark pair and then the quark pair through the strong interactions turns into several hadrons. The total cross section for the process II: $e^+e^- \to \mu^+\mu^-$ is

$$\sigma(e^+e^- \to \mu^+\mu^-) = \frac{4\pi\alpha^2}{3s},$$ (1.85)

where $\sqrt{s}$ is the energy in the centre of mass system. The cross section for the $e^+e^- \to q_i\bar{q}_i$ is obtained by replacing muons by quark and hence, the total cross section

$$\sigma(e^+e^- \to q_i\bar{q}_i) = \frac{4\pi\alpha^2}{3s} Z_{qi}^2,$$

where Z_{qi} is the charge of the quark with flavour i in units of the charge of the electron.

Of course, the quarks do not appear as free particles in the laboratory and they must fragment into colorless hadrons. Since, the cross section for the production of quark pair is proportional to the square of the charge, the total cross section, which is obtained by summing over all reactions should be proportional to the sum of all the quark charges squared. For example, for the Gell-Mann-Zweig (GW) model[27],

$$\frac{\sigma(e^+e^- \to q_i\bar{q}_i)}{\sigma(e^+e^- \to \mu^+\mu^-)} = Z_{qi}^2 = \left(\frac{2}{3}\right)^2 + \left(-\frac{1}{3}\right)^2 + \left(-\frac{1}{3}\right)^2$$

Hence, $R = \dfrac{\sigma(e^+e^- \to \text{hadrons})}{\sigma(e^+e^- \to \mu^+\mu^-)} = \dfrac{2}{3}$.

However, the experimental ratio is close to 2 below 4 GeV. If we consider three color degrees of freedom. Then $R = 3 \times (2/3) = 2$, which is in agreement with experiment. The factor 3 arises because of the three possible color combinations of the $q\bar{q}$ pair.

Colors of quark imply 9 different ways of coupling the gluons (G_i) to quarks. Following nine bicolored states exist: $R\bar{R}$, $R\bar{G}$, $R\bar{B}$, $B\bar{G}$, $G\bar{B}$, $G\bar{R}$, $B\bar{R}$, $G\bar{G}$, $B\bar{B}$. A quark emitting (or absorbing) a gluon changes its color (*see* Fig. 1.14). The quarks cluster together to form color singlet configuration because forces between them are all due to gluon exchange. One of the nine combination $(R\bar{R} + B\bar{B} + G\bar{G})$, is a color singlet, which lacks any net color charge and therefore can not play the roles of gluon carrying color from one quark to another. Thus, there are eight colored gluons.

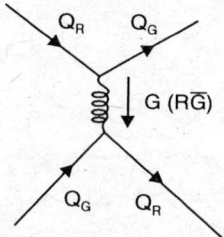

Fig. 1.14: Quark-Quark Interaction via Gluon Exchange

All hadrons are postulated to be colorless, i.e. they belong to the singlet representation of the $SU(3)$ color group. The color wave function of the baryon is therefore, a 3×3 Slater determinant of color singlet state of three quarks :

$$\Psi_{col\cdot sin\,glet} = \frac{1}{\sqrt{6}} \left[|RGB\rangle - |RBG\rangle + |BRG\rangle - |BGR\rangle + |GBR\rangle - |GRB\rangle\right]$$

$$= |Color\rangle_A |qqq\rangle_A \equiv \Psi_{total} = \Psi_{col} \cdot \Psi_{space} \cdot \Psi_{spin-flavour} \qquad (1.86)$$

Note the contrast with the state function for the three nucleons in ^3H or ^3H:

$$|NNN\rangle_A = |space,\ spin,\ isospin\rangle_A$$

For the ground state (s-state) baryons, Ψ_{space} is symmetric and therefore, in order that the total wave function is antisymmetric (for fermions) (Eq. 1.86), we require ($\Psi_{spin-flavour}$) symmetric since Ψ_{col} is antisymmetric. We write explicitly the $(3q)$ wave function for proton $(S = 1/2)$. We use the following spin-flavour notation for the proton: $|p\uparrow\rangle \equiv uud\,(\uparrow\uparrow\downarrow)$. For a proton with z spin component $m_s = +1/2$, we may write the spin wave function as a product of spin wave function of one quark and that of the remaining pair :

$$\chi_P\left(S = \frac{1}{2},\ m_s = \frac{1}{2}\right) = \sqrt{\frac{2}{3}}\,uud(\uparrow\uparrow\downarrow) - \sqrt{\frac{1}{3}}\,uud(\uparrow\downarrow\uparrow).$$

Here we have singled out the d-quark and coupled the u-quark pair. The factors in the above equation are the Clebs-Gordon coefficients for the coupling of spin 1 and spin 1/2. We should replace $uud\,(\uparrow\downarrow\uparrow)$ by the correct $S = 1$, $M_S = 0$ wave function, $\left[uud\,(\uparrow\downarrow\uparrow) + udd(\downarrow\uparrow\uparrow)\right]/2$. Then in one spin flavour notation

$$|p\uparrow\rangle = \sqrt{\frac{2}{3}}\,uud\,(\uparrow\uparrow\downarrow) - \sqrt{\frac{1}{6}}\,uud\,(\uparrow\downarrow\uparrow) - \sqrt{\frac{1}{6}}\,uud\,(\downarrow\uparrow\uparrow)$$

$$= \sqrt{\frac{1}{6}}\left[2uud\,(\uparrow\uparrow\downarrow) - uud\,(\uparrow\downarrow\uparrow) - uud\,(\downarrow\uparrow\uparrow)\right].$$

This expression is still symmetric in terms of the exchange of the first and second quarks, and not two arbitrary quarks as we need. With appropriate normalization factor and taking the permutation udu ($\uparrow\uparrow\downarrow$) and duu ($\uparrow\uparrow\downarrow$), finally we can write,

$$|p\uparrow\rangle = \frac{1}{\sqrt{18}} [2uud\,(\uparrow\uparrow\downarrow) - uud\,(\uparrow\downarrow\uparrow) - duu\,(\uparrow\downarrow\uparrow)$$

$$+ 2duu(\downarrow\uparrow\uparrow) - duu\,(\uparrow\uparrow\downarrow) - duu\,(\uparrow\downarrow\uparrow)$$

$$+ 2udu\,(\uparrow\downarrow\uparrow) - udu(\uparrow\uparrow\downarrow) - udu\,(\downarrow\uparrow\uparrow)] \qquad (1.87)$$

The neutron wave function is obtained by exchanging the u- to d-quarks. Employing (1.87) for $|p\uparrow\rangle$ we can calculate the magnetic moment of the proton. The magnetic moment operator is $\sum_i \mu_i(\sigma_3)_i$, where μ_i is the magnetic moment of the ith quark and summation is over all constituent quarks. σ_3 is the 3-component of quark spin.

More Quark Flavours: In the 1960's, the experimentally available energy was upto a few GeV and the value of R was measured to be close to 2, not 2/3. This was thought to clearly indicate the existence of colored fractionally charged quark. When the experiments were done at higher energy in a 8 GeV electron-positron collider at Stanford Linear Accelerator Centre (SLAC), the results were rather curious. It was found that $R = 4$ above 4 GeV.

In 1974, a spectacular occurrence took place: the discovery of the new particle J/ψ (J–psi)[28]. It was a colorless hadron made of quarks with a new flavour called "charm". As the energy of the beam is varied, the reaction shows a narrow peak (resonance) around 3.105 GeV. The narrow width indicates that J/ψ is much more stable than other mesons (ρ,ω,ϕ) observed earlier.

The charm quark designated by the symbol c is endowed with an additional quantum number $c = +1$, the anticharm has $c = -1$. The c-quark has a mass ≈ 1.55 GeV and charge 2/3 in units of e. Generalization of (1.84) to include c leads to

$$Q = T_3 + \frac{B + S + c}{2}. \qquad (1.88)$$

In 1977, at the Fermi National Argone Laboratory (FNAL) scientists discovered a new resonance state of mass 9.46 GeV using 400 GeV proton beam.[29] This resonance particle, known as upsilon particle, decay into $\mu^+\mu^-$ pair. It is made of a new type of quark flavour known as bottom quark designated by quantum number beauty b. Bottom quark has charge $-1/3$ and mass 4.72 GeV.

In addition to the five quarks (u, d, s, c and b) the existence of which has been confirmed experimentally, the existence of a sixth quark known as top quark, designated by the quantum number truth t ($=1$), was predicted theoretically, and has been recently confirmed by experiment.[30] Finally, the Gell-Mann-Nishijima formula takes the form:

$$Q = T_3 + \frac{B+S+c+b+t}{2}.$$ (1.89)

The properties of these three new quarks (c, b, t) are given in Table 1.7.

Table 1.7: Properties of Quarks beyond GW Model

Quark	Spin	Q	B	T_3	Y	Mass(GeV/c²)
c	1/2	2/3	1/3	0	1/3	1.55
b	1/2	−1/3	1/3	0	1/3	4.72
t	1/2	1/3	1/3	0	(+1/3)	199 ± 22

Exercise 1.14. Employing constituent *GW* quark model calculate the magnetic moment ratio $\mu_{neutron}/\mu_{proton}$ and compare the same with the experimental value.

The magnetic moment of a point like spin 1/2 particle of charge e and mass m in natural units ($\hbar = c = 1$) is $e/2m$. Thus, a structureless quark of charge $Q_i e$ and mass m_i has magnetic moment

$$\mu_i = Q_i \left(\frac{e}{2m_i} \right).$$

Hence, in the non-relativistic approximation, we may write the magnetic moment of proton as

$$\mu_p = \sum_{i=1}^{3} \langle p \uparrow | \mu_i (\sigma_3)_i \langle p \uparrow |.$$

Using the explicit wave function (1.87) we obtain

$$\mu_p = \frac{1}{18} \left[(\mu_u - \mu_u + \mu_d) + (-\mu_u + \mu_u + \mu_d) + 4(2\mu_u - \mu_d) \right] \times 3,$$

where the factor 3 takes care of the permutations. The magnetic moment of proton in terms of quark moment is then

$$\mu_p = \frac{1}{3} [4\mu_u - \mu_d].$$

The neutron magnetic moment is obtained by the interchange of u and d. Hence, $\mu_n = \left[4\mu_d - \mu_u\right]/3$. Assuming $m_u \approx m_d$ and employing (1), we get $\mu_u = -2\mu_d$. Therefore, the quark model predicts

$$\left(\frac{\mu_n}{\mu_p}\right)_{\text{quark model}} = -\frac{2}{3} = 0.66, \text{ which agrees with}$$

$$\left[\frac{\mu_n}{\mu_p}\right]_{\text{expt}} = -\frac{1.91}{2.79} \approx -0.68.$$

REFERENCES

1. Weisskopf VF. Proc. Intl. Conf. Nucl. Structure, Kingston Canada (1960) John Wiley (1960).
2. Rutherford E. Phil. Mag. **21**, 669 (1911).
3. Mott NF. Proc. Roy. Soc. (London) **A126**, 259 (1930).
4. Chadwick J. Proc. Roy. Soc. (London) **A136**, 692 (1932).
5. Heisenberg W, Z. Physik **77**, 1 (1932).
6. Felix Bloch, in Heismberg and Early Days of Quantum Mechanics', Physics Today, December 1976, page 23.
7. Gamow G, Z. Physik **51**, 204 (1928).
   Geiger H, Nuttall JM. Phil. Mag. 22, 613 (1911).
8. Fermi E. Z. Physik **88**, 161 (1934).
9. Dirac PAM. Proc. Roy. Soc. (London) **A117**, 610 (1928).
   Dirac PAM. Proc. Roy. Soc. **A232**, 435 (1971), Ibid. **A328**, 1 (1978).
10. Hofstadter H. Ann. Rev. Nucl. Sc. **7**, 231 (1957) and Nuclear and Nucleon Structure, Bengamin, New York (1963). H. Dejbaksh, Phys. Lett. **249**, 195 (1990).
11. Chamberlain O. et al. Phys. Rev. **100**, 947 (1955).
12. Lee TD, Yang CN. Phys. Rev. **104**, 254 (1956).
13. Particle Data Group, K. Hagiwara et al., Phys. Rev. **D66**, 01001 (2002).
14. Neils Bohr, Collected Works (Ed. L. Rosenfeld), North Holland, New York, 1976. N. Bohr, Nature **137**, 344 (1936).
15. Bethe HA, Bacher RF. Rev. Mod. Phys. **8**, 193 (1936).
16. Von Weizsacker CF. Z. Physik **96**, 431 (1935).
17. Ikeda K. et al. Prog. Theoret. Phys. (Japan) (Suppl) 464 (1968), also K. Wildermuth and Y.C. Tang in A Unified Theory of the Nucleus, Academic Press, N.Y. (1977).
18. Schiff LI. in Quantum Mechanics, McGraw-Hill, N.Y. (1955).
19. The Upanishads, Translated by Alistair Shearer and Peter Russell, Herper Colophon Books, New York, 1978.
20. Siegert AJF. Phys. Rev. **52**, 787 (1937).
21. Goldhaber M, Teller E. Phys. Rev. **74**, 1046 (1948).
22. Levinger JS, Bethe HA. Phys. Rev. **78**, 115 (1950).
23. Foldy LL. Phys. Rev. **107**, 1303 (1957).

24. Nishijima K. Prog. Theor. Phys. **13**, 285 (1955).
25. Sakata S. Prog. Theor. Phys. **6**, 636 (1956).
26. Gell-Mann M. Phys. Rev. **125**, 1067 (1962); Phys. Lett. **8**, 214 (1964), Neuvo Cimento **4**, 284 (1956). Y. Ne'eman, Nucl. Phys. **26**, 222 (1961).
27. Gell-Mann M. Phys. Lett. **8**, 214 (1964). G. Zweig CERN Reports TH401 and TH402 (1964).
28. Albert JJ. et al. Phys. Rev. Lett. **33**, 1404 (1974), J.E. Augustin et al., Phys. Rev. Lett. **33**, 1406 (1974).
29. Herb SW. et al., Phys. Lett. **39**, 252 (1977).
    Innes WR. et al. Phys. Rev. Lett. **39**, 1240 (1977).
30. Abachi S. et al. Phys. Rev. Lett. **74**, 2632 (1995).
    Abe A. et al. Phys. Rev. Lett. **74**, 2626 (1995).

SUGGESTED BOOKS FOR FURTHER READING

31. From Nucleons to the Atomic Nucleus
    (Perspective in Nuclear Physics)
    K. Heide, Springer Verlag, Heidelberg (1998).
32. Neutron, Nuclei and Matter
    J. Byrne
    Institute of Physics Publishing U.K. 1994.
33. Particles and Nuclei
    B. Povh, Klaus Rith, Christoph Scholz and Frank Zetsche,
    Springer Verlag, Berlin 1995.
34. Facts and Mysteries in Particle Physics
    Martinus J.G. Veltman
    World Scientific, Singapore and I.C.P. (U.K.) 2003.
35. Subatomic Physics, Third Edition
    Ernest M. Henley and Aljandro Garcia
    World Scientific, Singapore and I.C.P. (U.K.) 2007.

2

Two Nucleon Problem

"There is no other problem in physics in which so much owes so little".

–Henry Marshak[1]

2.1 GATEWAY TO NUCLEAR THEORY

Nuclear Force Concept: The force concept in nuclear physics in the beginning was all empirical. The general properties of nuclear force were deduced by Wigner[2] from the binding energy data of light nuclei. The constancy of binding energy per nucleon (*BE/A*) with mass number *A* (*see* Fig. 1.3) revealed the saturation property of nuclear force.

Yukawa[3] was the first to approach nuclear force problem from the first principle by considering analogies with more familiar electromagnetic inter-action between two charges mediated by the exchange of a virtual photon (Fig. 2.1a). Yukawa assumed that the nucleon-nucleon (*N-N*) interaction process can be viewed to be analogous to the electromagnetic interaction, except that the exchanged field quantum having rest mass, which later was identified as the pion rest mass (Fig. 2.1b). He obtained '*N-N* potential form' from the meson exchange process and connected the range of *N-N* force with the mass of the exchanged meson. We show below how this connection follows from the uncertainty relation.

Consider two nucleons each moving slowly, i.e. $v/c \ll 1$, where v is the velocity of either nucleon and c is the velocity of light. Upon emission of a meson by the one nucleon, the energy conservation is violated by an amount

$$\Delta E = \sqrt{p^2 c^2 + m^2 c^4},$$

where p is the momentum of the emitted meson and m the rest mass of the meson.

63

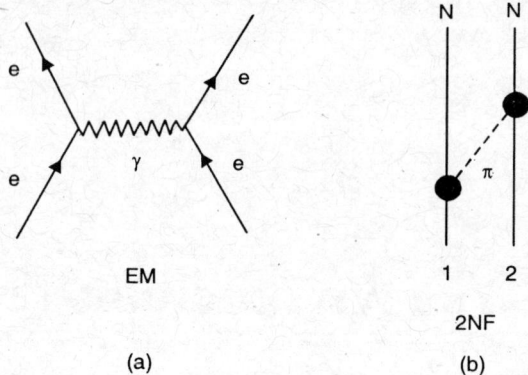

Fig. 2.1: Photon and one pion exchange processes: (*a*) Electromagnetic interaction (*b*) Two-nucleon force (2NF)

We shall assume here $pc << mc^2$ so that $\Delta E = mc^2$. The violation of energy conservation cannot last longer than the time Δt determined by the uncertainty relation

$$\Delta t = \frac{\hbar}{\Delta E} = \frac{\hbar}{mc^2}.$$

The maximum speed the virtual meson can attain is that of the speed of light and, therefore, the farthest distance it can travel is

$$r = c\Delta t = \frac{\hbar}{mc}. \tag{2.1}$$

If the exchanged particle is π-meson $(m \cong 139 \text{ MeV})$ then the range of *N-N* interaction, $r = 1.4$ fm.

The potential field $\phi_1(\vec{r})$ surrounding the first nucleon emitting a virtual meson can be calculated from the relativistic energy momentum relation, resulting the Klein-Gordon equation for a free particle of rest mass *m*:

$$\left(-p^2c^2 - m^2c^4 + E^2\right)\phi_1(\vec{r}) = 0, \tag{2.2}$$

where $p = -i\hbar\nabla$ and $E = i\hbar\dfrac{\partial}{\partial t}$. As the Coulombic potential $\phi_1(\vec{r})$ is time independent and surrounds a point source $g_1\delta(0)$ of coupling strength g_1 placed for convenience at the origin, the differential equation defining the field is

$$(\Box + \mu^2)\,\phi_1(\vec{r}) = 4\pi g_1\delta(0), \tag{2.3}$$

where $\mu = \dfrac{mc}{\hbar}$ = (pion Compton wave length)$^{-1}$, $\Box = \nabla^2 - \dfrac{1}{c^2}\dfrac{\partial}{\partial t^2}$ and the factor 4π has been added according to the normal convention*. Eq. (2.3) has a solution of the form (for the static case, $\Box = \nabla^2$)

$$\phi_1(r) = g_1 \frac{e^{-\mu r}}{r},$$

which satisfies the physical constraint that $\phi_1(\vec{r})$ vanishes at infinity. The virtual meson described by $\phi_1(\vec{r})$ can next be absorbed by the second nucleon at $\vec{r}$ to give the N-N interaction. $V(r) = g_2\phi_1(\vec{r})$ where g_2, the coupling constant for the second nucleon, describes the strength of absorption of the virtual meson. Hence, for the same strength of emission and absorption $(g_1 = g_2 = g)$, we get

$$V(r) = -g_1 g_2 \frac{e^{-\mu r}}{r} = -g^2 \frac{e^{-\mu r}}{r},$$

which is the Yukawa potential. The negative sign shows that the interaction is attractive. The interaction is also local, i.e. the emitted pion and the final state nucleon materialize at the same point where the initial state nucleon disappeared. In order to express central two-nucleon interaction $V(r)$, usually one employs two parameters namely, the depth V_0, range b, and a radial shape function $f(r/b)$, i.e.

$$V(r) = -V_0\, f(r/b). \tag{2.4a}$$

The depth gives the strength of the potential and the range gives the distance upto which the potential is felt. The commonly used radial shape functions are:

(a) Square well: $\quad f(r/b) = 1(0) \qquad r < b\ (r > b)$ $\hfill$ (2.4b)

(b) Yukawa: $\quad f(r/b) = \dfrac{e^{-r/b}}{r/b}.$ $\hfill$ (2.4c)

(c) Hulthen: $\quad f(r/b) = \dfrac{e^{-r/b}}{\left(1 - e^{-r/b}\right)}.$ $\hfill$ (2.4d)

---

* For example, inhomogeneous wave equation in electro-dynamics has the form $\nabla^2 A(r) = -4\pi Q\rho$. For a point charge located at the origin $A(r) = \dfrac{Q}{r}$ and $\nabla^2(1/r) = -4\pi\delta(0)$, where δ is the Dirac delta function.

A year after Yukawa had put forward his theory, Breit, Condon and Present[4] did a painstaking analysis of proton-proton scattering and they proposed the "hypothesis of charge independence". Much later discovery of pi-meson by Powell and Occhialini[5] combined with Yukawa's meson exchange idea made it possible to regard physically approximate charge independence of N-N system as a consequence of the basic symmetry principles involved in the classification of elementary particles and their interactions.

A. One Pion Exchange Potential

We shall now give a formal derivation of Yukawa's One Pion Exchange Potential (OPEP) following Brown and Jackson[32]. Let us consider N-N interaction arising from the coupling with a scalar meson ($J^\pi = 0^+$), i.e. boson. Non-relativistically this is described by a Lagrangian density for the coupling of meson field with the scalar nucleon current $\bar{\psi}\psi$ as

$$\mathcal{L}(x) = g\,\bar{\psi}(x)\,\phi(x)\,\psi(x), \tag{2.5a}$$

where g is the dimensionless coupling constant which is real, $\psi(x)$ is the nucleon wave function and $\phi(x)$ is the meson field. We use the second quantized formalism and the natural units ($\hbar = c = 1$), and expand the meson field as a plane wave:

$$\phi(r) = \sum_q \sqrt{\frac{2\pi}{\omega_q}}\, a_q e^{i\vec{q}\cdot\vec{r}}, \tag{2.5b}$$

where $a_q(a_q^\dagger)$ in the language of the second quantization (Appendix C2), is a meson annihilation (creation) operator. The factor $\omega_q^{-1/2}$ arises from the nomalization of $\phi(r)$ such that the density

$$\rho = 2\omega_q|\phi|^2.$$

Here, $\omega_q = \sqrt{q^2 + \mu^2}$ is the meson energy. Upon quantization of the meson field (*see* Appendix C2), one finds that a_q has matrix element

$$\langle n_q - 1|a_q|n_q\rangle = \langle n_q|a_q^\dagger|n_q - 1\rangle = \sqrt{n_q},$$

where $n_q = $ the number of quanta in oscillator state q. We shall take $n_q = 1$ for presence of a meson and $n_q = 0$ for no presence of meson. The interaction arising is shown in Fig. 2.2. The matrix element of the interaction is given by the second order perturbation theory. In Fig. 2.2a, for example, the initial state is that of two nucleons p_1 and p_2, it has an energy $E_0 = \dfrac{p_1^2}{2M} + \dfrac{p_2^2}{2M}$, where M is the nucleon mass. In the intermediate state a meson of energy ω_q

and momentum $\vec{q}$ is present, the matrix element of its emission equals $g\left(2\pi/\omega_q\right)^{1/2}$, where nucleon 1 has the energy $(\vec{p}_1 - \vec{q})^2/2M$. Thus, the energy of the intermediate state $E_i = \omega_q + (\vec{p}_1 - \vec{q})^2/2M + p_2^2/2M$. The meson is absorbed in transition to the final state with matrix element $g\left(2\pi/\omega_q\right)^{1/2}$. The entire expression for the process (Fig. 2.2a) is then

$$\frac{\langle f|\mathcal{L}|i\rangle \langle i|\mathcal{L}|o\rangle}{E_0 - E_i} = g^2 \left(\frac{2\pi}{\omega_q}\right) \frac{1}{\dfrac{p_1^2}{2M} - \omega_q - (\vec{p}_1 - \vec{q})^2/2M}. \tag{2.6a}$$

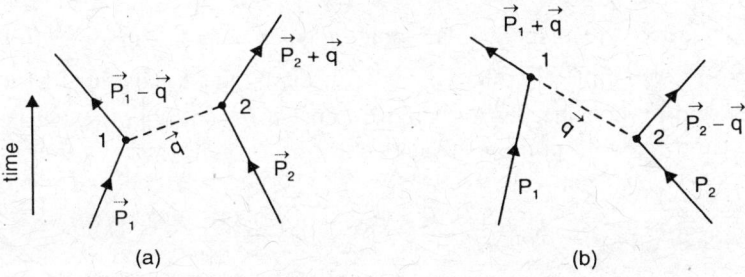

(a)   (b)

Fig. 2.2: The Yukawa interaction originating from meson exchange: (a) nucleon 1 (solid line) emits a meson (dotted line), which is absorbed by nucleon 2, (b) nucleon 2 (solid line) first emits a meson (dotted line), which is is absorbed by nucleon 1

Again for the process (Fig. 2.2b), we have

$$\frac{\langle f|\mathcal{L}|i\rangle \langle i|\mathcal{L}|o\rangle}{E_0 - E_i} = g^2 \left(\frac{2\pi}{\omega_q}\right) \frac{1}{\dfrac{p_2^2}{2M} - \omega_q - (\vec{p}_2 - \vec{q})^2/2M}. \tag{2.6b}$$

We shall consider p_1 and p_2 small and also $q \sim m$, where m is the meson mass. Thus, the terms $p_1^2/2M$ or $(p_1 - q)^2/2M$ will be of the order of m/M compared with ω_q. Since, π-meson mass $\mu \ll M$, we can neglect everything in the denominator except ω_q. Adding (2.6a) and (2.6b) we get

$$\frac{\langle f|\mathcal{L}|i\rangle \langle i|\mathcal{L}|o\rangle}{E_0 - E_i} = -g^2 \frac{4\pi}{\omega_q^2} = -g^2 \frac{4\pi}{(q^2 + \mu^2)} = V(q). \tag{2.6c}$$

The Fourier transform of $V(q)$ (*see* Ex. 2.1) gives the interaction in configuration space:

$$V(r) = -g^2 \frac{e^{-\mu r}}{r}. \tag{2.7}$$

The scalar meson exchange interaction Lagrangian (2.5a) needs modification. Since the pion, the mediator of N-N interaction is charged (isospin $T = 1$), a vector in isospin space and pseudoscalar in space (spin zero and negative intrinsic parity), its coupling with nucleon is pseudo scalar-isovector type. The pion field ϕ is a vector in isospin space. Because of pion's negative intrinsic parity, we have to include an operator with negative parity; but because pion is rotationally invariant for its zero spin, this negative parity operator has to be also rotationally invariant, so we take a scalar product of nucleon spin and pion momentum as $(\vec{\sigma} \cdot \vec{\nabla})$. Therefore, we replace Eq. (2.5a) by

$$\mathcal{L}(x) = \frac{f}{\mu} \, \overline{\psi}(x) \, \vec{\sigma} \cdot \vec{\nabla} \vec{\tau} \cdot \vec{\phi}(x) \, \Psi(x) \,, \qquad (2.8a)$$

where τ and σ are respectively nucleon isospin and spin operators and $4\pi f^2 = g^2$. A completely relativistic description, which is avoided here, will be achieved by considering a simple Lorentz invariant coupling of a Dirac field operator ψ with pseudo scalar-isovector pion field ϕ as

$$\mathcal{L} = \frac{f}{\mu} \, \overline{\psi} \, \gamma_5 \, \vec{\tau} \cdot \vec{\phi} \, \psi. \qquad (2.8b)$$

The scalar product in the isospin space ensures charge independence. The pion nucleon interaction Lagrangian $\mathcal{L}$ is a scalar in both configuration and isospin space. We note that $\vec{\tau} \cdot \vec{\phi} = \tau_1 \phi_1 + \tau_2 \phi_2 + \tau_3 \phi_3 = \tau_i \phi_i$ with $i = 1,2,3$, in the Cartesian coordinates.

We define the spherical tensors:

$$\tau_+ = -\frac{1}{\sqrt{2}} \, (\tau_1 + i\tau_2), \, \tau_- = -\frac{1}{\sqrt{2}} (\tau_1 - i\tau_2), \, \tau_0 = \tau_3$$

and similarly for $\phi_\pm, \phi_0$. Hence $\vec{\tau} \cdot \vec{\phi} = -[\tau_- \phi_+ + \tau_+ \phi_-] + \tau_0 \phi_0$.

Writing g for f/μ in (2.8a) and carrying out little algebra we get

$$\mathcal{L} = -\sqrt{2} \, g_{NN\pi} \left[\overline{\psi}_p \, \gamma_5 \, \overline{\psi}_n \, \phi_- - \overline{\psi}_n \, \gamma_5 \, \psi_p \, \phi_+ \right]$$
$$- g_{NN\pi} \left[\overline{\psi}_p \, \gamma_5 \, \psi_p \, \phi_0 - \overline{\psi}_n \, \gamma_5 \, \psi_n \, \phi_0 \right].$$

We conclude from the above that the charge independence demands that the strength for a charged pion coupling to a nucleon is $\sqrt{2}$ times that of a neutral pion. We may also replace scalar nucleon current $\overline{\psi} \psi$ by pseudovector nucleon current $\overline{\psi} \gamma_5 \gamma^\nu \psi$ due to which the Lorentz invariant Lagrangian takes the form:

$$\mathcal{L} = \frac{f}{\mu} \, \overline{\psi} \, \gamma^\nu \gamma_5 \, \psi \, \vec{\tau} \, \partial_\nu \, \vec{\phi} \,, \qquad (2.8c)$$

where ψ's are now the four component Dirac spinor for nucleon and γ's are the Dirac matrices. The pion field is described by three independent real fields ϕ_1, ϕ_2 and ϕ_3, which are three components of the vector in isospin space. The Lagrangian (2.8c) is called the derivative coupling Lagrangian. We have coupled the derivative $\partial_\nu \phi$ with pseudovector nucleon current.

The explicit form of OPEP is obtained as follows:

Employing the same arguments which led to Eq. (2.5) and writing $\vec{r} = |\vec{r}_1 - \vec{r}_2|, \nabla_1 = -\nabla_2 = \nabla$ we get N-N interaction from (2.8a) in the form:

$$V(r) = \frac{f^2}{\mu^2} (\vec{\tau}_1 \cdot \vec{\tau}_2) (\vec{\sigma}_1 \cdot \vec{\nabla})(\vec{\sigma}_2 \cdot \vec{\nabla}) \frac{e^{-\mu r}}{r}.$$

Since functions operated on are functions of r only, we can replace $\vec{\nabla}$ by $\hat{n} \dfrac{d}{dr}$, where $\hat{n}$ is the unit vector along the direction r. Hence,

$$(\vec{\sigma}_1 \cdot \vec{\nabla})(\vec{\sigma}_2 \cdot \vec{\nabla}) = (\vec{\sigma}_1 \cdot \hat{n})(\vec{\sigma}_2 \cdot \hat{n}) \left[\frac{d^2}{dr^2} - \frac{1}{r}\frac{d}{dr} \right] + \vec{\sigma}_1 \cdot \vec{\sigma}_2 \frac{1}{r}\frac{d}{dr}.$$

Finally, $(\vec{\sigma}_1 \cdot \vec{\nabla})(\vec{\sigma}_2 \cdot \vec{\nabla}) = \dfrac{1}{3} S_{12} \left(\dfrac{d^2}{dr^2} - \dfrac{1}{r}\dfrac{d}{dr} \right) + \dfrac{1}{3} \vec{\sigma}_1 \cdot \vec{\sigma}_2 \nabla^2,$

where $S_{12} = \dfrac{3}{r^2} (\vec{\sigma}_1 \cdot \vec{r})(\vec{\sigma}_2 \cdot \vec{r}) - \vec{\sigma}_1 \cdot \vec{\sigma}_2.$

Now $\left[\dfrac{d^2}{dr^2} - \dfrac{1}{r}\dfrac{d}{dr} \right] \dfrac{e^{-\mu r}}{r} = \left(\dfrac{3}{r^3} + \dfrac{3\mu}{r^2} + \dfrac{\mu^2}{r} \right) e^{-\mu r}$ and

$$\frac{1}{3} \vec{\sigma}_1 \cdot \vec{\sigma}_2 \nabla^2 \frac{e^{-\mu r}}{r} = \frac{1}{3} \mu^2 (\vec{\sigma}_1 \cdot \vec{\sigma}_2) \left[\frac{e^{-\mu r}}{r} - 4\pi\delta(r) e^{-\mu r} \right],$$

since $\nabla^2 \left(\dfrac{1}{|r|} \right) = -4\pi\delta(\vec{r})$. The N-N interaction is obtained in the form:

$$V(r) = f^2 \mu(\vec{\tau}_1 \cdot \vec{\tau}_2) \left[\frac{1}{3} S_{12} \left\{ \frac{3}{(\mu r)^3} + \frac{3}{(\mu r)^2} + \frac{1}{\mu r} \right\} e^{-\mu r} \right.$$

$$\left. + (\vec{\sigma}_1 \cdot \vec{\sigma}_2) \frac{1}{3} \left\{ \frac{e^{-\mu r}}{\mu r} - \frac{4\pi}{\mu^3} \delta(r) \right\} \right].$$

Since at large r, N-N interaction $V(r)$ is mediated by one pion exchange, we can disregard $\delta(r)$ term; because at small r the multiple pion exchange becomes important. Then OPEP is expressed in a simple form

$$V_{OPEP}(x) = \frac{1}{3}\left(\frac{g^2}{4\pi}\right)\mu(\vec{\tau}_1 \cdot \vec{\tau}_2)\left[(\vec{\sigma}_1 \cdot \vec{\sigma}_2) + S_{12}\left(1 + \frac{3}{x} + \frac{3}{x^2}\right)\right]\frac{e^{-x}}{x}, \quad (2.10)$$

where $x = \mu r$. It is remarkable that the present theory naturally yields the tensor interaction S_{12} (*see* 2.45). In application to phenomenological data analysis, the OPEP is applied almost exclusively to the first order and only to the calculation of high ℓ phase shifts.

B. Two Nucleon Systems at Low Energies

The deuteron is presumably the simplest nucleus consisting of a neutron and a proton and is the only bound state of two-nucleon systems. It might be expected, therefore, to occupy the same place in nuclear theory as occupied by the hydrogen atom in the atomic theory. From the simple inspection of deuteron data (Table 2.1) one can reach a number of conclusions concerning the qualitative nature of the nuclear force. The existence of a bound state of n-p system indicates that the n-p force is attractive.

The magnetic moment of deuteron indicates approximate additivity of the magnetic moments of neutron and proton in the deuteron: (2.97270 nm − 1.91316 nm = 0.87954 nm). At the first approximation, the ground state (g.s) of deuteron is taken to be the triplet s-state $(^3S_1)$. This assumption is partially justified because the total angular momentum of deuteron is 1, and experiments on magnetic moment show that the neutron and the proton spins add up to give spin angular momentum $S = 1$. However, the difference between the observed magnetic moment and that obtained from the above additive value, $\Delta\mu = 0.0231$ nm, is more than the experimental uncertainties. Furthermore, a purely central potential is spherically symmetric, and the quadrupole moment vanishes for spherically symmetric state. The deuteron, however, possesses a small and positive quadrupole moment. Hence, in an improved approximation we treat n-p interaction as non-central, i.e. an interaction which depends on the relative orientation of vectors joining neutron and proton with respect to their spins.

For predominantly $\ell = 0$ $(^3S_1$-state), the radial S. Eq. for deuteron (*see* Appendix B1) has the form:

$$\frac{d^2u}{dr^2} + \frac{M}{\hbar^2}\left[-\epsilon - V(r)\right]u = 0, \quad (2.11)$$

where ϵ is the numerical value of the deuteron binding energy and $M(=\mu/2)$ the nucleon mass. For a square well potential (2.11) takes the form:

$$\frac{d^2u}{dr^2} + k^2u = 0 \qquad r < b \quad (2.12)$$

$$\frac{d^2v}{dr^2} - \alpha^2 v = 0 \qquad\qquad r > b \qquad\qquad (2.13)$$

where $k^2 = \dfrac{M}{\hbar^2}(V_0 - \epsilon)$ and $\alpha^2 = \dfrac{M\epsilon}{\hbar^2}$.

The solutions of Eq. (2.12) and Eq. (2.13) which should respectively vanish at the origin and also at infinity are:

$$u = A \sin kr \qquad\qquad r < b$$

$$v = Be^{-\alpha r} \qquad\qquad r > b$$

where $\alpha = \sqrt{\dfrac{M\epsilon}{\hbar^2}} = \sqrt{\dfrac{Mc^2\,\epsilon}{\hbar^2 c^2}} = 0.232 \text{ fm}^{-1}.$ $\qquad\qquad (2.14)$

A and B are the normalization constants. Since u and $\dfrac{du}{dr}$ should be continuous at $r = b$, their logarithimic derivatives must match at $r = b$. Such an approach to study n-p system was proposed first by Bethe and Peirls.* Therefore, $k \cot kb = -\alpha$.

Hence $\quad x \cot x = -\alpha b,$ where $x = kb,$ $\qquad\qquad\qquad (2.15)$

If we assume $b \sim 2$ fm, then $\alpha b \sim 0.5$, so that $\cot x = -\dfrac{0.5}{x}$.

The solution of the above equation can be obtained from the intersection of the curves $y = \cot x$ and $y = -0.5/x$ (Fig. 2.3a). The roots are approximately equal to $\pi/2, 3\pi/2 \dots$ etc. For the ground state, $x = kb = \pi/2$. In other words,

$$k^2 b^2 = \frac{M}{2}(V_0 - \epsilon)b^2 = \frac{\pi^2}{4}. \text{ Since } \epsilon \text{ is small, } V_0 - \epsilon \approx V_0.$$

Hence, $\qquad V_0 b^2 = \dfrac{\pi^2 \hbar^2}{4M},$ $\qquad\qquad\qquad\qquad\qquad (2.16)$

which is the well known 'depth range' relation. If we take b to be roughly equal to the pion compton wavelength $\left(= \dfrac{\hbar}{m_\pi c} = 1.45 \text{ fm} \right)$ then

---

* H.A. Bethe, Rev. Mod. Phys. **8**, 325 (1936). This review is known in the literature as the "Bethe's Bible".

$V_0 \approx 50$ MeV, which is consistent with our assumption $V_0 >> \epsilon$. Evaluation of the normalization constants A and B is done by employing the boundary conditions, namely, the continuity of u, v, v' and u' at $r = b$ and the normalization condition: $\int_0^b u^2 dr + \int_b^\infty v^2 dr = 1$. The boundary conditions yield

$A \sin kb = Be^{-\alpha b}$ and $Ak \cos kb = -Be^{-\alpha b}$. The normalization condition yields

$$\left[\int_0^b A^2 \sin^2 kr\, dr + \int_b^\infty B^2 e^{-2\alpha r} dr \right] = 1, \text{ so that}$$

$$A = \left[2\alpha(1 - \alpha b) \right]^{1/2} \text{ and } B = \left[2\alpha(1 + \alpha b) \right]^{1/2}.$$

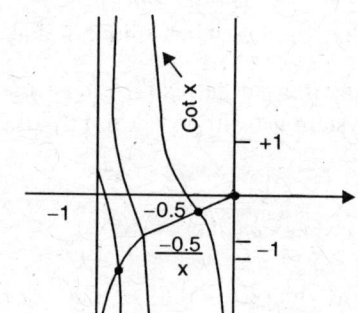

Fig. 2.3: (*a*) Intersection of curves if $y = \cot x$ and $y = -0.5/x$.

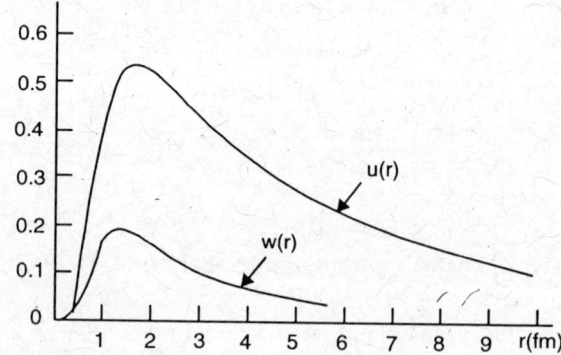

Fig. 2.3: (*b*) Complete deuteron wave function for the Reid soft core potential

The S-state deuteron wave function then reads as

$$u = \left\{ 2\alpha(1 - \alpha b) \right\}^{1/2} \sin kr \qquad r < b \tag{2.17}$$

$$v = \{2\alpha(1 + \alpha b)\}^{1/2} e^{-\alpha r} \qquad r > b \tag{2.18}$$

The shape of the complete deuteron wave function is shown in Fig. 2.3(*b*) for Reid soft core potential[6]. However, an approximate function mentioned above (2.18) of the form

$$u \approx \sqrt{2\alpha} \; e^{-2\alpha r}, \tag{2.19}$$

has been used at several places for the model calculations.

Experimentally there is no evidence for the existence of any excited states of the deuteron, which can be confirmed also theoretically. We shall start with *P*-state, a result which could then be generalized for higher angular momentum states. We employ a square well potential. The *P*-state radial equation is

$$\frac{d^2 u_1}{dr^2} - k^2 u_1 - \frac{2}{r^2} \, u_1 = 0 \quad r < b \tag{2.20}$$

and

$$\frac{d^2 v_1}{dr^2} - \alpha^2 v_1 - \frac{2}{r^2} \, v_1 = 0 \quad r > b \tag{2.21}$$

The solutions of the above equations are:

$$u_1 = A_1 \sqrt{\frac{2}{\pi k}} \left[\frac{\sin kr}{kr} - \cos kr \right] \text{ for } r < b$$

$$v_1 = B_1 \, e^{-\alpha(r-b)} \left[1 + \frac{1}{\alpha r} \right] \text{ for } r > b$$

Next we apply the boundary condition for the wave functions:

$$\left(\frac{u_1'}{u_1} \right)_{r=b} = \frac{1}{b} \frac{\cos x + \left(x - \dfrac{1}{x} \right) \sin x}{\dfrac{\sin x}{x} - \cos x}, \text{ where } x = kb \text{ and } r < b$$

$$\left(\frac{v_1'}{v_1} \right)_{r=b} = -\frac{1}{b} \frac{1 + \alpha b + \alpha^2 b^2}{1 + \alpha b} = -\frac{1}{b} \text{ for } r > b \text{ and } \alpha^2 b^2 \ll 0.$$

In the above equation we have put $\alpha^2 b^2 = 0$ to get the minimum potential energy for a bound state. Matching the logarithimic derivatives at $r = b$, we get

$$-\frac{1}{b} = \frac{1}{b} \frac{\cos x + \left(x - \dfrac{1}{x} \right) \sin x}{\dfrac{\sin x}{x} - \cos x} \text{ or } \sin x = 0. \tag{2.22}$$

Then $kb = \pi$ is the smallest positive root of (2.22) and, therefore, $V_0 = \dfrac{\pi^2 \hbar^2}{Mb^2}$.

Thus, V_0 is four times the potential depth for the ground state of the deuteron (*see* Eq. (2.16)). So in order to produce a bound *P*-state the potential depth must be considerably larger than that for the ground state.

Low Energy Nucleon-Nucleon Scattering: In addition to the study of bound state of *n-p* system, the study of *N-N* scattering is the second very important source of information of *N-N* interaction. In the case of bound state problem it was convenient to treat the problem in the c.m system. *N-N* scattering problem on the other hand, has to be referred to the laboratory system, since it is in that system, the experimental measurements are made. However, calculations are still made in the c.m system and they are then transformed to the laboratory system in order to compare theory with experiment (Appendix A4).

In order to calculate the total *n-p* scattering cross section we need to know the phase shift 'δ_ℓ' for each partial wave and the number of partial waves need to be taken into account in the infinite sum. Now we shall try to obtain phase shifts in terms of integrals over the radial wave functions. We begin with

$$\left[\frac{d^2}{dr^2} + k^2 - \frac{\ell(\ell+1)}{r^2} - U(r) \right] u_{\ell k}(r) = 0 \text{ for } r < b, \tag{2.23a}$$

and $\quad \left[\dfrac{d^2}{dr^2} + k^2 - \dfrac{\ell(\ell+1)}{r^2} \right] v_{\ell k}(r) = 0$, for $r > b.$ $\qquad$ (2.23b)

where $\quad U(r) = \dfrac{2\mu V(r)}{\hbar^2}.$

We multiply (2.23a) by $v_{\ell k}(r)$ and (2.23b) by $u_{\ell k}(r)$ and subtract the resulting equations to get

$$u''_{\ell k}(r) v_{\ell k}(r) - v''_{\ell k}(r) u_{\ell k}(r) = v_{\ell k}(r) U(r) u_{\ell k}(r).$$

Integrating the above equation from 0 to ∞, we obtain

$$\left| u'_{\ell k}(r) v_{\ell k}(r) - u_{\ell k}(r) v'_{\ell k}(r) \right|_0^\infty = \int\limits_0^\infty v_{\ell k}(r) U(r) u_{\ell k}(r) \, dr.$$

Now $u_{\ell k}(r = 0) = u'_{\ell k}(r = 0) = 0$ and

$$u'_{\ell k}(r \to \infty) v_{\ell k}(r \to \infty) - u_{\ell k}(r \to \infty) v'_{\ell k}(r \to \infty)$$

$$= \frac{1}{k} \sin(kr - \ell\pi/2) \cos(kr - \ell\pi/2 + \delta_\ell)$$

$$-\frac{1}{k}\cos(kr - \ell\pi/2)\sin(kr - \ell\pi/2 + \delta_\ell) = -\frac{1}{k}\sin\delta_\ell.$$

Hence, $\quad \sin\delta_\ell = \tan\delta_\ell \cong \delta_\ell = -k \int\limits_0^\infty v_{\ell k}(r)\, U(r)\, u_{\ell k}(r)\, dr.$ $\qquad\qquad$ (2.24)

The general solution of (2.23b) is a linear combination of regular and irregular solutions, which are spherical Bessel function $j_\ell(kr)$ and Neumann function $\eta_\ell(kr)$ respectively.

$$v_{\ell k}(r) = A_\ell\, j_\ell(kr) + B_\ell\, \eta_\ell(kr) \qquad r > b,$$

where $B_\ell/A_\ell = \tan\delta_\ell$, and δ_ℓ measures the amount by which phase of the radial function $v_{\ell k}(r)$ differs from the wave function in the absence of potential. The phase shift δ_ℓ can be computed by fitting the radial wave function $u_{k\ell}(r)$ for $r < b$ to the solution $v_{k\ell}(r)$ at $r = b$. The required boundary condition at $r = b$ is

$$\left[\frac{du_{\ell k}(r)}{dr} \bigg/ u_{\ell k}(r)\right]_{r=b} = \left[\frac{dv_{k\ell}(r)}{dr} \bigg/ v_{\ell k}(r)\right]_{r=b}.$$

$$RHS = \frac{k\, j'_\ell(kb)\, \cos\delta_\ell - k\, \eta'_\ell(kb)\, \sin\delta_\ell}{j_\ell(kb)\, \cos\delta_\ell - \eta_\ell(kb)\, \sin\delta_\ell}.$$

Denoting LHS by γ_ℓ, and using $r = kb$ we can write

$$\tan\delta_\ell = \frac{k\, j'_\ell(kb) - \gamma_\ell\, j_\ell(kb)}{k\, \eta'_\ell(kb) - \gamma_\ell\, \eta_\ell(kb)} = \frac{\rho\, j'_\ell(\rho) - b\gamma_\ell j_\ell(\rho)}{\rho\, \eta'_\ell(\rho) - b\gamma_\ell\, \eta_\ell(\rho)}, \qquad (2.25)$$

The phase shifts δ_0 as measured in *N-N* scattering is plotted (Fig. 2.4a) for singlet state against E_{Lab}. For E larger than 200 MeV, the *s*-wave phase shift δ_0 is negative and below this it is positive. This indicates that the nuclear force is repulsive at a short distance and attractive at larger separations. The wavelength of unscattered wave $\left(= \dfrac{2\pi}{k}\right)$ is changed to $2\pi/\sqrt{k^2 - U(r)}$ for the presence of the potential within its range b. For an attractive potential ($U(r)$ is −ve), the wavelength is reduced, resulting +ve phase shift ($\delta_0 > 0$) and for a repulsive potential, the wavelength is increased, resulting −ve phase shift ($\delta_0 < 0$). In the former case the wave function is pulled in by the potential whereas in the latter one the wave function is pushed out by the potential (Fig. 2.4b). From Eq. (2.23b) the asymptotic wave function

$$\frac{v_{\ell k}(r)}{r} = R_{k\ell}(r)$$

which is outside the range of the interaction for $\ell = 0$ takes the following form (dropping the subscript ℓk in (2.23b)):

$$v(r) = r\, R(r \to \infty) \sim \frac{e^{i\delta}}{k}\, \sin(kr + \delta). \tag{2.26}$$

For the scattering of low energy neutrons ($E_{cm} < 10$ MeV) by protons, all the phase shifts δ_ℓ (except s-wave phase shift δ_0 for $\ell = 0$) are zero and the total scattering cross section

$$\sigma = \frac{4\pi}{k^2}\, \sin^2\delta = \frac{4\pi}{k^2 + k^2 \cot^2\delta}, \tag{2.27a}$$

which follows from Eq.14 in Appendix B2.

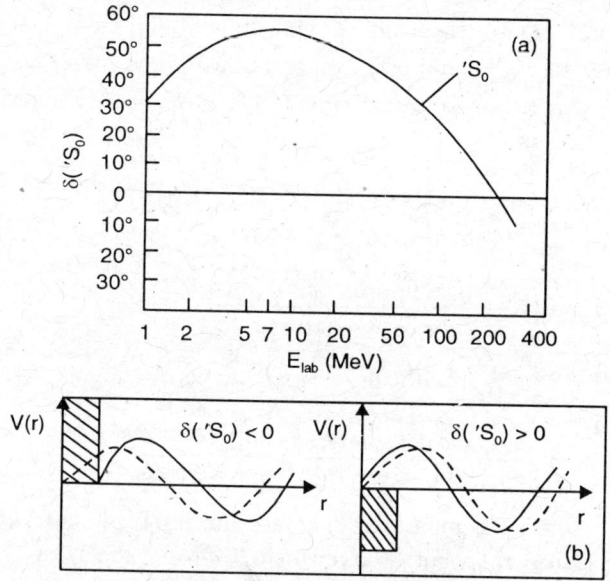

Fig. 2.4: (*a*) S-wave scattering phase shift $\delta({}^1S_0)$. (*b*) For an attractive potential the wave function is pulled in, resulting positive phase shift (right). For a repulsive potential the wave function is pushed out resulting negative phase shift (left)

We define σ in terms of a new quantity $-a^{-1} = k \cot d$ for $k \to 0$. "a" is called scattering length and $\sigma = 4\pi a^2$. Thus, in the limiting case of zero energy ($k \to 0$), the scattering cross section tends to a finite value. 'a', has a significance of a radius of a hard sphere from which a point neutron is scattered (note classically $\sigma = \pi a^2$). Since $\delta \to 0$ as k does and $\delta/k = -a$, we can rewrite (2.26) as

$$\lim_{k \to 0} v(r) = \frac{kr}{k} + \frac{\delta}{k} = r - a, \tag{2.27b}$$

which is a equation of a straight line.

·The scattering length a, is the intercept on the r axis and is obtained by extrapolating the radial wave function inside the well beyond the range of force b. Figure 2.5 illustrates the significance of scattering length. The positive scattering length indicates a bound state and negative scattering length indicates an unbound state.

Since the deuteron binding energy is very small, we can equate the logarithimic derivative of the deuteron wave function outside the range of interaction (2.13) to the logarithmic derivative of the asymptotic n-p scattering wave function (2.26) and get

$$k \cot (kr + \delta) = -\alpha \quad \text{or} \quad \sin^2 (kr + \delta) = \frac{k^2}{k^2 + \alpha^2}$$

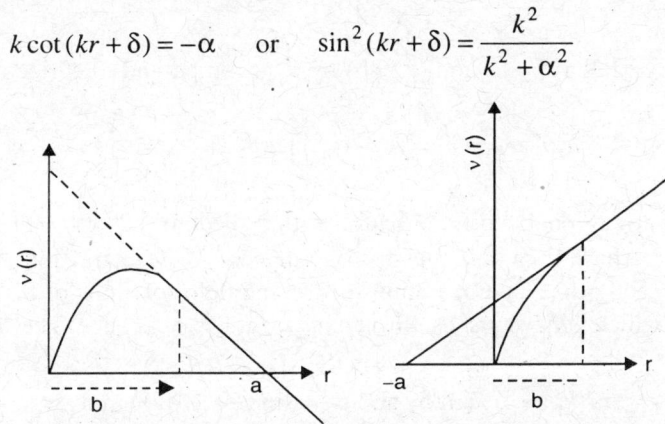

Fig. 2.5: The positive and negative scattering lengths indicating respectively, bound and unbound two-nucleon systems

The cross section for zero energy neutrons is then given by

$$\sigma_{theo} = \frac{4\pi}{k^2 + \alpha^2} = \frac{4\pi\hbar^2}{M(E+\in)} \approx 2.4 \text{ barn, whereas}$$

$$\sigma_{exp} = (20.36 \pm 0.1) \text{ barn.}$$

The disagreement between the values shows that there is a fundamental error in the assumption of nuclear forces used to derive σ_{theo}. Realizing that in the deuteron the spins of neutron and proton are parallel while np scattering cross section measures uncorrelated spins, we can write $\sigma = \frac{1}{4}\sigma_s + \frac{3}{4}\sigma_t$, where subscripts s and t indicate singlet and triplet np states respectively. Using σ_{theo} for σ_t and σ_{expt} for σ, we get $\sigma_s = 74$ barn. Thus we conclude that n-p forces are strongly spin dependent.

Effective Range Theory: The treatment of low energy n-p and p-p data is often presented in terms of so called "effective range theory". This is not a theory in the sense of offering an explanation of how scattering takes place but is merely a convenient parametrization of results of theories. We shall now show that the function $k \cot \delta$ is a linear function of the energy, independent of the shape and depth of the potential. In order to derive equation for $k \cot \delta$, we write the radial S. Eq. for two states of energies E_1 and E_2:

$$\frac{d^2 u_1}{dr^2} + \frac{M}{\hbar^2} \left[E_1 - V(r) \right] u_1 = 0, \tag{2.28a}$$

$$\frac{d^2 u_2}{dr^2} + \frac{M}{\hbar^2} \left[E_2 - V(r) \right] u_2 = 0. \tag{2.28b}$$

Multiply (2.28a) by u_2 and (2.28b) by u_1, subtract and integrate to get

$$u_2 u_1' - u_1 u_2' \Big|_0^R = (k_2^2 - k_1^2) \int_0^R u_1 u_2 \, dr, \tag{2.29}$$

where the primes denote differentiation with respect to r, $\hbar^2 k_{1,2}^2 = 2\mu E_{1,2}$ and R is some arbitrary radius outside the range of N-N interaction. We now perform exactly the same operation for free particle solutions of S. Eq. with $V(r) = 0$; which behave just as $u_1(r)$ and $u_2(r)$ at large distance. The functions (Eq. 2.26) are:

$$v_1 = \sin(k_1 r + \delta_1)/\sin \delta_1 \text{ and } v_2 = \sin(k_2 r + \delta_2)/\sin \delta_2, \tag{2.30}$$

These functions are normalized to unity at $r = 0$. This is different from the usual normalization. We obtain the same way

$$v_2 v_1' - v_1 v_2' \Big|_0^R = (k_2^2 - k_1^2) \int_0^R v_1 v_2 \, dr. \tag{2.31a}$$

Allowing $R \to \infty$ in (2.31a) leads to the well-known orthogonality of eigen functions belonging to two different energies E_1 and E_2.

We subtract (2.29) from (2.31a). On the left hand side of the resulting equation there is no contribution from the upper limit because v_1 and u_1, v_2 and u_2 are equal there, so

$$v_2'(0) - v_1'(0) = (k_2^2 - k_1^2) \int_0^R (v_1 v_2 - u_1 u_2) \, dr \tag{2.31b}$$

or $\qquad k_2 \cot \delta_2 - k_1 \cot \delta_1 = (k_2^2 - k_1^2) \int_0^R (v_1 v_2 - u_1 u_2) \, dr. \tag{2.31c}$

Let $k_1 \to 0$, and $k_2 = k$. Then $(\sin k_1 \cot \delta_1)_{k_1 \to 0} = -a^{-1}$, where a is the scattering length, so that

$$k\cot\delta = -a^{-1} + k^2 \int_0^R (v_1 v_2 - u_1 u_2)\, dr = -a^{-1} + \frac{1}{2} k^2 \rho(0, E), \quad (2.32\text{a})$$

with

$$\rho(E_1, E_2) = \int_0^R (v_1 v_2 - u_1 u_2)\, dr. \tag{2.32b}$$

The function u differs from v only inside the range of interaction. But just in this region the dependence of u on energy is slight, because the potential energy $V(r)$ is much larger than k^2 in the entire low energy region up to 5-10 MeV. Therefore, it is a good approximation in the low energy region to replace v and u in (2.32b) by the corresponding function at zero energy. We define the effective range r_0 by

$$\rho(0,0) = \frac{1}{2} r_0 = \int_0^\infty (v_0^2 - u_0^2)\, dr \qquad \text{so that (2.32a) becomes}$$

$$k\cot\delta = -a^{-1} + \frac{1}{2} k^2 r_0. \tag{2.33}$$

Employing (2.27a) and (2.33) we get

$$\sigma = \frac{4\pi}{k^2 + \left(\dfrac{1}{a} - \dfrac{r_0 k^2}{2}\right)^2}. \tag{2.34}$$

Besides being a function of k^2, the cross section is expressed in terms of two parameters, the effective range r_0 and scattering length 'a'. It is clearly independent of the form and shape of the potential. We can determine r_0 and 'a' by measuring cross sections for different low energy neutrons. Since the neutron proton scattering is spin dependent, actually four parameters are involved: the effective range and scattering length in the triplet and singlet states respectively (r_t, a_t) and (r_s, a_s). Hence, the average cross section for scattering of unpolarized neutrons by unpolarized free protons is finally expressed as

$$\sigma = \frac{3\pi}{k^2 + \left(\dfrac{1}{a_t} - \dfrac{r_t k^2}{2}\right)^2} + \frac{\pi}{k^2 + \left(\dfrac{1}{a_s} - \dfrac{r_s k^2}{2}\right)^2}. \tag{2.35}$$

The scattering cross sections for scattering of neutrons by free protons in triplet and singlet states are separately

$$\sigma_t = \frac{4\pi}{\left(a_t^{-1} - \frac{1}{2}k^2 r_t\right)^2 + k^2} \text{ and } \sigma_s = \frac{4\pi}{\left(a_s^{-1} - \frac{1}{2}k^2 r_s\right)^2 + k^2}. \qquad (2.36)$$

The result (2.31a) applies to both bound-state and scattering states. We can use (2.31b) and (2.31c) and take for deuteron wave function $v_2 = e^{-\alpha r}$ with

$$\alpha^2 = \frac{M \in}{\hbar^2}. \text{ Then } v_2'\big|_{r=0} = -\alpha \text{ and } (k_2^2 - k_1^2)\int_0^R (v_1 v_2 - v_1 v_2)\, dr = -\frac{1}{2}\,\alpha^2\, r_t\,.$$

In the triplet spin, neutron-proton interaction forms the ground state of deuteron. When (2.31b) and (2.33) are applied, we get

$$\alpha = a_t^{-1} + \frac{1}{2}\,\alpha^2\, \rho(0, -\in) = a_t^{-1} + \frac{1}{2}\,\alpha^2\, r_t, \qquad (2.37)$$

which is an interesting relationship between the deuteron binding energy, the triplet scattering length and the triplet effective range.

Knowing the relation between the scattering length in triplet state and the deuteron binding energy (Eq. 2.37), we can determine the effective range in the triplet state with a high degree of accuracy: $r_t = (1.759 \pm 0.005)$ fm. Using the experimental dependence of the averaged cross section on the neutron energy for the scattering of slow neutrons by free protons, we find the singlet effective range $r_s = (2.75 \pm 0.05)$ fm. The error in the determination of r_s turns out to be much greater than the error in r_t.

Proton-proton Scattering at Low Energy

The proton-proton (p-p) scattering differs from the neutron-proton scattering for several reasons:

1. p-p scattering measurements are more accurate than n-p scattering measurements because it is possible to obtain highly monoenergetic beam of protons from an accelerator and its ionization ability makes its detection easy.

2. p-p interaction differs from n-p interaction because of the Coulomb repulsion.

3. Identical particles must have antisymmetric wave function in accordance with the Pauli principle. Therefore, for $T = 1$ pp system, symmetric spatial wavefunction combines with an antisymmetric spin wave function and vice-versa. For example, S-state is space symmetric, there is no $^3S_1$-state in $\ell = 0$ p-p scattering.

Therefore, to find the proton-proton scattering cross section we must take into account the identity of the two particles by employing the relation between the isotopic spin and space-spin symmetries for states of two-nucleon systems.

Interchange of their coordinates is equivalent to replacing θ by $\pi - \theta$. For spin zero particles the differential cross section will be

$$\sigma_s(\theta) = |f(\theta) + f(\pi - \theta)|^2 \qquad (2.40a)$$

because of the symmetry of the wave function. At $\theta = \pi/2$

$$\sigma(\pi/2) = |2f(\pi/2)|^2 = 4 |f(\pi/2)|^2,$$

twice the classical value of $2 |f(\pi/2)|^2$. For the spin $\dfrac{1}{2}$ case the space part of the wave function will be symmetric or antisymmetric depending upon the spin state of the two-particle system. Hence, the singlet and triplet scattering cross sections, $\sigma_s(\theta)$ and $\sigma_t(\theta)$ are respectively given by

$$\sigma_s(\theta) = |f(\theta) + f(\pi - \theta)|^2 \qquad (2.40b)$$

for antisymmetric spin $(S = 0)$ and $\sigma_t(\theta) = |f(\theta) - f(\pi - \theta)|^2$ for symmetric spin $(S = 1)$. If the incident proton beam is unpolarized, then the relative weights of triplet and singlet states are 3:1, moreover their contributions add incohorently. We thus obtain

$$\sigma(\theta) = \frac{3}{4} |f(\theta) - f(\pi - \theta)|^2 + \frac{1}{4} |f(\theta) + f(\pi - \theta)|^2$$

$$= |f(\theta)|^2 + |f(\pi - \theta)|^2 - \mathrm{Re} \left[f^*(\theta) \; f(\pi - \theta) \right] \qquad (2.41)$$

The first term describes the direct scattering, the second the exchange scattering and the third the interference between direct and exchange processes. With the presence of Coulomb field, S. Eq. takes the form:

$$\left(\nabla^2 + k^2 - \frac{\beta}{r} \right) \psi = 0, \qquad (2.42)$$

where $\qquad \beta = \dfrac{2\mu e^2}{\hbar^2} = \dfrac{Mc^2}{(\hbar c)} \left(\dfrac{e^2}{\hbar c} \right) = \dfrac{938 \text{ MeV}}{197 \text{ MeV}} \text{fm}^{-1} \left(\dfrac{1}{137} \right) \approx \dfrac{1}{2.8} \text{ fm}^{-1}$

is the inverse of characteristic Coulomb length for protons (proton Compton wavelength). For large r the solution of (2.42) is the Coulomb distorted plane wave. Employing $f_c(\theta)$, the amplitude for pure Coulomb scattering, we obtain the famous Mott scattering formula:

$$\sigma_{Mott}(\theta) = \left(\frac{e^2}{4E} \right)^2 [\operatorname{cosec}^4 \theta/2 + \sec^4 \theta/2$$

$$- \operatorname{cosec}^2 \theta/2 \, \sec^2 \theta/2 \, \cos(n \log \tan \theta/2)],$$

where all quantities are measured in the c.m. system. When nuclear forces are present, we may write

$$\sigma(\theta) = \sigma_{Mott}(\theta) + \sigma_N(\theta) + \sigma_{NC}(\theta),$$

where $\sigma_N(\theta)$ involves only nuclear phase shifts δ_ℓ. By straightforward calculation we obtain

$$\sigma(\theta) = \left(\frac{e^2}{4E}\right)^2 [\cosec^4\theta/2 + \sec^4\theta/2$$

$$- \cosec^2\theta/2 \ \sec^2\theta/2\cos(n\log\tan\theta/2)]$$

$$- \left(\frac{e^2}{4E}\right)^2 \frac{1}{k}\sin\delta\ [\cosec^2\theta/2\ \cos(\delta+n\log\sin\theta/2)$$

$$+ \sec^2\theta/2\cos(\delta+n\log\cos\theta/2)] + \frac{1}{k^2}\sin^2\delta, \qquad (2.43)$$

where $n = \dfrac{\beta}{2k} = \dfrac{e^2}{2\hbar v}$ and $\delta = \delta_0$ is the S-wave nuclear phase shift.

We have mentioned earlier that the analysis of n-p and p-p scattering data provides a test for the charge independence of nuclear forces. We may ask: Are isotropic p-p differential cross section and n-p differential cross section, which are approximately symmetrical about $\pi/2$, compatible with the hypothesis of charge independence of N-N force? To address the above question let us examine the behaviour of both $\sigma_{pp}(\theta)$ and $\sigma_{n-p}(\theta)$ at $\theta = \pi/2$. While both $T = 1$ and $T = 0$ states contribute to $\sigma_{np}(\theta)$, only $T = 1$ states contribute to $\sigma_{pp}(\theta)$. At $\pi/2$, $P_\ell(\cos\theta) = 0$ for odd ℓ and only even ℓ spin singlet ($S = 0$) states contribute to $\sigma_{pp}(\theta)$. Thus, $\sigma_{pp}(\pi/2) = [\sigma(\pi/2)]_{T=1,S=0}$.

$$\sigma_{np}(\pi/2) = \frac{1}{4}[\sigma(\pi/2)]_{T=1,\ S=0} + \frac{3}{4}[\sigma(\pi/2)]_{T=0,\ S=1}.$$

If the charge independence holds, it follows that $4\sigma_{np}(\pi/2) \geq \sigma_{pp}(\pi/2)$.

The above inequality is always satisfied even at the highest incident energies where $\sigma_{np}(\pi/2)$ is very small. A generic n-p state with a $|v(1)\pi(2)\rangle$ isospin component is half $T = 0$ and half $T = 1$ states and corresponding scattering amplitude is

$$f_{np}(\theta) = \frac{1}{\sqrt{2}}[f_{T=1}(\theta) + f_{T=0}(\theta)].$$

Thus, $\sigma_{np}(\theta) = \dfrac{1}{2}\Big[\sigma_{T=1}(\theta) + \sigma_{T=0}(\theta) + 2\mathrm{Re}\, f_{T=1}(\theta) f_{T=0}^*(\theta)\Big]$

Due to the generalized Pauli principle, $f_{T=1}(\theta)$ and $f_{T=0}(\theta)$ have opposite parity for given value of S, i.e. $S < \begin{smallmatrix}T = 0,\, \ell = \text{even}\\ T = 1,\, \ell = \text{odd}\end{smallmatrix}$ and $S = 0 < \begin{smallmatrix}T = 0,\, \ell = \text{odd}\\ T = 1,\, \ell = \text{even}\end{smallmatrix}$.

Therefore, $2\,\mathrm{Re}\Big[f_{T=1}(\theta)\, f_{T=0}^*(\theta)\Big] = -2\,\mathrm{Re}\Big[f_{T=1}(\pi - \theta)\, f_{T=0}^*(\pi - \theta)\Big].$

Using this property and assuming the charge independence of nuclear forces, that is, $\sigma_{pp}(\theta) = \sigma_{T=1}(\theta)$, one obtains

$$\sigma_{np}(\theta) + \sigma_{np}(\pi - \theta) = \sigma_{T=1}(\theta) + \sigma_{T=0}(\theta) = \sigma_{pp}(\theta) + \sigma_{T=0}(\theta)$$

Hence, $\quad \sigma_{T=0}(\theta) = \sigma_{np}(\theta) + \sigma_{np}(\pi - \theta) - \sigma_{pp}(\theta).$ \hfill (2.44)

At high energies $\sigma_{pp}(\theta) \approx 4$ mb, independent of incident proton energy. Using Eq. (2.44) one finds that $\sigma_{T=0}$ is always positive around $\pi/2$ with a forward and backward values of 30 mb and a value of about 3 mb at $\theta = \pi/2$. This is shown in Fig. 2.6.

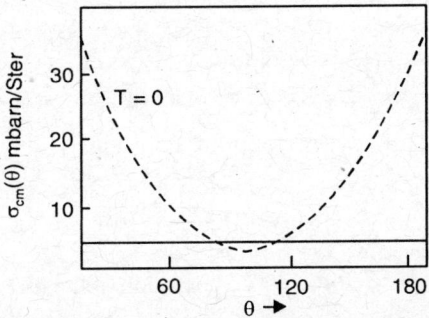

Fig. 2.6: Differential elastic scattering cross section for nucleon-nucleon scattering

Exercise 2.1. Give the underlying steps for Eq. (2.7), which is obtained from Eq. (2.6c).

$$\int e^{iq\cdot r}\frac{e^{-\mu r}}{r}\,d^3r = \iint \sum_{\ell m} 4\pi\, i^\ell j_\ell(qr)\, Y_\ell^{m*}(\Omega_q) Y_\ell^m(\Omega)\frac{e^{-\mu r}}{r} r^2\, dr\, d\Omega$$

$$= \sum_{\ell m} 4\pi i^\ell Y_\ell^{m*}(\Omega_q) \int d\Omega\, Y_\ell^m(\Omega)\, Y_0^o(\Omega)\sqrt{4\pi}\int e^{-\mu r} r\, j_\ell(qr)\, dr$$

$$= \sum_{\ell m} 4\pi i^\ell\, Y_\ell^{m*}(\Omega_q)\sqrt{4\pi}\int e^{-\mu r} r\, j_\ell(qr)\,\delta_{\ell 0}\,\delta_{m0}\, dr$$

$$= 4\pi \int e^{-\mu r} \frac{\sin(qr)}{q} dr = \frac{4\pi}{(q^2 + \mu^2)} = \frac{4\pi}{\omega_q^2}.$$

Exercise 2.2. Suppose we have the nucleon in the isodoublet representation $\psi = (\psi_p, \psi_n)$ and the pion field in the isotriplet representation $(\varphi_1, \varphi_2, \varphi_3)$ with $\phi_3 = \pi^0, \frac{1}{\sqrt{2}}(\phi_1 - i\phi_2) = \pi^+$ and $\frac{1}{\sqrt{2}}(\phi_1 + i\phi_2) = \pi^-$. Construct the *SU* (2) invariant pion nucleon (πNN) coupling.

Employing (1.15b) we get

$$\tau \cdot \phi = \begin{pmatrix} 0 & 1 \\ 1 & 0 \end{pmatrix} \phi_1 + \begin{pmatrix} 0 & -i \\ i & 0 \end{pmatrix} \phi_2 + \begin{pmatrix} 1 & 0 \\ 0 & -1 \end{pmatrix} \phi_3$$

$$= \begin{pmatrix} \phi_3 & \phi_1 - i\phi_2 \\ \phi_1 + i\phi_2 & -\phi_3 \end{pmatrix} = \begin{pmatrix} \pi^0 & \sqrt{2}\pi^+ \\ \sqrt{2}\pi^- & -\pi^0 \end{pmatrix}$$

$$\mathcal{L}_{\pi NN} = g \, \bar{\psi}(\tau \cdot \varphi) \, \psi = g \, (\bar{\psi}_p, \bar{\psi}_n) \begin{pmatrix} \pi^0 & \sqrt{2}\pi^+ \\ \sqrt{2}\pi^- & -\pi^0 \end{pmatrix} (\psi_p, \psi_n)$$

$$= g \left[(\bar{\psi}_p \, \psi_p - \bar{\psi}_n \, \psi_n) \, \pi^0 + \sqrt{2} (\bar{\psi}_p \, \psi_n \, \pi^+ - \sqrt{2} \, \bar{\psi}_n \, \psi_p \, \pi^-) \right]$$

and, thus, the relation between coupling constants are:

$$g_{pp\pi^0} = -g_{nn\pi^0} = \frac{1}{\sqrt{2}} g_{pn\pi^+} = \frac{1}{\sqrt{2}} g_{np\pi^-} = g.$$

Exercise 2.3. Employing Hulthen n-p interaction of the form $V(r) = -V_0 \, e^{-r/b}$, where $V_0 = 32.7$ MeV and $b = 2.18$ fm, calculate the s-state deuteron binding energy using variational method.

The Hamiltonian
$$H = -\frac{\hbar^2}{M} \left[\frac{d^2}{dr^2} + \frac{2}{r} \frac{d}{dr} \right] - V_0 \, e^{-r/b}.$$

We choose the trial function $\phi(r) = N \, e^{-\beta r/b} Y_{00}$, where β is the variational parameter and the normalization constant, $N^2 = (2\beta/b)^3$.

The energy, $E(\beta) = \int \phi^* H \phi \, d\tau = \frac{\hbar^2}{M} \left[\frac{\beta^2}{b^2} - \frac{8V_0 \beta^3}{(1 + 2\beta)^3} \right]$.

Setting $\dfrac{\partial}{\partial\beta}\left[\dfrac{\hbar^2}{M}\left(\dfrac{\beta^2}{b^2}-\dfrac{8V_0\beta^3}{(1+2\beta)^3}\right)\right]=0$, we get $\dfrac{(1+2\beta)^4}{\beta}=\dfrac{12M}{\hbar^2}V_0b^2$,

which gives $\beta = 0.67$ and hence the required s-state binding energy = 2.2432 MeV.

Exercise 2.4. Calculate the n-p scattering differential cross section at $\theta = 90°$ from the following data: $E_{c.m} = 5\,\text{MeV}$, $\delta_0 = 50°$, $\delta_1 = 20°$ and $\delta_2 = 5°$.

$$f(\theta) = \frac{1}{ik}\sum_{\ell=0}^{\infty}(2\ell+1)\,[2^{2i\delta_\ell}-1]\,P_\ell(\cos\theta)$$

$$f(90°) = \frac{1}{2i}\left(\frac{20.7}{5}\right)^{1/2}[\cos 100° - 1 + i\sin 100°)$$

$$+3\,(\cos\ 40° - 1 + i\sin 40)\cos(90°)$$

$$+5(\cos 10° - 1 + i\sin 10°)\frac{1}{2}(3\cos\ 90° - 1)] = (1.94\,i - 0.26)\,\text{fm}\,.$$

Hence, $\dfrac{d\sigma(90°)}{d\Omega} = 38.3\,\text{mb/Sr.}$

Exercise 2.5. The total n-p and p-p scattering cross section at $E_{tab} = 10$ MeV are respectively 900 mb and 370 mb. Assuming pure s-wave scattering calculate the singlet phase shift $\delta(^1S_0)$ and the triplet phase shift $\delta(^3S_1)$.

For n-p scattering we have

$$90\ \text{fm}^2 = \frac{4(3.14)\,2(197)^2}{(939)E_L}\left[\frac{1}{4}\sin^2\delta(^1S_0)+\frac{3}{4}\sin^2\delta(^3S_1)\right]\frac{MeV^2\,fm^2}{MeV^2}.$$

or $\quad \dfrac{1}{4}\sin^2\delta(^1S_0)+\dfrac{3}{4}\sin^2\delta(^3S_1) = 0.87$.

For p-p scattering, we have

$$37\ \text{fm}^2 = \frac{4(3.14)\,2(197)^2}{(939)E_L}\times\frac{1}{2}\sin^2\delta\,(^1S_0)\frac{MeV^2\,fm^2}{MeV^2}.$$

Solving we get $\delta(^1S_0)=1.0$ radians and $\delta(^3S_1)=1.85$ radians for $E_L = 10$ MeV.

2.2 SPIN DEPENDENCE OF N-N INTERACTION

As pointed out in the beginning, the existence of an electric quadrupole moment of deuteron and the non-additivity of magnetic moments of neutron and proton in the deuteron indicate that the nuclear force is not purely central. Because in the case of central forces, the ground state of deuteron should be S-state described by a spherically symmetric wave function.

The ground state of deuteron can not be a combination of S-state ($\ell = 0$) and P-state ($\ell = 1$) due to the parity consideration, because an isolated system is not allowed to switch back and forth from states of opposite parity. The next possibility is that the ground state of deuteron is an admixture of S and D-states. Such complex ground state is possible only when the interaction between the particles is non-central. Allowance for these non-central or tensor forces enables us to explain the existence of the electric quadrupole moment and extra magnetic moment in the deuteron.

N-N force should be translation invariant, which implies that interaction depends only on the relative separation $\vec{r} = \vec{r}_1 - \vec{r}_2$. Further, it should be invariant under reflection (parity conserving for strong force) and invariant under rotation of the coordinate system.

While constructing a tensor interaction, which depends on the relative position vector $\vec{r}$ between the nucleons and the spin vectors $\vec{\sigma}_1$ and $\vec{\sigma}_2$, we must ensure that the nuclear interaction potential V is invariant with respect to spatial rotation and reflection. Therefore, only the invariant combination that can be formed from the quantities $\vec{r}, \vec{\sigma}_1, \vec{\sigma}_2$ can appear in the potential V. Of the quantities $V(\vec{r}), \vec{r}, \vec{\sigma}_1$ and $\vec{\sigma}_2$, only the following terms are true scalars: $V(\vec{r}), \vec{r}, \vec{\sigma}_1 \cdot \vec{\sigma}_2, (\vec{\sigma}_1 \cdot \vec{r})(\vec{\sigma}_2 \cdot \vec{r}), (\vec{\sigma}_1 \times \vec{r}) \cdot (\vec{\sigma}_2 \times \vec{r})$ and their products. The last term also reduces to simpler components, e.g.,

$$(\vec{\sigma}_1 \times \vec{r}) \cdot (\vec{\sigma}_2 \times \vec{r}) = r^2 \, \vec{\sigma}_1 \cdot \vec{\sigma}_2 - (\vec{\sigma}_1 \cdot \vec{r})(\vec{\sigma}_2 \cdot \vec{r}).$$

It is usual to define the non-central potential in such a way that its average over all the directions equals to zero. We define the tensor operator,

$$S_{12} = \frac{3}{r^2}(\vec{\sigma}_1 \cdot \vec{r}) \cdot (\vec{\sigma}_2 \cdot \vec{r}) - \vec{\sigma}_1 \cdot \vec{\sigma}_2, \tag{2.45}$$

and show that the average value of S_{12} over all directions vanishes.

Using the Cartesian coordinates we get:

$$(\vec{\sigma}_1 \cdot \vec{r})(\vec{\sigma}_2 \cdot \vec{r}) = (\sigma_{1x} x + \sigma_{1y} y + \sigma_{1z} z)(\sigma_{2x} x + \sigma_{2y} y + \sigma_{2z} z)$$

$$= (\sigma_{1x} \sigma_{2x} x^2 + \sigma_{1y} \sigma_{2y} y^2 + \sigma_{1z} \sigma_{2z} z^2)$$

$$+ (\sigma_{1x}\sigma_{2y} + \sigma_{1y}\sigma_{2x}) xy + (\sigma_{1z}\sigma_{2x} + \sigma_{1x}\sigma_{2z}) xz$$

$$+ (\sigma_{1y}\sigma_{2z} + \sigma_{1z}\sigma_{2y}) yz$$

$$\overline{xy} = \overline{yz} = \overline{xz} = 0 \text{ and } \overline{x^2} = \overline{y^2} = \overline{z^2} = \frac{r^2}{3}, \text{ so that}$$

$$\overline{(\vec{\sigma}_1 \cdot \vec{r})(\vec{\sigma}_2 \cdot \vec{r})} = \frac{r^2}{3}(\vec{\sigma}_1 \cdot \vec{\sigma}_2). \text{ Hence, } \langle S_{12} \rangle = 0.$$

We can derive the same result by using spherical coordinates and show that

$$\langle (\vec{\sigma}_1 \cdot \vec{r})(\vec{\sigma}_2 \cdot \vec{r}) \rangle = \frac{1}{4\pi} \iint (\vec{\sigma}_1 \cdot \vec{r})(\vec{\sigma}_2 \cdot \vec{r}) \sin \theta \, d\theta \, d\phi = \frac{1}{3} r^2 \vec{\sigma}_1 \cdot \vec{\sigma}_2.$$

Next we show that the tensor potential (2.45) vanishes in the singlet state. For singlet state we set $\vec{\sigma}_1 = -\vec{\sigma}_2$. Hence,

$$S_{12} = \frac{3(\vec{\sigma}_1 \cdot \vec{r})(-\vec{\sigma}_1 \cdot \vec{r})}{r^2} - (\vec{\sigma}_1) \cdot (-\vec{\sigma}_2) = -\frac{3(\vec{\sigma}_1 \cdot \vec{r})}{r^2} + \sigma_1^2 = -3 + 3 = 0.$$

S_{12} satisfies the following commutation relations:

$$[S^2, S_{12}] = [j^2, S_{12}] = [j_z, S_{12}] = 0, \tag{2.46}$$

where S, j and ℓ are the total spin, total angular momentum and relative orbital angular momentum operators of two-nucleon system:

$$\vec{S} = \frac{1}{2}(\vec{\sigma}_1 + \vec{\sigma}_2), \quad \vec{j} = \vec{\ell} + \vec{S} \quad \text{and} \quad S^2 = \frac{1}{2}(3 + \vec{\sigma}_1 \cdot \vec{\sigma}_2).$$

Equations (2.46) imply that in the presence of tensor interaction only the total angular momentum j, its Z-component m_j and the total spin S are the constants of motion. Therefore, they are conserved. The most general N-N local interaction which is momentum independent (except for the ℓ dependence), translation invariant, Galilian-invariant, Hermitian (for real functions $V(r)$), parity conserving, time reversal invariant and scalar under rotation can be expressed as $V_{NN} = V_{OPEP} + V(r)$, where

$$V(r) = V_c(r) + V_\sigma(r) \, \vec{\sigma}_1 \cdot \vec{\sigma}_2 + V_{so} \, \vec{\ell} \cdot \vec{S} + V_T(r) \, S_{12}$$

$$+ V_{qso}(r) \left[\{ \delta_{\ell j} - \vec{\sigma}_1 \cdot \vec{\sigma}_2 \} \, \ell^2 - (\ell \cdot S)^2 \right] \tag{2.47}$$

The five terms in the N-N potential are respectively, the spin-independent and spin-spin central forces, a linear spin-orbit potential, the two nucleon tensor force and a quadratic spin-orbit interaction.

A. Exchange Potentials

An alternative way of characterizing the central part of the potential (2.47) is in terms of exchange operators. In some sense this is more in keeping with the meson exchange model. Four central exchange potentials are required:

(I) Wigner Exhange

The ordinary potential, written to exhibit formally its nature as an operator, is

$$\tilde{V}(r) = V_W(r)\,\tilde{P}^W \quad \text{with } \tilde{P}^W = I, \text{ the identity operator.}$$

(II) Bartlett or spin exchange potential

$$\tilde{V}(r) = V_B(r)\,\tilde{P}^B \quad \text{with } \tilde{P}^B = P^\sigma = \frac{1}{2}(1 + \vec{\sigma}_1 \cdot \vec{\sigma}_2).$$

(III) Heisenberg or isospin exchange potential

$$\tilde{V}(r) = V_H(r)\,\tilde{P}^H \quad \text{with } \tilde{P}^H = \tilde{P}^\tau = \frac{1}{2}(1 + \vec{\tau}_1 \cdot \vec{\tau}_2).$$

(IV) Majorana or space exchange potential

$$\tilde{V}(r) = V_M\,\tilde{P}^M \quad \text{with } {}^M = P^r \text{ and } P^r\,\phi(\vec{r}) = \phi(-\vec{r}).$$

For two identical fermions

$$\tilde{P}^\sigma\,\tilde{P}^\tau\,P^r[\phi(\vec{r})\,\chi_\sigma(1,2)\,\xi_\tau(1,2)] = -\phi(\vec{r})\,\chi_\sigma(1,2)\,\xi_\tau(1,2),$$

where χ_σ and ξ_τ are respectively spin and isospin functions of two fermions.

Therefore, $\tilde{P}^\sigma\,\tilde{P}^\tau\,P^r = -I$. Since $(P^r)^2 = I$, the above relation gives

$$\tilde{P}^M = \tilde{P}^r = -\frac{1}{4}(1 + \vec{\sigma}_1 \cdot \vec{\sigma}_2)(1 + \vec{\tau}_1 \cdot \vec{\tau}_2).$$

It is straight forward to show that $P^\sigma = \frac{1}{2}(1 + \vec{\sigma}_1 \cdot \vec{\sigma}_2)$ and

$P^\tau = \frac{1}{2}(1 + \vec{\tau}_1 \cdot \vec{\tau}_2)$ exchange spin and isospin of two nucleons respectively.

We define $\sigma_{1+} = (\sigma_{1x} + i\sigma_{1y})$, $\sigma_{1-} = (\sigma_{1x} - i\sigma_{1y})$ and so on.

Therefore, $\frac{1}{2}(\sigma_{1+}\sigma_{2-} + \sigma_{1-}\sigma_{2+}) = (\sigma_{1x}\sigma_{2x} + \sigma_{1y}\sigma_{2y})$

or $P^\sigma = \frac{1}{2}(1 + \vec{\sigma}_1 \cdot \vec{\sigma}_2) = \frac{1}{2}\left[1 + \frac{1}{2}(\sigma_{1+}\sigma_{2-} + \sigma_{1-}\sigma_{2+}) + \sigma_{1z}\sigma_{2z}\right].$

Let us consider a two nucleon spin state $\chi = \chi_{1/2}^{1/2}(1)\,\chi_{1/2}^{-1/2}(2)$, where (1) and (2) denote particle labels. We also note:

$$\sigma_{1z}\,\sigma_{2z}\,\chi_{1/2}^{1/2}(1)\chi_{1/2}^{-1/2}(2) = -\chi_{1/2}^{1/2}(1)\chi_{1/2}^{-1/2}(2).$$

$$P^\sigma \chi_{1/2}^{1/2}(1)\chi_{1/2}^{-1/2}(2)$$

$$= \frac{1}{2}\left[1 + \frac{1}{2}(\sigma_{1+}\sigma_{2-} + \sigma_{1-}\sigma_{2+}) + \sigma_{1z}\,\sigma_{2z}\right]\chi_{1/2}^{1/2}(1)\,\chi_{1/2}^{-1/2}(2)$$

$$= \frac{1}{2}\left[\chi_{1/2}^{1/2}(1)\,\chi_{1/2}^{-1/2}(2) + 2\chi_{1/2}^{1/2}(2)\,\chi_{1/2}^{-1/2}(1) - \chi_{1/2}^{1/2}(1)\,\chi_{1/2}^{-1/2}(2)\right]$$

$$= \chi_{1/2}^{1/2}(2)\,\chi_{1/2}^{-1/2}(1).$$

Further, we assume all potentials to have the same r-dependence with an universal shape $V_0(r)$ and setting $V_W(r) = aV_0(r)$, $V_B(r) = b\,V_0(r)$,

$$V_H(r) = c\,V_0(r) \text{ and } V_M(r) = d\,V_0(r), \text{ we can write}$$

$$V(r) = V_0(r)\left[a\,\tilde{P}^W + b\tilde{P}^B + c\tilde{P}^H + d\tilde{P}^M\right].$$

Classification of States of the N-N System

We have shown that in a singlet state $S_{12} = 0$. This means that for $S = 0$ the nuclear forces are central. Therefore, in a singlet state the orbital angular momentum ℓ is conserved and states with $S = 0$ can be classified by the quantity ℓ.

The triplet state ($S = 1$), which can be characterized by determining the value of the total angular momentum j, is a superposition of states with values of ℓ determined by the rules of the addition of angular momenta. For example, the triplet state with $j = 1$ is a superposition of states $^3S_1$, $3P_1$ and $^3D_1$.

The parity of the state is connected with the value of the orbital angular momentum by the relation $p = (-)^\ell$. Therefore, the states $^3S_1$ and $^3D_1$ with even values of ℓ and the state $^3P_1$ with an odd value of ℓ are characterized by the opposite parity.

It follows from the experimental data that the total angular momentum of deuteron, $j = 1$. As pointed out earlier also, the deuteron ground state is $^3S_1 + ^3D_1$. Thus, the ground state of deuteron is characterized by $j = 1$, $S = 1$ and parity $\pi = $ even, i.e. $\psi_0 = \psi_S + \psi_D$, where ψ_S and ψ_D respectively describe the S and D-states and they are products of radial functions and functions depending on the angles and spin variables. The spin angle variables

$$y_{\ell s j}^{m_j} = \sum_{m + m_s = m_j} \langle \ell S m m_s | j m_j \rangle\, Y_\ell^m\, \chi_S^{m_s},$$

where ℓ and S are the quantum numbers of the squares of the orbital and spin angular momenta and m and m_s are their components along Z-axis, and $\langle \ell S m m_s | j m_j \rangle$ are the Clebs Gordon coefficients.

Hence, $\phi_S = \dfrac{u(r)}{r}\, y_{011}^{m_j}$ and $\phi_D = \dfrac{w(r)}{r}\, y_{211}^{m_j}$,   (2.48a)

where u and w are respectively the radial functions of the S and D-states. The spin angle wave functions have the form: $y_{011}^{m_j} = Y_0^o\, \chi_1^{m_j}$ and

$$y_{211}^{m_j} = \sqrt{\frac{(3+m_j)\,(2+m_j)}{4\times 5}}\; Y_2^{m_j+1}\, \chi_1^{-1}$$

$$+\sqrt{\frac{(2-m_j)\,(2+m_j)}{2\times 5}}\; Y_2^{m_j}\, \chi_1^0 + \sqrt{\frac{(2-m_j)\,(3-m_j)}{4\times 5}}\; Y_2^{m_j-1}\, \chi_1^1 \; . \quad (2.48b)$$

Since, $y_{\ell s j}^{m_j}$ are orthonormal functions, it follows from the normalization condition for the wave function of the deuteron ground state, i.e. $\int |\psi_0|^2 d^3 r = 1$,

that the relation $\int_0^\infty (u^2 + w^2)\,dr = 1$ holds. Clearly, the quantity $P_S = \int_0^\infty u^2 dr$ and

$P_D = \int_0^\infty w^2 dr$ are interpreted respectively as the probabilities of finding the

deuteron in S- and D-states $(P_S + P_D = 1)$.

Since, S_{12} commutes with j^2, S^2, j_z then

$$S_{12}\, y_{011}^1 = a y_{011}^1 + b y_{211}^1 \qquad (2.48c)$$

Since, y_{011}^1 is independent of θ and ϕ, $\int S_{12}\, d\Omega = 0$ and $\int y_{211}^1\, d\Omega = 0$,

then the integration of (2.48c) over $d\Omega$ requires that $a = 0$. Further, the resulting equation must hold for any θ and ϕ. Consider the special case $\theta = 0$ or $\hat{r} = \hat{z}$, then $S_{12} = (3\sigma_{1z}\,\sigma_{2z} - \vec{\sigma}_1 \cdot \vec{\sigma}_2)$. Therefore,

$$S_{12}y_{011}^1 = (3\sigma_{1z}\,\sigma_{2z} - \vec{\sigma}_1 \cdot \vec{\sigma}_2)\frac{1}{\sqrt{4\pi}}\alpha(1)\alpha(2) = \frac{2}{\sqrt{4\pi}}\alpha(1)\alpha(2)\,. \quad (2.48d)$$

Now $b\, y_{211}^1 = b\,\dfrac{1}{\sqrt{10}}\left(\dfrac{5}{4\pi}\right)^{1/2}\alpha(1)\,\alpha(2).$ (2.48e)

Comparing (2.48d) and (2.48e) we get $b = \sqrt{8}$. Similarly for $m_j = 0$ and $m_j = -1$, we again get $b = \sqrt{8}$. Therefore, $S_{12}\, y_{011}^{m_j} = \sqrt{8}\, y_{211}^{m_j}$. In a similar manner we find $S_{12}y_{211}^{m_j} = \sqrt{8}\,y_{011}^{m_j} - 2y_{211}^{m_j}$. Therefore, the wave function of the deuteron

$$\psi_0 = \frac{1}{\sqrt{4\pi}}\left[\frac{u(r)}{r} + \frac{1}{\sqrt{8}}\frac{w(r)}{r}S_{12}\right]\chi_1^{m_j}, \tag{2.49}$$

where $\chi_1^{m_j}$ is the spin function of the system with the spin component coinciding with the component m_j of the total angular momentum of the deuteron. The Schrodinger equation for deuteron with tensor interaction then reads as

$$\left[\frac{-\hbar^2}{M}\nabla^2 + V_C(r) + V_T(r)S_{12}\right]\psi_0 = 0.$$

We should note that ∇^2 operating on states with $P_\ell^m(\cos\theta)\,e^{im\phi}$ reduces to $\nabla^2 = \frac{1}{r}\frac{d^2}{dr^2}r - \frac{\ell(\ell+1)}{r^2}$. Employing the explicit form of ψ_0 we resolve the S. Eq. into a system of coupled differential equations of radial waves u and w:

$$\frac{-\hbar^2}{M}\frac{d^2u}{dr^2} + V_C u + \epsilon u = -\sqrt{8}\,V_T w \tag{2.50a}$$

$$\frac{-\hbar^2}{M}\left(\frac{d^2w}{dr^2}\right) - \frac{6}{r^2}w + V_C w - 2V_T w + \epsilon w = -\sqrt{8}\,V_T u, \tag{2.50b}$$

where ϵ is the binding energy of the deuteron. The above pair of coupled equations can not be solved analytically but entail numerical integration by computer. Outside the range of nuclear forces ($r > b$), the equations for the radial functions u and w are uncoupled. The D-state radial wave function w for the Reid soft core potential is shown in Fig. 2.3b.

B. Static Moments of Deuteron

Electromagnetic moments of the deuteron enable us to estimate the amount of mixing of the individual $^3S_1$ and $^3D_1$ states in the ground state of deuteron.

$$\mu_{op} = \mu_p\vec{\sigma}_p + \mu_n\vec{\sigma}_n + \frac{1}{2}\vec{\ell},$$

where μ_p and μ_n are respectively magnetic moments of the proton and the neutron in units of nuclear magneton. The neutron being uncharged does not contribute except through its magnetic moment due to spin. In c.m system the orbital angular momentum associated with the proton is half of the relative orbital angular momentum ℓ. The observed value of the deuteron magnetic moment μ_d, is the quantum mechanical expectation value of the Z-component of μ_{op} in a state in which the total angular momentum $(\vec{j} = \vec{\ell} + \vec{s})$ has its maximum projection value along Z-direction. Hence,

$$\mu_d = \frac{\langle j_z \rangle \, \langle \vec{\mu}_{op} \cdot \vec{j} \rangle}{\langle j^2 \rangle} = \frac{1}{j+1} \langle \psi_{\ell s j m_j = j} \mid \vec{\mu}_{op} \cdot \vec{j} \mid \psi_{\ell s j m_j = j} \rangle. \quad (2.51a)$$

In order to evaluate the above quantity we express μ_{op} as

$$\mu_{op} = (\mu_p + \mu_n)\vec{s} + \frac{1}{2}(\mu_n - \mu_p)\,(\vec{\sigma}_n - \vec{\sigma}_p) + \frac{1}{2}\,\vec{\ell}. \quad (2.51b)$$

We calculate μ_d for the different ground states namely $^3S_1$ and $^3D_1$ defined earlier by Eq. (2.48a). We also note that the triplet spin state is even and $(\vec{\sigma}_n - \vec{\sigma}_p)$ is odd, hence, the expectation value of the middle term in (2.51b) is zero so that

$$\mu_d = \frac{1}{2}\langle \psi_{\ell s j m_j = j} \mid (\mu_n + \mu_p)\vec{s} \cdot \vec{j} + \frac{1}{2}\,\vec{\ell} \cdot \vec{j} \mid \psi_{\ell s j m_j = j} \rangle.$$

Employing the explicit form of ψ and carrying out the spin-angle algebra, we get

$$(\mu_d)_{3_{S_1}} = (\mu_n + \mu_p)\,P_S \text{ and } (\mu_d)_{3_{D_1}} = \frac{1}{2}\left[\frac{3}{2} - (\mu_n + \mu_p)\right]P_D.$$

For pure S-state $\mu_d = \mu_p + \mu_n = 2.7927 - 1.91316 = 0.87954$ nm.

However, as pointed out earlier $(\mu_d)_{\text{expt}}$ differs from the above value by an amount $(\Delta \mu_d) = 0.0221$ nm. Combining these results, we get

$$(\mu_d)_{\text{theory}} = (\mu_n + \mu_p)(1 - P_D) + \frac{1}{2}\left[\frac{3}{2} - (\mu_n + \mu_p)\right]P_D$$

$$= \mu_n + \mu_p - \frac{3}{2}\left[(\mu_n + \mu_p) - \frac{1}{2}\right]P_D. \quad (2.52)$$

Finally we get $(\mu_d)_{\text{theory}} = (\mu_d)_{\text{expt}}$ for $P_D = 4$ %.

Quadrupole Moment: In the case of deuteron, Eq. (1.21c) for a single proton gives $\hat{Q} = \sqrt{\dfrac{\pi}{5}}\, r^2\, Y_{2,0}$, where $r\left(= \dfrac{r_1}{2}\right)$ is the distance of the proton from the centre of mass. The observed quadrupole moment

$$Q_d = \langle \psi_S \mid Q_{zz} \mid \psi_S \rangle + 2\langle \psi_S \mid Q_{zz} \mid \psi_D \rangle + \langle \psi_D \mid Q_{zz} \mid \psi_D \rangle,$$

where

$$\psi_S = \frac{u}{r} Y_0^0 \chi_1^1 \text{ and } \psi_D = \frac{w(r)}{r} \left[\sqrt{\frac{3}{5}} Y_2^2 \chi_1^{-1} - \sqrt{\frac{3}{10}} Y_2^1 \chi_1^0 + \sqrt{\frac{1}{10}} Y_2^0 \chi_1^1 \right].$$

Substituting the deuteron wave functions (2.48) we get

$$Q_d = 2 \int \frac{uw}{r^2} Y_0^{0*} \sqrt{\frac{\pi}{5}} r^2 Y_2^0 \sqrt{\frac{1}{10}} Y_2^0 r^2 \, dr \, d\Omega$$

$$+ \frac{w^2}{r^2} \sqrt{\frac{\pi}{5}} Y_2^0 \left\{ \frac{3}{5} Y_2^{2*} Y_2^2 + \frac{3}{10} Y_2^{1*} Y_2^1 + \frac{1}{10} Y_2^{0*} Y_2^2 \right\} r^2 \, dr \, d\Omega. \quad (2.53a)$$

Employing $\int Y_\ell^{m*} Y_2^0 Y_\ell^m \, d\Omega = \sqrt{\frac{5}{4\pi}} \frac{\ell(\ell+1) - 3m^2}{(2\ell-1)(2\ell+3)}$ \quad (2.53b)

we get $\quad Q_d = \frac{1}{\sqrt{50}} \int_0^\infty \left(uw - \frac{w^2}{\sqrt{8}} \right) r^2 \, dr.$ \quad (2.54)

Employing the normalization condition for u and w and the fact that we need 4% D-state to fit the deuteron magnetic moment, we get $\langle w \rangle = \frac{\langle u \rangle}{5}$. Therefore,

$$Q_d = \frac{1}{5} \frac{1}{\sqrt{50}} \int_0^\infty u^2 r^2 \, dr, \text{ neglecting the second term.}$$

Substituting u from Eq. (2.17), we finally get

$$Q_d = \frac{1}{5} \frac{1}{\sqrt{50}} (2\alpha) \int_0^\infty e^{-2\alpha r} r^2 \, dr = \frac{1}{10} \frac{1}{\sqrt{50}} \frac{1}{\alpha^2} = 2.2627 \text{mb} \cdot$$

This value is in nice agreement with the experimental value (Table 2.1).

Corrections to the Deuteron Magnetic Moment: If the entire discrepancy between $(\mu_d)_{\text{expt}}$ and $(\mu_p + \mu_n)$ is ascribed to the D-state, we have a well known estimate of $P_D = 4\%$. But almost all local potential models which fit the deuteron quadrupole moment and nucleon-nucleon scattering data give a value of P_D between (6-7)%. This is because, the above model of magnetic moment is subject to a large number of well known corrections. Generally, following corrections are considered: (1) correction due to the presence of velocity dependent terms in N-N interaction (minimal substitution),

(2) corrections due to meson exchange currents and (3) corrections due to relativistic effects.

Hence, $(\mu_d)_{\text{expt}} = (\mu_p + \mu_n) - \dfrac{3}{2} \left[(\mu_n + \mu_p) - \dfrac{1}{2} \right] P_D + \Delta\mu_d,$ (2.55a)

where $\Delta\mu_d = (\Delta\mu_d)_{MS} + (\Delta\mu_d)_{ME} + (\mu_d)_{\text{Rel}}$ (2.55b)

and 'MS', 'ME' and 'Rel' stand for the minimal substitution, meson exchange and relativistic effects respectively. Now we shall see how the magnetic moment operator for deuteron gets modified due to the presence of linear spin orbit term in N-N potential. The form of linear spin-orbit force is

$$V_{LS} = V(r)\, \vec{L} \cdot \vec{S} = \frac{1}{\hbar} V(r) \left[\vec{r} \times \vec{p} \right] \cdot \vec{S} = \frac{1}{2\hbar} V(r) \left[\vec{r} \times \vec{p} \right] \cdot \vec{S}$$

where $\vec{p} = \vec{p}_1 - \vec{p}_2$; $\vec{p}_1 = $ proton momentum and $p_2 = $ neutron momentum. Gauze invariance requires us to replace p_1 by $\left(p_1 - \dfrac{e\vec{A}}{c} \right)$ in the presence of electromagnetic field. This replacement is also called minimal substitution. Therefore,

$$V_{LS} = \frac{1}{2\hbar} V(r) \left[\vec{S} \times \vec{r} \right] \cdot \left[\vec{p}_1 - \vec{p}_2 \right] - \frac{e}{2\hbar c} V(r) \left[\vec{S} \times \vec{r} \right] \cdot \vec{A},$$

where $\vec{A}$ is the vector potential. The additional interaction energy,

$H_{e\cdot m} = -\dfrac{e}{2\hbar c} V(r)(\vec{S} \times \vec{r}) \cdot \vec{A}.$ For a uniform magnetic field, the vector potential $\vec{A} = -\dfrac{1}{4} \vec{r} \times \vec{H}$ so that $H_{e\cdot m} = \dfrac{e}{8\hbar c} V(r)(\vec{S} \times \vec{r}) \times \vec{r} \cdot \vec{H} = (\Delta\mu)_{LS} \cdot \vec{H},$

where $(\Delta\mu)_{LS}$ represents the required correction to the magnetic moment of deuteron arising due to the presence of spin-orbit force in N-N interaction. We now estimate the numerical value of $(\Delta\mu)$ for the ground state of deuteron.

$$(\Delta\mu)_{LS} = \frac{1}{j+1} \langle jm_j = j \mid \Delta\vec{\mu} \cdot \vec{j} \mid jm_j = j \rangle$$

$$= -\frac{e}{16\hbar c} \langle \mid V(r) \left[\frac{1}{4} (\vec{\sigma}_1 \cdot \vec{r})^2 + (\vec{\sigma}_1 \cdot \vec{r})^2 \right.$$

$$\left. + 2(\vec{\sigma}_1 \cdot \vec{r})(\vec{\sigma}_2 \cdot \vec{r}) - \frac{1}{2} r^2 \left\{ j^2 + s^2 - \ell^2 \right\} \right] \rangle$$

$$= -\frac{e}{32\hbar c} \langle V(r)\, r^2 \left\{ 1 + \frac{1}{3} S_{12} + \frac{1}{3}\, \vec{\sigma}_1 \cdot \vec{\sigma}_2 - \left(j^2 + s^2 = \ell^2 \right) \right\} \rangle.$$

Employing the deuteron wave function (2.49) we get

$$(\Delta\mu)_{LS} = -\frac{e}{32\hbar c} \left[\frac{u(r)}{r} \,|r^2 V(r)|\, \frac{u(r)}{r} \rangle \left(\frac{8}{3} \right) \right.$$

$$\left. + \langle \frac{u(r)}{r} \,|r^2 V(r)|\, \frac{w(r)}{r} \rangle \left(\frac{2\sqrt{8}}{3} \right) + \langle \frac{w(r)}{r} \,|r^2 V(r)|\, \frac{w(r)}{r} \rangle \left(\frac{8}{3} \right) \right]$$

$$= -\frac{e}{12\hbar c} \left[\langle S\,|r^2 V(r)|\, S \rangle - \frac{1}{\sqrt{2}} \langle S\,|r^2 V(r)|\, D \rangle - \langle D\,|r^2 V(r)|\, D \rangle \right] nm$$

$$(2.56)$$

If $V(r)$ is known then $<S|S>$, $<S|D>$ and $<D|D>$ integrals can be evaluated either analytically or numerically[7].

Exercise 2.6. Show that for a charge independent and parity conserving N-N interaction, S^2, the square of the total two nucleon spin is a good quantum number.

For charge independent N-N interaction we can consider only space reflection and spin exchange through operator $P = P^\pi P^\sigma$ where P^π is the parity operator and P^σ is the spin exchange operator.

We have, $[H, P^\sigma] = 0$ and $[H, P^\pi] = 0$. Therefore, $[H, P^\pi P^\sigma] = 0$.

Now $[H, P^\pi P^\sigma] = P^\pi [H, P^\sigma] + [H, P^\pi] P^\sigma = P^\pi [H, P^\sigma]$

$$P^\pi [H, S^2 - 1] = P^\pi [H, S^2] = [H, S^2] = 0,$$

so S^2 is a good quantum number.

2.3 PHENOMENOLOGY OF NUCLEON-NUCLEON INTERACTION

The nucleon-nucleon potential is called semi-phenomenological or realistic when it is obviously not a pure phenomenological one, but when it is based on some underlying theory, mostly some form of the meson exchange. The important ingredients of all such potential models are parameters of these models. At present there exists no theory which allows us to calculate in a parameter free way, a N-N potential, which predicts realistic phase shifts obtained by a multi energy phase shift analysis of p-p and n-p scattering data below the meson production threshold ($E < 350$ MeV). It is obvious that more parameters one has at one's disposal, the easier it will be to get satisfactory fit to the data. Mainly there are two sets of parameters. The first set will contain the physical parameters, the parameters that can be checked independently, some where else in physics. An example is the pion nucleon coupling constant. Second set of parameters contains purely phenomenological parameters. These are introduced to cover up our ignorance of N-N interaction and are used at the same time to improve fit to the data. For example, the hard core radii and the description of short range forces.

Scattering Matrix and Observables

In the case of nucleon-nucleon scattering (scattering of spin ½ particles by spin ½ target) we need nine independent experiments to determine nine independent parameters. This comes about as follows:

Consider for simplicity p-p scattering which occurs in $T = 1$ states only at fixed angular momentum j, which is conserved. There are two spins $S = 0$ and $S = 1$, singlet and triplet respectively. Because of the Pauli principle there can not be a singlet triplet transition. Then corresponding to $S = 1$ and fixed j, the total orbital angular momentum ℓ takes the values $j \pm 1, j$. Hence, we shall encounter nine scattering amplitudes with the following combinations of $(\ell\ell')$ for $S = 1$:

$(\ell = j-1, \ell' = j),$ $(\ell = j, \ell' = j-1),$ $(\ell = j, \ell' = j+1),$

$(\ell = j+1, \ell' = j-1),$ $(\ell = j-1, \ell' = j+1),$ $(\ell = j+1, \ell' = j-1),$

$(\ell = j, \ell' = j),$ $(\ell = j-1, \ell' = j-1)$ and $(\ell = j+1, \ell' = j+1).$

Because of the parity conservation, four of these amplitudes having $(\ell = j-1, \ell' = j), (\ell = j, \ell' = j-1), (\ell = j, \ell' = j+1)$ and $(\ell = j-1, \ell' = j)$ are zero and because of time reversal invariance, the amplitudes corresponding to $(\ell = j-1, \ell' = j+1)$ and $(\ell = j+1, \ell' = j-1)$ are equal. Hence, there remains only four amplitudes corresponding to $S = 1$ and for $S = 0$ there is only one amplitude for $(\ell = j, \ell' = j)$. So there are five amplitudes which are also complex. If we eliminate an overall phase factor, the specification of these amplitudes at one energy and angle will require nine parameters and hence nine independent experiments. If, however, measurements are made over all angles from $0°$ to $90°$, then the unitarity relation shows that in principle at least five kinds of experiments at one energy and angle suffice to give p-p scattering matrix completely.

Polarization Measurements: Experiments are first done by scattering nucleons, once to measure the differential cross section $\dfrac{d\sigma(\theta)}{d\Omega}$ with unpolarized beam of nucleons by unpolarized target, twice to measure the polarization of the beam which before scattering was unpolarized and three times to measure the polarization produced in a beam which is already polarized (Fig. 2.7) before scattering.

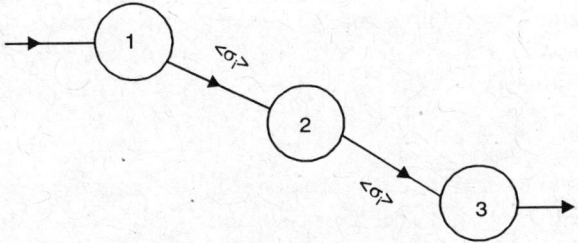

Fig. 2.7: Triple scattering diagram

The cross-section for the unpolarized beam is the single scattering parameter and the polarization $P(\theta)$ of the beam once scattered by the target is the double scattering parameter. The triple scattering parameters namely, depolarization $D(\theta)$, rotation parameter $R(\theta)$ and $A(\theta)$ describe how the second scatterer changes the direction or magnitude or both of the polarization. The first scatterer acts as a polarizer and the final scatterer as a analyzer.

Therefore, to measure polarization, we do double scattering measurements in which an unpolarized beam is first scattered by an unpolarized target and then a part of the scattered beam undergoes a second scattering for which the differential cross section is measured. As a result of the first scattering the beam incident on the second target is partially polarized, because the scattering force is spin dependent. So the measured cross section has left right asymmetry about the direction of the intermediate beam. Due to the experimental difficulties associated with using polarized targets, following measurements are generally made:

(*i*) Unpolarized differential cross section, measured in single scattering

$$\frac{d\sigma(\theta)}{d\Omega}.$$

(*ii*) Polarization $P(\theta)$ upon single scattering, determined by a subsequent

second scattering defined (for spin $-\frac{1}{2}$ particles) as

$$P(\theta) = \frac{N_+(\theta) - N_-(\theta)}{N_+(\theta) + N_-(\theta)},$$

where $N_\pm$ are the number particles whose spin components are parallel (antiparallel) to a preferred direction (usually taken as Z-direction).

(*iii*) Depolarization $D(\theta)$ produced by a second scatterer in a coplanar triple scattering. $D(\theta)$ determines the amount of polarization perpendicular to the scattering plane converted to the same after the second scattering.

(*iv*) Rotation of polarization $R(\theta)$ in triple scattering using successive scattering planes at right angles.

(*v*) Longitudinal polarization in triple scattering $A(\theta)$ where transverse polarization resulting from the first scattering is rotated to longitudinal polarization using a magnetic field prior to second scattering.

The reduction of all the two-nucleon scattering data to phase shift contribution has been carried out notably by the Yale Group (Breit et al., 1960)[8] and by Livermore group (Mac Gregor et al.[9]). A general discussion of these points as well as the entire two nucleon problem can be found in an article by Breit and Haracz[8]. In the single energy phase shift analysis, one includes partial waves upto a maximum $\ell = \ell_{max}$. For $\ell > \ell_{max}$ phases are generated by OPEP. The phase shift analysis is performed by determining phase parameter in such a way that the least square sum

$$\chi^2 = \sum_{i=1}^{n} \left| \frac{Y_i(\delta) - Y_i(\exp)}{\Delta Y_i(\exp)} \right|^2 ,$$

for n experimental data $Y_i(\exp) \pm \Delta Y_i(\exp)$ has a minimum. The expression $Y_i(\delta)$ is the value of the corresponding observables calculated from given phase parameter, generated by a chosen potential obtained from meson theory of nuclear forces. $Y_i(\exp)$ is the experimental value and $\Delta Y_i(\exp)$ is the probable error in $Y_i(\exp)$. The determination of g^2 in p-p and n-p scattering, has been used as a source of information regarding g^2 and its approximate independence on the Isospin component T_3, furnish tests of charge independence of N-N interaction[10].

A. Realistic Potentials

The most general form of N-N potential is given by (2.47). Since the phenomenological potentials have been obtained by fitting experimental data over a certain energy range usually below meson production threshold, it is a general practice to require the determination of as few parameters in the potential as possible. The high $\ell(\ell \geq 5)$ phase shifts δ_ℓ are calculated employing (2.25) by substituting spherical Bessel function $j_\ell(kr)$ for both $u_{\ell k}(r)$ and $v_{\ell k}(r)$ for the one pion exchange potential:

$$\tan \delta_\ell = -k \int_0^\infty [j_\ell(kr)]^2 V_{OPEP}(r) r^2 \, dr. \tag{2.57}$$

The above expression is also called the first order Born approximation expression for the phase shift.

The low ℓ phase shift are searched to fit the experimental data. Detailed efforts to derive potentials using theoretical guidance that would generate experimental phase shifts have been vigorously pursued by several workers for the last thirty years. Several phenomenological potential models that are rather based on the multi energy phase shift analysis, employ N-N potential of the form (2.47) in which V_{OPEP} is the one pion exchange potential with $\dfrac{g^2}{4\pi} = 13.16$ in singlet even states and 14.0 otherwise. The pion mass m_π is the neutral pion mass used for singlet even and triplet odd states. For singlet-odd and triplet-even states, a mass of the pion was statistically weighted to include a neutral and charged pions in the ratio 1:2. The hard core radius was taken to be $x_c = 0.35$. V's other than V_{OPEP} has the form

$$V = \sum_n a_n \frac{e^{-2x}}{x^n}, \quad \text{where} \quad x = \mu r. \tag{2.58}$$

The coefficients a_n are obtained by the multi energy phase shift analysis of data by the Yale group (Yale potential)[8]. Hamada and Johnston (HJ)[11] employ a potential form of different spin orbit dependence. The numerical values of the parameters of HJ potential are given in Ref. 6. The parameters of Yale potential developed by Breit et al. are given in Ref. 8.

The Reid potentials[6] are similar to the HJ Potentials particularly for $r \geq 1$ fm, but notably they include a variant in which the hard core is replaced by repulsive Yukawa form terms. For example, the soft core version of the potential operating in the ^1S state includes a repulsive component having a behaviour e^{-7x}/x and a rather large relative strength. Soft core implies small but non-vanishing wave functions within the core region. A somewhat larger number of parameters are used than for HJ potential, and no particular attempt is made to associate x with any particular meson mass.

The Paris potential[12] derives the contributions from meson exchange into three distinct types: (1) π-exchange (long range), (2) resonant and non-resonant two-pion exchange (medium range) and (3) 3-pion exchange mediated primarily by ω-exchange (short range). The Paris group resorts to pure phenomenology to describe the short range (SR) part of the N-N interaction[12]. The complete potential is written as a sum of theoretical part plus a phenomenological part, together with a weighing function $f(r)$:

$$V = V_{theo}(r) f(r) + V_{pheno}(r) \left[1 - f(r)\right]. \tag{2.59}$$

The function $f(r)$ in (2.59) is designed to sharply suppress V_{pheno} for $r > r_0$, where $r_0 = 0.8$ fm. The weighing function has the form:

$$f(r) = \frac{(xr)^{\alpha}}{1 + (xr)^{\alpha}},$$

with $x = 1.25$ fm^{-1} and $\alpha = 10$. The parameters in V_{pheno} are then fitted to the properties of the deuteron and N-N phase shifts for $\ell < 6$ upto 350 MeV incident nucleon Lab. energy. The phase shift analysis carried by the Paris group, a total 2239 data points were considered. χ^2 per degree of freedom for Paris potential is 1.99 for p-p scattering and 2.17 for n-p scattering, which is one of the best fit presently available.

One way to gauge the quality of the different NN potentials is to see how well they describe the experimental data (Table 2.1). For that one needs first of all a good phase shift analysis of all the data.

Table 2.1: Comparison of two-nucleon data with theoretical values obtained by various potentials

	Quantity	Experiment	Bonn[13]	Paris[12]	Reid[6]
Deuteron	B_d (MeV)	−2.224575(9)	−2.22458	−2.2249	−2.2246
	P_d (%)	−	4.38	5.77	6.47
	Q_d (mb)	2.8590(3)	2.74	2.79	2.796
	μ_d (nm)	0.857406(1)	0.8548	0.853	0.875
	R_e^d (fm)	1.9650(68)	1.9684	1.9716	1.9567
$^3S_1$	a_t (fm)	5.424(4)	5.424	5.427	5.39
n-p scattering	r_t (fm)	1.759(5)	1.760	1.766	1.720
$^1S_0$					
n-p scattering	a_S (fm)	−23.75(1)	−23.74	−	−
	r_s (fm)	2.75(5)	2.70	−	−
p-p scattering	a_S (fm)	−7.828(8)	−7.810	−7.78	−
	r_s (fm)	2.80(2)	2.797	2.72	

B. Three-Nucleon Systems with Two-Nucleon Inputs

Few-Body systems are understood as consisting of small number of well defined constituent structures, which can be either elementary, quasi-elementary or composites behaving collectively in many particle system. Examples of such systems in Nuclear Physics range from few-nucleon nuclei ($^3H, ^3He, ^4He$, etc.) or heavier nuclei treated as collection of clusters of nucleons, hyper nuclei and quark clusters representing baryons.

Many techniques have been used to solve time independent S. Eq. for three nucleon system. Earliest technique used is the "variational method". Many different procedures exist for developing the trial function used in these calculations. The variational method is specially applicable for determination of lowest energy state of the system. The expectation value of energy $\langle E \rangle = \int \psi^* H \psi \, d\tau$. Here ψ is any normalized function of the coordinates of the system.

Table 2.2: Properties of ^3H and ^3He

Property	^3H	^3He
Ground state spin ($\hbar$)	1/2	½
Ground state parity	+	+
Magnetic moment (nm)	2.9786	−2.1274
Binding energy (MeV)	8.482	7.718
rms charge radius (fm)	1.70 ± 0.05	1.87 ± 0.05
Half life (Y)	12.26	Stable

This integral is known as variation integral and it gives an upper limit to the energy E_0 of the lowest state of the system. The function ψ is the variation function and its choice may be quite arbitrary, but more wisely it is chosen such that E approaches closely to E_0.

The mathematical formulation for the bound state calculation of three nucleon system starts with a suitable choice of coordinate system. We neglect here spin-degree of freedom and use spin independent N-N interaction. Let R be the position vector of c.m. of three particles defined as

$$\vec{R} = \frac{\vec{r}_1 + \vec{r}_2 + \vec{r}_3}{3} \, .$$

We define $\rho = \dfrac{\vec{r}_2 + \vec{r}_3}{2}$ and $\vec{r} = \dfrac{\sqrt{3}}{2}(\vec{r}_2 - \vec{r}_3)$

and express kinetic and potential energy operators in terms of $\vec{R}$, $\vec{\rho}$ and $\vec{r}$:

$$KE = \frac{-\hbar^2}{2M}(\nabla_1^2 + \nabla_2^2 + \nabla_3^2) = \frac{-3\hbar^2}{4M}(\nabla_\rho^2 + \nabla_r^2) - \frac{-\hbar^2}{6M}\nabla_R^2 \, , \quad (2.60)$$

where M is the nucleon mass assumed same for both neutron and proton. If there is no external force, we have to deal only with the first two terms describing the relative motion. We also assume that the two body interaction $V(r_{ij})$ is not affected by the presence of the third particle. Further, neglecting the three-body force we can assume:

$$V(r_{12}) = V(r_{13}) = V(r_{23}) \tag{2.61a}$$

We take a simple two-body force of Yukawa form:

$$V(r_{12}) = \frac{1}{2}(1+q)V_0 \frac{e^{-br_{12}}}{br_{12}}, \tag{2.61b}$$

where b (= 0.855 fm^{-1}) is the usual range parameter and V_0 (= 67.3 MeV) is the depth of the potential and $q = 0.69$ is the ratio of singlet to triplet depth.

Since $r_{23} = r_{12} = r_{13} = \frac{2}{\sqrt{3}}r$, the potential energy operater

$$= 3V(r_{12}) = \frac{3}{2}(1+q)V_0 \frac{e^{-\frac{2}{\sqrt{3}}br}}{\frac{2}{\sqrt{3}}br} \quad \text{and} \quad E_{Coul} = \frac{e^2}{r_{23}} = \frac{\sqrt{3}}{2}\frac{e^2}{r}. \tag{2.62}$$

Next step is to choose a suitable trial wave function. The wave function is assigned three basic properties: (1) It can interfere with itself. This property explains the phenomena of diffraction. (2) It is large in magnitude where particles are likely to be located and small elsewhere. $|\psi|^2$ gives the probability density of finding the physical system at a particular place at a given time.(3)The wave function is single valued, positive definite, continuous, non singular and differentiable. The one suggested by Irving[14] is

$$\psi = N^{1/2} \exp\left[-\alpha(r_{12}^2 + r_{13}^2 + r_{23}^2)^{1/2}\right], \tag{2.63}$$

where N is the normalization constant and α is the variational parameter, which is used to minimize the binding energy.

Now $r_{12}^2 + r_{13}^2 + r_{23}^2 = (\vec{r}_1 - \vec{r}_2)^2 + (\vec{r}_1 - \vec{r}_3)^2 + (\vec{r}_2 - \vec{r}_3)^2 = 2(\rho^2 + r^2)$.

Then the normalization constant is obtained from the relation

$$N \int \exp\left\{-2\sqrt{2\alpha}(\rho^2 + r^2)^{1/2}\right\}\rho^2 \, d\rho r^2 \, dr \, (4\pi)^2 = 1.$$

In order to evaluate the normalization integral we substitute $r = s \sin \phi$ and $\rho = s \cos \phi$. Carrying out the integration we get:

$$\frac{1}{N} = (4\pi)^2 \int_0^\infty e^{-2\sqrt{2\alpha s}} s^5 \, ds \int_0^{\pi/2} \sin^2 \phi \cos^2 \phi \, d\phi$$

We use the standard integral (6.46b) and get

$$N = \frac{2^9 \alpha^6}{\pi^3 5!}.$$

$$KE = -\frac{3\hbar^3}{4M} N \int e^{-\sqrt{2}\alpha(\rho^2+r^2)^{1/2}} \left[\nabla_\rho^2 + \nabla_r^2\right] e^{-\sqrt{2}\alpha(\rho^2+r^2)^{1/2}} \, d\vec{r} \, d\vec{\rho}$$

$$= -\frac{3\hbar^3 N}{4M}(4\pi)^2 \int \frac{e^{-\sqrt{2}\alpha(r^2+\rho^2)^{1/2}}}{(r^2+\rho^2)} [2\alpha^2(\rho^2+r^2)$$

$$- 5\sqrt{2}\alpha(r^2+\rho^2)^{1/2}] \, r^2\rho^2 \, dr \, d\rho$$

$$= -\frac{3\hbar^3 N}{4M} \frac{\Gamma 3/2}{2\sqrt{3}} \int_0^\infty e^{-2\sqrt{3}\alpha s} \left[2\alpha^2 s^5 - 5\sqrt{2}\alpha^4\right] ds = \frac{3}{2}\frac{\hbar^2}{M}\alpha^2 .$$

$$PE = -\frac{(1+q)V_0}{5\pi a(1-a^2)^3} \left[8 + 9a^2 - 2a^4 - \frac{15a}{\sqrt{1-a^2}} \cos^{-1}a\right], \text{ where } a = \frac{b}{\sqrt{6}\alpha}.$$

$$E_{Coul} = \frac{16\sqrt{2}}{5\sqrt{3}\pi} e^{2\alpha}. \tag{2.64}$$

With spin independent central interaction of the type (2.61b) and the adjustable variational parameter $\alpha = 0.92$ fm^{-1}, one gets reasonable value of BE:

$$KE(MeV) = 52.3 \, ; \, PE(MeV) = -60.2 \, ; \, BE(^3H) = -7.9 \text{ MeV}.$$

$$E_{Coul}(MeV) = 1.10 \text{ and } BE(^3He) = -6.8 \text{ MeV}.$$

The above values are in reasonable agreement with experimental values (Table 2.2). However, Irving wave function gives too small a value for root mean square radius of charge distribution of H^3 compared to electron scattering measurements. For example, for Irving wave function (2.63) and r_1 as proton coordinate, we get

$$R_c(H^3) = [\{(\vec{r_1} - \vec{R})^2\}_{00}]^{1/2} = \left[\frac{7}{6\alpha^2}\right]^{1/2} = 1.17 \text{ fm}, \tag{2.65}$$

which is $\sim 2/3$ of the experimental value (Table 2.2).

Mathur, Mukherjee and Rustgi[15] used a two-body spin dependent force of exponential type with hard core and found that the presence of hard core gives rise to reasonable values for the binding energy as well as r.m.s. charge radius of ^3H. With the inclusion of spin degrees of freedom, the ground state function which must be antisymmetric between two like nucleons can be written as

$$\Psi = \chi_a \Psi_s + \chi_s \Psi_a,$$

where $\chi_a = \frac{1}{\sqrt{2}} [\alpha(1)\beta(2) - \alpha(2)\beta(1)]\alpha(3)$

and $\chi_s = \left[\alpha(1)\,\beta(2) + \alpha(2)\,\beta(1) \right] \dfrac{\alpha(3)}{\sqrt{6}}$

$\qquad = \dfrac{2}{\sqrt{3}} \alpha(1)\,\alpha(2)\,\beta(3),$

where χ_a and χ_s are the antisymmetric and symmetric spin wave functions respectively, and ψ_s and ψ_a are the space wave functions which are symmetric and antisymmetric respectively with regard to two like nucleons. With the approximation $\psi_a = 0$, the wave function ψ becomes

$$\psi = \chi_a\,\psi_s$$

The space part of the wave function is of exponential type:

$$\psi_s = \prod_{ij} \left\{ \exp\left[-\mu(r_{ij} - r_0) - \exp\left[-v(r_{ij} - r_0) \right] \right\} \text{ for } r_{ij} \geq r_0 \right.$$

$$= 0 \qquad\qquad\qquad\qquad\qquad \text{for } r_{ij} \leq r_0 \qquad (2.66)$$

Hence, $R_c = \left[\displaystyle\int \psi^* \left\{ \dfrac{\vec{r}_{12} + \vec{r}_{13}}{3} \right\}^2 \psi\, r_{12}\, r_{13}\, r_{23}\, dr_{12}\, dr_{13}\, dr_{23} \right]^{\frac{1}{2}}.$

Employing (2.66) for ψ and the following parameters: $\mu = 0.4$ fm^{-1}, $v = 4.5$ fm^{-1} and $r_0 = 0.4$ fm, which fit the binding energy of ^3H reasonably, we get $R_c(^3H) = 1.71$ fm, in agreement with experiment.

Mixed Symmetric Spin States: Three spin 1/2 identical particles may give rise to four $S = 3/2$ states which are symmetric under the interchange of any two spins and in addition there are two types of mixed symmetric states with total spin $S = 1/2$. Let α and β denote, respectively, spin-up and spin-down Pauli spinors. Then $S = 3/2$ and $S_z = +3/2$ state is

$$\chi^s_{3/2,3/2} = \alpha(1)\,\alpha(2)\,\alpha(3),$$

the superscript s on the spin function χ stands for symmetric. The remaining three symmetric $S = 3/2$ states are:

$$\chi^s_{3/2,-3/2} = \beta(1)\,\beta(2)\,\beta(3),$$

$$\chi^s_{3/2,1/2} = \frac{1}{\sqrt{3}} \left[\alpha(1)\,\alpha(2)\,\beta(3) + \alpha(1)\,\beta(2)\,\alpha(3) + \beta(1)\,\alpha(2)\,\alpha(3) \right],$$

$$\chi^s_{3/2,-1/2} = \frac{1}{\sqrt{3}} \left[\alpha(1)\,\beta(2)\,\beta(3) + \beta(1)\,\alpha(2)\,\beta(3) + \beta(1)\,\beta(2)\,\alpha(3) \right].$$

The mixed symmetric states may be formed by combining the spins of 1 and 2 to give $S_{(12)} = 0$ and then coupling the spin S_3 of the third particle to yield the total spin $S = 1/2$. The wave functions which have mixed symmetry with respect to permutation of (1,2,3) but are antisymmetric under the permutation $1 \leftrightarrow 2$ are :

$$\chi_0 = \chi^{ms}_{1/2,1/2}(1) = \frac{1}{\sqrt{2}} \left[\alpha(1)\beta(2) - \beta(1)\alpha(2) \right] \alpha(3)$$

and

$$\chi_0 = \chi^{ms}_{1/2,-1/2}(2) = \frac{1}{\sqrt{2}} \left[\alpha(1)\,\beta(2) - \beta(1)\alpha(2) \right] \beta(3),$$

which are antisymmetric between particles 1 and 2 but there are no overall symmetry under the exchange of $1 \leftrightarrow 3$ or $2 \leftrightarrow 3$. The superscript 'ms' on χ stands for mixed symmetric wave function which is antisymmetric between 1 and 2. The second mixed symmetric state is constructed by coupling spins of particles 1 and 2 to give $S_{(12)} = 1$ and then coupling spin S_3 to yield $S = 1/2$. The wave functions which have mixed symmetry under the permutation of (1,2,3) but are symmetric w.r.t. the permutation $1 \leftrightarrow 2$ are :

$$\chi_1 = \chi^{ms}_{1/2,1/2}(1) = \sum_{m_s} \left\langle 1\frac{1}{2} m\, m_s \,| \frac{1}{2}\, m_s \right\rangle |1\, m\rangle \left| \frac{1}{2} m_s \right\rangle$$

$$= \sqrt{\frac{2}{3}}\, |1\,1\rangle \left| \frac{1}{2} -\frac{1}{2} \right\rangle - \sqrt{\frac{1}{3}}\, |1\, 0\rangle \left| \frac{1}{2}\, \frac{1}{2} \right\rangle$$

$$= \frac{1}{\sqrt{6}} \left[2\alpha(1)\, \alpha(2)\, \beta(3) - \alpha(1)\, \beta(2)\, \alpha(3) - \beta(1)\, \alpha(2)\, \alpha(3) \right]$$

$$\chi_1 = \chi^{ms}_{1/2,-1/2}(2) = \frac{1}{\sqrt{6}} \left[2\beta(1)\, \beta(2)\, \alpha(3) - \alpha(1)\beta(2)\, \beta(3) - \beta(1)\, \alpha(2)\, \beta(3) \right].$$

In the same way we can find mixed symmetry isospin states ϕ_0 and ϕ_1.

Besides ψ_s and ψ_a, there is a possibility of a state of mixed symmetry (ms) arising due to the exchange of both space and spin-isospin coordinates of three nucleon system. The total wave function with spatially symmetric Irving wave function (2.63) can be written as an admixture of mixed symmetric state:

$$\psi = \left[1 + \delta_{S'}^2 \right]^{-1/2} \left[\psi_s \Phi_a + \frac{\delta_{S'}}{\sqrt{2}} \left(\psi(2)\, \varphi(1) - \psi(1)\, \varphi(2) \right) \right],$$

where ψ's and ϕ's are respectively orthonormal position and spin-isospin functions having symmetric or antisymmetric character. The free parameter that can be adjusted to produce a fit to experimental data are S'/S mixing ratio $\delta_{S'}$ and the size parameter α.

Mathur and Lagu[16] were able to attain good agreement with the values of the rms charge radius of ^3H by using the hard core mixed symmetry wave function with a 2 to 4 percent S' state admixture.

Among the exact formalism, Faddeev's equation and its generalization by Yakuboski are most popular.[17] These are respectively the three and many body generalizations of the Lippan Schwinger equation. In solving Faddeev's equations, separable interactions introduced by Mitra[18] were used. These interactions are fitted to N-N scattering phase shifts. Another very powerful approach is the hyper spherical harmonics (HH) or K-harmonics expansion method, which reduces the N-body Schrodinger equation to a more convenient form. The HH are the angular part of the 3N-demensional harmonic polynomials. In brief, in this technique one expands the few-body wave function in the HH, which constitute of a complete set of functions in the 3N-1 dimensional space. Ultimately, the HH technique approximates the S. Eq. by a truncated set of coupled differential equations in a single variable called hyperradius. The application of this technique consists of three parts. First, we choose a set of coordinates and derive hyperspherical harmonics. Second, we express S. Eq. in terms of these functions. This entails the calculation of so called potential matrix elements. Lastly, the resulting equations are solved.[19-20]

In order to describe the nuclear properties, starting from a basic *N-N* force, one has to make the following postulates:

(1) In nuclei, *N-N* force may be replaced by a potential between nucleons.
(2) Nucleons within the nucleus interact primarily via two-body forces. Hence,

the nuclear Hamiltonian can be expressed as $H = \sum_i T_i + \sum_{i<j} V_{ij}$,

where T_i is the kinetic energy of the ith nucleon and V_{ij} the two-nucleon interaction.

(3) Nucleons within the nucleus move slowly and the non-relativistic dynamics prevails.
(4) Nucleus consists of only nucleons and the subnucleonic degrees of freedom are suppressed.

This is an enormous simplification of the physical problems but, in fact, accounts amazingly well for much of the experimental data. Studies in the past decade searching new physics, have focused attention on examining the validity of potential approach, the presence of three body forces, the relativistic effects and the role of subnucleon degrees of freedom.

Three-body Forces: By early sixties there was already a growing realization that the nuclear theory with the two-nucleon potential fails. The method of solving three body equations both for the bound states and scattering problems and two nucleon potential inputs were made reliable enough to suggest what is missing is the inclusion of the three-nucleon force. Therefore, the three and four nucleon systems became a testing ground for the presence of few-body forces.

With the inclusion of three-body force the non-relativistic Hamiltonian for a many nucleon system is given by

$$H = \sum_{i<j} T_i + \sum_{i<j} V_{ij} + \sum_{i<j<k} V_{ijk} + \cdots \qquad (2.67)$$

The one body part, T_i is determined by the property of a single nucleon, the two-body part V_{ij} in principle, is determined by the analysis of two-nucleon systems. There is a indication from a precise calculation of binding energy of ^3H that one needs a three-body force V_{ijk} as well. The origin of three nucleon force (3NF) lies in the fact that nucleons are treated as point particles interacting via two-body local potential. This approximation neglects the contribution of the mechanism, whereby one of the nucleons emits a meson that scatters off a second nucleon and then gets absorbed on to a third nucleon (Fig. 2.8 left). In the Hamiltonian picture (2.67) based on nucleon degrees of freedom, a true 3NF depends in an irreducible way upon the simultaneous coordinates of three interacting nucleons, i.e. one has V_{123} and not ($V_{12}+V_{23}$) as assumed in (2.61a).

Motivation for studying three nucleon system and solving exactly corresponding S. Eq. within the frame work of Faddeev's theory are the following:(1) Can we explain observed data from our knowledge of two-nucleon interaction? (2) Should we introduce three-body forces into the calculation in order to explain observed data? (3) Can we derive any information about two-nucleon interaction which can not be derived from two-nucleon interaction alone?

There are two methods of including three nucleon force (3NF) in the three-body calculation. First, we start from any N-N realistic interaction which reasonably reproduce the N-N observables, and then add 3NF with sufficient adjustable parameter to fit the binding energy of ^3H or ^3He. The second method exploits the common origin of two nucleon force (2NF) and three nucleon force (3NF). Effect of 3NF on the binding energy of H^3 has been studied extensively by many workers[21-22]. Table 2.3 summarizes calculations of Sasakawa[21].

Calculations of charge densities and electron scattering form factors of ^3He with realistic two nucleon force (2NF) alone did not meet much success. They fail to improve fit to the size of secondary maximum in the charge form factor and the related central depression in the charge density. Inclusion of 3NF, however, did not improve the situation as shown in Fig. 2.8. The inclusion of 3NF increases the binding energy and brings it close to experiment

(Table 2.3) but it concomitantly compresses the density and stretches the form factor as against experimental behaviour[22].

Table 2.3: Binding energy of 3-nucleon system

N-N Potentials	BE (^3H) (MeV)
Reid	7.35
Pairs	7.53
Pairs + 3NF	8.47
Expt.	8.48

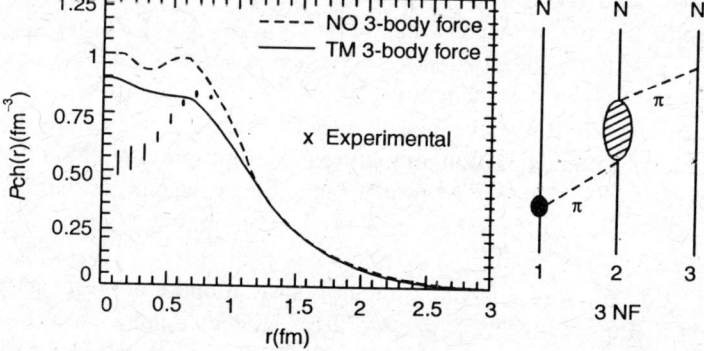

Fig. 2.8: The charge density distribution of ^{3}He with (solid line) and without (dotted line) three nucleon force (3NF)

Exercise 2.7. Derive an expression for the n-p scattering differential cross section for the scattering of unpolarized neutron with unpolarized proton including spin degree of freedom.

The scattering amplitudes are:

$$f_{\nu\mu}(\theta,\varphi)\,\chi_s^\nu = \frac{1}{2ik}\Bigg[\sum_{j\ell m} e^{2i\delta_{\ell j}}\,(4\pi(2\ell+1))^{1/2}$$

$$\times\langle\ell s0\mu|j\,m_j\rangle\langle\ell sm\nu|j\,m_j\rangle\left\{\frac{(2\ell+1)\,(\ell-m)!}{4\pi(\ell+m)!}\right\}^{1/2}$$

$$\times(-)^m\,e^{m\varphi}\,P_\ell^m(\cos\theta)\chi_s^\nu - \sum_\ell(2\ell+1)P_\ell^0(\cos\theta)\chi_s^\mu\Bigg],$$

where S is the total spin of the neutron proton system which takes the values 1 and 0. For $S = 1$, μ or $\nu = \pm 1, 0$ and for $S = 0$, $\mu = \nu = 0$. The explicit form of $f_{\nu\mu}(\theta, \phi)$ are given below :

$$f_{11}(\theta, \phi) = \frac{1}{2ik}\left[\sum_{\ell}\{(2\ell+1)\,e^{2i\delta_+}\,\langle\ell\,1\,0\,1\,|\,\ell+1\,1\rangle^2\right.$$

$$+(2\ell+1)e^{2i\delta_0}\,\langle\ell\,1\,0\,1\,|\,\ell\,1\rangle^2$$

$$\left.+(2\ell+1)\,e^{2i\delta_-}\,\langle\ell\,1\,0\,1\,|\,\ell-1\,1\rangle^2 - (2\ell+1)\}\,P_\ell^0(\cos\theta)\right]$$

$$= \frac{1}{2ik}\left[\sum_{\ell=0}\frac{1}{2}\{(\ell+2)\,e^{2i\delta_+}+(2\ell+1)e^{2i\delta_0}\right.$$

$$\left.+(\ell-1)\,e^{2\delta_-}-(2\ell+1)]\,P_\ell^0(\cos\theta)\right],$$

where $\delta_+ = \delta_{\ell\,\ell+1}$, $\delta_0 = \delta_{\ell\ell}$ and $\delta_- = \delta_{\ell\,\ell-1}$.

$$f_{00} = \frac{1}{2ik}\left[\sum_{\ell}\{(2\ell+1)\,e^{2i\delta_+}\,\langle\ell\quad1\quad0\quad0\,|\,\ell+1\quad0\rangle^2\right.$$

$$+(2\ell+1)\,e^{2i\delta_0}\langle\ell\quad1\quad0\quad0\,|\,\ell\quad0\rangle^2$$

$$\left.+(2\ell+1)\,\langle\ell\,1\quad0\quad0\,|\,\ell-1\,0\rangle^2\,e^{2i\delta_-}-(2\ell+1)\right]P_\ell^0(\cos\theta)$$

$$= \frac{1}{2ik}\left[\sum_{\ell=0}\{(\ell+1)\,e^{2i\delta_+}+\ell\,e^{2i\delta_-}-(2\ell+1)\}\right]P_\ell^0(\cos\theta).$$

$$f_{01} = \frac{1}{2ik}\left[\sum_{\ell=1}^{\infty}(-)\frac{(2\ell+1)}{\sqrt{\ell(\ell+1)}}\,e^{i\phi}\{\langle\ell\,1\,0\,1\,|\,\ell+1\,1\rangle\right.$$

$$\times\langle\ell\,1\,1\,0\,|\,\ell+1\,1\rangle\,(2\ell+1)\,e^{2i\delta_+}+\langle\ell\,1\,0\,1\,|\,\ell\,1\rangle$$

$$\times\langle\ell\,1\,1\,0\,|\,\ell\,1\rangle\,e^{2i\delta_0}+\langle\ell\,1\,0\,1\,|\,\ell-1\,1\rangle$$

$$\left.\langle\ell\,1\,0\,1\,|\,\ell-1\,1\rangle^2\,e^{2i\delta_-}\}\times P_\ell^1(\cos\theta)\right].$$

$$= \frac{1}{2ik}\left[\sum_{\ell=1}\frac{1}{\sqrt{2}}\left\{-\frac{(\ell+2)}{(\ell+1)}\,e^{2i\delta_+}+\frac{(2\ell+1)}{\ell(\ell+1)}e^{2i\delta_0}+\frac{\ell-1}{\ell}\,e^{2i\delta_-}\right\}P_\ell^1(\cos\theta)e^{i\phi}\right].$$

$$f_{10} = \frac{1}{2ik}\left[\sum_{\ell=1}\frac{1}{\sqrt{2}}\{e^{2i\delta_+}-e^{2i\delta_-}\}\right]P_\ell^1(\cos\theta)e^{i\phi}.$$

The spin singlet amplitude $f_s = \dfrac{1}{k}\left[\displaystyle\sum_{\ell=0}^{\infty}(2\ell+1)\,e^{2\delta_\ell}\sin\delta_\ell\,P_\ell(\cos\theta)\right]$.

Here $P_\ell^m(\cos\theta) = \sin^m\phi\left[d^m/d(\cos\theta)^m\right]P_\ell(\cos\theta)$. The n-p differential scattering cross section for unpolarized projectile and target ($g = 1/4$) finally is expressed as

$$\left(\frac{d\sigma}{d\Omega}\right) = \frac{1}{4}\left[\sum_{\nu\mu}|f_{\nu\mu}|^2 + |f_s|^2\right],$$

and the total cross sections are obtained by integrating over the angles.

Exercise 2.8. Assuming charge symmetry of N-N force, calculate the charge radius of ^3He.

The binding energies of ^3H and ^3He can be interpreted in the light of the charge symmetry of internucleon forces. If we take the principle to be valid, we should be able to account for the difference in binding energy

$$\Delta E_B = BE(^3H) - BE(^3He) = 8.48 - 7.718 = 0.764 \text{ MeV}$$

exclusively through Coulomb repulsion between the two protons in ^3He and thereby arrive at an estimate of charge radius of this nucleus:

$$\Delta E_B = \Delta E_C = \frac{6}{5}\frac{e^2}{R_C}(Z-1) = \frac{6}{5}\frac{e^2}{R_C},$$

which gives the value of $R_C(^3He) = 2.31$ fm.

Exercise 2.9. Estimate the percentage of D-state from the magnetic moments of ^3H and ^3He systems.

$$\vec{\mu}(^3H) + \vec{\mu}(^3He) = (\mu_p + \mu_n)\sum_{k=1}^{3}\vec{\sigma}_k + \sum_{k=1}^{3}\vec{L}_k = 2(\mu_p + \mu_n)\vec{S} + \vec{L},$$

where $\vec{J} = \vec{L} + \vec{S}$. Building the expectation value of the z-component when $J_z = J = \dfrac{1}{2}$:

$$\left[\mu(^3H) + \mu(^3He)\right]_{expt} = \frac{1}{J+1}\langle LSJ\,|\,(\vec{\mu}(^3H) + \vec{\mu}(^3He))\cdot\vec{J}\,|LSJ\rangle$$

$$= \frac{1}{2}\left(\mu_p + \mu_n + \frac{1}{2}\right) + \frac{2}{3}\left(\mu_p + \mu_n - \frac{1}{2}\right)[S(S+1) - L(L+1)]. \tag{1}$$

By weighting the S-state $(S = \dfrac{1}{2}, L = 0)$ with a weight $(1 - p_D)$ and the

D-state $(S = \dfrac{3}{2}, L = 2)$ with the weight p_D, we get

$$\left[\mu(^3H) + \mu(^3He)\right]_{expt} = \frac{1}{J+1}\langle LSJ \mid (\bar{\mu}(^3H) + \bar{\mu}(^3He)) \cdot \vec{J} \mid LSJ \rangle$$

$$= \frac{1}{2}\left(\mu_p + \mu_n + \frac{1}{2}\right) + \frac{2}{3}\left(\mu_p + \mu_n - \frac{1}{2}\right)\left[\frac{3}{4}\right](1 - p_D) + \left[-\frac{9}{4}\right]p_D \quad (2)$$

Substituting the experimental value of $\mu(^3H)$ and $\mu(^3He)$ from Table 2.3 and that of μ_p and μ_n in (2) we get $p_D = 0.0375$. The calculation therefore indicates that the ground states of ^3H and ^3He comprise of a mixture of 96% S-state with 4% D-state.

2.4 ELECTROMAGNETIC PROBES FOR NUCLEON-NUCLEON SYSTEMS

The process of photodisintegration of the deuteron arises when the energy of the incident gamma ray (γ-photon) is greater than the binding energy (= 2.224 MeV) of the deuteron. The deuteron breaks up into a proton and a neutron and the excess gamma ray energy is shared equally by the outgoing proton and the residual neutron. The reaction may be represented in the following way:

$$d + \hbar\omega \longrightarrow n + p$$

The inverse reaction is called n-p capture reaction in which neutron captures a proton to form a bound deuteron and a gamma ray is emitted. As the neutron is uncharged, the incident gamma photon interacts with the proton only. The incoming photon has an energy $\hbar\omega$ and momentum $\hbar\omega/c$. The detected particle (proton) has momentum k relative to the neutron. Energy conservation requires:

$$\hbar\omega = (KE)_{proton} - (BE)_{deuteron} = \frac{\hbar^2}{M}(k^2 + \alpha^2),$$

where α is related to the deuteron binding energy by Eq. (2.14). With the threshold energy of $d\,(\gamma, p)\,n$ reaction at 2.224 MeV, the first peak of the total cross-section occurs at nearly 4.4 MeV.

A. Dipole Cross-Section

The cross section for the electric dipole absorption will now be calculated by employing the standard perturbation theory, used for the interaction of electromagnetic radiation with nuclei[24-26]. The selection rule for electric dipole transition demands $\Delta I = \pm 1$ and $\Delta \pi = +$. The ground state of deuteron is predominantly $^3S_1$ state (we neglect the $^3D_1$ state contribution), therefore, the final state must be ^3P. It can't be ^1P, since the electric dipole operator does not operate on the spin part of the wave function, which must therefore, remain unchanged. Hence, we consider $^3S_1 \rightarrow {}^3P_1$ transition without spin flip. In accordance to theory outlined before (1.78), the E1-γ-absorption cross section is given by

$$d\sigma = 4\pi^2 \left(\frac{e^2}{\hbar c} \right) \hbar \omega_\gamma \left| \int \psi_f^* \frac{1}{2} r \cos \theta \, \psi_i \, d\tau \right|^2 \rho(E_f),$$

where ψ_f stands for final state wave function and ψ_i stands for deuteron bound state ($^3S_1$) wave function. θ is the angle between the vector joining neutron and proton with the polarization direction, which is taken as Z-direction, i.e. the angle between electric dipole moment and the electric field vector of the incident gamma radiation. $\rho(E_f)$ is the energy density in the final state and is given by (1.32):

$$\rho(E_f) = \frac{dn}{dE_f} = \frac{dn}{dk} \cdot \frac{dk}{dE_f}.$$

The number of states dn in the energy interval is calculated as follows:

$$dn = \frac{\text{Volume of phase space}}{\text{Volume of phase cell}} = \frac{L^3 p^2 dp \, d\Omega}{(2\pi\hbar)^3},$$

where $\dfrac{dn}{dk} = \dfrac{L^3 d\Omega}{8\pi^3} k^2$, and hence, $\rho(E_f) = \dfrac{L^3}{8\pi^3} k^2 d\Omega \cdot \dfrac{dk}{dE_f}$.

The final energy is of course the proton energy $\left(\dfrac{\hbar^2 k^2}{M} \right)$ in the relative system. Then $\dfrac{dk}{dE_f} = \dfrac{M}{2k\hbar^2}$. Finally we get $\rho(E_f) = \dfrac{L^3 kM}{16\pi^3 \hbar^2} d\Omega$. The function ψ_f ought to describe the free proton in the relative system and because proton and neutron in the P-state weakly interact, we can set ψ_f equal to a plane wave $\psi_f = L^{-3/2} e^{ikz}$, where $z = r \cos \theta$ and L^3 denotes the normalization volume.

Employing the plane wave expansion of e^{ikz} (*see* Appendix B2), the P-state ($\ell = 1$) wave function can be written as

$$\psi_f = L^{-3/2} \frac{3i}{kr} \left(\frac{\sin kr}{kr} - \cos kr \right) \cos \theta',$$

where θ′ is the angle between the direction of ejection of the proton and the radius vector *r*. If χ is the angle between the direction of ejection of the proton and that of polarization of the proton, which we take as Z-axis, then

$$\cos\theta' = \cos\chi\cos\theta + \sin\chi\sin\theta\cos\phi$$

The integration over cosφ will make the later term vanish and we are left with

$$\frac{d\sigma}{d\Omega} = \left(\frac{e^2}{\hbar c}\right)\hbar\omega_\gamma \frac{kM}{4\pi\hbar^2}\left|\iiint \frac{3i}{kr}\left(\frac{\sin kr}{kr} - \cos kr\right)\cos\chi\cos\theta\right.$$

$$\left.\times\frac{1}{2}r\cos\theta\left\{\frac{\alpha}{2\pi}(1+\alpha b)\right\}^{1/2}\frac{e^{-\alpha r}}{r}\cdot r^2\sin\theta\, dr\, d\theta\, d\varphi\right|^2,$$

where we have used the radial deuteron wave function (2.17-2.18) and $\frac{1}{\sqrt{4\pi}}$ for the angular part. Simplifying we get

$$\frac{d\sigma}{d\Omega} = \frac{9}{32\pi^2}\left(\frac{e^2}{\hbar c}\right)\frac{(k^2+\alpha^2)}{k^3}\alpha(1+\alpha b)\cos^2\chi$$

$$\times\left|\iiint (\sin kr - kr\cos kr)\, e^{-\alpha r}dr\cos^2\theta\sin\theta\, d\theta\, d\varphi\right|^2$$

$$= \frac{9}{32\pi^2}\left(\frac{e^2}{\hbar c}\right)\frac{(k^2+\alpha^2)}{k^3}\alpha(1+\alpha b)\cos^2\chi$$

$$\times\left|\left\{\int_0^\infty \sin kr\, e^{-\alpha r}dr - \int_0^\infty kr\cos kr\, e^{-\alpha r}dr\right\}\right.$$

$$\left.\times\int_0^\pi \cos^2\theta\sin\theta\, d\theta\int_0^{2\pi}d\varphi\right|^2 = \left(\frac{e^2}{\hbar c}\right)\frac{2k^3}{(k^2+\alpha^2)^3}\alpha(1+\alpha b)\cos^2\chi. \qquad (2.68a)$$

The angular distribution of the ejected protons for an unpolarized incident beam is obtained by averaging $\cos^2\chi$ over all directions of polarization perpendicular to the direction of propagation of the incident beam. So for the unpolarized photons, we integrate over photon direction β and integration over the solid angle Ω yields 2π. Therefore,

$$d\sigma = \left(\frac{e^2}{\hbar c}\right) \frac{k^3 \alpha(1+\alpha b)}{(k^2+\alpha^2)^3} \sin^2\beta \, \sin\beta \, d\beta \, (2\pi),$$

where β is the angle between the direction of photon propagation and the direction of ejected proton.

Hence, $\sigma = 2\pi\left(\frac{e^2}{\hbar c}\right)\frac{k^3 \alpha(1+\alpha b)}{(k^2+\alpha^2)^3} \int\limits_0^\pi \sin^3\beta \, d\beta = \frac{8\pi}{3}\left(\frac{e^2}{\hbar c}\right)\frac{k^3 \alpha(1+\alpha b)}{(k^2+\alpha^2)^3}.$

Denoting $\dfrac{\hbar\omega_\gamma}{(BE)_{deut.}} = \eta$ and using (energy conservation requirement) we

get, $\eta = \left(1 + \dfrac{k^2}{\alpha^2}\right)$ and

$$\sigma = \frac{8\pi}{3}\left(\frac{e^2}{\hbar c}\right)\frac{1}{\alpha^2}\frac{(\eta-1)^{3/2}}{\eta^3}(1+\alpha b). \qquad (2.68b)$$

On substitution $\dfrac{e^2}{\hbar c} = \dfrac{1}{137}$ and α^2 from (2.14) we can estimate σ_{max} as follows. If the gamma photon energy just exceeds the threshold, so that $(\eta - 1) \ll 1$ the cross section increases roughly as $(\hbar\omega_\gamma - BE)_{deut.})^{3/2}$. The function $\dfrac{(\eta-1)^{3/2}}{\eta^3}$ has its maximum at $\eta = 2$, that is, $\hbar\omega_\gamma = 2(BE)_{deut.}$ and takes on value $\dfrac{1}{8}$. Therefore, $\sigma_{max} = \dfrac{\pi}{3}\,(1+\alpha b)(1.32)$ mb. Taking $b = 2$ fm and $\alpha b = 0.5$, we get $\sigma_{max} \cong 2.0$ mb.

For the large photon energies $(\eta \gg 1)$, the cross section is proportional to $(\hbar\omega_\gamma)^{-3/2}$. The measured cross section behaves almost the same way as described by this simple theory.

The theoretical cross section for forward produced protons in the deuteron photodisintegration, i.e. $\left(\dfrac{d\sigma}{d\Omega}\right)_{\theta=0}$ shows serious disagreement with experimental value. The experimental cross section in the photon energy range

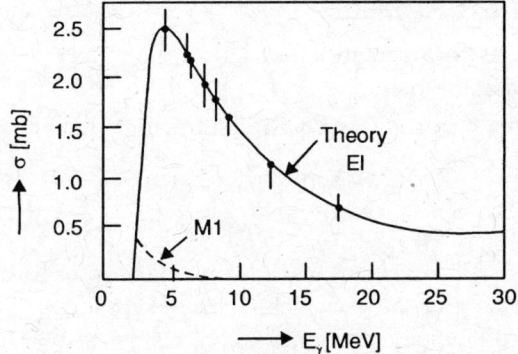

Fig. 2.9: Photodisintegration cross section for deuteron at low energies

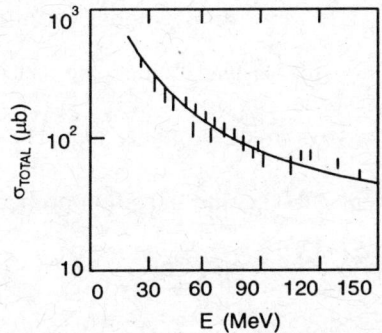

Fig. 2.10: The total cross section for the photodisintegration of deuteron at high energies. Solid line represents calculated values of Vyas et al. (Ref. 25)

(20-130) MeV lies between (1.5-2) $\mu b/Sr$, whereas the theoretical cross section lies between (5-7) μb. Lomon[23] found that the theoretical estimates are moderately sensitive to the D-state of the deuteron. Decreasing percentage of D-state from 7.53% to 4.58% decreases the forward cross sections at low energy by nearly 17%. Conservation of linear momentum during the process of γ absorption demands that the nucleus as a whole must recoil with momentum equal and opposite to that of photon. This recoiling nucleus gives additional currents that have been ignored so far.

Vyas et al.[24-25] have calculated the total cross section for photodisintegration of deuteron for photon energy of (20-150) MeV. They included all the transitions induced by the electromagnetic multipoles upto and including the fourth order. All effects of recoil and nucleonic magnetic moments on electric multipole transitions were taken into account. In Fig. 2.10 the total cross section for deuteron photodisintegration with unpolarized gamma for Paris potential is shown along with the experimental values.

n-p Capture Cross Section

In order to calculate *n-p* capture cross section initiated by the thermal neutrons we again use the perturbation method described earlier. The random motion of a thermal neutron is in equilibrium with its surroundings, at which, the kinetic energy $E = \dfrac{3}{2} kT$, where $k = 8.62 \times 10^{-5} \, \text{eV}^0 \, \text{K}^{-1}$ is the Boltzman constant. At room temperature $T = 300°K$; $E = 0.025$ eV. By virtue of extremely low relative kinetic energy, the wave function for the initial unbound n-p system must be all s-wave, which is space symmetric. The spin part of n-p system should then be antisymmetric (spin singlet; $S = 0$) in accordance to the Pauli principle. We do not consider isospin. The final state is the ground state of deuteron which is predominantly in $^3S_1$ (triplet $\ell = 0$) state. Hence, we consider spin-flip $^1S_0 \rightarrow {}^3S_1$ transition without change in parity, which is magnetic dipole transition (*M1*). The incident neutron flux $v = \dfrac{p}{\mu} = \dfrac{2\hbar K}{M}$,

where K is the relative wave number for n-p system. The capture cross section

$$\sigma_C = \frac{M}{2\hbar K} T(M1),$$ where $T(M1)$ is the transition probability (*see* Eq. 1.76b)

for the magnetic dipole transition given by

$$T(M1) = \frac{8\pi}{\hbar} \frac{2}{9} k^3 \left| \left\langle {}^3S_0 \mid M_{10} \mid {}^1 S_0 \right\rangle \right|^2 .$$

The magnetic dipole operator M_{10} is written by Calling nucleon 1 as proton and nucleon 2 as neutron as

$$M_{10} = \sqrt{\frac{3}{4\pi}} \; \mu_0 \left[\frac{1}{2}(\mu_1 + \mu_2) \, (\vec{\sigma}_1 + \vec{\sigma}_2)_z + \frac{1}{2} (\mu_1 - \mu_2)(\vec{\sigma}_1 - \vec{\sigma}_2)_z \right],$$

where $\mu_0 = \dfrac{e\hbar}{2Mc}$

Now $\langle x_1^0 \mid (\vec{\sigma}_1 + \vec{\sigma}_2) \mid x_0^0 \rangle = 0$ for singlet state; the total spin being zero. The initial state of n-p wave function is approximated by zero energy wave function (2.27b) as

$$\varphi_i(r) = \frac{v(r)}{r} \approx \left[1 - \frac{a_s}{r} \right],$$

where a_s = singlet scattering length (= −23.75 fm). In the final state we have deuteron in $^3S_1$ state (Eq. 2.19), i.e.

$$\varphi_f(r) = \frac{1}{\sqrt{4\pi}} \frac{u(r)}{r} = \frac{1}{\sqrt{4\pi}} \sqrt{2\alpha} \frac{e^{-\alpha r}}{r}, \qquad \alpha = 0.232 \text{ fm}^{-1}.$$

Therefore,

$$\langle {}^3S_1 \mid M_{10} \mid {}^1 S_0 \rangle = \sqrt{\frac{3}{4\pi}} \mu_0 (\mu_1 - \mu_2) \int_0^\infty \sqrt{\frac{\alpha}{2\pi}} \frac{e^{-\alpha r}}{r} \left(1 - \frac{a_s}{r}\right) 4\pi r^2 dr$$

$$= \sqrt{6\alpha} \ \mu_0 (\mu_1 - \mu_2) \int_0^\infty e^{-\alpha r} (r - a_s) \, dr$$

$$= \frac{6}{\alpha} \mu_0 (\mu_1 - \mu_2)(\alpha^{-1} - a_s).$$

Hence,
$$T(M1) = \frac{16\pi}{9\hbar} k^3 \frac{6}{\alpha} \mu_0^2 (\mu_1 - \mu_2)^2 (\alpha^{-1} - a_s)^2.$$

The total transition probability from singlet to the triplet is three times this, since there are three magnetic dipole modes with different projections ($\pm 1, 0$). The contributions of all three must be the same because of rotational symmetry.

In addition, the probability that the initial n-p system is in the singlet state is $\frac{1}{4}$. Hence,

$$T(M1) = \frac{3}{4} \frac{16\pi}{9\hbar} k^3 \frac{6}{\alpha} \mu_0^2 (\mu_1 - \mu_2)^2 (\alpha^{-1} - a_s)^2$$

$$= \frac{8\pi}{\hbar} k^3 \mu_0^2 (\mu_1 - \mu_2)^2 \frac{1}{\alpha} (\alpha^{-1} - a_s)^2.$$

At the threshold, $k = \dfrac{E_\gamma}{\hbar c} = \dfrac{\epsilon}{\hbar c}$, so that

$$\sigma_C = \frac{M}{2\hbar K} \frac{8\pi}{\hbar} \left(\frac{\epsilon}{\hbar c}\right)^3 \left(\frac{e^2 \hbar^2}{4M^2 c^2}\right)(\mu_1 - \mu_2)^2 \frac{\hbar}{\sqrt{M \epsilon}} (\alpha^{-1} - a_s)^2$$

$$= \frac{2\pi}{(v/c)} \left(\frac{e^2}{\hbar c}\right) \left(\frac{\epsilon}{M}\right)^{5/2} (\mu_1 - \mu_2)^2 (\alpha^{-1} - a_s)^2, \qquad (2.69)$$

where v is the relative neutron velocity and M is the nucleon mass expressed in MeV. It is clear from Eq. (2.69) that for thermal neutrons $\sigma_c \propto \dfrac{1}{v}$. Now,

$$(\mu_1 - \mu_2)^2 = (4.7)^2; \quad (\alpha^{-1} - a_s)^2 = \frac{(6.5)^2}{0.54} \text{fm}, \left(\frac{\epsilon}{M}\right)^{5/2} = \left(\frac{2.238}{938}\right)^{5/2}. \text{ For an}$$

incoming thermal neutron of energy $E_n \cong 2 \, \text{eV}$, the relative $\dfrac{v}{c} = \dfrac{1}{2}\sqrt{\dfrac{E_{cm}}{Mc^2}} =$

30×10^{-6}, since, $\dfrac{v}{c} = \dfrac{1}{2}\sqrt{\dfrac{E_{cm}}{Mc^2}}$. Therefore, $(\sigma_C)_{IA} \cong 300 \, \text{mb}$.

Considering the various approximations made in the calculation, it is rather remarkable to see that the theoretical estimate is low only by 10% from the experimental value. Calculations done with more accurate wave functions obtained by employing realistic potential could not explain this difference and the discrepancy existed for many years until Riska and Brown[27] explained it by considering meson exchange current effects.

B. Role of Meson Exchange Currents

A nucleus is a composite system of neutrons and protons and the electromagnetic interaction depends upon the structure of this composite system. For the low energy nuclear physics, nucleon can be treated as structureless particle.

So far we have treated the interaction of electromagnetic radiation with nuclei in impulse approximation (*IA*), which means photons directly interact with free nucleons; the presence of strongly interacting mesons was ignored. The success of nucleon description of the nucleus can be seen by a comparison of the total cross section per nucleon for photon absorption. Such a comparison of $\dfrac{\sigma(r)}{A}$ versus E_γ reveals that it has a universal shape throughout the periodic system. The cross section per nucleon is simply the free one, broadened by nucleon motion, indicating that the impulse approximation (*IA*) works reasonably well (Fig. 2.11).

In the above example, sufficient care has been given to the two-nucleon potential input to justify the point of view that the origin of discrepancies may lie in the incomplete nature of the impulse approximation[28,33].

Nucleon itself is a composite system, bare nucleon dressed by mesons that are continuously emitted and reabsorbed by its constituent nucleons. The nucleus may also contain their various excited states in addition to nucleon and antinucleon pair states. Therefore, the electromagnetic interaction of the nucleus can be described in terms of two classes of currents interacting with electromagnetic fields: the single nucleon current and the meson exchange current (MEC) due to the hidden components in the nucleus.

Sometimes the predictions of *IA* do not agree with experimental data. Such discrepancies are generally regarded as strong evidence for the presence of

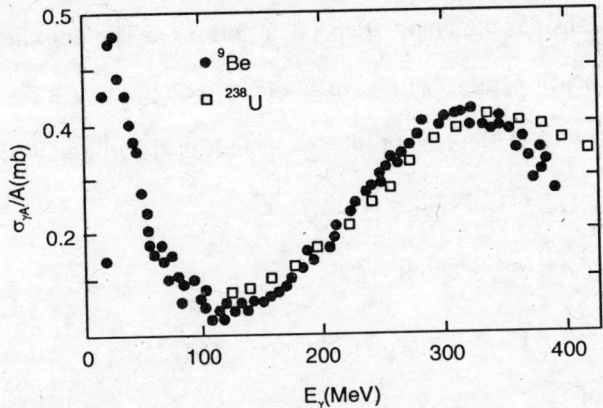

Fig. 2.11: Universal shape of photon nucleus cross section per nucleon ($\sigma_{\gamma A}/A$) versus photon energy (E_g). (Adopted from J. Ahrens Ref. 28)

meson exchange current (MEC) in the nucleus. There are some important discrepancies:

Radiative capture of thermalized neutrons by proton is one of the simplest electromagnetic nuclear reactions

$$n + p \rightarrow d + \hbar\omega_\gamma \,.$$

The measured n-p capture cross section of thermal neutrons

$$\sigma_{n-p}^{expt} = (334.2 \pm 0.5) \text{ mb} \,,$$

while the most accurate *IA* result is

$$\sigma_{n-p}^{IA} = (302.5 \pm 0.4) \text{ mb} \,.$$

It was in the early seventies, Riska and Brown[27] calculated the effects due to meson exchange current on n-p capture cross section. The additional interaction term appearing in the form of an exchange moment is calculated non-relativistically as follows. In the presence of electromagnetic field gauge invariance requires $\vec{\nabla} \rightarrow \vec{\nabla} \pm i\, e\, \vec{A}$, where the $-$ve and $+$ve sign refer to π^+ and π^- emission respectively and we use $\hbar = c = 1$. This means that in the presence of electromagnetic field we shall have an additional term in the non-relativistic Lagrangian density (2.8a):

$$\delta\mathcal{L} = \frac{f}{\mu}\, \psi^\dagger(x)\, \frac{1}{\sqrt{2}}\, (\mp i\, e\, \vec{\sigma} \cdot \vec{A})\, \tau_\mp\, \phi_\pm(x)\, \psi(x),$$

where $\quad \varphi_\pm(x) = \dfrac{1}{\sqrt{2}} \left[\phi_1(x) \pm i\phi_2(x) \right] \quad$ and $\quad \sigma_\mp = \left[\sigma_1 \mp i\sigma_2 \right].$

We first consider the charge dependent term: (1) The emission of π^+ at 1 and its absorption at 2 will give a factor equal to $-\frac{1}{2} i e (\tau_{1x} + i\tau_{1y})$ $(\tau_{2x} - i\tau_{2y})$. (2) The emission of π^- at 1 and its absorption at 2 will give a factor $\frac{1}{2} i e (\tau_{1x} - i\tau_{1y}) (\tau_{2x} + i\tau_{2y})$.

The sum of factors 1 and 2

$$= -\frac{1}{2} i e (\tau_{1x}\tau_{2x} - i\tau_{1x}\tau_{2y} + i\tau_{1y}\tau_{2x} + \tau_{1y}\tau_{2y})$$

$$+ \frac{1}{2} i e (\tau_{1x}\tau_{2x} + i\tau_{1x}\tau_{2y} - i\tau_{1y}\tau_{2x} + \tau_{1y}\tau_{2y})$$

$$= -e (\tau_{1x}\tau_{2y} - \tau_{1y}\tau_{2x}) = -e(\vec{\tau}_1 \times \vec{\tau}_2)_z.$$

We see that the exchange current correction is isovector, this fact being connected with the isovector nature of the pi-meson. Considering the pion emission and absorption process leading to Yukawa interaction, we can get the additional correction term

$$H_{12} = \frac{f^2}{4\pi\mu^2} \vec{\sigma}_1 \cdot \vec{A}(r_1) \vec{\sigma}_2 \cdot \vec{\nabla}_2 \frac{e^{-\mu r}}{r}.$$

We must add the term H_{21} in which particle 2 interacts with γ-ray. We have

$$H_{21} = \frac{f^2}{4\pi\mu^2} \vec{\sigma}_2 \cdot \vec{A}(r_2) \vec{\sigma}_1 \cdot \vec{\nabla}_1 \frac{e^{-\mu r}}{r}.$$

For static magnetic field H, $A(r) = \frac{1}{2}(\vec{r} \times \vec{H})$ so that

$$H_{12} + H_{21} = -(\vec{\tau}_1 \cdot \vec{\tau}_2)_z\, e\, \frac{f^2}{4\pi\mu^2} \frac{1}{2} [\vec{\sigma}_1 \cdot (\vec{r}_1 \times \vec{H})\, \vec{\sigma}_2 \cdot \vec{\nabla}_2 \frac{e^{-\mu r}}{r}$$

$$+ \vec{\sigma}_2 \cdot (\vec{r}_2 \times \vec{H})\, \vec{\sigma}_1 \cdot \vec{\nabla}_1 \frac{e^{-\mu r}}{r}.$$

$$= -(\vec{\tau}_1 \cdot \vec{\tau}_2)_z\, \frac{ef^2}{4\pi\mu^2} \frac{1}{2} [\vec{\sigma}_1 \cdot (\vec{r}_1 \times \vec{H})\, \vec{\sigma}_2 \cdot \vec{r} + \vec{\sigma}_2 \cdot (\vec{r}_2 \times \vec{H})\vec{\sigma}_1 \cdot \vec{r}]$$

$$\times \frac{1}{r} \frac{\partial}{\partial r}\left(\frac{e^{-\mu r}}{r}\right).$$

Then the exchange moment $(\vec{\mu}_{12})_{ex} = -\dfrac{e}{2M}(\vec{\mu}_{12})_{ex} \cdot \vec{H} = H_{12} + H_{21}$,

where M is the nucleon mass and $(\hbar = c = 1)$. By rearranging terms we get

$$(\vec{\mu}_{12})_{ex} = (\vec{\tau}_1 \cdot \vec{\tau}_2)_z \frac{f^2}{4\pi} \frac{M}{\mu^2} \left[(\vec{\sigma}_1 \times \vec{r}_1) \ \vec{\sigma}_2 \cdot \vec{r} + (\vec{\sigma}_2 \times \vec{r}_2) \ \vec{\sigma}_1 \cdot \vec{r} \right] (\mu r + 1) \frac{e^{-\mu r}}{r^3}.$$

(2.70a)

If we define $\vec{R} = \dfrac{\vec{r}_1 + \vec{r}_2}{2}$, then we can break up the term in square bracket

in (2.70a) into the sum of a R dependent term:

$$(\vec{\sigma}_1 \times \vec{R})(\vec{\sigma}_2 \times \vec{r}) + (\vec{\sigma}_1 \cdot \vec{r})(\vec{\sigma}_2 \times \vec{R})$$

and a term which depends only on r:

$$\frac{1}{2}\left\{ (\vec{\sigma}_1 \times \vec{r})(\vec{\sigma}_2 \cdot \vec{r}) - (\vec{\sigma}_1 \cdot \vec{r})(\vec{\sigma}_2 \times \vec{r}) \right\}; \text{ which can be written as}$$

$$\frac{1}{3}(\vec{\sigma}_1 \times \vec{\sigma}_2)r^2 - \frac{1}{2}\left\{ (\vec{\sigma}_1 \times \vec{\sigma}_2)\vec{r} \cdot \vec{r} - \frac{1}{3}(\vec{\sigma}_1 \times \vec{\sigma}_2)r^2 \right\},$$

where we have decomposed it into irreducible spherical tensors.

The R-dependent term appears to violate translation invariance. Such a term occurs in the magnetic moment because $\vec{A}$ depends on absolute, not only on relative position. This R-dependent term give an exchange current correction to the orbital gyromagnetic ratio, it will not enter into capture process. Collecting terms which depend only on r we get exchange moment

$$(\vec{\mu}_{12})_{ex} = \frac{1}{2}(\vec{\tau}_1 \cdot \vec{\tau}_2)_z \frac{f^2}{4\pi} \frac{M}{\mu^2} \left\{ \frac{2}{3}(\vec{\sigma}_1 \times \vec{\sigma}_2)r^2 - (\vec{\sigma}_1 \times \vec{\sigma}_2)\vec{r} \cdot \vec{r} \right.$$

$$\left. - \frac{1}{3}(\vec{\sigma}_1 \times \vec{\sigma}_2)r^2 \right\} \frac{e^{-\mu r}}{r^3}.$$

(2.70b)

The term $\dfrac{2}{3}(\vec{\sigma}_1 \times \vec{\sigma}_2)\,r^2$ in (2.70b) gives transitions from incoming $^1S_0$ state of n-p system to the $^3S_1$ state of the deuteron and the other term, to the $^3D_1$ state of the deuteron. Employing the deuteron $^1S_0$ state wave function from a standard N-N interaction, we can calculate the matrix element of $(\vec{\mu}_{12})_{ex}$.

The exchange current contribution to $n + p \rightarrow d + \gamma$ is obtained by calculating the enhancement δ of the single-particle amplitude:

$$\delta = \frac{\langle {}^3S_1 + {}^3D_1 \,|(\vec{\mu}_{12})_{ex}\,|{}^1S_0\rangle}{\langle {}^3S_1 \,|\,\vec{\mu}_{s\cdot p}\,|{}^1S_0\rangle}.$$

where $\vec{\mu}_{s\cdot p} = \frac{1}{2}(\mu_1 - \mu_2)\,(\vec{\sigma}_1 - \vec{\sigma}_2)_z$ is the single particle operator; it does not give rise to S-D state transition but contributes for ${}^3S_0 \to {}^1S_0$ transition (Exercise 2.11). In Eq. 2.70: $\frac{1}{2}(\mu_1 - \mu_2) = 2.353$.

For the deuteron states $|{}^3S_1$ and $|{}^3D_1\rangle$, Riska and Brown[27] used the hard core wave function of Reid and for $|{}^1S_0\rangle$ wave function they used the solution of Schrodinger equation for the Reid hard core $|{}^1S_0\rangle$ state potential. This potential has a hard core $x_c\,(=\mu_\pi\, r) = 0.296$ and gives a singlet state scattering length in agreement with experiment. Evaluation of the integrals entering into the above equation then gives $\delta^{SS} = 1.90\%$ and $\delta^{SD} = 1.38\%$.

Exercise 2.10. Derive an expression for the magnetic dipole contribution to the bremstrahlung weighted cross section σ_b^m for the photo-disintegration of deuteron.

$$M_{11} = \frac{1}{2}\,\vec{L} + \frac{1}{2}\,(\vec{\sigma}_1 + \vec{\sigma}_2)\,(\mu_1 + \mu_2) + \frac{1}{2}\,(\mu_1 - \mu_2)(\vec{\sigma}_1 - \vec{\sigma}_2).$$

Energy weighted cross section for $M1$ transition to the singlet state is

$$\sigma_b^{ms} = \frac{\pi^2 e^2}{\hbar c}\left(\frac{\hbar}{Mc}\right)^2 \frac{(\mu_1 - \mu_2)^2}{4}\left\{\frac{4}{3}\left|\int_{x_0}^{\infty} u^2(x)\,dx\right| + \frac{2}{15}\left|\int_{x_0}^{\infty} w^2(x)\,dx\right|\right\}$$

$$= \frac{\pi^2 e^2}{3\hbar c}\left(\frac{\hbar}{Mc}\right)^2 (\mu_1 - \mu_2)^2\left\{\int_{x_0}^{\infty} u^2(x)\,dx + \frac{2}{10}\int_{x_0}^{\infty} w^2(x)\,dx\right\}$$

The bramstrahlung weighted cross section for M1 transition to the triplet states becomes

$$\sigma_b^{mt} = \left(\frac{\pi^2 e^2}{3\hbar c}\right)\left(\frac{\hbar}{Mc}\right)^2\left(\mu_1 - \mu_2 - \frac{1}{2}\right)^2 \times \left(\int_{x_0}^{\infty} u^2(x)\,dx + \int_{x_0}^{\infty} w^2(x)\,dx\right).$$

Exercise 2.11. Evaluate $\langle x_1^0 \,|\,(\vec{\sigma}_1 - \vec{\sigma}_2)_z\,|x_0^0\rangle$

$$= \left[\left\{ \frac{\alpha(1)\,\beta(2) + \beta(1)\,\alpha(1)}{\sqrt{2}} \right\}^{\dagger} (\sigma_{1z} - \sigma_{2z})_{z} \left\{ \frac{\alpha(1)\,\beta(2) - \beta(1)\,\alpha(1)}{\sqrt{2}} \right\} \right]$$

$$= \frac{1}{2} [\alpha^{\dagger}(1)\beta^{\dagger}(2)(\sigma_{1z} - \sigma_{2z})\alpha(1)\beta(2) - \beta^{\dagger}(1)\,\alpha^{\dagger}(2)\,(\sigma_{1z} - \sigma_{2z})\,\beta(1)\,\alpha(2)]$$

$$= \frac{1}{2} \left[2\alpha^{\dagger}(1)\,\beta^{\dagger}(2)\,\alpha(1)\,\beta(2) + 2\beta^{\dagger}(1)\,\alpha^{\dagger}(2)\beta(1)\alpha(2) \right] = 2.$$

2.5 THE QUARK PICTURE OF NUCLEAR FORCE

As mentioned earlier, quarks possess color charges and these charges appear to be source of forces that clusters quarks together to form the bound states that we collectively can call nucleons. Quark-quark interaction is mediated by exchange of massless spin 1 neutral particles called gluons, which are responsible for binding quarks to form nucleons. With the proposal of treating nucleons as bound state of three quarks[37,38], soon models of mesons and baryons[30] and nuclear forces[34] were proposed. One such model, the MIT bag model[29], based on picture of confinement, visualizes two nucleons as a system of two interacting clusters of three quarks inside a bag confined by assumed boundary conditions imposed on the bag surface. The potential energy of this 'six quark' configuration is interpreted as *N-N* potential.

A. Quantum Chromodynamics

Considering the scope of the book, we have no intention to describe the color gauge field theory here. There are standard text books on the subject, for examples, Fields, Symmetries and Quarks by Mosel[35] and Quantum Chromodynamics by Muta.[36]

Color plays an "observable" role in determining the spin structure of nucleon. The hadrons have no net color (white) but feel the strong forces because of their color constituents. This is in its way analogous to the way that covalent or Van der Walls forces arise between electrically neutral atoms. The strong nuclear force is but a remnant of the more powerful forces between quarks. This powerful color force is neutralized within protons and neutrons with the result that its remnant acts only over short distances - the nucleus. In the quark model, NN scattering is just a collision of two nucleon clusters of quarks.

Let us consider quark-antiquark interaction energy:

$$V_{q\bar{q}} = \frac{\lambda_{i}^{q}}{2} \cdot \frac{\lambda_{i}^{\bar{q}}}{2} \cdot \frac{g_{s}^{2}}{4\pi r},$$

where $\lambda^q = \{\lambda_1^q, \lambda_2^q \cdots \lambda_8^q\}$ is the set of eight generators of the color SU(3) group (Appendix D1), acting on the color wave function of the quark q and $\lambda^{\bar{q}}$ is the analogous set of eight generators for antiquark. The product $\lambda^q \cdot \lambda^{\bar{q}} = \frac{1}{2}\left[(\lambda^q + \lambda^{\bar{q}})^2 - (\lambda^q)^2 - (\lambda^{\bar{q}})^2\right]$. From the definition of λ-matrices:

$$(\lambda^q)^2 = \frac{16}{3}\begin{pmatrix} 1 & 0 & 0 \\ 0 & 1 & 0 \\ 0 & 0 & 1 \end{pmatrix} = (\lambda^{\bar{q}})^2 \text{ or } \lambda^q \cdot \lambda^{\bar{q}} = -\frac{16}{3}.$$

Of all color states that can be formed by q and $\bar{q}$, we are only interested in the color singlet states of $q\bar{q}$ (colorless) system (mesons for example) as all observed isolated particles are in color singlet states.

Hence, $V_{q\bar{q}} = -\frac{4}{3}\frac{g_s^2}{4\pi r} = -\frac{4}{3}\frac{\alpha_s}{r}$, where $\alpha_s = \frac{g_s^2}{4\pi}$. \hfill (2.71)

Asymptotic Freedom and Confinements: We know that in a polarizable medium the potential energy $V(r)$ of two static charges q_1 and q_2 is

$$V(r) = \frac{q_1 q_2}{4\pi \in r}, \hspace{2cm} (2.72a)$$

where r is the distance between two charges and $\in$ is the dielectric constant of the medium, which in vacuum takes the value 1. Ordinarily, the polarizability of the medium causes the screening of interaction between charges, meaning $\in > 1$. Antiscreening on the other hand corresponds to $\in < 1$.

A relativistic quantum field theory will have a vacuum that in many respects behaves like a polarizable medium. However, Lorentz invariance condition demands $\mu\in = 1$ (in unit $c = 1$). Therefore, ordinary screening means $\mu < 1$ and antiscreening means $\mu > 1$. The magnetic permeability μ is related to magnetic susceptibility χ by $\mu = 1 + 4\pi\chi$ and the energy density of the medium in the presence of external magnetic field H is

$$E = -\frac{1}{2}4\pi\chi H^2. \hspace{2cm} (2.72b)$$

If the medium contains magnetic dipole, they tend to align themselves with magnetic field leading to negative energy density, so that in that case $\chi > 0$ and $\mu > 1$ and the material is called paramagnetic. In accordance to Lorentz invariance this implies antiscreening. In QED we view two heavy static charges at a distance r as surrounded by virtual electron-positron pairs. In other words, a virtual photon emitted from an electron does not only see the charge of another electron but also charges of virtual electron positron pair. Since the electron under consideration attracts particles of opposite charges and repel those of same charge it appears to be surrounded by a cloud of charges of opposite sign, consequently, net charge, seen from certain distance is less than the base charge. Thus, the $e^+ e^-$ polarization bubble in the virtual photon in QED gives rise to charge renormalization. These clouds will be polarized by test charges producing a value of $\in$ different than 1. But if test charges are close together they can penetrate each other particles cloud and thus not feel any screening. This means that when adopting Eq. (2.72b) to QED we have to let $\in \rightarrow 1$ for $r \rightarrow 0$, so the dielectric constant is a function r for this application. Correcting (2.72a) for r dependence of $\in$, we get

$$V(r) = \frac{q_1 q_2}{4\pi \in (r) \, r}. \tag{2.73}$$

We see that the effective strength of the interaction measured by the effective charge of test charge 1

$$q_1^2(r) = \frac{q_1^2}{\in (r)}, \tag{2.74}$$

in the screened case, where $\in(r) > 1$, $\in(r) \rightarrow 1$ for $r \rightarrow 0$. This what happens in QED, the strength of interaction increases as $r \rightarrow 0$. Also $\in > 1$ means $\mu < 1$; and QED vacuum is called diamagnetic. In the antiscreening case $\in (r) < 1$ so that $\in (r) \rightarrow 1$ for $r \rightarrow 0$, i.e. the effective strength of the interaction decreases as $r \rightarrow 0$ ($q_1^2(r) \rightarrow 0$ as $\in (r) \rightarrow \infty$ if $r \rightarrow 0$) and if it decreases to zero in the limit we call the theory asymptotically free. This is what happens in QCD vacuum as discussed below:

In QCD, quarks have color charge and interact by exchange of gluons. Gluons like photons are spin 1 objects, but unlike photons they carry color charges. In QCD vacuum, a gluon can produce virtual $q\bar{q}$ pairs which would screen the interaction as in the case QED. However, since gluons have color charges as well as spin they can cause color magnetization of the medium and this effect actually overcomes the diamagnetic property of $q\bar{q}$ pairs and the overall result is that $\mu_c > 1$ for QCD vacuum (where subscript c stands for color) and hence $\in_c < 1$, so that color electric interaction between charged objects become stronger for larger distances and $\in_c \rightarrow 1$ for $r \rightarrow 0$ making the interaction weaker. The latter is called asymptotic freedom and this is what happens in QCD. Thus, the QCD perturbative vacuum is antiscreening, since $\mu > 1$.

Because of the QCD vacuum polarization antiscreening effect the coupling constant α_s varies with Q^2 or r so that α_s is not a constant at all but function of Q^2 or r.

$$\alpha_s(Q^2) = \frac{12\pi}{(33 - 2N_f)\log\left(\dfrac{Q^2}{\Lambda^2}\right)}, \qquad Q^2 >> \Lambda^2. \tag{2.75}$$

where α_s is a dimensionless momentum dependent quantity, it has to be a function of a dimensionless variable (Q^2/Λ^2), where Λ is called the fundamental QCD scale parameter and N_f is the number of flavors. As long as $N_f < 33/2$, $\alpha_s(Q^2)$ is positive, so that self coupling of gluons results antiscreening. From large Q^2 experiments and perturbative QCD calculations to fit the data, it is found that $\Lambda \cong 200$ MeV. These experiments include deep inelastic scattering process and e^+e^- annihilation cross section.

The principle of "asymptotic freedom" determines that the renormalized QCD coupling is small only at high energies, and it is only in this domain high precision tests-similar to those in QED - can be performed using perturbation theory. For low energy hadron physics, $Q \sim \dfrac{1}{R}$, where R is the size of the hadron. For $R \sim 1$ fm, Q is of the same order as Λ (200 MeV $\equiv 1$ fm^{-1}) and Eq. (2.75) should not be used. The Fourier transform of (2.75) is given by

$$\alpha_s(r) = \frac{6\pi}{(33 - 2N_f)\log\left(\dfrac{1}{\Lambda r}\right)}, \tag{2.76}$$

Hence, $\alpha_s \to 0$ as $r \to 0$, which gives rise to so called "asymptotic freedom", i.e. the fact that quarks and gluons inside a hadron behaves like free particles when they are very close together. On the other hand, $\alpha_s(r)$ apparently diverges as $r \to R \equiv 1/\Lambda$. This divergence simply heralds breakdown of perturbation theory, of course, but nonetheless it leads us to expect that the strength of the force between quarks will increase if they are pulled apart. As a result the

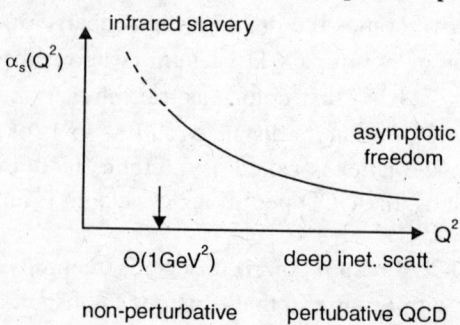

Fig. 2.12: Coupling constant in QCD

quarks and gluons are "Confined" inside hadrons whose size is of the order of R. This is so called infra-red slavery (*see* Fig. 2.12). The crucial parameter characterizing the strong interaction is the energy scale that appears in Eq. (2.76). It turns out that $\Lambda \approx 0.2$-0.3 GeV. Not surprisingly, it is of the same order of magnitude as the masses of lightest hadrons, for example, pion.

We also notice that instead of giving a numerical value for dimensionless parameter α_s, we end up specifying an energy scale Λ to quantify the strength of the interaction. This phenomena is known as "dimensional transmutation" and is characteristic of renormalizable non-Abelian gauge field theories. QCD does not predict the value of α_s; rather it determines how it varies with Q^2. Finally, we write

$$V(r) = -\frac{4}{3}\frac{\alpha_s(r)}{r}, \text{ where } \alpha_s(r) \text{ is given by } (2.76). \tag{2.77}$$

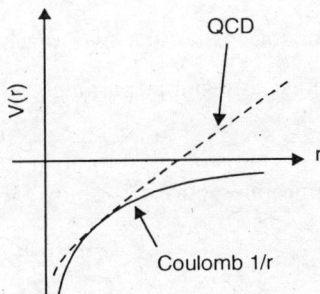

Fig. 2.13: QCD based Color Coulombic potential compared with the Coulomb potential

At short distance QCD interactions are in asymptotic freedom, one gluon exchange interaction become dominant with rather small coupling constant (α_s) and potential turn out Coulombic type. This is schematically shown in Fig. (2.13). To represent better the behaviour at large r, we write

$$V(r) = -\frac{4}{3}\frac{\alpha_s(r)}{r} + A(r) + V_{OGEP} + V_{conf}, \tag{2.78}$$

where A is called string tension for linear confinement.

Hyperfine Interaction: In QED, the spin-spin interaction gives rise to hyperfine splitting, which is expressed in terms of the point interaction of magnetic moments $V_{HF} = \frac{2}{3}\vec{\mu}_1 \cdot \vec{\mu}_2\, \delta(r_{12})$ where $\mu_i = \frac{e_i}{2m_i}\sigma_i$ with $\hbar = c = 1$ and e_i, m_i, σ_i denote charge, mass and spin of the ith particle respectively and r_{12} is the distance between particles ($i = 1,2$).

Hence,
$$\Delta E_{HF} = \frac{2}{3} \frac{e^2}{4} |\psi(0)|^2 \frac{\vec{\sigma}_1 \cdot \vec{\sigma}_2}{m_1 m_2} = \frac{2\pi}{3} \propto |\psi(0)|^2 \frac{\vec{\sigma}_1 \cdot \vec{\sigma}_2}{m_1 m_2}, \qquad (2.79)$$

where $e_1 = e_2 = e$ and $e^2 = 4\pi\alpha$.

To obtain analogous result for QCD we have to replace electric charges e_1 and e_2 by the appropriate color charges, which will lead to a spin-spin interaction called the chromomagnetic or color magnetic interaction. Employing the Gellmann matrices, we can write

$$V_{HF} = \frac{2\pi}{3} \alpha_s \left(\frac{1}{4} \lambda_1 \cdot \lambda_2 \right) \frac{\vec{\sigma}_1 \cdot \vec{\sigma}_2}{m_1 m_2} \delta(r_{12}),$$

Therefore,
$$(\Delta E)_{HF} = \frac{2\pi}{3} \left(\frac{1}{4} \lambda_1 \cdot \lambda_2 \right) \frac{\vec{\sigma}_1 \cdot \vec{\sigma}_2}{m_1 m_2} \alpha_s |\psi(0)|^2 .$$

Here $|\psi(0)|^2$ is the probability that the two quarks are at the same place. We treat $\alpha_s |\psi(0)|^2$ amongst adjustable parameters in a phenomenological model.

Therefore, hadron masses are supposed to arise from a sum of constituent quark masses and hyperfine interactions[30]. Thus, for baryons

$$m(q_1 q_2 q_3) = m_1 + m_2 + m_3 + a \sum_{i,j>i} \frac{\vec{\sigma}_i \cdot \vec{\sigma}_j}{m_i m_j}, \qquad (2.80a)$$

and for mesons
$$m(q\bar{q}) = m_q + m_{\bar{q}} + b \frac{\vec{\sigma}_q \cdot \vec{\sigma}_{\bar{q}}}{m_q m_{\bar{q}}}. \qquad (2.80b)$$

We regard the constants a and b and quark masses as free parameters and try to reproduce hadron mass spectrum with a consistent set of values. In terms of the spins of the constituent quarks the baryon spin J is given by

$$J^2 = (\vec{s}_1 + \vec{s}_2 + \vec{s}_3)^2 = s_1^2 + s_2^2 + s_3^2 + 2(\vec{s}_1 \cdot \vec{s}_2 + \vec{s}_1 \cdot \vec{s}_3 + \vec{s}_2 \cdot \vec{s}_3)$$

Hence,
$$\sum_{i,j>i} \vec{\sigma}_i \cdot \vec{\sigma}_j = 2\left[j(j+1) - \frac{9}{4} \right]$$

For $j = 1/2$ baryon octet, $\sum_{i,j>i} \vec{\sigma}_i \cdot \vec{\sigma}_j = -3$. For nucleon Eq. (2.80a) with $m_u = m_d$ yields

$$m_N = 3m_u - 3\frac{a}{m_u^2}.$$

From the above equation, we can fix the value of the parameter a, by taking $m_N = 939 \cdot \dfrac{\text{MeV}}{c^2}$ and $m_u = 363 \cdot \dfrac{\text{MeV}}{c^2}$ so as to give $\dfrac{a}{m_u^2} = 50 \cdot \dfrac{\text{MeV}}{c^2}$.

Bag Model: The bag model, suggested by M.I.T. group[29], based on a picture for confinement is likely to be more realistic than that for potential models[30], especially for light hadrons. In the static spherical bag model there are two basic assumptions[31,34];

(1) Hadrons can be visualized as composed of weakly interacting quarks and gluons confined in a bag by assumed boundary conditions imposed on the bag surface. For computational convenience, one assumes spherical bag of radius R.

(2) The outward pressures of quarks and gluons in the bag are counter balanced by an inward pressure exerted by the bag surface, generated by the assumption that the bag interior contains a constant positive energy density B.

The quarks obey the free Dirac equation inside the bag but they are subject to a linear boundary condition. We can get an estimate of the magnitude of the bag pressure B by considering massless free fermions in a spherical cavity of radius R. The Dirac equation for a massless fermion in the cavity is $\gamma \cdot p\psi = 0$,

where $\gamma^0 = \begin{pmatrix} I & 0 \\ 0 & -I \end{pmatrix}$ and $\gamma = \begin{pmatrix} 0 & \sigma \\ -\sigma & 0 \end{pmatrix}$ are in the Dirac representation. We

write the four component wave function of the massless fermion as

$$\psi = \begin{pmatrix} \psi_+ \\ \psi_- \end{pmatrix},$$

where ψ_+ and ψ_- are two dimensional spinor so that the Dirac Eq. becomes

$$\begin{pmatrix} p_0 & -\vec{\sigma} \cdot \vec{p} \\ \vec{\sigma} \cdot \vec{p} & -p_0 \end{pmatrix} \begin{pmatrix} \psi_+ \\ \psi_- \end{pmatrix} = 0. \tag{2.81}$$

Writing the above matrix equation into two equations and eliminating the ψ_- component, we obtain

$$\left[p^2 - (p^0)^2 \right] \psi_+ = 0.$$

The lowest energy solution for the above equation ($s_{1/2}$ state) is given by

$$\psi_+(r,t) = N\, e^{-ip^0 t} j_0(p^0 r)\chi_+,$$

where j_0 is the spherical Bessel function of order zero, χ_+ the two dimensional spinor and N the normalization constant, which is obtained from the relation

$$1 = \int d^3r\, \psi^\dagger \psi = N^2 \int_0^R r^2 dr \left[\{j_0(p^0 r)\}^2 + \{j_1(p^0 r)\}^2 \right]$$

$$= 2N^2 R^3 (p^0 R - 1) \sin^2(p^0 R) / (p^0 R)^3,$$

so that $N = \left[\dfrac{(p^0 R)^3}{2R^3(p^0 R - 1)\, \sin^2(p^0 R)} \right]^{\frac{1}{2}}$. From Eq. (2.81) the other component is

$$\psi_-(r,t) = N e^{-ip^0 t} \vec{\sigma} \cdot \hat{r}\, j_1(p^0 r)\, \chi_-.$$

The confinement of the quarks is equivalent to the requirement that the normal component of the vector current $j_\mu = \overline{\psi}\, \gamma_\mu\, \psi$ vanishes at the surface, i.e.

$$[\overline{\psi}\, \gamma_\mu\, \psi]_{r=R} = 0.$$

The condition is the same requirement that the scalar density $\overline{\psi}\, \psi$ of the quarks vanishes at the bag surface $r = R$. This leads to

$$\overline{\psi}\psi\big|_{r=R} = \left[j_0(p^0 R) \right]^2 - \vec{\sigma} \cdot \hat{r}\, \vec{\sigma} \cdot \hat{r} \left[j_1(p^0 R) \right]^2 = 0.$$

where $\hat{r}$ is the unit vector in the direction r and $\left[j_0(p^0 R) \right]^2 = \left[j_1(p^0 R) \right]^2$.

We find from the tabulated values of spherical Bessel function that this equation is satisfied for

$$p^0 R = 2.04 \text{ or } p^0 = \frac{2.04}{R}. \tag{2.82}$$

For a system of N quarks in a bag, the total kinetic energy ($= 2.04\, N/R$) of the confined quark is inversely proportional to R. It decreases with an increase in the bag radius.

The contents of the bag (the quarks and gluons) form an overall color singlet object so that static color fields do not exist in the physical vacuum. In fact, the MIT bag model defines the boundary condition of the gluon field at the bag boundary in such a way that the color field do not penetrate the physical vacuum and this leads to color neutrality of the bag. That is, the antisymmetric field tensor $F_a^{\mu\nu}$ describing a gluon field must also satisfy a surface boundary condition for their confinement in the MIT bag model:

$$n_\mu\, F_a^{\mu\nu}\bigg|_{r=R} = 0,$$

where a is gluon color label $(1...8)$ and n_μ is the four normal.

In the bag model, the effects of confinements are represented phenomeno-logically by the presence of bag pressure directed by the region outside the bag towards the region inside the bag. The energy density of the vacuum inside the bag is higher than that outside the bag. The difference being the bag pressure B, the energy of the system of N confined quarks in a bag of radius R is proportional to its volume and kinetic energy of its contents

$$E = p^0 N + \frac{4\pi}{3} R^3 B = \frac{2.04}{R} N + \frac{4\pi}{3} R^3 B. \qquad (2.83)$$

We observe that the tendency to increase the radius due to kinetic energy of the quarks is counterbalanced by this inward pressure B directed from the region outside the bag. The equilibrium radius of the system is located at the radius R determined by $\frac{dE}{dR} = 0$, which relates the bag pressure constant B to the bag radius. The value of the bag pressure $B^{1/4}$ is in the range between (145-235) MeV. The bag model mass formula for baryon masses can be ob-tained by putting three quarks in the cavity to yield;

$$M_b = \frac{3(2.04)}{R} + \frac{4\pi}{3} BR^3, \qquad (2.84)$$

which is the eigen energies of 3-quarks and the energy required to create a cavity in the vacuum. We have assumed quarks have been placed in the lowest eigen states of the cavity. The pressure balance condition at the bag surface is equivalent to energy minimization with respect to the bag radius which yields

$$\frac{4\pi}{3} R^4 B = 2.04. \qquad (2.85)$$

B. Dynamics of Six-Quark System

We consider exchange of two gluons between two color singlet objects (Fig. 2.18) and this leads to a color octet baryonic intermediate state of hidden color. Let us put all six quarks into the same orbital 1s state, which can accommodate 12 quarks because of 3 colors, 2 spins and 2 flavors. We show below the Young tableaux which are tabular presentation of the antisymmetrized states for color singlet of 6q system. There are as many boxes as there are particles in the system. Three distinct colors red, green and blue are denoted by 1, 2, 3 (or 4, 5, 6) respectively. Boxes placed in row, label symetric function. Boxes placed in column, label antisymmetric function. Also by convention, quarks 1, 2 and 3 are said to make up one baryon, while quarks 4, 5, 6 make the other. The color singlet state of 6q system is then obtained by various permissible permutation of the quark labels:

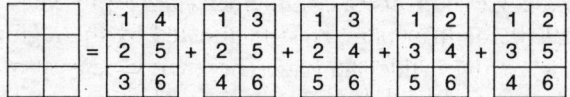

Fig. 2.18: Young tableaux for color singlet states of six-quark system

The first standard tableaux contain two color octet baryons. We therefore conclude that two completely overlapping nucleons in the $(1s)^6$ configuration contain 80% hidden color components. The two $SU(3)_c$ octet states cannot be physical baryons.

As an example we can describe the deuteron wave function in terms of six-quark state as

$$|D\rangle = a|(uud)_1 (ddu)_1\rangle + b|(uud)_8 (ddu)_8\rangle \qquad (2.86)$$

The first term corresponds to the usual np structure of deuteron. The second term corresponds to hidden color configurations where the three quark clusters are in color octets, but the overall state is a color singlet.

In Fig. 2.19 we show that the lowest-order diagram of one gluon exchange for nucleon-nucleon interaction is zero. Both nucleons are color neutral, and thus they cannot interact by color forces. One cannot couple the color singlet of a nucleon with color octet of the gluon to a color singlet of the second nucleon.

If one includes the antisymmetrization between the three quarks in one nucleon and the three quarks in the other nucleon, one gets quark exchange as shown in Fig. 2.19. Thus, a gluon exchange between two nucleons is accompanied by a quark-exchange. This diagram can be different than zero.

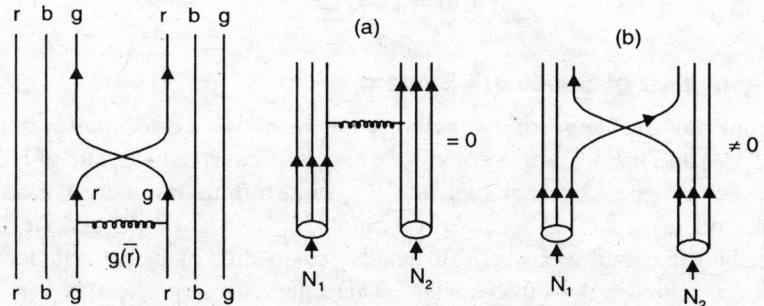

Fig. 2.19: One gluon exchange diagram for N-N interaction

Quarks are fermions, so that N–N wave function must be totally anti-symmetric with respect to the exchange of any two quarks. We assume that the $6q$ wave function can be written as a product of two three cluster wave functions and a relative motion wave function:

$$\psi_6 = A\left[\sum_{ab} \psi_a(1,2,3)\,\psi_b(4,5,6)\,\chi_{ab}(r)\right]. \tag{2.87}$$

$\psi_a(\psi_b)$ is a well defined internal wave function for a three quark baryon a (b) and $\chi_{ab}(r)$ is the relative ab wave function where a and b are separated by r. The operator A ensures that the total wave function for six quarks is antisymmetric under interchange of any two quarks. The six quark wave function satisfies the S. Eq. $(H-E)\psi_6(r)=0$, where H is the six quark Hamiltonian given by

$$H = \sum_{i=1}^{6}\left[6m + \frac{p_i^2}{2m}\right] + \sum_{i,<j}\left(V_{ij}^{conf} + V_{ij}^{HF}\right) - E_{c.m}, \tag{2.88}$$

where all the six quarks in two baryons have the same mass m. To solve the above equation one employs the standard three-body technique and integrate out the relative quark coordinates in known baryon cluster $\psi_a(\psi_b)$ (assumed relative s-state) and arrives at a set of coupled integral equation for $\chi_{ab}(r)$. Then the coupled equations are solved to find the scattering phase shift for N-N scattering. For details see the work of Oka and Yazaki.[31]

We only examine here the symmetry properties of 6q wave function to elucidate the quark perspective of short range repulsion of NN interaction. First let us construct the total six quark wave function. We use curly brackets to denote SU(4) spin flavor (spin-isospin) symmetries. A nucleon has a symmetric spin flavour wave function

$$\{3\} = \boxed{}\boxed{}\boxed{}$$

and when it is multiplied by another symmetric wave function $\{3\}$, we get four possible spin-flavor wave function of the six quarks:

One is completely symmetric under the interchange of any two quarks and we denote this by $\psi_{\{6\}}^{SF}$. The other three have mixed symmetry.

B. Short-range Repulsion

The color singlet six quark state has a definite symmetry in the color space. Six quark color wave function is not completely antisymmetric under the interchange of any two quarks (it has mixed symmetry). {51} symmetric spin flavor state can not couple [6] space symmetric state. We use square bracket to denote the symmetry of the spatial wave function. Therefore, for two nucleons in relative s-state, the six-quark wave function has the following form:

$$\psi_6(r) = \psi_6^{color}\left[a\ \psi_{[6]}^{space}\ \psi_{\{33\}}^{SF} + b\ \psi_{[42]}^{space}\ \psi_{\{51\}}^{SF} + c\ \psi_{[42]}^{space}\ \psi_{\{33\}}^{SF}\right]. \quad (2.89)$$

Since, the two nucleons are in relative s-state, the allowed spin and isospin combinations are $S = 1$, $T = 0$ and $S = 0$, $T = 1$ ($S + T =$ odd). Here a, b and c are Clebs Gordon coefficients of the permutation group of 6 objects. Their squares give .the probability to find completely symmetric spatial representation [6] or mixed orbital space symmetry [42].

When $r \to \infty$, we have two well separated free nucleons in relative S-state. The orthogonal transformation coefficients between the six-quark symmetry basis and the physical N-N basis have the following values:

$$a = \frac{1}{3},\ b = \frac{2}{3}\ \text{and}\ c = -\frac{2}{3}.$$

Accordingly Eq. (2.89) is written as

$$\psi_{NN}(r \to \infty) = \psi_6^{col}\left[\frac{1}{3}\ [6]\ \{33\} + \frac{2}{3}\ [42]\ \{33\} - \frac{2}{3}[42]\ \{51\}\right]. \quad (2.90)$$

Faessler[34] has explained the short range repulsion of the N-N interaction. Figure 2.20 shows the spatial symmetries of the 6-quarks at small distances. We see that it is more probable by the weight 8/9 to have at short distance the [42] symmetry compared with the completely symmetric orbital wave function [6] which has only the weight 1/9. Figure 2.20 indicates also the lowest energy realization of the orbital symmetry [6] and mixed orbital symmetry [42]. The last configuration requires that at zero distance of the two nucleons ($r = 0$), the two quarks are in $1p$ state of the lowest energy realization of this configuration. The usual way of representing two nucleon wave function by 6 quarks in $1s$ state is only contained with probability 1/9.

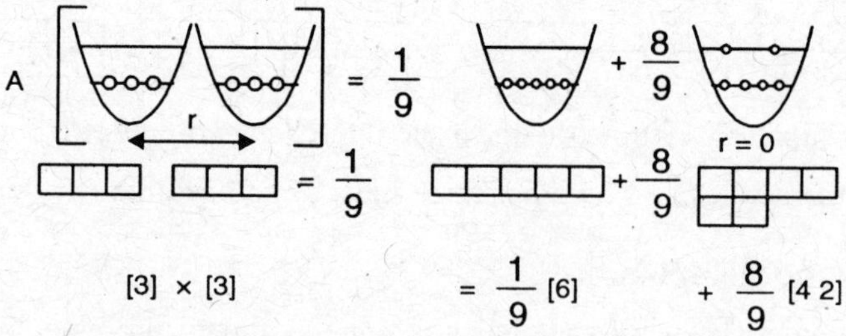

Fig. 2.20: Orbital [6] and mixed orbital [42] symmetries

It is obvious that the [42] orbital symmetry can not be neglected. If for a moment we neglect 1/9 compared to 8/9 we have at small distances $r = 0$ at least two harmonic oscillator quanta excited. Or at least two quarks have not to be in 1s state. For the lowest configuration they are in 1p state. That means at short distances this configuration with the probability 8/9 has at least two harmonic oscillator quanta excited. Since one has to conserve the number of harmonic oscillator quanta, the two quanta must be contained in the relative motion.

Exercise 2.12. Get an estimate of effective coupling strength for $Q^2 = 4$ GeV^2 and $Q^2 = 10$ GeV^2.

At $Q^2 = 4$ GeV^2, $N_f = 3$ and at $Q^2 = 10$ GeV^2, $N_f = 5$.

Therefore, $\alpha_s[(4 \text{ GeV})^2] = \dfrac{12 \times 3.141}{(33 - 2 \times 3)} \dfrac{1}{2 \log_e 20} \cong 0.33$

$$\alpha_s[(10 \text{ GeV})^2] = \dfrac{12 \times 3.141}{(33 - 2 \times 5)} \dfrac{1}{2 \log_e 50} \cong 0.26.$$

Exercise 2.13. Show that $m_{\Lambda^0} = 2m_u + m_s - \dfrac{3a}{m_u^2}$ and obtain the value of m_{Λ^0}.

For Λ^0, $J_{ud} = 0$ so that $\vec{\sigma}_u \cdot \vec{\sigma}_d = -3$.

Hence, $m_{\Lambda^0} = 2m_u + m_s + a \left[\dfrac{\vec{\sigma}_u \cdot \vec{\sigma}_d}{m_u^2} + \dfrac{\vec{\sigma}_u \cdot \vec{\sigma}_s + \vec{\sigma}_d \cdot \vec{\sigma}_s}{m_u m_s} \right]$

$$= 2m_u + m_s - \dfrac{3a}{m_u^2},$$

since $\vec{\sigma}_u \cdot \vec{\sigma}_s + \vec{\sigma}_d \cdot \vec{\sigma}_s = 2 \left[\dfrac{1}{2} \times \dfrac{3}{2} - \dfrac{9}{4} \right] - \vec{\sigma}_u \cdot \vec{\sigma}_d = -3 + 3 = 0.$

Substituting the values of m_u, m_s and $\dfrac{a}{m_u^2}$ we get $m_{\Lambda^0} = 1114$ MeV/c^2,

which agrees with the experimental value 1115.6 MeV/c^2.

Exercise 2.14. Assuming the equilibrium radius of the bag to be the proton radius calculate $B^{1/4}$.

$$\frac{dE}{dR} = -\frac{2.04N}{R^2} + \frac{4\pi}{3} 3R^2 B = 0 \quad \text{or} \quad B^{1/4} = \left(\frac{2.04N}{4\pi} \right) \cdot \frac{1}{R}.$$

If we take the confinement radius to be the radius of proton (= 0.8 fm) for a 3-quark system (*uud*), we can obtain an estimate of the bag pressure constant

$$B^{1/4} = \left(\frac{2.04 \times 3}{4\pi}\right)^{1/4} \frac{1}{0.8 \text{ fm}} \approx 206 \text{ MeV.}$$

Exercise 2.15. Assuming that three massless quarks are put in $s_{1/2}$ orbital for nucleon, calculate various static properties of nucleon in the bag model.

For proton (neutron), we have two u-quarks (d quarks) and one d-quark (u-quark) in the bag. As described earlier, we construct the wave function of the nucleon as an antisymmetric combination of three-quark wave function. The MSCR for proton

$$\langle r_p^2 \rangle_{ch} = \int d^3r \; r^2 \psi^+ \psi = N^2 \int_0^R r^4 dr \left[\{j_0(p^0 r)\}^2 + \{j_1(p^0 r)\}^2 \right]$$

$$= R^2 \frac{2(p^0 R)^3 - 2(p^0 R)^2 + 4(p^0 R) - 3}{6\left[(p^0 R)^3 - (p^0 R)^2\right]} = 0.53 \; R.$$

Thus for a bag radius of 1 fm the MSCR is about 0.53 fm. We can compare this with the values given in Table 1.1. One can calculate

The magnetic moment of a nucleon is defined as the expectation value of the 3-component of the magnetic moment operator in a nucleon state with spin $\frac{1}{2}$. Calculating the space part of the matrix element, we get

$$\langle (\mu_z)_i \rangle_{space} = \frac{2}{3} \; e_i(\sigma_z)_i \; N^2 \int_0^R r^3 dr \; j_0(p^0 r) \; j_1(p^2 r)$$

$$= \frac{4(p^0 R) - 3}{12[(p^0 R) - 1] p^0 r} = R \; e_i(\sigma_z)_i \approx 1.93 \; R \; e_i(\sigma_z)_i.$$

Here R is in Fermi and magnetic moments is in units of nuclear magnetons. To calculate the magnetic moment, we must calculate the expectation value of $e_i(\sigma_z)_i$ in nucleon wave function:

$$\langle e_i(\sigma_z)_i \rangle = 1(-2/3) \qquad \text{proton (neutron)}$$

Thus, for a bag radius of 1 fm, the proton neutron magnetic moments are 1.93 and −1.29 nuclear magnetons repectively. The experimental values are 2.79 and −1.9 respectively. So bag model underestimates magnetic moments.

However, $\left[\dfrac{\mu_n}{\mu_p}\right]_{Bag \; model} = -\dfrac{2}{3}$, which agrees well with the experimental value.

REFERENCES

1. E. Burhop, High Energy Physics, Ed. E. Burhop Academic Press, New York 1962.
2. E.P. Wigner, Phys. Rev. **43**, 252 (1933).
3. H. Yukawa, Proc. Phys. Math. Soc. (Japan) **17**, 467 (1935).
4. G. Breit, E.V. Condon and R.D. Present, Phys. Rev. **50**, 825 (1936).
5. C.F. Powel and G.P.S. Ochialini, Nature **159**, 694 (1947).
6. R.V. Reid, Ann. Phys. (N.Y.) **50**, 411 (1968).
   R. Machbidt, K. Holinde and C. Elster, Phys. Rep. 149, 1 (1987).
7. S.N. Mukherjee and R. Shyam, Phys. Rev. **C8**, 1149 (1973).
   R.R. Scheerbaum, Phys. Rev. **C11**, 255 (1975).
8. G. Breit, M.H. Hull, K.E. Lassila and K.D. Pyatt, Phys. Rev. 120; 2227 (1960),
   Phys. Rev. **128**, 826 (1962).
   R.E. Seamon, K.A. Friedman, G. Breit, R.D. Haracz, J.M. Holt and A. Prakash,
   Phys. Rev. **165**, 1579 (1988).
9. M.H. Mac Gregor, R.A. Ardnt and R.M. Wright, Phys. Rev. **169**, 1128 (1968).
   Phys. Rev. **182**, 11714 (1969).
10. G. Breit, M. Tischler, S.N. Mukherjee and J. Lucas, Proc. Natl. Acad. Sc. U.S.A.
    68, 897 (1971), **70**, 2178 (1973).
11. T. Hamada and I.D. Johnston, Nucl. Phys. **34**, 382 (1962).
12. M. Lacombe et al., Phys. Rev. **C21**, 861 (1980); Ibid. Phys. Rev. **C23**, 2405
    (1981).
13. R.A. Bergervoet et. al., Phys. Rev. Lett. **59**, 2255 (1987).
14. J. Irving, Phil. Mag. **42**, 338 (1951).
15. V.S. Mathur, S.N. Mukherjee and M.L. Rustgi, Phys. Rev. **127**, 1663 (1962).
16. V.S. Mathur and A.V. Lagu, Nucl. Phys. **A118**, 369 (1968).
17. L.D. Faddeev, Z. Eksperim 1. Theo. Fiz. **39**, 1459 (1961), A.N. Mitra, Nucl. Phys.
    32, (1962); O.A. Yakubovski, Sov. J. Nucl. Phys. **5**, 937 (1967).
18. H.T. Cohelo et. al., Phys. Rev. **C28**, 1812 (1983).
19. T.K. Das and H.T. Coelho, Phys. Rev. **C26**, 697, 754 (1982).
20. S. Sanyal and S.N. Mukherjee, Phys. Rev. **C31**, 33 (1985), **C36**, 67 (1987).
21. T. Sasakawa and S. Ishikawa, Few-Body System **1**, 3 (1986).
    T. Sasakawa and S. Ishikawa, Phys. Rev. Lett. **54**, 1875 (1990).
    T.K. Das et. al., Phys. Rev. **C26**, 2288 (1982).
22. B.F. Gibson, Nucl. Phys. **A543**, 1C (1992).
23. E. Lomon, Phys. Lett. **68B**, 419 (1977).
24. Reeta Vyas, M. Chopra and M.L. Rustgi, Phys. Rev. **C25**, 1801 (1982).
25. R. Vyas and M.L. Rustgi, Phys. Rev. **C26**, 1399 (1982).
26. R. Shyam and S.N. Mukherjee, Prog. Theoret. Phys. **53**, 1846 (1975).
27. D.O. Riska and G.E. Brown, Phys. Lett. 38B, 1993 (1972).
    D.O. Riska, Phys. Reports, **181**, 207 (1989).
28. J. Ahrens, Nucl. Phys. **A446**, 229C (1985).
29. J. Ashman et. al., Phys. Lett. **B206**, 364 (1988), Nucl. Phys. **B368**, 1 (1989), Nucl.
    Phys. **B337**, 509 (1990).

30. S.N. Mukherjee, R.Nag, S.Sanyal, T.Morii, J.Morishita and M.Tsuge in Quark Potential approach to mesons and baryons, Phys.Reports. **231**, 201(1993).

N. Isgur and G. Karl, Phys. Lett. **B72**, 109 (1977).

A. De Rujula, H. Georgi and S.L. Glashow Phys. Rev. **D12**, 147 (1975).

31. M. Oka and K. Yazaki, Quarks in Nuclei, Int. Rev. Nucl. Phys., Vol. I, Ed. W. Weise, World Scientific 1984.

K. Yazaki, Nucl. Phys. **416**, 87 (1984).

SUGGESTED BOOKS FOR FURTHER READING

32. The Nucleon-Nucleon Interaction, G.E. Brown and A.D. Jackson, North Holland, 1976.

33. Mesons in Nuclei Ed. M. Rho and D.H. Wilkinson, North Holland, 1979.

34. Quarks and Nuclear Forces, Editors : D. Fries and B. Zeitnitz, Springer Tracts, Vol. 100 (1982). See also A. Faessler in "How should or will QCD influence nuclear physics", chapter 6, page 214.

35. Fields, Symmetries and Quarks, U. Mosel, Springer.35. Foundation in Quantum Chromodynamics, T. Muta 2nd Ed. World Scientific Publication (1998).

36. Foundation in Quantum Chromodynamics, T. Mutta 2nd Ed. World Scientific Publication (1998).

37. G. Zweig, CERN Reports TH401, TH402 (1964).

38. Models of the Nucleon from Quarks to Solitons, R. K. Bhadhuri, Addition Wesley Reading, M. A. 1988.

Nuclear Shell Theory

"Trying to explain magic by miracles"

– Oppenheimer[1]

3.1 SINGLE PARTICLE ORBITS

The periodicity in atomic properties, such as valence and ionization potential forms the basis of the familiar periodic table. It is explained in terms of filling of successive levels of a screened Coulomb potential by electrons whose numbers are determined by the Pauli principle. Similar periodicity in the nuclear properties was first pointed out by Elasser[2]. For example, nuclides

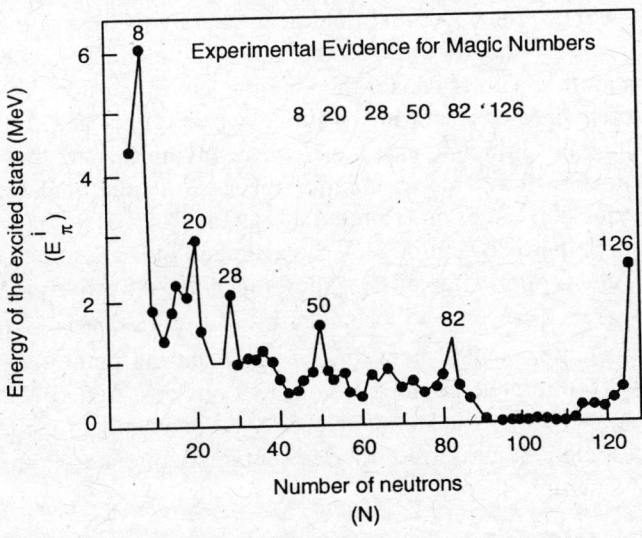

Fig. 3.1: Energy of the first excited state E_x^1 versus neutron number N

having neutron or proton number (N or Z) equal to the 'magic number' (2, 8, 20, 28, 50, 82, 126) have following observed properties:

1. The energy of the first excited state in even N or even Z nuclei is at an anomalously high energy when N or Z equals the magic number. This is illustrated in Fig. 3.1.
2. They possess large BE/A values.
3. The separation energy for the last bound nucleon is very low for nuclei with N or Z = magic number + 1.
4. The number of stable and long lived isotopes for magic nuclei is greater than for nearby elements.
5. The nuclei, having excited states with a long half life for gamma emission (isomeric states) are abundant for the magic nuclei.
6. They possess zero intrinsic quadrupole moment in the ground state.
7. There is definite evidence of strong systematic irregularities in binding energies as compared to the smooth values obtained from the semi-empirical mass formula.

It seems that inside the nucleus certain nucleon numbers (magic numbers) complete shells and have extra stability in analogy to noble gas electronic structure. However, there is no direct evidence of dominant central interaction inside the nucleus.

A simple nuclear model, which provides a reasonable accounting for many observed properties related to the shell closure, is the single particle shell model. The main assumption of this model is that, each nucleon moves independently in an average potential field generated by the rest of the nucleons. In other words, interaction between the individual nucleons is weak and the nucleon mean free path is greater than the nuclear dimension. This seems to be, however, in direct conflict with the strong nature of nuclear force. Weisskopf has used the Pauli Principle as a means of resolving this contradiction. The strong interaction is unable to manifest because all the quantum states to which this nucleon might be scattered are already occupied.

In the single particle shell model, the average field is assumed to be local and central and is approximated by a short range and attractive potential $V_c(r)$ such that $V_c(r) \leq 0$; $V_c(r \to \infty) = 0$; and $\mathrm{Lim}_{r \to 0} \, r^2 \, V_c(r) \leq 0$. It is hard to understand why central potential may form a good starting point because unlike atom, no fixed reference point exists inside a nucleus. Actually, the average field is generated by the Nucleon-Nucleon (N-N) interaction. In order to obtain the average field explicitly from the N-N interaction, one uses the 'Hartree-Fock' self consistent theory.

A. Nucleon Motion in a Central Potential

Commonly used forms of the average single particle potential are: the harmonic oscillator, the square well and the Saxon-Woods. One solves an eigen value problem (Appendix B1) to study the motion of a single nucleon in a potential field and considers the radial Schrodinger Equation (S. Eq.) for the bound state $(E < 0)$ namely,

$$\frac{d^2R}{dr^2} + \frac{2}{r}\frac{dR}{dr} + \left[\frac{2\mu}{\hbar^2\cdot}\{E - V_c(r)\} - \frac{\ell(\ell+1)}{r^2}\right]R = 0. \tag{3.1}$$

For the harmonic oscillator (h.o.) potential $V_c(r) = -V_0 + \frac{1}{2}\mu\omega^2r^2$ (Fig. 3.2)

one obtains

$$\frac{d^2u}{dr^2} + \left[\frac{2\mu}{\hbar^2}\left(V_0 + E - \frac{1}{2}\mu\omega^2r^2\right) - \frac{\ell(\ell+1)}{r^2}\right]u(r) = 0, \tag{3.2}$$

where $R(r) = \dfrac{u(r)}{r}$ is substituted in (3.1) to get an equation in u without the

first order differential. The boundary conditions which the wave function u should satisfy are as follows:

1. The wave function and its derivative must be everywhere finite and continuous.
2. The wave function must vanish at infinity.

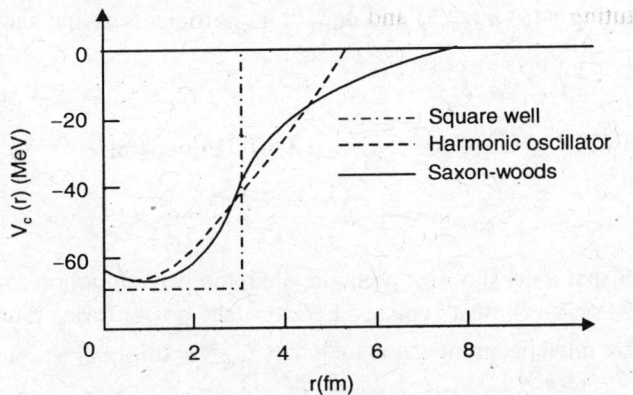

Fig. 3.2: Shape of the single-particle central potential

To solve Eq. (3.2) we introduce a dimensionless variable $\rho = r/b$, where

$$b = \sqrt{\frac{\hbar}{\mu\omega}}.$$

Hence,
$$\frac{d^2u}{d\rho^2} + \left[\frac{2(E + V_0)}{\hbar\omega} - \rho^2 - \frac{\ell(\ell+1)}{\rho^2}\right]u = 0. \tag{3.3}$$

For small ρ, $\dfrac{d^2 u}{d\rho^2} - \dfrac{\ell(\ell+1)}{\rho^2} u = 0$, which is satisfied by $u \sim \rho^{\ell+1}$.

For large ρ, $\dfrac{d^2 u}{d\rho^2} - \rho^2 u = 0$, which is satisfied by $u \sim e^{-(1/2)\rho^2}$. Therefore,

we try the solution $u = \rho^{\ell+1} e^{-(1/2)\rho^2} v(\rho)$ and on substitution in (3.3) get

$$\left[\frac{d^2}{d\rho^2} + 2\left(\frac{\ell+1}{\rho} - \rho \right) \frac{d}{d\rho} + 2(\lambda - \ell) \right] v(\rho) = 0, \tag{3.4}$$

where $\lambda = \dfrac{(E + V_0)}{\hbar\omega} - \dfrac{3}{2}$. We introduce a new variable $x = \rho^2$ and get

$$\left[x \frac{d^2}{dx^2} \left(\ell + \frac{3}{2} - x \right) \frac{d}{dx} + \frac{\lambda - \ell}{2} \right] v(x) = 0 . \tag{3.5}$$

Since ℓ is an integer variable, we try a series solution $v(x) = \displaystyle\sum_{m=0}^{\infty} c_m \, x^m$.

Substituting $v(x)$ in (3.5) and equating coefficients of the same power of x, we get:

$(2\ell + 3)C_1 = -(\lambda - \ell)\, C_0$, $2(2\ell + 3 + 2)C_2 = -(\lambda - \ell - 2)\, C_1$ and

$3[(2\ell + 3) + 4]C_3 = -(\lambda - \ell - 4)\, C_2$. In general,

$$C_{m+1} = -\frac{(\lambda - \ell - 2m)}{(m+1)\,[(2\ell + 3) + 2m]} C_m. \tag{3.6}$$

In order that $u(x)$ signifies a bound state, the wave function must be finite and for $x \to \infty$, $v(x)$ must vanish. Therefore, the power series must break off, that is, there must be an integer n such that $C_{m+1} = 0$ for $m = n-1$. Therefore, $\lambda - \ell - 2(n-1) = 0$, or $E = -V_0 + \dfrac{3}{2}\hbar\omega + \{2(n-1) + \ell\}\hbar\omega$. Here $\dfrac{3}{2}\hbar\omega$ is the zero point energy for the 3-dimensional harmonic oscillator. Denoting $(-V_0 + 3/2\,\hbar\omega) = E_{10}$, we get

$$E_{(n\ell)} - E_{10} = N\hbar\omega, \text{ where } 2(n-1) + \ell = N. \tag{3.7}$$

The closed shells occur for N or $Z = \displaystyle\sum_{\ell} 2(2\ell + 1)$. We see that the single

particle energy and the single particle states of the harmonic oscillator do not depend upon the magnetic quantum number m. Hence, the level is $2(2\ell+1)$ fold degenerate. The factor 2 corresponds to spin states of a nucleon. The levels appear in groups such as $1s$; $1p$; $(1d,2s)$; $(1f,2p)$; $(1g,2d,3s)$ etc. These grouped levels are degenerate and the closed shells occur at N or $Z = 2, 8, 20$. However, for N or Z beyond 20, the predicted shell closure occur at 40 and 70 whereas the actual values are 50 and 82. The level separation $\hbar\omega \approx 41 \, A^{-(1/3)}$, is estimated by setting $\left|V_c(r)\right|_{r=r_0} = 0$, where r_0 is the range of interaction, usually taken as the nuclear radius.

The energy $E_{n\ell}$ is generally an increasing function of n and ℓ. Same energy is obtained by increasing ℓ and decreasing n by appropriate integer; this leads to degeneracy and the shell structure. Bohr and Mottelson[3] generalized this idea by expanding $E_{n\ell}$ around a given point (n_0, ℓ_0):

$$E_{n\ell} = E_{n_0\ell_0} + (n-n_0)\left(\frac{\partial E_{n\ell}}{\partial n}\right)_{n=n_0} + (\ell-\ell_0)\left(\frac{\partial E_{n\ell}}{\partial \ell}\right)_{\ell=\ell_0} + \text{higher order terms.}$$

Approximate degeneracy is obtained by setting

$$c\left(\frac{\partial E_{n\ell}}{\partial n}\right)_{n=n_0} = a\left(\frac{\partial E_{n\ell}}{\partial \ell}\right)_{\ell=\ell_0},$$

where a and c are small integers. We therefore obtain the condition

$$an + c\ell = an_0 + c\ell_0.$$

So the levels with constant value of $N = an + c\ell$, only slightly differ in energy due to the neglect of higher order derivatives. We may redefine N as

$$N_{shell} = a(n-1) + c\ell,$$

which now looks similar to the value obtained for h.o. (Eq. 3.7) with $a = 2$, and $c = 1$.

It is also customary to denote the function $v(x)$ by $v_{n\ell}(x)$ for the state $n\ell$, so that (3.5) reads as

$$\left[x\,\frac{d^2}{dx^2} + \left(\ell+\frac{3}{2}-x\right)\frac{d}{dx} + (n-1)\right]v_{n\ell}(x) = 0.$$

The above equation resembles the Laguerre equation:

$$\left[x\,\frac{d^2}{dx^2} + \left(m+1-x\right)\frac{d}{dx} + k\right]L_k^m(x) = 0.$$

Hence, $v_{n\ell}(x) \sim L_{n-1}^{\ell+(1/2)}(x)$. The complete wave function

$$u_{n\ell}(\rho) = A_{n\ell}\,\rho^{\ell+1}\,e^{(-1/2)\rho^2}\,L_{n-1}^{\ell+(1/2)}(\rho^2), \tag{3.8a}$$

where the normalization $A_{n\ell}$ is determined by the condition

$$\int_0^\infty |R_{n\ell}(r)|^2 \, r^2 \, dr = 1, \text{ or } b^3 \int_0^\infty u_{n\ell}^2(\rho) \, d\rho = 1. \tag{3.8b}$$

Let us consider Eq. (3.8b) in which

$$\text{L.H.S.} = A_{n\ell}^2 \, b^3 \int_0^\infty \rho^{2\ell} e^{-\rho^2} \left| L_{n-1}^{\ell+(1/2)}(\rho^2) \right|^2 \rho^2 d\rho$$

$$= \frac{1}{2} A_{n\ell}^2 \, b^3 \int_0^\infty e^{-x} x^{\ell+1/2} \left| L_{n-1}^{\ell+1/2}(x) \right|^2 dx.$$

Employing the orthogonality relation for the Laguerre functions

$$\int_0^\infty x^{\ell+1/2} \left| L_{n-1}^{\ell+1/2}(x) \right|^2 e^{-x} \, dx = \frac{(2n+2\ell-1)!!}{2^{n+\ell}(n-1)!} \sqrt{\pi}, \text{ we get}$$

$$A_{n\ell}^2 \, b^3 \frac{(2n+2\ell-1)!!\sqrt{\pi}}{2^{n+\ell+1}(n-1)!} = 1 \text{ or } A_{n\ell}^2 = \frac{2^{n+\ell+1}(n-1)!}{(2n+2\ell-1)!! \sqrt{\pi} \, b^3}. \tag{3.9}$$

Here, the double factorial $(p)!! = p(p-2)\,(p-4)\ldots\ldots$, and n is the number of radial nodes including the one at the origin. Some authors use n one less and in that case it starts with the value zero instead of 1 used here.

Employing the recurrence relation for the Laguerre functions,

$$L_n^a(x) = \frac{a+n}{n} \left[a + 2n - 1 - x \right] L_{n-1}^a(x) - (a+n-1)^2 \, L_{n-2}^a(x);$$

we can calculate harmonic oscillator radial functions (Table 3.1).

Table 3.1: Harmonic Oscillator radial functions: $R_{n\ell}(r)$

State	n	ℓ	$R_{n\ell}(r)$
s	1	0	$(4/\sqrt{\pi}b^3)^{1/2} e^{-r^2/2b^2}$
p	1	1	$(8/3\sqrt{\pi}b^5)^{1/2} r e^{-r^2/2b^2}$
d	1	2	$(16/15\sqrt{\pi}b^7)^{1/2} r^2 e^{-r^2/2b^2}$
f	1	3	$(32/105b^9 \sqrt{\pi})^{1/2} r^3 e^{-r^2/2b^2}$

Spin-Orbit Interaction: To reproduce the higher magic numbers Mayor, Haxel, Suess and Jensen (1949)[4] suggested to add a spin-orbit interaction which would remove degeneracy of h.o. states and predict magic numbers accurately. The entire origin of the spin-orbit interaction can not be electromagnetic because neutron is uncharged. However, its form is guessed from the electromagnetic interaction energy of an electron bound in an atom (Thomas form). By analogy, for the nucleus,

$$V_{LS} = -\frac{\lambda}{\mu^2 c^2}\left(\frac{1}{r}\frac{\partial V(r)}{\partial r}\right)\vec{\ell}\cdot\vec{s}. \qquad (3.10)$$

Here λ is a dimensionless constant characterizing the magnitude of spin-orbit coupling. It follows from the analysis of experimental data, that $\lambda \approx 10$.

When $V(r) = -V_0 + \frac{1}{2}\mu\omega^2 r^2$, then

$$V_{LS} = -\frac{\lambda}{\mu^2 c^2}\frac{1}{r}\frac{\partial}{\partial r}\left(\frac{1}{2}\mu\omega^2 r^2\right)\vec{\ell}\cdot\vec{s} = -\lambda\frac{\hbar\omega}{\mu c^2}\frac{\vec{\ell}}{\hbar}\cdot\frac{\vec{s}}{\hbar}\quad \hbar\omega = -\alpha\,\vec{\ell}\cdot\vec{s}, \quad (3.11)$$

where α is positive and has dimension of energy if ℓ and s are expressed in units of $\hbar$.

In a phenomenological shell model approach α (or λ) is used as an adjustable parameter to achieve agreement with experimental energy levels. The S. Eq. with spin-orbit term then takes the form:

$$\left[-\frac{\hbar^2}{2\mu}\nabla^2 - V_0 + \frac{1}{2}\mu\omega^2 r^2 - \alpha\,\vec{\ell}\cdot\vec{s}\right]\phi(\vec{r},\vec{\sigma}) = E\phi(\vec{r},\vec{\sigma}),$$

where, $\quad \phi(\vec{r},\vec{\sigma}) = \frac{1}{r}u_{n\ell}(r)\,Y_\ell^m(\theta\ \varphi)\chi_{1/2}^{m_s}. \qquad (3.12)$

Y_ℓ^m and $\chi_{1/2}^{m_s}$ are the simultaneous eigen functions of $\hat{\ell}^2$, $\hat{\ell}_z$, $\hat{s}_z$ and $\hat{s}^2$ but not of the operator $\vec{\ell}\cdot\vec{s}$. It is necessary to couple the spin and orbital angular momentum to a total angular momentum $\vec{j}$. So

$$\phi(\vec{r},\vec{\sigma}) \equiv \frac{u_{n\ell}(r)}{r}\left|j\,m_j\right\rangle, \qquad (3.13)$$

where $\quad \left|j\,m_j\right\rangle = \sum_{m_s = \pm 1/2}\langle\ell\frac{1}{2}\,m_j - m_s\,m_s\,|\,j\,m_j\rangle Y_\ell^{m_j - m_s}\chi_{1/2}^{m_s}$

$$= \left(\frac{\ell + \frac{1}{2}\mp m_j}{2\ell+1}\right)^{1/2}Y_\ell^{m_j + \frac{1}{2}}\chi_{1/2}^{-1/2} \pm \left(\frac{\ell + \frac{1}{2}\pm m_j}{2\ell+1}\right)^{1/2}Y_\ell^{m_j - \frac{1}{2}}\chi_{1/2}^{1/2}$$

for $j = \ell \pm \dfrac{1}{2}$. $\langle \ell \dfrac{1}{2} \ m_j - m_s \ m_s \ | \ j \ m_j \rangle$ are the Clebsch-Gordon (C.G.) coefficients (Appendix D2). $| j \ m_j \rangle$ are simultaneous eigen functions of $\hat{j}^2 \ (\vec{j} = \vec{\ell} + \vec{s})$, $\hat{\ell}^2, \hat{s}^2$ and $\hat{j}_z = \hat{\ell}_z + \hat{s}_z$ with corresponding eigen values $j \ (j + 1), \ \ell(\ell + 1), \dfrac{3}{4}$ and m_j. The operator $\vec{\ell} \cdot \vec{s}$ is diagonal in $| j \ m_j \rangle$ basis so that

$$\langle j \ m_j \ | \ \vec{\ell} \cdot \vec{s} \ | \ j \ m_j \rangle = \frac{j(j+1) - \ell(\ell+1) - 3/4}{2}$$

$$= \begin{cases} \ell / 2 & \text{for } j = \ell + 1/2 \\ -(1/2)(\ell + 1) & \text{for } j = \ell - 1/2. \end{cases} \qquad (3.14a)$$

The radial equations are obtained by multiplying (3.12) by $\phi^*(\vec{r}, \vec{\sigma})$ and then integrating over angles and summing over spins. For $j = \ell \pm 1/2$, we get,

$$\frac{d^2 u_{n\ell}^{\pm}}{dr^2} + \left[k^2 - \frac{2\mu}{\hbar^2} \left\{ -V_0 + \frac{1}{2} \ \mu\omega^2 r^2 \right\} \right.$$

$$\left. - \frac{\ell(\ell+1)}{r^2} - \alpha \begin{cases} \ell / 2 & \text{or} \\ -(1/2) \ (\ell + 1) \end{cases} \right] u_{n\ell}^{\pm}(r) = 0. \qquad (3.14b)$$

The superscript plus on u_{nk} corresponds to $j = \ell + 1/2$ and the superscript minus on $u_{\ell k}$ corresponds to $j = \ell - 1/2$ and $k = (2\mu E / \hbar^2)$. The energy degeneracy found earlier in ℓ quantum number is removed by $\vec{\ell} \cdot \vec{s}$ force, e.g.,

$$E_{n\ell j = \ell + 1/2} = E_{10} + N\hbar\omega - \frac{\alpha}{2} \ \ell, \qquad \text{pushed down since } \alpha > 0 \text{ and}$$

$$E_{n\ell j = \ell - 1/2} = E_{10} + N\hbar\omega + \frac{\alpha}{2}(\ell + 1), \qquad \text{pushed up since } \alpha > 0.$$

Hence, the spin-orbit doublet splitting, $(E_{n\ell j = \ell - 1/2} - E_{n\ell j = \ell + 1/2})$,

$$\Delta E = \frac{\alpha}{2}(2\ell + 1). \qquad (3.15)$$

Thus, with the introduction of spin-orbit interaction, the level $n\ell j = \ell + 1/2$ comes down from the unpertubed position $E_{n\ell}$ and the level $n\ell j = \ell - 1/2$ goes up from $E_{n\ell}$ and separation between them increases with ℓ.

The $g_{9/2}$ level comes down from the rest of the levels in $N = 4$ to the vicinity of the levels in $N = 3$, and behaves as an intruder level of opposite parity. Similarly, $i_{13/2}$ level of $N = 6$ is moved sufficiently below to the vicinity of $N = 5$. The $f_{7/2}$ level is shifted least and hence 20 remains a magic number and an additional magic number occurs at 28 (Fig. 3.3). Within each shell the level order will depend upon the specific properties of potential assumed. In general, one can write

$$\Delta E = (2\ell + 1) \int_0^\infty \left| R_{n\ell}(r) \right|^2 V_{LS}(r) r^2 \, dr \,, \tag{3.16}$$

which can be readily evaluated for a known single particle potential. In the nuclear spectra ΔE is observed to be of the order of 1-6 MeV. The complete single particle level scheme is given in Fig. 3.3. Bohr and Mottelson[3] found that the radial integral can be approximated by

$$\int_0^\infty V_{LS}(r) \left[R_{n\ell}^2(r) \right] r^2 dr = 20 \, A^{-2/3} \text{ MeV} \cdot$$

The oscillator potential is further improved by adding a term $\beta \ell^2$. This term produces a level scheme which is in better agreement with experiment.

One of the most fundamental properties of a nucleus is the spatial distribution of its neutrons and protons. The saturation property of nuclear forces ensures that nuclei have a central region of fairly constant density, and a surface region where density falls rapidly to zero (Fig. 1.5). The distribution of protons in the nucleus gives its charge distribution and this can be determined by elastic scattering of high energy electrons. The radial distribution of these quantities may be represented by a Fermi distribution (1.28). Therefore, the single particle potential for the core nucleus is mostly chosen in a more realistic form like Saxon-Woods:

$$V(r) = V_{Coul} + V_0 \, f(r) + V_{so}(r) \,, \text{ where}$$

$$f(r) = \frac{1}{1 + \exp \left[(r - R)/a \right]}.$$

The Coulomb potential $V_{Coul}(r)$ (for protons only) and spin orbit potential $V_{so}(r)$ have their usual forms. The eigen value problem is solved numerically. The potential depth V_0 is then searched for standard values of 'R' and 'a' to fit the last nucleon binding energy. This is done by searching numerically until potential depth that causes the logarithmic derivatives of the interior solution and the exterior solution, satisfying decaying boundary condition at $r \to \infty$ and signifying a bound state, to match at the boundary.

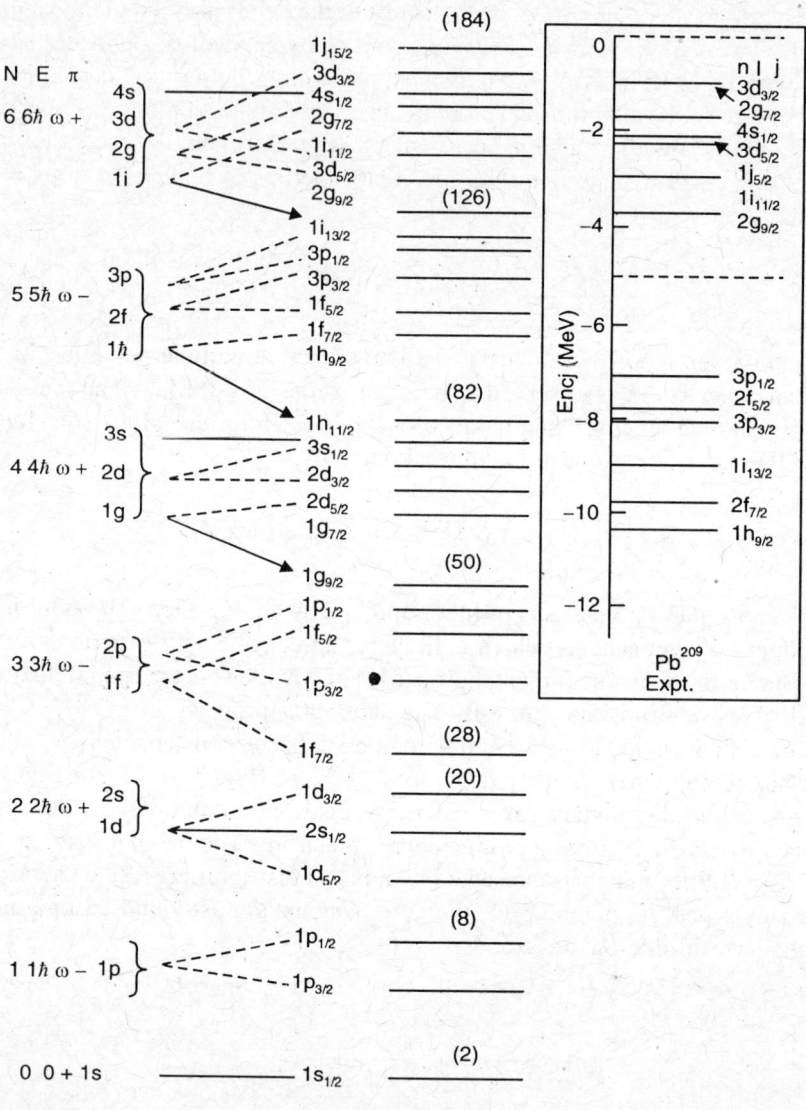

Harmonic oscillator

Fig. 3.3: The energy levels in the simple shell model (Inset: experimentally determined single particle level sequence for ^{209}Pb)

The neutron and proton potentials for the same nuclei are often appreciably different. Guided by the nuclear asymmetry term, the dependence of the single particle potential on isospin can be found as follows. Let us assume N-N interaction (without Coulomb interaction) in the form $v_{ij} = v_0 + (\vec{\tau}_i \cdot \vec{\tau}_j) v_1$, where v_0 and v_1 are depths.

The potential experienced by the particle j is

$$V_j = \sum_{i \neq j} v_0 + \vec{\tau}_j \cdot \sum_{i \neq j} \vec{\tau}_i v_1 = -V_0 + \frac{4V_1}{A} \, \vec{t} \cdot \vec{T},$$

where $\vec{\tau} = 2\vec{t}, \sum_{i \neq j} v_0 = -V_0, \frac{1}{2} \sum_{i \neq j} \vec{\tau}_i v_1 = \frac{\vec{T}V_1}{A}$ and $\vec{T}(= \frac{1}{2} \sum_{i \neq j} \vec{\tau}_i)$ is the total isospin of the nucleons other than the one labelled by the subscript j. The nucleon isospin $\vec{t}$ under consideration can couple with the total isospin $\vec{T}$ of the A-1 nucleus to form $\vec{T}' = \vec{T} + \vec{t}$.

Now, $\quad \vec{t} \cdot \vec{T} = \frac{1}{2} \left[T'(T'+1) - T(T+1) - \frac{3}{4} \right] = T/2 \qquad$ for $T' = T + \frac{1}{2}$

$$= -\frac{1}{2}(T+1) \qquad\qquad\qquad\qquad \text{for } T' = T - \frac{1}{2}$$

We define a T dependent interaction for V as follows :

$$\acute{V}(T,T') = -V_0 + \frac{2T}{A} V_1 \qquad\qquad\qquad \text{for } T' = T + \frac{1}{2}$$

$$= -V_0 - \frac{2(T+1)}{A} V_1. \qquad\qquad \text{for } T' = T - \frac{1}{2}.$$

Since the coupling of the isospin of the $A-1$ nucleons may result in a configuration that is a linear combination of several possible T values, we take this configuration to be a mixture of pure isospin states with a weight factor $|a_T|^2$. Hence, average interaction $\overline{V}$ is defined as,

$$\overline{V} = \sum_T |a_T|^2 \sum_{T'} \langle T \, \tfrac{1}{2} \, Mm \mid T'M' \rangle^2 V(T,T').$$

We substitute the values of C-coefficients and use $\overline{V} = \sum_T |a_T|^2 \sum_T |a_T|^2 = 1$,

to get $\overline{V}_Z$ and $\overline{V}_N$ as follows:

Taking the nucleon under consideration to be a proton, we have $m = +\frac{1}{2}$

so that

$$M' = M + \frac{1}{2} = \frac{Z-1-N}{2} + \frac{1}{2} = -\frac{\Delta}{2}, \text{ where } \Delta = N\text{-}Z \text{ and } M = -\frac{(\Delta+1)}{2}.$$

Employing

$$\langle T \frac{1}{2} M \frac{1}{2} \mid T + \frac{1}{2} M + \frac{1}{2} \rangle = [(T + M + 1)/(2T + 1)]^{1/2}$$

and $\langle T \dfrac{1}{2} M \dfrac{1}{2} \mid T - \dfrac{1}{2} M + \dfrac{1}{2} \rangle = [(T - M)/(2T + 1)]^{1/2}$, we get

$$\bar{V}_Z = \sum_T |a_T|^2 \left\{ \frac{T + M + 1}{2T + 1} V \left(T, T + \frac{1}{2} \right) + \frac{T - M}{(2T + 1)} V \left(T, T - \frac{1}{2} \right) \right\}$$

$$= \sum_T |a_T|^2 \left\{ \frac{T + M + 1}{2T + 1} \left(-V_0 + \frac{2V_1 T}{A} \right) + \frac{T - M}{2T + 1} \left(-V_0 - \frac{2V_1(T + 1)}{A} \right) \right\}$$

$$= \sum_T |a_T|^2 \left(-V_0 + \frac{2M}{A} V_1 \right).$$

Hence, $\bar{V}_Z = [-V_0 - V_1(1 + \Delta)/A]$ and $\bar{V}_N = [-V_0 - V_1(1 - \Delta)/A]$.   (3.17)

B. Predictions of Ground State Properties

The extreme single particle model is the simplest form of the shell model. The coupling rules giving the arrangement of the particles in the unfilled shells are as follows:

(1) An even number of identical particles in equivalent j orbital couple to total $I = 0$ in *gs*. Equivalent 'j' orbital are those with $(2j+1)$ fold degeneracy.

(2) An odd number of identical particles in equivalent j orbital couple to total $I = j$.

(3) The state into which a given proton goes is independent of number of neutrons in the nucleus and vice versa.

The extreme single particle model is rather successful in predicting spin-parity of nuclear *gs* for odd-*A* nuclides in a straight forward manner.

For example: $^{33}_{16}S_{17} \left[(1s_{1/2})^2 (1p_{3/2})^4 (1p_{1/2})^2 (id_{5/2})^6 (1d_{3/2})^3 \right]$ configuration

predicts $(I)_{sp} = 3/2$ and $(I)_{expt} = 3/2$.

The cases *N*-odd and *Z*-odd are more complicated particularly when the odd neutron and proton are in different orbitals. A reliable prediction of the *gs* spin of odd-odd nuclei requires accurate calculation of effect of residual interaction. However, Nordheim proposed the following empirical rule for predicting the *gs* spin of odd-*A* nuclides. The possible values of total angular momentum *I* is given by

$$|j_1 - j_2| \leq I \leq j_1 + j_2.$$

The problem is to decide about I value which would correspond to gs spin. We define the Nordheim number $N = j_p - \ell_p + j_n - \ell_n$ and give the rules:

When $N = 0$, then $I = | j_n - j_p |$; this is called the strong rule.

When $N = \pm 1$, then $I = j_n + j_p$ or $| j_n - j_p |$; this is called the weak rule.

Example: In $^{14}_{7}N_7$ both the odd-proton and the odd-neutron configuration is

$(s_{1/2})^2 (p_{3/2})^4 (p_{1/2})^1$.

Hence, $N = 1/2 - 1 + 1/2 - 1 = -1$. Therefore, we apply the weak rule, $I = 1/2 + 1/2 = 1$ or $I = 1/2 - 1/2 = 0$; whereas, $I_{expt.} = 1^+$.

The gs spin of large number of nuclei are well predicted by the extreme single particle model. However, there are several exceptions. For example:

$$^{47}_{22}Ti_{25} \text{ and } ^{55}_{25}Mn_{30} \text{ have } I_{sp} = 7/2 \text{ but } I_{expt} = 5/2$$

$$^{75}_{33}As_{42} \text{ and } ^{61}_{28}Ni_{33} \text{ have } I_{sp} = 5/2 \text{ but } I_{expt} = 3/2$$

$$^{103}_{45}Rh_{58} \text{ and } ^{107}_{47}Ag_{60} \text{ have } I_{sp} \text{ but } I_{expt} = 1/2.$$

Again in most even-even nuclei, the first excited state is 2^+ with few exceptions, for example $^{16}_{8}O_8(3^-)$, $^{208}_{82}Pb_{126}(3^-)$, $^{77}Ge(0^+)$.

Many such exceptions have been eliminated by modifying coupling rules with the inclusion of **pairing interaction**. The evidence of pairing energy comes from the binding energy (BE) data. Consider a nucleus with binding energy $B(N)$. When two neutrons are added to it, one finds

$$\{B(N + 2) - B(N)\} > 2 \{B(N + 1) - B(N)\} \qquad (3.18)$$

In other words, there exists extra pairing energy (2Δ) to pair up two neutrons in the same shell over and above the "individual" binding energy. Thus, nuclei with even Z or even N are more stable indicating stronger binding energies. It has also been established experimentally that the pairing energy of two similar nucleons confirm on the average with the rule

$$\Delta = \frac{11.2}{\sqrt{A}} \text{ MeV}, \qquad (3.19)$$

where Δ decreases from 3.5 MeV to 0.75 MeV between $A = 10$ and $A = 225$. If we disregard the small difference in the symmetry and Coulomb energies, the *BE* of even-even nucleus is greater by 2Δ than the *BE* of odd-odd nucleus with the same A. Therefore, the fact that even-even nuclei are more abundant in nature than odd-odd nuclei is related to the pairing energy gap Δ.

The pairing interaction, which exists between nucleons of equal and opposite angular momentum projection is defined as

$$V_{pair} = -G(-)^{j_1-m_1+j_2-m_2+j_3-m_3+j_4-m_4} \, \delta_{j_1j_2} \, \delta_{j_3j_4} \, \delta_{m_1-m_2} \, \delta_{m_3-m_4}, \quad (3.20a)$$

where G is the strength of pairing. Two identical particles in the same angular momentum state interact most strongly when the orbital overlap of their wave function is maximum, i.e. when they are coupled to total angular momentum $I = 0$.

To obtain pairing energy E_{pair}, we calculate the matrix element $\langle j_1m_1j_2m_2 \,|V_{pair}|\, j_3m_3j_4m_4 \rangle$ by employing the coupled wave function $|IM\rangle$ instead of product wave function $|j_3m_3\rangle \,|\, j_4m_4\rangle$ etc. for two particles in the 'j' orbit. Therefore,

$$\langle j^2 IM \,|V_{pair}|\, j^2 IM \rangle = -\sum_{mm'} \langle jjm-m\,|\,I0\rangle$$

$$\langle j'j'm'-m'\,|\,I0\rangle (-)^{j-m+j'-m'} G$$

Now $\displaystyle\sum_m (-)^{j-m} \langle jj\,m-m\,|\,I0\rangle = \sqrt{2j+1}\,\delta_{I0}$ so that *g.s.* $(j = j')$ pairing energy,

$$E_{pair} = -\delta_{I0}(2j+1)\,G. \quad (3.20b)$$

Hence, for a given nucleus, the pairing energy of nucleons in equivalent orbits increases with increasing j. Thus, if two levels of different j are close together in energy and $E_{j_1} < E_{j_2}$ with $j_2 < j_1$, the paring interaction may be large enough to push E_{j_2} below E_{j_1} so that pairs of like particles prefer to occupy j_2 orbital leaving j_1 for an odd particle.

In $_{40}Zr_{50}^{90}$, the proton fills $2p_{1/2}$ level at $Z = 40$, after this subsequent odd proton nuclei should have $j = 9/2$, because the last odd proton must go to $1g_{9/2}$ level according to single particle model. Instead, the nuclei $_{45}Rh_{58}^{103}$ and $_{47}Ag_{60}^{107}$ are known to have *g.s.* spin 1/2. However, if pairing interaction is included $|(1g_{9/2})^2, I = 0\rangle$ is pushed below $|(2g_{1/2})^2, I = 0\rangle$ by an amount $8G$ because

$$\left[E_{pair}(2p_{1/2})^2 - E_{pair}(g_{9/2})^2 \right] = -2G + 10G = 8G.$$

Therefore, $g_{9/2}$ levels will be first filled in pairs and the last odd proton should occupy $2p_{1/2}$ thus predicting the correct *g.s.* spins of Rh^{103} and Ag^{107} mentioned above.

Magnetic moments: According to the extreme single particle model, the magnetic moment of the whole nucleus is due to the last odd particle. We assume $I = j$. The expectation value of the magnetic moment operator in a single particle state

$| jm_j = j \rangle$ is given by

$$\mu = \frac{\langle j_z \rangle \langle \vec{\mu} \cdot \vec{j} \rangle}{j^2} = \frac{1}{j+1} \left[g_\ell \langle jj | \vec{\ell} \cdot \vec{j} | jj \rangle + g_s \langle jj | \vec{s} \cdot \vec{j} | jj \rangle \right] nm.$$

Hence, $\quad \mu(j = \ell + 1/2) = \left[(j - 1/2)g_\ell + 1/2g_s \right] nm,$ $\qquad$ (3.21a)

and $\quad \mu(j = \ell - 1/2) = \dfrac{j}{j+1} \left[g_\ell (j + 3/2) - (1/2)g_s \right] nm.,$ $\qquad$ (3.21b)

For odd Z: $\mu \begin{cases} = \left[(j - 1/2) + 2.792 \right] nm & \text{for } j = \ell + 1/2 \quad (3.22a) \\[2mm] = \dfrac{j}{j+1} \left[(j + 3/2) - 2.792 \right] nm & \text{for } j = \ell - 1/2 \quad (3.22b) \end{cases}$

For odd N : $\mu \begin{cases} = -1.913 \text{ nm} & \text{for } j = \ell + 1/2 \quad (3.23a) \\[2mm] = \dfrac{j}{j+1}(1.913) \, nm & \text{for } j = \ell - 1/2 \quad (3.23b) \end{cases}$

The numerical values of μ computed from (3.21-3.23) are plotted with respect to j. These curves are known in the literature as the Schmidt's curves. They have the following characteristics: (1) All experimental moments lie close to or in between Schmidt's curves. (2) The general trend of deviations between theory and experiment is quite similar on odd Z and odd N diagrams. The measured values, in large majority of cases, permit us to guess whether for the nucleus the relation $I = j = \ell + 1/2$ or $I = j = \ell - 1/2$ is valid. Thus, from the measurement of magnetic moment and from its proximity from the Schmidt's curve we can get the value of the orbital angular 'ℓ' of the state in question if 'j' is known and hence can determine the parity $(= (-)^\ell)$. This result is important, since the parity of nuclear state can be measured in other ways with great difficulty. A comparison of g.s. spin and magnetic moment of some light nuclei is shown in Table 3.2.

Quadrupole Moment: The quadrupole moment in the extreme shell model picture can arise from the last incomplete proton shell. If there is only one proton in such a shell, then employing the single particle wave function given by (3.13) and the expression for quadrupole moment operator (dropping k in (1.21c) we get

Table 3.2: Comparison between predictions of the extreme single particle model (sp) and experimental values

$^{A}_{Z}X_{N}$	I^{π}		μ	
	sp.	*expt.*	*sp. (nm)*	*expt. (nm)*
$^{3}_{2}He_{1}$	$\frac{1}{2}^{+}$	$\frac{1}{2}^{+}$	-1.91	-2.13
$^{7}_{3}Li_{4}$	$\frac{3}{2}^{-}$	$\frac{3}{2}^{-}$	3.79	3.03
$^{11}_{6}B_{6}$	$\frac{3}{2}^{-}$	$\frac{3}{2}^{-}$	3.79	2.69
$^{13}_{6}C_{7}$	$\frac{1}{2}^{-}$	$\frac{1}{2}^{-}$	0.64	0.70
$^{15}_{7}N_{8}$	$\frac{1}{2}^{-}$	$\frac{1}{2}^{-}$	-0.26	-0.28
$^{17}_{8}O_{9}$	$\frac{5}{2}^{+}$	$\frac{5}{2}^{+}$	-1.913	-1.89
$^{25}_{12}Mg_{13}$	$\frac{5}{2}^{+}$	$\frac{5}{2}^{+}$	-1.913	-0.85
$^{27}_{13}Al_{14}$	$\frac{5}{2}^{+}$	$\frac{5}{2}^{+}$	4.79	3.40

$$Q = \int u_{n\ell}^{*}\, r^{2}\, u_{n\ell}\, dr \sqrt{\frac{16\pi}{5}} \langle j\, m_{j} \mid Y_{2}^{0} \mid j\, m_{j} \rangle d\Omega. \qquad (3.24)$$

Employing (3.13) and the orthogonality of spin function, one gets

$$\langle j\, \ell\, m_{j} \mid Y_{2}^{0} \mid j\, \ell\, m_{j} \rangle = \frac{\ell + (1/2) + m_{j}}{2\ell + 1} \int \left| Y_{\ell}^{m_{j}+1/2} \right|^{2} Y_{2}^{0}\, d\Omega$$

$$+ \frac{\ell + (1/2) - m_{j}}{2\ell + 1} \int \left| Y_{\ell}^{m_{j}+1/2} \right|^{2} Y_{2}^{0}\, d\Omega.$$

Now employing (2.53b) we get

$$Q(m_{j}) = \frac{2\langle n\ell \mid r^{2} \mid n\ell \rangle}{(2\ell + 1)(2\ell - 1)(2\ell + 3)} \left[(2\ell + 1)\left\{ \ell(\ell + 1) - 3\left(m_{j}^{2} + \frac{1}{4} \right) \right\} \pm 6m_{j}^{2} \right]$$

$$\text{for } j = \ell \pm \frac{1}{2} \qquad (3.25)$$

One finally gets from (3.25)

$$Q(m_j) = \langle r^2 \rangle_{n\ell} \left[\frac{j(j+1) - 3m_j^2}{2j(j+1)} \right], \text{ where } \langle r^2 \rangle_{n\ell} = \int_0^\infty r^2 u_{n\ell}^2 \, dr,$$

also $\bar{Q}(m_j) = \frac{1}{2} \left[\{Q(m_j)\}_{j=\ell+1/2} + Q(m_j) \}_{j=\ell-1/2} \right].$

Hence, $Q(m_j = j) = Q_{sp} = -\dfrac{2j-1}{2j+2} \langle n\ell \mid r^2 \mid n\ell \rangle = -\dfrac{2j-1}{2j+2} \overline{r^2}$ \hfill (3.26)

$Q_{sp} = 0$ for $j = 1/2$ and $Q_{sp} < 0$ for $j > 1/2$. $\overline{r^2} = <r^2> n\ell$.

It immediately follows that the quadrupole moment of a nucleus with closed proton shell is zero:

$$\sum_{m_j=-j}^{+j} Q(m_j) = \frac{\langle n\ell \mid r^2 \mid n\ell \rangle}{2j(j+1)} \left[\sum_{m_j} j(j+1) - 3 \sum_{m_j=-j}^{+j} m_j^2 \right]$$

$$= \frac{\langle n\ell \mid r^2 \mid n\ell \rangle}{2j(j+1)} \left[(2j+1)\, j(j+1) - 6 \sum_{m_j=+1/2}^{j} m_j^2 \right].$$

Since, $\displaystyle\sum_{m_j=1/2}^{j} m_j^2 = \left[(1/2)^2 + (3/2)^2 + \cdots\cdots j^2 \right],$

then $\displaystyle\sum_{m_j=-j}^{+j} Q(m_j) = \frac{\langle n\ell \mid r^2 \mid n\ell \rangle}{2j(j+1)} \left[(2j+1)\, j(j+1) - \frac{1}{6} 6j(2j+1)(j+1) \right] = 0.$

A rough estimate of $\overline{r^2}$ is r_0^2 where r_0 is the nuclear radius ($r_0 = a_0 A^{1/3}$ fm with a_0 ranging from 1.1 to 1.4). It is clear that Q_{sp} is non-vanishing when $j > 1/2$. Since $\overline{r^2}$ is positive definite, and $2j - 1 > 0$ for $j > 1/2$, it follows that quadrupole moment of single proton state is negative. This result can also be interpreted semi-classically as follows : In the present case angular momentum points in the Z direction because $m_j = j$. Hence, the particle orbit plane is perpendicular to the Z axis and the charge distribution of the particle will obviously be flattened (oblate) and give rise to negative quadrupole moment. The single particle estimates of quadrupole moment Q_{sp} obtained from Eq. (5.38) are given in Table 3.3 for the last odd proton (or proton hole) nuclides.

Table 3.3: Single particle quadrupole moment

Nucleus	ℓj	$\langle r^2 \rangle_{n\ell}$ (fm)	Q_{sp} (barn)	Q_{expt} (barn)
$^{17}_{9}F_8$	$d_{5/2}$	8.0*	−0.46	–
$^{39}_{19}K_{20}$	$d^{-1}_{3/2}$	13.0**	−0.052	0.057
$^{121}_{51}Sb_{70}$	$d_{5/2}$	35.3*	−0.201	−0.530
$^{209}_{83}Bi_{126}$	$h_{9/2}$	35.0**	−0.254	−0.350

* Estimated from $\overline{r^2} = r_0^2$ with $r_0 = a_0 \, A^{1/3}$ *fm* and $a_0 = 1.1\text{-}1.2$.

** Values of the radial matrix elements are taken from ref. 3.

According to the extreme single particle model, the odd neutron nuclei should have zero quadrupole moment. This is in contradiction with the observed quadrupole moment of the odd neutron nuclei. For example, the observed quadrupole moment of $^{17}_{8}O_9$ is −0.026 barn and is comparable to the quadrupole moment of odd proton nucleus $^{17}_{9}F_8$. Therefore, the observed quadrupole moments of odd neutron nuclei strongly suggests that (3.26) must be modified for the recoil effects of the core and polarizing deformation of the core induced by the single particle. The recoil effect arises from the fact that Q should be properly referred to the center of mass. The required recoil correction is $(\Delta Q)_{\mathrm{Rec}} \approx (Z_c/A_c^2) \, Q_{sp}$, where Z_C and A_C are respectively the charge and mass number of the core. A rough estimate of the polarization effect can be obtained by noting the volume eccentricity dependence on A. An additional quadrupole moment for each core proton results and the required contribution to quadrupole moment due to the core polarization is $(\Delta Q)_{pol} \approx \left(\dfrac{Z_c}{A}\right) Q_{sp}$. Non zero quadrupole moment due to the polarization of otherwise symmetric core of $Q_n \approx (Z/A) \, Q_p$, with the same sign of Q_p calculated for a single proton.

For a single proton hole, the quadrupole moment is $-Q_{sp}$, the opposite sign of quadrupole moment for a particle; hole corresponds to the fact that quadrupole moment operator, being a function of position coordinates, transforms under particle hole conjugation with phase $= -1$. The quadrupole moment of $^{39}_{19}K_{20}$ should be therefore $-(-0.052) = +0.052$ in good agreement with observed value (Table 3.3).

Although the periodic dependence of observed quadrupole moments shows the evidence of shell structure, single particle model gives magnitude of Q far less than the observed value. Comparison of observed Q_{expt} with Q_{sp} value indicates the following:

(1) Agreement of Q_{sp} with experiment is achieved for some nuclei which are immediately above or below closed shells (magic number ± 1)

(2) Odd neutron nuclei possess large quadrupole moment comparable to odd-proton nuclei instead of zero value predicted by single particle model.

(3) Most of nuclei possess +ve quadrupole moment contrary to single particle model predictions.

(4) Observed Q is large, sometime by a factor of ten from Q_{sp} for many nuclei, particularly those occurring in the middle of the shell.

(5) Large Q occurs for both odd Z and odd N.

Gamma Transitions: The total radiative transition rate for the electric mode of multiple order λ is given by (1.75a). Usually we are not interested in the orientation of either the initial or final nucleus, so we sum over M_f and averages over M_i.

We define the reduced matrix element for $\hat{Q}_{\lambda\mu}$ (1.75a) as

$$B\left(E\lambda, I_i \to I_f\right) = (2I_i + 1)^{-1} \sum_{M_i M_f} |\langle f | \hat{Q}_{\lambda\mu} | i \rangle|^2. \qquad (3.27a)$$

Hence,
$$T(E, \lambda\mu) = \frac{8\pi(\lambda+1)}{\lambda[(2\lambda+1)!!]^2} \frac{k^{2\lambda+1}}{\hbar} B(E\lambda). \qquad (3.27b)$$

Of course, only one value of $\mu = M_f - M_i$ contributes to each term.

Magnetic multipole transition probabilities are calculated in a manner quite similar to the electric multipole transition probabilities. For example,

$$T(M, \lambda\mu) = \frac{8\pi(\lambda+1)}{\lambda[(2\lambda+1)!!]^2} \frac{k^{2\lambda+1}}{\hbar} |\langle f | \hat{M}_{\lambda\mu} | i \rangle|^2, \qquad (3.28a)$$

where $\hat{M}_{\lambda\mu}$ is given by (1.76). When a single proton participates in the transition, then

$$M_{\lambda\mu} = \mu_0 \vec{\nabla}(r^\lambda Y_\lambda^\mu) \cdot \left(\frac{2}{\lambda+1} g_\ell \ \vec{\ell} + \mu_p \ \vec{\sigma}\right)$$

where μ_p is the magnetic moment of proton.

Now, $\vec{\nabla}(r^\lambda Y_\lambda^\mu) \cdot \vec{A} = [\lambda(2\lambda+1)]^{1/2} r^{\lambda-1} (Y^{\lambda-1}, A)_\mu^\lambda$, where A stands for $\vec{\ell}$ or $\vec{\sigma}$.

Hence, $\hat{M}_{\lambda\mu} = \mu_0 \left[\lambda(2\lambda+1)\right]^{1/2} r^{\lambda-1} \left[\frac{2}{\lambda+1} g_\ell (Y^{\lambda-1}, \ell)_\mu^\lambda + \mu_p (Y^{\lambda-1}, \sigma)_\mu^\lambda\right]$.

$$(3.28b)$$

Here, $g_\ell = 1$ for the proton and $g_\ell = 0$ for the neutron. Substituting $\lambda = 1$, in (3.28b) we get the expression for single particle magnetic moment operator for $\mu = 0$.

Weisskopf Estimate: In the extreme single particle model, we assume that only proton participates in the transition and the nucleus goes from a state having angular momentum λ to a state having angular momentum zero. The transition matrix element (dropping suffix i) is

$$\langle f \mid r^\lambda \ Y_\lambda^{\mu^*}(\theta, \varphi) \mid i \rangle = \int \sum \frac{w}{r} Y_0^0 \ \chi_{1/2}^{m_s} \ r^\lambda \ Y_\lambda^\mu (\theta, \phi) \ \frac{u}{r} Y_\lambda^m (\theta, \phi) \ \chi_{1/2}^{m_s} \ r^2 \ dr \, d\Omega .$$

$$(3.29)$$

where (r, θ, φ) are the coordinates of the single participating proton and the single particle model wave functions are expressed in terms of them.

The gamma ray energy $\hbar\omega$ is equal to the energy difference ΔE of two nuclear levels connected by the γ-transition. Therefore, the wave number $k \ (= w/c)$ of photon is given by $k = (\hbar c)^{-1} \Delta E$, where ΔE is in MeV. Thus, the γ-ray energy of 1 to 10 MeV yields value of k in the range 0.005 fm to 0.05 fm. The dimensionless radial coordinate kr should be of the order of kr_0, r_0 being the nuclear radius. Using $r_0 = 1.2 \, A^{1/3}$ fm, $E = 5$ MeV, $A = 216$, we get $kr_0 = 0.20$. Hence, $kr_0 < 1$. The main determining factors for transition rate are therefore multi polarity λ, the energy of the radiation $\hbar\omega = c\hbar k$ and the dimension of the system which enters through $(kr_0)^{2\lambda}$. The nuclear structure informations come through the reduced matrix element.

The radial integral is evaluated by assuming that the radial functions have constant value up to radius r_0 and zero outside. We denote $[u(r)/r] = [w(r)/r] = c$ (say) so that the condition

$$\int_0^{r_0} u^2 dr = \int_0^{r_0} w^2 dr = 1 \text{ yields } c = \left(\frac{3}{r_0^3}\right)^{1/2} .$$

Therefore, $\langle f \mid r^\lambda \ Y_\lambda^{\mu^*}(\theta, \phi) \mid i \rangle = \frac{1}{\sqrt{4\pi}} \frac{3}{r_0^3} \int_0^{r_0} r^{\lambda+2} \, dr = \frac{3}{\sqrt{4\pi}} \frac{r_0^\lambda}{\lambda+3}.$ (3.30)

The angular integration yields $\delta_{\mu m}$ and the spin summation yields $\delta_{m_s m_s'}$. Substituting the value of the matrix element, we get

$$T(E,\lambda\mu) = \frac{8\pi(\lambda+1)}{\lambda[(2\lambda+1)!!]^2} \frac{k^{2\lambda+1}}{\hbar} \frac{e^2}{4\pi} \left(\frac{3}{\lambda+3}\right)^2 r_0^{2\lambda}.$$

$$= \frac{2(\lambda+1)}{[(2\lambda+1)!!]^2}(kc)\left(\frac{e^2}{\hbar c}\right)(kr_0)^{2\lambda}\left(\frac{3}{\lambda+3}\right)^2. \tag{3.31a}$$

The single particle transition estimated by Weisskopf in this way is known as the Weisskopf estimate. The very strong dependence on λ values makes this estimate useful even though overlap integral may be off by a large factor. Sometime $T(E,\lambda\mu)$ is conveniently expressed as

$$T(E,\lambda\mu) = \frac{4.4(\lambda+1)(1.2)^{2\lambda}}{\lambda[(2\lambda+1)!!]^2}\left(\frac{3}{\lambda+3}\right)^2\left(\frac{E_\gamma}{197}\right)^{2\lambda+1} A^{2\lambda/3}(10^{21} \ sec^{-1}),$$

$$\tag{3.31b}$$

and the reduced transition probability according to (1.75a) and (3.30) is

$$B(E\lambda, I_i \rightarrow I_f) = \frac{(2I_f+1)}{(2I_i+1)}e^2\left(\frac{9}{4\pi}\right)\frac{(1.2)^{2\lambda}}{(\lambda+3)}A^{2\lambda/3} \ fm^{2\lambda}. \tag{3.32a}$$

Example: For $\lambda = 2$, $\left[B(E2, I_i \rightarrow I_f)/(e^2 \ fm^4)\right]_{sp} \approx 0.059 \frac{(2I_f+1)}{(2I_i+1)} A^{4/3}$

$$\tag{3.32b}$$

$$\left[B(E2, \ 0 \rightarrow 2)/(e^2 \ fm^4)\right]_{sp} \approx 0.295 \ A^{4/3}.$$

Table 3.4: Single particle estimates of reduced $E2$ transition rate

Nucleus	$(\ell j)_i$	$(\ell j)_f$	$B(E2)_{sp}$ $(e^2 \ fm^4)$	$B(E2)_{expt}$ $(e^2 \ fm^4)$
$^{17}_{8}O_9$	$s_{1/2}$	$d_{5/2}$	7.9	6.3
$^{41}_{20}Ca_{21}$	$p_{3/2}$	$f_{7/2}$	17.0	66.0
$^{209}_{82}Pb_{127}$	$s_{1/2}$	$d_{5/2}$	223.2	150.0

The upward transition rate $\left[B(E2, \ 0 \rightarrow 2)\right]_{up} = 5\left[B(E2, \ 2 \rightarrow 0)_{down}\right]$. In Table 3.4 we compare $B(E2, j_i \rightarrow j_f)_{sp}$ with the observed values for few odd A nuclides, assuming $I_i = j_i$ and $I_f = j_f$. In Table 3.5 we compare

$[B(E2, 0 \rightarrow 2)]_{sp}$ with the observed value for several rare earth and actinide nuclides.

We see from Table 3.4 that for closed shell plus one nucleon nuclei the single particle estimates of $E2$ transitions compare reasonably with the observed values. However, Table 3.5 shows that for the rare earth and actinide nuclei, $[B(E2, 0 \rightarrow 2)]_{sp}$ values are too small and mostly smaller by a factor of 100 with respect to the observed values. This indicates the collective participations of nucleons in the gamma transitions.

Table 3.5: Reduced single particle transition rates for the rare earth and actinide nuclei

Nucleus	Energy (KeV)	$[B(E2, 0 \rightarrow 2)]_{expt}$ $(e^2 \, fm^4)$	$[B(E2, 0 \rightarrow 2)]_{sp}$ $(e^2 \, fm^4)$
^{152}Sm	121.85	$(3.4 \pm 1.2) \times 10^4$	2.43×10^2
^{160}Gd	75.3	$(5.67 \pm 1.25) \times 10^4$	2.60×10^2
^{178}Hf	93.3	$(4.5 \pm 0.20) \times 10^4$	3.0×10^2
^{230}Th	53.0	$(7.9 \pm 0.8) \times 10^4$	4.2×10^2
^{238}U	44.0	$(11.9 \pm 1.5) \times 10^4$	4.42×10^2

To obtain Weisskopf estimate for the magnetic transition, we again assume that the nucleus goes from a state having angular momentum λ to a state having angular momentum zero. The radial matrix element

$$\langle f \mid r^{\lambda-1} \mid i \rangle = \frac{3}{\sqrt{4\pi}} \frac{r_0^{\lambda-1}}{\lambda+2}$$

is obtained by substituting $\lambda-1$ for λ in (3.30). The required expression

$$T(M, \lambda\mu) = \frac{2(\lambda+1)}{\lambda \left[(2\lambda+1)!! \right]^2} \frac{k^{2\lambda+1}}{\hbar} \frac{e^2}{4\pi} \left(\frac{3}{\lambda+3} \right)^2 r_0^{2\lambda-2}$$

$$\lambda \mid \mu_0 \left[\lambda(2\lambda+1) \right]^{1/2} \langle f \mid \{ \frac{2}{\lambda+1} g_\ell (Y^{\lambda-1},\sigma)_\mu^\lambda + \mu_p (Y^{\lambda-1},\sigma)_\mu^\lambda \} \mid i \rangle \mid^2$$

$$= \mu_0^2 \left[\lambda(2\lambda+1) \right] \frac{2(\lambda+1)}{\lambda \{ (2\lambda+1)!! \}^2} \left(\frac{3}{\lambda+2} \right)^2 \frac{k^{2\lambda+1}}{\hbar} r_0^{2\lambda-2}$$

$$\times \mid \langle f \mid \frac{2g_\ell}{\lambda+1} (Y^{\lambda-1}, j)_\mu^\lambda + \mu_p - \frac{g_\ell}{\lambda+1} (Y^{\lambda-1},\sigma)_\mu^\lambda \} \mid i \rangle \mid^2 \qquad (3.33)$$

writing $\vec{\ell} = \vec{j} - (1/2)\vec{\sigma}$.

$$T(M,\lambda \to 0) = \frac{2(\lambda+1)\,\lambda(2\lambda+1)}{\lambda\,[(2\lambda+1)!!]^2}\,(kr_0)^{2\lambda}\left(\frac{3}{\lambda+2}\right)^2\frac{k}{\hbar}\frac{1}{r_0^2}$$

$$\times\frac{e^2\hbar^2}{(2Mc)^2}\frac{1}{\lambda(2\lambda+1)}\left[\left(\mu_p\,\lambda-\frac{\lambda}{\lambda+1}\right)^2\right]\frac{4(\lambda+2)}{3}$$

$$=\frac{2(\lambda+1)}{\lambda[(2\lambda+1)!!]^2}(kr_0)^{2\lambda}\left(\frac{3}{\lambda+2}\right)^2\frac{e^2}{\hbar c}\left(\frac{\hbar/Mc}{r_0}\right)^2$$

$$\left(\frac{\omega\,r_0}{c}\right)^{2\lambda}\omega\times\left(\mu_p^\lambda-\frac{\lambda}{\lambda+1}\right)^2\,\sec^{-1}. \qquad (3.34)$$

The comparison of electric and magnetic radiation of same multipolarity yields

$$\frac{T(M\lambda)}{T(E\lambda)}\approx 10\left(\frac{\hbar}{Mr_0c}\right)^2\approx 0.3\,A^{-2/3}.$$

Similarly,
$$\frac{T(E\lambda+1)}{T(M\lambda)}=\frac{T(E\lambda+1)}{T(E\lambda)}\cdot\frac{T(E\lambda)}{T(M\lambda)}=\frac{10^{-4}}{(2\lambda+3)^2}\,A^{4/3}$$

and
$$\frac{T(E,\lambda+1\mu)}{T(E,\lambda\mu)}\approx\frac{(Kr_0)^2}{(2\lambda+3)^2}=\frac{3.6\times10^{-5}}{(2\lambda+3)}A^{2/3},$$

for $r_0 = 1.2\,A^{1/3}\,fm$ and $\hbar\omega = 1$ MeV (or $k = 0.005$ fm^{-1}).

Nuclear Isomerism: There is a large number of nuclei that are found in excited states with a half life ranging from about a microsecond to many years. These are of extremely long deviation on the nuclear time scale ($\sim 10^{-22}$ sec). These nuclei are known as isomers. They deexcite by the emission of γ-rays. The question arises, how γ-rays for which there is no Coulomb barrier can remain confined for such a long time within the nucleus. The phenomenon of nuclear isomerism can be understood within the framework of the single particle model.

We mentioned earlier that in the spin-orbit level scheme, a high spin state is depressed and is grouped with the low spin states, opposite in parity, but differing in energy only by a small amount. For example, $g_{9/2}$ belonging to $N = 4$ is pushed below and it comes in the vicinity of $p_{1/2}$, belonging to $N = 3$ and similarly $f_{7/2}$ belonging to $N = 3$ comes near the top of $d_{3/2}$, belonging to $N = 2$. If a gamma ray is now emitted from an excited state that has a high spin value to a lower state, usually the ground state, with low spin value, there is a large spin change involved in the low energy transition. According to (3.31) and (3.34), the large spin change leads to low transition probability and hence long life time.

Exercise 3.1. Calculate the mean square radius of a particle in the oscillator state with quantum numbers, n, ℓ.

For the classical harmonic oscillator potential $\langle KE \rangle = \langle PE \rangle$.

Hence, $E = 2\langle PE \rangle = 2\left(\dfrac{1}{2} \mu\omega^2 \langle r^2 \rangle \right)$. Employing (3.7) we get

$$\langle r^2 \rangle = \left(N + \frac{3}{2} \right) \frac{\hbar}{\mu\omega} = 2\left(2n + \ell - \frac{1}{2} \right) b^2 \text{, where } b = \sqrt{\frac{\hbar}{\mu\omega}}.$$

Exercise 3.2. Find the total number of nucleons from all shells $N = 0$ to $N = N_{max}$ being filled and, hence, calculate the spacings between oscillator levels.

We use the relation $N = 2(n-1) + \ell$, in which for N even, ℓ is even and for N odd, ℓ is odd. We set $N = 2i$ for $N = $ even and $N = 2i + 1$ for $N = $ odd. Now

$$N_{even} = \sum_{\ell} 2(\ell+1) = \sum_{i=0}^{N/2} 2\left\{2(2i)+1\right\} \quad \text{for } N = \text{even}$$

$$= 2\left[2\{1+5+\cdots 2N+1\} \right] = (N+1)\,(N+2).$$

$$N_{odd} = \sum_{i=0}^{(N-1)/2} 2\{2(2i+1)\ +1\} = \sum_{i=0}^{(N-1)/2} 2(4i+3) \quad \text{for } N \text{ odd}$$

$$= 2\left[3+7+\cdots 2N+1 \right] = (N+1)(N+2).$$

Hence, $$A = \sum_{N=0}^{N_{max}} 2(N+1)(N+2) = \sum_{N=0}^{N_{max}} (2N^2+6N+4)$$

$$= \frac{2}{3}\left[(N_{max}+1)(N_{max}+2)(N_{max}+3) \right]$$

$$= \frac{2}{3}(N_{max}+2)^3 - (N_{max}+2)] \approx \frac{2}{3}(N_{max}+2)^3.$$

The total energy

$$E = \hbar\omega \sum_{N=0}^{N_{max}} 2(N+1)(N+2)(N+3/2)$$

$$= \frac{1}{2}\hbar\omega(N_{max}+2)^4 = \frac{1}{2}\left(\frac{3}{2}A\right)^{4/3}\hbar\omega.$$

Alternatively,

$$E = A\left[\mu\cdot\omega^2\langle r^2\rangle\right] = A^{5/3}\frac{938 \text{ MeV } (\hbar\omega)^2}{(197)^2 \text{ MeV}^2 fm^2}\times\frac{3}{5}\times 1.44 \text{ fm}^2.$$

Hence, $\qquad \hbar\omega = 41 \, A^{-1/3} \text{ MeV}.$

Exercise 3.3. The single particle potential is expressed as a mixture of oscillator potential plus spin-orbit and ℓ^2 terms in the form

$$V = V_{h\cdot o}(r) - \alpha \, \vec{\ell}\cdot\vec{s} + \beta\ell^2.$$

Show that in a given oscillator shell N, the single particle energies are

$$E_{j=\ell+1/2}(\alpha,\beta) = E_{h\cdot o}(N) - \alpha\frac{\ell}{2} + \beta\ell(\ell+1)$$

and

$$E_{j=\ell+1/2}(\alpha,\beta) - E_{j'=\ell+2+1/2}(\alpha,\beta) = [\alpha - \beta(4\ell+6)].$$

Apply the above relations to the ^{17}O spectrum given below:

$E_{d_{5/2}} = -4.15 \text{ MeV}$, $E_{d_{3/2}} = 0.93 \text{ MeV}$ and $E_{s_{1/2}} = -3.28 \text{ MeV}$, and determine the value of α and β.

Using the relation (3.14a) and that ℓ^2 is diagonal in $|jm_j\rangle$ representation, we can write:

$$E_{j=\ell+1/2}(\alpha,\beta) = E_{h\cdot o}(N) - \alpha\frac{\ell}{2} + \beta\ell(\ell+1), \qquad (1)$$

$$E_{j=\ell+2+1/2}(\alpha,\beta) = E_{h\cdot o}(N)$$

$$- \frac{\alpha}{2}\left[\left(\ell+\frac{5}{2}\right)\left(\ell+\frac{7}{2}\right) - (\ell+3)(\ell+2) - \frac{3}{4}\right] + \beta(\ell+2)(\ell+3). \qquad (2)$$

Employing 3.15 and the ^{17}O levels we get

$$E_{j=\ell+1/2} - E_{j=\ell-1/2} = -\frac{\alpha}{2}\,(2\ell+1) = -\frac{5\alpha}{2} = -4.15 - 0.93 = -5.08 \text{ MeV},$$

which yields $\alpha \cong 2$ MeV. Subtracting Eq. 2 from Eq. 1 we get

$$E_{j=\ell+1/2}(\alpha,\beta) - E_{j'=\ell+2+1/2}(\alpha,\beta) = [\alpha - \beta(4\ell+6)].$$

According to the given data: $\alpha - 6\beta = -3.28 - (-4.15) = 0.87$ so that $\beta \cong 0.19$.

Exercise 3.4. Assuming n-p residual interaction to have a mixture of Wigner and Bartlett forces, show that $I = |j_n - j_p|$ corresponds to the lower state confirming Nordheim's strong rule.

Let n–p interaction has the form:

$$V_{np} = -V_0\left[(1-\alpha) \;+\; \alpha\,\vec{\sigma}_n \cdot \vec{\sigma}_p\right]\delta(r_n - r_p),$$

where α is a positive constant lying between 0 and 1. If the n-p system is in the triplet state ($S = 1$), then

$$\langle V_{np}\rangle = -V_0\left[(1-\alpha) + \alpha\langle x_1^1 |\,\vec{\sigma}_n \cdot \vec{\sigma}_p\, | x_1^1\rangle\right]\delta(r_n - r_p) = -V_0(r_n - r_p).$$

If the n-p system is in the singlet state $(S=0)$

$$\langle V_{np}\rangle = -V_0(1-4\alpha)\delta(r_n - r_p).$$

Therefore, if the net spin exchange force is attractive, then $S = 1$ (triplet) interaction is stronger than $S = 0$ (singlet) interaction. Now, $S = 1$ state is obtained when j_n and j_p are oppositely directed in accordance with the Pauli principle. Hence, $I = |j_n - j_p|$ assignment conforms Nordheim's strong rule.

Exercise 3.5. Show that the magnetic moment of a 'hole' state is same as that of a single particle state.

If $| jm\rangle$ represents a single particle state then $|j-m\rangle$ represents the hole state. Had the state $|j-m\rangle$ be present in the occupied state we would have obtained zero value for the magnetic moment because for closed levels it is zero. Since the contribution of this state is absent, the results must be given by its negative. We employ the Wigner-Eckert theorem with μ as a tensor of rank 1. Hence,

$$-\langle n\ell j - m |\,\mu_z\,| n\ell j - m\rangle = -\langle j1 - m0 | j - m\rangle\,\langle n\ell j \,\|\mu_z\| n\ell j\rangle$$

$$= -(-)^{j+1-j}\langle j1m0 | jm\rangle\,\langle n\ell j \,\|\,\mu_z\,\| n\ell j\rangle = \langle n\ell jm |\,\mu_z\,| n\ell jm\rangle.$$

Thus, the magnetic moment of the hole state is the same as that of a particle state. Following the same argument, we can write

$$-\langle n\ell j - m \mid \hat{Q} \mid n\ell j - m \rangle$$

$$= -\langle j2 - m0 \mid j - m \rangle = \langle n\ell j \parallel Q \parallel n\ell j \rangle, \text{ since } \hat{Q} \text{ is tensor of rank 2}$$

$$= -(-)^{j+2-j} \langle j2m0 \mid jm \rangle \langle n\ell j \parallel Q \parallel n\ell j \rangle = -\langle n\ell jm \mid Q \mid n\ell jm \rangle.$$

Exercise 3.6. Find the quadrupole moment of a nucleus which has 'k' protons in the unfilled shell.

For an odd number of k protons, the angular momentum of $(k-1)$ protons saturate. The unpaired proton then determine the angular momentum of the nucleus. We choose $m_j = j$ for this proton. Since the wave function must be antisymmetric in all the protons and $m_j = j$ is no longer to be considered for the paired protons, then these protons must be distributed with equal weights over the quantum numbers $m_j = -j+1, \cdots\cdots j-1$, that is, over $2j-1$ states (excluding $m_j = j$ and $m_j = 0$). Hence,

$$Q = Q_{sp} + \frac{k-1}{2j-1} \frac{\langle n\ell \mid r^2 \mid n\ell \rangle}{2j(j+1)} \left[\sum_{m_j=-j+1}^{j-1} j(j+1) - 3 \sum_{m_j=-j+1}^{j-1} m_j^2 \right],$$

following (3.25). Now, $\displaystyle\sum_{m_j=-j+1}^{j-1} m_j^2 = \frac{2}{3} j(j+1/2)(j+1) - 2j^2$. Therefore,

$$Q = Q_{sp} + \frac{k-1}{2j-1} \langle r^2 \rangle_{n\ell} \left[j(j+1)(2j-1) - 3 \cdot \frac{2}{3} j(j+1)(j+1/2) + 6j^2 \right]$$

$$= Q_{sp} + \frac{k-1}{2j-1} \langle r^2 \rangle_{n\ell} [2j(2j-1)]/2j(j+1)$$

$$= Q_{sp} + \frac{k-1}{2j-1}(-2Q_{sp}) = Q_{sp} \left[1 - \frac{2(k-1)}{2j-1} \right].$$

3.2 WAVE FUNCTION OF A NUCLEUS

We have seen the inadequacies of the extreme single particle shell model. We shall now consider some refinements by allowing interaction between particles in a partially filled shell and treating completely filled shells as inert core. We thus attribute all the low energy properties of nuclei to their valence nucleons outside the closed shell. However, we assume that the interactions of these valence nucleons amongst themselves do not perturb them appreciably from the single particle orbits described by quantum numbers $n\ell j$. Hence, this is called the extended single particle model.

Let us denote k particles in $(n\ell j)$ shell by $(n\ell j)^k$. When we disregard the interaction between particles, all the different states which can be formed by k-particles with the same $n\ell j$, have the same energy. This degeneracy is removed by considering the interaction between particles. This interaction is assumed to be strong enough to remove degeneracy but not so strong compared to spin-orbit interaction that 'j' ceases to be a good quantum number. It is sometime possible to relax this approximation taking the particle state to be a superposition of wave function of few $(n\ell j)$ states whose energies are close to one another. Such a procedure is called the "configuration mixing". The force which removes the degeneracy of shell model states is generally referred as the "residual interaction". The bulk of the nuclear properties indicates that the interaction between nucleons inside nucleus is predominantly a two-body interaction V_{ij} and we are required to solve the non-relativistic many-body Schrodinger equation

$$H\,\psi_n(\vec{r}_1, \vec{r}_2 \cdots\cdots \vec{r}_A) = E_n\,\psi_n(\vec{r}_1, \vec{r}_2 \cdots\cdots \vec{r}_A) \qquad (3.35)$$

where $H = \sum_{i=1}^{A} T_i + \sum_{i<j} V_{ij}$, and solution of (3.35) should yield infinite number

of discrete levels E_n. However, (3.35) can not be solved in a straight forward manner. We rewrite H as

$$H = \sum_i (T_i + V_c) + \sum_{i<j}(V_{ij} - V_c), \qquad (3.36a)$$

where V_C is generally the average (mean field) of V_{ij} operating between a particular valence nucleon and nucleons in the closed shells. We again write

$$H = H_0 + H_1, \qquad (3.36b)$$

where H_1 is the residual interaction and H_0 is the single particle Hamiltonian, which contains single particle potential (also called auxiliary potential), for which S. Eq. gives a set of single particle states ϕ_α, ϕ_β, ϕ_γ, $\cdots$ ϕ_ξ with corresponding eigenvalues $\in_\alpha$, $\in_\beta$, $\in_\gamma$, $\cdots$ $\in_\xi$ respectively, satisfying

$$H_0\,\phi_\alpha = \in_\alpha \phi_\alpha \text{ and } \langle \phi_\alpha | \phi_\beta \rangle = \delta_{\alpha\beta}. \qquad (3.37)$$

The A-particle product wave function

$$\phi_{\alpha\beta\gamma\ldots}(\vec{r}_1, \vec{r}_2, \vec{r}_3 \cdots r_A) = \phi_\alpha(\vec{r}_1)\phi_\beta(\vec{r}_2) \cdots\cdots \phi_\xi(\vec{r}_A), \qquad (3.38)$$

where α, β, $\gamma \cdots \xi$ are the single particle states and $\vec{r}_1$, $\vec{r}_2$, $\vec{r}_3 \cdots \vec{r}_A$ are the coordinates of A particles. We have assumed that there are as many states as there are number of particles. The product wave function satisfies many-body S. Eq.

$$H_0 \, \phi_{\alpha,\beta,\gamma \,\ldots}(\vec{r}_1,\vec{r}_2,\cdots\vec{r}_A) = E_0 \; \phi_{\alpha,\beta,\cdots\xi}(\vec{r}_1,\vec{r}_2,\cdots\vec{r}_A),$$

where
$$E_0 = \epsilon_\alpha + \epsilon_\beta \cdots \epsilon_\xi. \tag{3.39}$$

A. Determinental Wave Function

Now we incorporate in the many-body unperturbed wave function the requirement of antisymmetry, i.e. the wave function must change sign under the exchange of any two nucleons, as demanded by the Pauli principle. First, we consider the case of two nucleons, for which the normalized antisymmetric wave function can be written as

$$\Phi_{\alpha\beta}(\vec{r}_1,\vec{r}_2) = \frac{1}{\sqrt{2}}\left[\phi_\alpha(\vec{r}_1)\phi_\beta(\vec{r}_2) - \phi_\alpha(\vec{r}_2)\phi_\beta(\vec{r}_1)\right]$$

$$= \frac{1}{\sqrt{2}}\left[1 - P(1,2)\right]\phi_\alpha(\vec{r}_1)\phi_\beta(\vec{r}_2),$$

where $P(1,2)$ exchanges the coordinates associated with the nucleons. $\phi_{\alpha\beta}$ also can be expressed in the form of a determinant and $P(1,2)$ amounts to interchanging two rows of the determinant:

$$\Phi_{\alpha\beta}(\vec{r}_1,\vec{r}_2) = \frac{1}{\sqrt{2}}\begin{vmatrix}\phi_\alpha(r_1) & \phi_\alpha(r_2) \\ \phi_\beta(r_1) & \phi_\beta(r_2)\end{vmatrix}.$$

The normalization of $\phi_{\alpha\beta}(\vec{r}_1,\vec{r}_2)$ is guaranteed by the orthogonality of the single-particle wave function ϕ_α and ϕ_β (*see* Eq. 3.37). Similarly, a normalized A particle wave function is defined by the Slater determinant

$$\Phi_{\alpha\beta\cdots\xi} = \frac{1}{\sqrt{A!}}\begin{vmatrix}\phi_\alpha(\vec{r}_1) & \phi_\alpha(\vec{r}_2) & \cdots & \phi_\alpha(\vec{r}_A) \\ \phi_\beta(\vec{r}_1) & \phi_\beta(\vec{r}_2) & \cdots & \phi_\beta(\vec{r}_A) \\ \vdots & \vdots & \vdots & \vdots \\ \phi_\xi(\vec{r}_1) & \cdots & \cdots & \phi_\xi(\vec{r}_A)\end{vmatrix}$$

$$= \frac{1}{\sqrt{A!}}\sum_P \pi(P)\left[\phi_\alpha(\vec{r}_1)\,\phi_\beta(\vec{r}_2)\cdots\phi_\xi(\vec{r}_A)\right], \tag{3.40}$$

where p stands for the number of permutation ($= A!$) and the sum is over all permutations of A objects: $\left[\phi_\alpha(\vec{r}_1)\cdots\phi_\xi(\vec{r}_A)\right] \cdot \pi(P) = +1$ for even permutations and -1 for odd permutations.

B. Expectation Value of Single-Particle Operator

We consider one-body operators. Due to indistinguishability of nucleons, these operators are symmetric in all the nucleon variables. A general one-body operator can be expressed as $\hat{O} = \sum_i f(r_i)$. We now consider

$$\langle \phi_{\alpha\beta\gamma\cdots\xi} \mid \sum_i f(r_i) \mid \phi_{\alpha\beta\gamma\cdots\xi} \rangle = \frac{1}{A!} \langle \phi_\alpha(\vec{r}_1)\phi_\beta(\vec{r}_2)\cdots \phi_\xi(\vec{r}_A) \mid \sum_p \pi(P)P\hat{O}$$

$$\sum_{p'} \pi(P')P' \mid \phi_\alpha(\vec{r}_1)\,\phi_\beta(\vec{r}_2)\cdots\phi_\xi(\vec{r}_A)\rangle.$$

$$\sum_p \langle \phi_\alpha(\vec{r}_1)\phi_\beta(\vec{r}_2)\cdots\phi_\xi(\vec{r}_A)\,)\pi(P)P\hat{O}$$

$$\sum_{p'} \pi(P')P' \mid \phi_\alpha(\vec{r}_1)\phi_\beta(\vec{r}_2)\cdots\phi_\xi(\vec{r}_A)\rangle.$$

Now employ the standard result: $\sum_{pp'} \pi(P)\pi(P')\,PP' = \sum_{p'} \pi(P')\,P'$ and get

$$\langle \phi_{\alpha,\beta,\cdots\xi} \mid \hat{O} \mid \phi_{\alpha,\beta,\cdots\xi} \rangle = \langle \phi_\alpha(\vec{r}_1)\,\phi_\beta(\vec{r}_2)\cdots\phi_\xi(\vec{r}_A) \mid \sum_i f(r_i) \mid \sum_{p'} \pi(P')\,P' \mid$$

$$\phi_\alpha(\vec{r}_1)\,\phi_\beta(\vec{r}_2)\cdots\phi_\xi(\vec{r}_A)\rangle. \tag{3.41}$$

Here, non-vanishing result can be obtained only for $P' = 1$ term, since all the single particle states are orthogonal. Thus,

$$\langle \phi_{\alpha,\beta,\cdots\xi} \mid \hat{O} \mid \phi_{\alpha,\beta,\cdots\xi} \rangle = \sum_{\alpha=1}^{A} \langle \alpha \mid f \mid \alpha \rangle, \tag{3.42}$$

where α stands for any of the single particle states of the set. As an example, let us consider the density operator $\hat{\rho}(r) = \sum_{i=1}^{A} \delta(\vec{r} - \vec{r}_i)$. Then the nucleon density in a nucleus

$$\rho(r) = \langle \phi_{\alpha,\beta\cdots\xi} \mid \sum_{i=1}^{A} \delta(\vec{r} - \vec{r}_i) \mid \phi_{\alpha,\beta\cdots\xi} \rangle$$

$$= \int \phi_\alpha^*(r_1)\phi_\beta^*(r_2)\cdots\phi_\xi^*(r_A) \left| \sum_{i=1}^{A} \delta(\vec{r} - \vec{r}_i) \right| \phi_\alpha(r_1)\phi_\beta(r_2)\cdots\phi_\xi(r_A)\,d\tau_A.$$

$$= \sum_{\alpha=1}^{A} \mid \phi_\alpha(r)\mid^2, \text{ using properties of the delta function.} \tag{3.43}$$

Next, employing the harmonic oscillator radial wave functions given in Table 3.1, we find the density of $_2^4He_2$, $_6^{12}C_6$. The radial and the spin-angle components of the single particle wave function (ϕ) are respectively given in Table 3.1 and Eq. (3.13). The explicit expansions are:

$$^4He: \varphi_{1/2\ 1/2} = \frac{u_{10}}{r}\ Y_0^0\ \chi_{1/2}^{1/2}, \qquad \varphi_{1/2-\ 1/2} = \frac{u_{10}}{r}\ Y_0^0\ \chi_{1/2}^{-1/2},$$

$$^{12}C: \varphi_{3/2\ 3/2} = \frac{u_{11}}{r}\ Y_1^1\ \chi_{1/2}^{1/2}, \qquad \varphi_{3/2-\ 3/2} = \frac{u_{11}}{r}\ Y_0^{-1}\ \chi_{1/2}^{-1/2},$$

$$\varphi_{3/2\ 3/2} = \frac{u_{11}}{r}\left[\sqrt{\frac{2}{3}}\ Y_1^0\ \chi_{1/2}^{1/2} + \frac{1}{\sqrt{3}}\ Y_1^1\ \chi_{1/2}^{-1/2}\right].$$

Hence, $[\rho(r)]_{4_{He}} = 2\left(\frac{4}{\sqrt{\pi}b^3}\right)e^{-r^2/b^2}\left(\frac{1}{2\pi}\right)$, (*see* Table 3.1 for $\frac{u_{n\ell}(r)}{r}$; the

factor 2 accounts for both the neutrons and protons).

$$[\rho(r)]_{12_C} = 4\left(\frac{1}{\pi b^2}\right)^{3/2} + 2\left(\frac{8}{3\sqrt{\pi}b^5}\right)$$

$$\left\{|Y_1^1|^2 + \frac{2}{3}\ |Y_1^0|^2 + \frac{1}{3}\ |Y_1^{-1}|^2 + \frac{2}{3}\ |Y_1^0|^2\ r^2 e^{-r^2/b^2}\right\},$$

the cross terms vanish due to the orthogonality of the spin functions. We use the general expression for summing the spherical harmonics,

i.e. $\displaystyle\sum_{m=-\ell}^{+\ell} |Y_\ell^m|^2 = \frac{2\ell+1}{4\pi}.$

Hence, $\qquad [\rho(r)]_{12_C} = 4\left(\frac{1}{\pi b^2}\right)^{3/2} + e^{-r^2/b^2}\left[1 + \frac{4}{3}\ \frac{r^2}{b^2}\right].$ \hfill (3.44)

Configuration Mixing

One physically plausible approximation employed so far is to treat closed shell as inert core, the presence of which simply provides a one-body potential for each of the remaining 'k' valence nucleons. When the number of possible shell configurations available for the valence nucleons is limited, one speaks of truncating the space of configuration. When more than just immediate unfilled shell (and occasionally subshell) is considered, one speaks of configuration mixing.

H_1 is the residual interaction (3.36b), which is small because V_C should cancel in an average way V_{ij} as much as possible. Hence, it is the left over part of the two-body interaction between particles. The effect of the residual force can be considered through perturbation approach. We are interested in some low lying energy levels of the nucleus which are observed experimentally. The wave function ψ_n of a nuclear state energy E_n could be written

$$\psi_n = \sum_\alpha C_\alpha^n \, \varphi_\alpha^n , \tag{3.45}$$

where ϕ's are the eigen states of eigen value (ϵ_α) of the unperturbed average Hamiltonian H_0. The C_α^n are the configuration mixing coefficients. One could get the eigen values and eigen states of the system from the secular equation:

$$\sum_\beta (\epsilon_\alpha \, \delta_{\alpha\beta} + \langle \phi_\alpha | H_1 | \phi_\beta \rangle) C_\beta^n = E_n C_\alpha^n . \tag{3.46}$$

Reasonable amount of configuration mixing may alter the energy of neighbouring states. But the prediction of the ground state spin arrived in the previous section (Sec. 3.1), normally does not change. However, the ground state moments may appreciably change when we give due consideration to both the neutrons and protons outside the closed core by simply considering the admixture of various magnetic substates. (*see* Exercise 3.8).

Exercise 3.7 Calculate the mean square radius of the charge distribution of the helium nucleus and compare it with the value obtained from the electron scattering measurements :

$$\langle r^2 \rangle_{4_{He}} = \frac{Z}{A} \int \left[\rho(r) \, r^2 \right] r^2 \, dr \, d\Omega$$

$$= \frac{2}{4} \cdot 4 \left(\frac{1}{\pi b^2} \right)^{3/2} b^5 \int_0^\infty x^4 \, e^{-x^2} \, dx, \text{ where } \frac{r}{b} = x.$$

We use the standard integral $\displaystyle\int_0^\infty x^{2n} \, e^{-ax^2} \, dx = \frac{1.3.5 \cdots\cdots (2n-1)}{2^{n+1} \, a^n} \sqrt{\frac{\pi}{a}}$.

Therefore,
$$\langle r^2 \rangle = \frac{3}{2} \, b^2 = \frac{3}{2} \frac{\hbar}{\mu\omega} = \frac{1.5(\hbar c)^2}{\mu c^2 \hbar\omega}.$$

Substituting $(\hbar c) = 197$ MeV fm, $\mu c^2 = 938$ MeV, $\hbar\omega = 25.625$, we get:

$$\langle r^2 \rangle_{sp} = 2.42 \ \text{fm}^2 \text{ whereas, } \langle r^2 \rangle_{expt} = 2.56 \, \text{fm}^2.$$

The agreement between the theory and experiment is reasonably good.

Exercise 3.8. Calculate ^7Li ground state magnetic moment and compare the same with Schmidt's and observed values.

The outermost orbit of $^7_3Li_4$ has two neutrons and a proton and the configuration is $(1s_{1/2})^2 (1p_{3/2})^2; (1s_{1/2})^2 (1p_{3/2})^1$. The g.s. spin of (^7Li) according to the extreme single particle model is 3/2, which is also the observed value.

The explicit form of g.s. wave function for which the total angular momentum I (=3/2) has maximum projection M (=3/2), can be expressed in terms of single particle wave functions:

$$\Phi_{3/2,3/2} = a\,|\,\varphi_{3/2\ 3/2}(\pi)\varphi_{3/2\ 3/2}(\nu)\varphi_{3/2-\ 3/2}(\nu)\rangle$$

$$+\,b\,|\,\varphi_{3/2\ 3/2}(\pi)\ \varphi_{3/2\ 1/2}(\nu)\ \varphi_{3/2-\ 1/2}(\nu)\rangle$$

$$+\,c\,|\,\varphi_{3/2\ 3/2}(\nu)\ \varphi_{3/2\ 1/2}(\pi)\ \varphi_{3/2-\ 1/2}(\nu)\rangle$$

$$+\,d\,|\,\varphi_{3/2\ 3/2}(\nu)\ \varphi_{3/2\ 1/2}(\nu)\ \varphi_{3/2-\ 1/2}(\pi)\rangle,$$

where, π and ν denote proton and neutron states respectively. Φ satisfies two conditions:

$$J_+\,\Phi_{3/2,3/2} = \sum_{i=1}^{3} j_+^{(i)}\,\Phi_{3/2,\ 3/2} = 0\ ,$$

$$T_+\,\Phi_{3/2,3/2} = \sum_{i=1}^{3} t_+^{(i)}\,\Phi_{3/2,\ 3/2} = 0.$$

We employ the property of the raising operator namely,

$$j_+\,\varphi_{jm_j} = \left\{ j(j+1) - m_j(m_j+1) \right\}^{1/2}\,\varphi_{jm_j+1}$$

and a similar relation for t_+. Hence,

$$[j_+^{(1)} + j_+^{(2)} + j_+^{(3)}]\,\{a\,|\,\phi_{3/2\ 3/2}(\pi)\phi_{3/2\ 3/2}(\nu)\phi_{3/2-\ 3/2}(\nu)\rangle$$

$$+\,b\,|\,\phi_{3/2\ 3/2}(\pi)\ \phi_{3/2\ 1/2}(\nu)\phi_{3/2-\ 1/2}(\nu)\rangle$$

$$+\,c\,|\,\phi_{3/2\ 3/2}(\nu)\ \phi_{3/2\ 1/2}(\pi)\ \phi_{3/2-\ 1/2}(\nu)\rangle$$

$$+\,d\,|\,\phi_{3/2\ 3/2}(\nu)\phi_{3/2\ 1/2}(\nu)\ \phi_{3/2-\ 1/2}(\pi)\rangle\} = 0$$

or $\qquad \sqrt{3}(a+b-c)\,|\,\phi_{3/2\ 3/2}(\pi)\ \phi_{3/2\ 3/2}(\nu)\ \phi_{3/2-\ 1/2}(\nu)$

$$+\,2(c-d)\,|\,\phi_{3/2\ 3/2}(\nu)\ \phi_{3/2\ 1/2}(\pi)\ \phi_{3/2\ 1/2}(\nu)\rangle = 0.$$

Therefore, $a + b - c = 0$ and $c - d = 0$. $T = 1/2$ gives another condition, $b + c + d = 0$. The normalization of ψ demands: $a^2 + b^2 + c^2 + d^2 = 1$. Thus, we have four equations to determine four constants. We find $a = \dfrac{3}{\sqrt{15}}$, $b = \dfrac{-2}{\sqrt{15}}$, $c = d = \dfrac{1}{\sqrt{15}}$ and express

$$\psi_{3/2\ 3/2}\left(T = 1/2\right) = \frac{1}{\sqrt{15}}\left[3\,|\,\varphi_{3/2}\left(\pi\right)\,\varphi_{3/2\ 3/2}\left(\nu\right)\,\varphi_{3/2 - 3/2}\left(\nu\right)\right.$$

$$+ 2\,|\,\varphi_{3/2\ 3/2}\left(\pi\right)\varphi_{3/2\ 1/2}\left(\nu\right)\,\varphi_{3/2 - 1/2}\left(\nu\right)\rangle$$

$$+ |\,\varphi_{3/2\ 3/2}\left(\nu\right)\,\varphi_{3/2\ 1/2}\left(\pi\right)\varphi_{3/2 - 1/2}\left(\nu\right)\rangle$$

$$\left. + |\,\varphi_{3/2\ 3/2}\left(\nu\right)\,\varphi_{3/2\ 1/2}\left(\nu\right)\,\varphi_{3/2 - 1/2}\left(\pi\right)\rangle\right].$$

We employ (Eq. 3.21a) and the above wave function to calculate the magnetic moment of ^7Li:

$$\mu_{theo} = \frac{1}{15}[(9 + 4)\,\mu_j^p + \mu_j^n],$$

where $\mu_j^p = [(j - 1/2) + \mu_p]\,nm = 3.79$ and $\mu_j^n = \mu_n = -1.91$ nm.

Finally, $\mu_{theo.} = 3.15$ nm, $\mu_{expt.} = 3.03$ nm and $\mu_{schmidt} = 3.79$ nm.

3.3 SHELL MODEL SPECTROSCOPY

In the preceding section we treated at length single particle shell model and made attempts to describe some ground state properties of nuclei. Now we shall examine the excited states of nuclei.

A starting point for the shell model spectroscopy for nucleus with A nucleons is to consider the non-relativistic S. Eq. (3.35), in which E_n, ψ_n are the eigen values and eigen functions for a state n in the Hilbert space. The first step towards a solution of A nucleon problem is to reduce the infinitely many degrees of freedom of the full Hilbert space to those represented by a physically motivated subspace. In shell model calculation we take closed shell as inert core and consider the remaining valence nucleons (or holes) restricted to the model space of tractable size. More formally one defines the projection operators: $P = \displaystyle\sum_{i=1}^{m} |\psi_i\rangle\,\langle\psi_i\,|,\ Q = \displaystyle\sum_{i=1+m}^{\infty} |\psi_i\rangle\,\langle\psi_i\,|.$

The operator P projects onto the chosen reduced shell model space (or P-space), $|\psi_i\rangle$ being the basis functions spanning this space and m the dimension of the space. The operator Q defines the complement of P, $Q = 1 - P$, which projects out of the shell model space. The basic shell model problem is then to obtain an effective Hamiltonian $H_{eff} = H_0 + H_1$, acting on P-space defined by

$$P H_{eff} P \, |\psi_n\rangle = P(H_0 + H_1) \, P \, |\psi_n\rangle$$

$$= P\{(H_0 + H_1)\} \sum_{i=1}^{m} |\psi_i\rangle \, \langle \psi_i | \psi_n \rangle$$

$$= P\left\{ \sum_{i=1}^{m} E_i \, \psi_i \, \delta_{in} \right\} = P \, E_n \, \psi_n = E_n \, P \, \psi_n \qquad (3.47)$$

where again E_n is the nth eigen state of the original Hamiltonian H in (3.35).

Thus, the eigen functions of H_{eff} are projections of the true eigen functions of H on to the P-space. The H_{eff} includes a one-body term PH_0P, which represents the mean field degrees of freedom of the chosen P-space and an effective interaction PH_1P acting among the nucleons in P-space.

In order to solve eigen value problem (3.47) we need to find the effective interaction PH_1P. However, in actual cases the dimension m could be very large. With the advent of high speed computers 'large' shell model calculations can be performed in which upto 10,000 basis states can be included.

Now we shall describe the procedure to calculate binding and excitation energies of a nucleus within the framework of the extended single particle shell model. We begin with a phenomenological approach and use the standard first order perturbation theory. The nucleus is described by an inert closed shell core and few nucleons outside the core. Generally the following procedure is adopted[5]:

(1) We first define a set of active single particle orbits based on either physical requirement or on technical limitations. This restricted set of states of nucleons are guided by the single particle shell model. A judicious choice of the model space may result in a big benefit in reducing the computational labour for the calculation of nuclear excited states.

(2) Next, we assume a coupling scheme to couple the angular momentum of the particles in the unfilled shell. Mostly j-j coupling scheme is used. However, for light nuclei L-S coupling may be used.

(3) Thereafter, we choose a reasonable form of the nucleon-nucleon interaction, $\sum_{i<j} V_{ij}$. However, for simplicity we do not deal with problem separating residual interaction from the two-nucleon interaction.

(4) Next, we calculate the Two-Body Matrix Elements (TBME), construct the energy matrix and diagonalize it to give the excited states.

(5) Finally, we extend the calculation for more number of states and more number of particles and check the consistency of procedure.

Example: Let there be 'k' active nucleons outside the close shell and they be represented by $(n\ell j)^k$. Neutrons and protons are treated as one and the same particles in "isospin" formalism. We first consider $k = 2$ system, i.e. two particles outside the close shell. For example, in ^{18}O, there are two neutrons which could be restricted to $(1d_{5/2})$ levels as model space outside the close core of ^{16}O. Or one may also choose a large model space which corresponds to ^{16}O core and two nucleons in the entire $(2s\ 1d)$ shell. The second choice is physically more meaningful, because spectrum of ^{17}O indicates that $2s_{1/2}$ level is very close to the g.s. level $1d_{5/2}$ (Fig. 3.4).

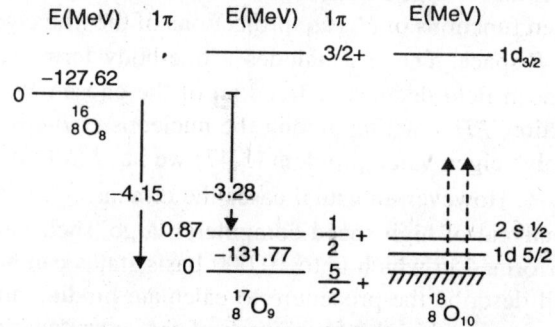

Fig. 3.4: Low lying levels of oxygen isotopes

Coupling rules for $(n\ell j)^k$ configuration for $k = 2$ will now be described. The Pauli principle requires that the wave function for a system of identical particles be antisymmetric under the interchange of all coordinates of any pair of particles. This antisymmetrization put certain restriction on the state which can be formed for $(n\ell j)^k$ configuration. We use j-j coupling and construct the antisymmetric wave function for a system of two particles.

Let the particles occupy the single particle states $|j_1 m_1|$ and $|j_2 m_2\rangle$. The coupled wave function of total angular momentum I and projection M is

$$|j_1 j_2 IM\rangle = \sum_{m_1 m_2} \langle j_1 j_2 m_1 m_2 | IM\rangle \, |j_1 m_1\rangle \, |j_1 m_2\rangle \text{ where } \langle 1 \rangle \text{ is the 'C' coefficient.}$$

Two particles may have $T = 0$ or 1. Since the radial function is symmetric, it is sufficient to consider spin-isospin function in the discussion of symmetry. To identify the particles we adopt the convention that quantum numbers of 'particle 1' are written to the left of the quantum number of the 'particle 2'. When particles are exchanged

$$| j_2 j_1 IM \rangle = | j_1 j_2 IM \rangle_{ex} = \sum_{m_1 m_2} \langle j_2 j_1 m_2 m_1 | IM \rangle \, |j_2 m_2 \rangle \, |j_1 m_1 \rangle.$$

Employing the symmetry properties of 'C' coefficients, we get

$$(j^2), | j \, j \, IM \rangle_{ex} = (-)^{2j-I} | j \, j \, IM \rangle.$$

For the equivalent particles $(j^2), | j \, j \, IM \rangle_{ex} = (-)^{2j-I} | j \, j \, IM \rangle$. Since $2j$ is odd, I should be even in order that the wave function for two identical equivalent particles be antisymmetric in accordance to the Pauli principle. Similarly,

$$| 1/2 \, 1/2 \, T \, T_3 \rangle_{ex} = (-)^{1-T} | 1/2 \quad 1/2 \quad T \quad T_3 \rangle. \qquad (3.48)$$

In order that the combined spin-isospin wave function be antisymmetric,

$$I + T = \text{odd}. \qquad (3.49)$$

For identical particles, $T = 1$ so that $I = $ even.

For more than 2 particles in the unfilled shell, the coupling rule is complicated and construction of antisymmetric wave function for $k \geq 3$ is more involved and an additional quantum number is required for the classification of states.

Example: Consider the case of ^{18}O, the allowed values of I for the two neutrons in the configuration $(d_{5/2})^2, (d_{3/2})^2, (2s_{1/2})^2$ are respectively (0, 2, 4), (0, 2) and (0). If the two neutrons are in different j orbit, both the even and odd values of I are allowed. The allowed values of I for the configuration $(d_{5/2})^1 (d_{3/2})^1, (d_{5/2})^1 (2s_{1/2})^1, (d_{3/2})^1 (2s_{1/2})^1$ are respectively (4, 3, 2, 1), (3, 2) and (2, 1).

In the case of ^{18}F, having a neutron and a proton in $(d_{5/2})$ shell, we have both the states $T = 0$ and $T = 1$. The ground state of ^{18}F is to arise from $T = 0$ (because $N = Z$). Therefore, the allowed values of I for $(d_{5/2})^2$ configuration in ^{18}F are $I = 1$, 3 and 5.

A. Two Body Matrix Elements

A simple two-body interaction used extensively in shell model calculation is the "surface delta interaction", which assumes that the interaction takes place on the nuclear surface only. It also assumes that the probability of finding a particle at the nuclear surface is independent of the shell model orbit in which the particle moves. The Surface Delta Interaction (SDI) is conveniently written as[5]

$$V_{SDI}(\vec{r}_{12}) = -4\pi V_0 \, \delta(\vec{r}_1 - \vec{r}_2) \delta(r_1 - r_0), \qquad (3.50)$$

where, $\vec{r}_1$ and $\vec{r}_2$ are position vectors of interacting particles and V_0 is the strength of the interaction. Here, r_0 is the nuclear radius. The factor 4π is used for convenience. The spin dependence of surface delta interaction is introduced by a new strength parameter, A_T so that

$$V_{SDI} = -4\pi A_T \, \delta(\vec{r}_1 - \vec{r}_2) \, \delta(r_1 - r_0).$$

The above interaction is separated into angular and radial parts as follows:

$$V_{SDI}(\vec{r}_{12}) = -A_T \, v(r_1, r_2) S(\Omega_{12}),$$

where $\qquad v(r_1, r_2) = \dfrac{\delta(r_1 - r_2)}{r_1 r_2} \, \delta(r_1 - r_0)$

and $\qquad S(\Omega_{12}) = 4\pi \displaystyle\sum_{\lambda\mu} Y_\lambda^{*\mu}(\hat{r}_1) Y_\lambda^\mu(\hat{r}_2).$

We assume j-j coupling scheme for the evaluation of TBME and employ $V_{SDI}(\vec{r}_{12})$ for that purpose.

Now, $\qquad \langle j_1 j_2 \mid V_{SDI}(r_{12}) \mid j_3 j_4 \rangle = -A_T \langle M_{Rad} \rangle \langle M_{Ang} \rangle_{IT},$

where $\langle M_{Rad} \rangle$ is the radial matrix element, which can be evaluated easily:

$$\langle M_{Rad} \rangle = \int R_{n_1 \ell_1}(r_1) R_{n_2 \ell_2}(r_2) \delta(r_1 - r_2) \delta(r_1 - r_0)$$

$$\times R_{n_3 \ell_3}(r_1) R_{n_4 \ell_4}(r_2) r_1 r_2 \, dr_1 \, dr_2$$

$$= R_{n_1 \ell_1}(r_0) R_{n_2 \ell_2}(r_0) R_{n_3 \ell_3}(r_0) R_{n_4 \ell_4}(r_0) r_0^2,$$

which follows from the properties of the delta function. If we assume that all radial functions for a given mass number have the same absolute value at $r = r_0$, then

$$\langle M_{Rad} \rangle = (-)^{n_1 + n_2 + n_3 + n_4} \, R_{n_1 \ell_1}^4(r_0) \, r_0^2$$

$$= (-)^{n_1 + n_2 + n_3 + n_4} \, C(r_0), \text{ where } C(r_0) = R_{n_1 \ell_1}^4(r_0) \, r_0^2.$$

It turns out that the agreement between the spectra obtained by SDI and experiment is not so good. The deviations occur for the spacings of the $T = 0$ and $T = 1$ centroids of level energies. The Modified Surface Delta Interaction (MSDI) is given by

$$V^{MSDI}(\vec{r}_{12}) = -4\pi A_T \, \delta(\vec{r}_1 - \vec{r}_2) \delta(r_1 - r_0) + B'(\tau(1) \cdot \tau(2)) + C'. \quad (3.51)$$

Now, $\langle \overrightarrow{\tau(1) \cdot \tau(2)} \rangle_T = 2T(T+1) - 3$. Thus, the contribution of the additional term is given by

$$B[\overrightarrow{\tau(1) \cdot \tau(2)}] + C = \begin{cases} -3B' + C' & \text{for } T = 0 \\ B' + C' & \text{for } T = 1 \end{cases}$$

We set $\qquad B'C(r_0) = B$ and $C'C(r_0) = C$.

The parameters of MSDI namely, A_T, B, C are functions of the mass number A. For the mass number in the range $A = 17$–22, their values are[5]

$$A_0 = A_1 = \frac{25}{A} \text{ MeV and } C = 0.$$

Evaluation of $\langle M_{Ang.} \rangle$ is more complicated and involves some Clebs Gordon algebra. We give the final expression. The diagonal matrix element is given by[5]

$$\langle j_1 j_2 \mid V^{MSDI} \mid j_1 j_2 \rangle_{IT} = \frac{A_T}{2(2I+1)} \frac{[j_1][j_2]}{(1+\delta_{12})} \{ (\langle j_2 j_1 - \frac{1}{2} \frac{1}{2} \mid I0 \rangle)^2$$

$$\times (-)^{2\ell_2 + 2 j_2} (1 - (-)^{\ell_1 + \ell_2 + I + T})$$

$$- (\langle j_2 j_1 \frac{1}{2} \frac{1}{2} \mid I1 \rangle) (1 + (-)^T) \} + [2T(T+1) - 3] B + C \qquad (3.52)$$

where the symbol $[x] = (2x + 1)$ and the phase factor $(-)^{2\ell_2 + 2 j_2} \equiv -1$.

For two equivalent particles, i.e. j^2 configuration (Eq. 3.20b):

$$\langle j^2 \mid V^{MSDI} \mid j^2 \rangle_{I,T=1} = -A_1 \frac{(2j+1)^2}{2(2I+1)} (\langle jj - 1/2 \ \ 1/2 \mid 0 \rangle)^2 + B + C.$$

For $\qquad I = 0, (\langle jj - 1/2 \ \ 1/2 \mid 00 \rangle)^2 = \frac{1}{2j+1}.$

Hence, $\langle j^2 \mid V^{MSDI} \mid j^2 \rangle_{I=0,T=1} = \frac{-A_1}{2} (2j+1) + B + C. \qquad (3.53a)$

This means that the absolute value of the binding of two particles in a j^2 configuration coupled to $I = 0$ and $T = 1$ increases with j (*see* 3.20b). Substituting the values of the C-coefficients, we get various TBME:

$$\langle j^2 \mid V^{MSDI} \mid j^2 \rangle_{I=2,T=1} - \frac{A_1 (2j+1)^2 \{ m_j^2 - j(j+1) \}^2}{2(2I+1)(2j-1) \ j(j+1)(2j+3)} + B + C \qquad (3.53b)$$

and

$$\langle j^2 \mid V^{MSDI} \mid j^2 \rangle_{I=4,T=1} = -0.145 \ A_1 \frac{(2j+1)^2}{2(2I+1)} + B + C. \qquad (3.53c)$$

We are now in a position to calculate the ground state (g·s.) binding energy and excited states of a nucleus. Presently, we shall give an illustration for the calculation of ground state binding energy and excited states of ^{18}O nucleus by considering ^{16}O as an inert core and two extra neutrons in $1d_{5/2}$ orbit not occupied by core nucleons (Fig. 3.4). The various terms contributing to the total binding energy of the nucleus can be written as

$$E_{j\pi}^b(\text{core} + j^2) = 2e_j + E^b(\text{core}) + E_{I\pi}^b(j^2), \qquad (3.54)$$

where e_j represents the negative value of the energy needed to remove two neutrons from the potential well in which they are assumed to move independently in the orbit j. It is usually assumed that this potential well does not depend upon the particles outside the core. The values of $\langle V^{MSDI}(r_{12}) \rangle$ are obtained from Eq. (3.53) by substituting $j = 5/2$:

$$E_{01}(1d_{5/2}^2) = \langle (5/2)^2 \mid V^{MSDI} \mid (5/2)^2 \rangle = -3A_1 + B + C$$

$$E_{21}(d_{5/2}^2) = \langle (5/2)^2 \mid V^{MSDI} \mid (5/2)^2 \rangle = -0.82A_1 + B + C$$

$$E_{41}(d_{5/2}^2) = \langle (5/2)^2 \mid V^{MSDI} \mid (5/2)^2 \rangle = -0.29A_1 + B + C .$$

The absolute value of the binding energy is largest for the nucleus in the ground state. The excitation energy $E_x(n)$ of the nth excited state follows from the binding $E^b(n)$ of the nucleus in that state taken with respect to the ground state binding energy, $E^b(0)$ i.e. $E_x(n) = E^b(n) - E^b(0)$. For ^{18}O, we have the relation:

$$E_{0+}^b(^{18}O \text{ g.s.}) = 2e_{d_{5/2}} + E^b(^{16}O) + (-3A_1 + B + C).$$

Now $\quad e_{d_{5/2}} = E^b(^{17}O) - E^b(^{16}O) = -131.77 - (-127.62) = -4.15 \text{ MeV}$

Substituting $A_1 = B = 1.4$ MeV and $C = 0$, we get $E_0^b(^{18}O \text{ g.s.}) = 138.7$ MeV.

Similarly, $E_{2+}^b(^{18}O) = 2e_{d_{5/2}} + E^b(^{16}O) - 0.82A_1 + B + C = -135.67 \text{ MeV}.$

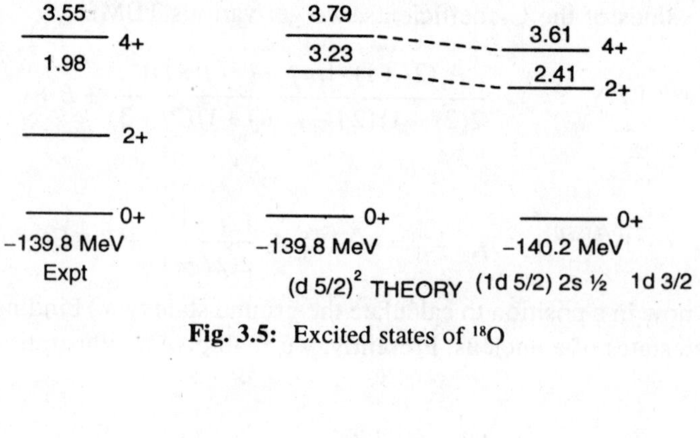

Fig. 3.5: Excited states of ^{18}O

Hence, $\qquad E_x(2+) = E_{2+}^b - E_{0+}^b = 3.234$ MeV,

and $\qquad E_{4+}^b(^{18}O) = 2e_{d_{5/2}} + E^b(^{16}O) - 0.29A_1 + B + C = -134.926,$

so that $\qquad E_x(4+) = 3.78$ MeV.

Comparison between the theory and experiment for the binding energy and excitation energies in units of MeV are shown in Fig. (3.5).

Configuration mixing: Comparison between the theory and experiment shows that the configuration space assumed for 2 neutrons outside ^{16}O core, i.e. $(2e_{d_{5/2}})^2$ is not adequate and we must allow the "configuration mixing" to give better agreement. Configuration mixing may arise from the virtual excitation and allow matrix elements which are off diagonal.

Suppose there are two states in a nucleus, both having the same spin I and isospin T to be described by the orthonormal wave functions $\phi_1^{(0)} \equiv |(1d_{5/2})^2\rangle$ and $\phi_2^{(0)} \equiv |(2s_{1/2})^2\rangle$. Without configuration mixing the energy of the two states are given by

$$H_{11} = \langle \phi_1^{(0)} | H | \phi_1^{(0)} \rangle \text{ and } H_{22} = \langle \phi_2^{(0)} | H | \phi_2^{(0)} \rangle.$$

In order to consider configuration mixing, we introduce two wave functions

$$\psi_p = a_{1p} \, \varphi_1^{(0)} + a_{2p} \, \varphi_2^{(0)}, \, p = 1, 2 \text{ with } a_{1p}^2 + a_{2p}^2 = 1.$$

Then the energies are obtained from the eigen value equation

$$\begin{pmatrix} H_{11} & H_{12} \\ H_{21} & H_{22} \end{pmatrix} \begin{pmatrix} a_{1p} \\ a_{2p} \end{pmatrix} = E_p \begin{pmatrix} a_{1p} \\ a_{2p} \end{pmatrix},$$

with the matrix $H_{k\ell} = H_{\ell k}$. The eigen values E_p^0 and E_p^1 follows from

$$\begin{vmatrix} H_{11} - E_p & H_{12} \\ H_{12} & H_{22} - E_p \end{vmatrix} = 0.$$

or $(H_{11} - E_p)(H_{22} - E_p) - H_{12}^2 = 0$, with the solution given by

$$E_p = \frac{1}{2} \left[(H_{11} + H_{22}) \pm \{(H_{11} - H_{22})^2 + (2H_{12})^2\}^{1/2} \right]. \qquad (3.55a)$$

For two possible roots E_1 (lower sign) and E_2 (upper sign) one obtains two different wave functions ψ_1 and ψ_2 respectively. The corresponding amplitude $a_{\ell p}$ can be calculated as follows:

$$\psi_1 = a_{11}\,\phi_1^{(0)} + a_{21}\,\phi_2^{(0)} = \pm\frac{H_{12}\,\phi_1^{(0)} + (E_{11} - H_{11})\,\phi_2^{(0)}}{\{(E_1 - H_{11})^2 + H_{12}^2\}^{1/2}}. \qquad (3.55b)$$

$$\psi_2 = a_{12}\,\phi_1^{(0)} + a_{22}\,\phi_2^{(0)} = \pm\frac{(E_2 - H_{22})\,\phi_1^{(0)} + H_{12}\,\phi_2^{(0)}}{\{(E_2 - H_{22})^2 + H_{12}^2\}^{1/2}}. \qquad (3.55c)$$

Since $(E_1 - H_{11}) = -(E_2 - H_{22})$, the wave functions ψ_1 and ψ_2 are indeed orthogonal with $a_{12} = \pm a_{21}$ and $a_{11} = \mp a_{22}$. It also follows from these equations that for a vanishing residual interaction, i.e. $H_{12} = 0$, each perturbed state goes over into one specific unperturbed sate, i.e. $\psi_k \to \varphi_k^{(0)}$ $(k = 1, 2)$. The energy separation between the states originally described by $\phi_1^{(0)}$ and $\phi_2^{(0)}$ changes when configuration mixing between these two states is taken into account. $\Delta = |H_{11} - H_{22}|$ be their energy separation without configuration mixing (Fig. 3.6). With configuration mixing, energy separation is given by

$$\Delta' = \sqrt{\Delta^2 + (2H_{12})^2}.$$

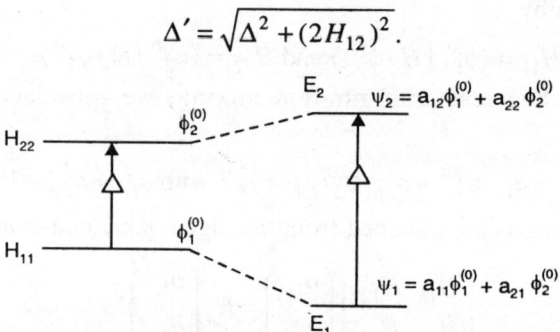

Fig. 3.6: Energy separation between states with and without configuration mixing

The level separation increases as a result of configuration mixing since $\Delta' \geq D$.

Now, $$H_{11} = -3A_1 + B + C + 2e_{1d_{5/2}}$$

and $$H_{22} = -A_1 + B + C + 2e_{1s_{1/2}}$$

Since H_0, the s.p Hamiltonian (3.37) only contributes to the diagonal matrix elements, we have $e_{1d_{5/2}} = -4.15$ and $2e_{2s_{1/2}} = -3.28$. The off diagonal element

$$H_{12} = \langle (2s_{1/2})^2 | V^{MSDI} | (1d_{5/2})^2 \rangle = -1.7321A_1 + B + C.$$

Hence, by substituting the value of A_1, B, C we evaluate H_{11}, H_{22} and H_{12} and then from (3.55a) we get $E_p^{(1)} = -6.33$ MeV.

Therefore, $E_{0+}^b(^{18}O) = -127.62 - 6.33 = -133.95$ and $E_p^{(2)} = -11.33\,\text{MeV}$.

Hence, $E_{0+}^b(^{18}O\ g.s.) = -127.62 - 11.33 = -138.95$.

The first excited

$$E_x^{0^1+}(^{18}O) = -133.95 + 138.95 \approx 5.0\,\text{MeV}.$$

We also find that extending shell model space from $(1d_{5/2})^2$ to $(1d_{5/2})$ $(2s_{1/2})$ leads to an improvement of g.s. binding energy of ^{18}O from -138.7 MeV to -138.95.

B. Classification of States and Seniority

We have seen earlier that for $|\ j^n IT\rangle$ configuration with $n > 2$, the angular momentum I and isospin T do not suffice to specify nuclear levels uniquely, there may be many distinct wave function having specification $|(j)^n I : T\rangle$ and we may again look for new methods of describing all the properties of wave functions. A new quantum number "seniority" has been introduced to specify distinction between wave functions.

Let us first consider configurations containing only protons or only neutrons outside of a closed shell. The shell is defined by $(2j + 1)$ levels. A pair of particles in the same shell may be coupled to total $I = 0$. Then because $\langle jj\ m - m\ |\ I0\rangle = (-)^{2j}\langle jj - mm\ |\ I0\rangle$ the zero coupled (zc) pair is anti-symmetric in spin and space and therefore must be symmetric in isospin $(T = 1)$. Since we deal here with identical nucleons, we need not specify the isospin. $|\ j\bar{m}\rangle = (-)^{2j-m}|\ j - m\rangle$ is the state obtained from $|\ jm\rangle$ by time reversal.

We can express the wave function of 'n' identical particles in j shell by an antisymmetrized product of a two particle wave function having $I = 0$ and a wave function of $n - 2$ particles coupled to I. For example,

$$|(j)^n I\rangle = \mathcal{A}\{|j^{n-2}I\rangle\ j^2 I = 0\rangle\}.$$

We may imagine this expansion may be considered for the wave function $|\ j^{n-2}I\rangle$, and continue in successive steps until the process is no longer possible; we are decoupling two-particles at a time, the pair having $I = 0$ and antisymmetrizing by operator $\mathcal{A}$ the resulting product. The process eventually fails either because antisymmetrized product vanishes or because the total I is large enough that it could not be constructed from the remaining particles.

The component v of the configuration $|(j)^v I\rangle$ when the decoupling process is no longer possible, is the "seniority" of the original wave function $|\ j^n I\rangle$. Hence, the seniority quantum number is the number of unpaired particles. The value of I imposes condition on the possible values of seniority v.

For example, If $I = 0$ $v = 0$, If $I \neq 0$, $v \neq 0$.

For even-even nuclei, $v =$ even. For odd A nuclei, $v =$ odd.

If $I \geq 2j + 1$ then $v \geq 3$. Consider two equivalent particles (j^2):

$$I = 0, \qquad\qquad\qquad v = 0$$
$$I = 2,4,6 \ldots (2j\text{-}1); \qquad v = 2, T = 1$$
$$I = 1,3,5 \ldots 2j; \qquad v = 2, T = 0$$

For $I = j$, $v = 1$, which is the lowest possible seniority. In Table 3.6 we give classification of states when identical fermions are in a single j shell.

Table 3.6: Classification of $(5/2)^n$ states

n	v	I	n	v	I
0	0	0	2	2	2,4
1	1	5/2	3	1	5/2
2	0	0	3	3	3/2, 9/2

In order to derive explicit form of the wave function of n-particles forming a $I = 0$ and $v = 0$ state, we introduce quasi-spin operators as follows:

$$s_+ = \sum_{m>0} (-)^{j-m} a^\dagger_{jm} a^\dagger_{j-m},$$

$$s_- = \sum_{m>0} (-)^{j-m} a_{jm} a_{j-m},$$

and
$$s_z = \frac{1}{2}\sum_m a^\dagger_{jm} a_{j-m} - \frac{2j+1}{2}, \qquad (3.56a)$$

where s_+ and s_- operating on a vacuum state respectively creates and destroys a zc pair. The operator part of s_z is just the number operator $\hat{n}$ (3.62), which tells us the number of particles in the state:

$$s_z = \frac{1}{2}(n - \Omega) \text{ where } \Omega = \frac{2j+1}{2}.$$

Further,

$$[s_+, s_-] = 2s_z = (n - \Omega) = [s_-, s_+]. \qquad (3.56b)$$

The term quasi-spin comes from the fact that the components s_x, s_y and s_z satisfy the usual angular momentum relations:

$$\vec{s} \times \vec{s} = i\vec{s}, \quad [s^2, s_+] = [s^2, s_-] = 0, \quad s^2 = s_+ s_- - s_z(s_z - 1) = s_x^2 + s_y^2 + s_z^2 \text{ etc.}$$

The ground state is characterized by all particles being zero coupled. Thus, for the ground state wave function we can write

$$\psi \sim s_+^{n/2}|0\rangle \tag{3.56c}$$

With the above choice, we solve the eigen value equation

$$H\psi = E\psi,$$

where the pairing interaction is defined in terms of s_+ and s_- as

$$H = -2G\, s_+\, s_-. \tag{3.56d}$$

Employing (3.56c) and (3.56d) we get

$$H\psi = -2Gs_+s_-s_+^{n/2}|0\rangle$$

$$= -2Gs_+(s_-s_+ - s_+s_- + s_+s_-)\, s_+^{(n-2)/2}|0\rangle$$

We use (3.56b) and the fact $s_+^{(n-2)/2}|0\rangle = \dfrac{2(n-2)}{2} = n-2$, and get

$$-2Gs_+s_-\ s_+^{n/2}|0\rangle = \{-2Gs_+^{n/2}(\Omega - n + 2) - 2Gs_+s_-s_+^{(n-2)/2}\}\,|0\rangle$$

Hence, $H(s_+^{n/2}|0\rangle) = -2Gs_+^{n/2}\left\{\dfrac{n}{2}(\Omega - n) + 2\displaystyle\sum_{x=1}^{n/2} x\right\}|0\rangle$

$$= -2G\left\{\dfrac{n}{2}\ (\Omega - n) + \dfrac{2\cdot n}{4}\left(\dfrac{n}{2}+1\right)\right\}\, s_+^{n/2}\ |0\rangle$$

Hence, the pairing energy eigen value for the ground state is given by

$$E_{v=0} - \dfrac{Gn}{2}\ (2\Omega - n + 2) = -\dfrac{Gn}{2}\ (2j + 3 - n), \tag{3.57a}$$

where $v = 0$ characterizes a ground state system in which all pairs of particles are zero coupled.

Similarly n-particle wave function that corresponds to v unpaired particle is

$$\psi = s_+^{(n-v)/2}|j^v\alpha I\rangle, \tag{3.57b}$$

where the function $|j^v\alpha I\rangle$ denotes a state of v particles coupled to angular momentum I and α is an extra label required to specify a j^n state and $\dfrac{n-v}{2}$ is the number of zc pairs.* Hence, the eigen value equation is

$$Hs_+^{n-v/2}|j^v\alpha I\rangle = E_v\ s_+^{n-v/2}|j^v\alpha I\rangle, \tag{3.57c}$$

We follow the same procedure as before and carry out the iterative procedure $(n - v)$ times. The resulting eigen values are found to be

---

* A relationship between the quasi-spin quantum number and the seniority quantum number is obtained in Exercise 3.10.

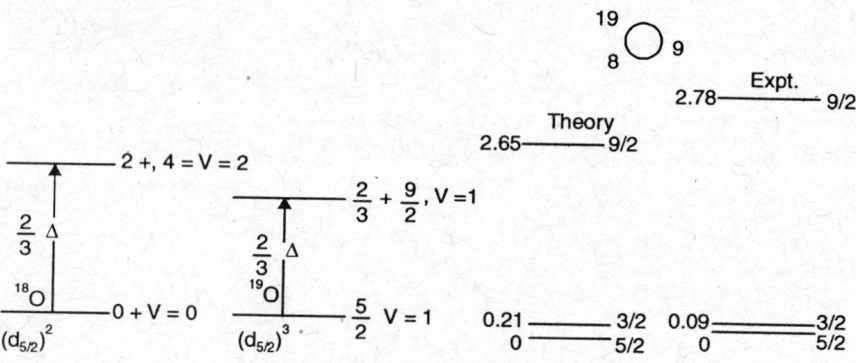

Fig. 3.7: Low lying levels of ^{19}O

$$E_v = -G\frac{n-v}{2}(2\Omega - n - v + 2) = -G\frac{n-v}{2}(2j + 3 - n - v) \qquad (3.57d)$$

These $(j)^n, v = 1$ levels have a stronger single particle character than others with $I = j$ but $v > 1$. This reveals that 'the lowest seniority forms the ground state of nuclei'.

Talmi[6] tried to obtain effective interaction from the actual spectra of nuclei. Shell model wave functions along with residual two-body interaction give definite relation between spectra of related configuration which can be verified comparing with experimental energy levels.

The matrix element of the two-body effective interaction in j^n configuration can be expressed as linear combination of matrix elements of effective interaction in j^2 configuration. He considered a simple two-body interaction in the following form:

$$H_{int} = a + 2b\sum_{i<k}\vec{j}_i \cdot \vec{j}_k + c\sum_{i<k}q_{ik}, \qquad (3.58)$$

where the second and third term respectively represent spin-spin and pairing interaction.

$$\langle j^2 I \mid H_{int}\, j^2 I \rangle = a + 2b\langle j^2 I \mid \vec{j}_1 \cdot \vec{j}_2 \mid j^2 I \rangle + \langle j^2 I \mid q_{12} \mid j^2 I \rangle$$

$$= a + b\left[I(I+1) - 2j(j+1)\right] + c(2j+1)\,\delta_{I0} \qquad (3.59a)$$

The matrix of H_{int} in j^n configuration is given by[6] (Eq. 3.57d) as

$$\langle j^n\, v \mid IM \mid H_{int} \mid j^n\, v\, IM \rangle$$

$$= a\,\frac{n(n-1)}{2} + b\left[I(I+1) - nj(j+1)\right] + c\frac{n-v}{2}\{2j + 3 - n - v\}, \qquad (3.59b)$$

where v is the seniority quantum number. For illustration, let us consider oxygen isotopes namely, ^{16}O, ^{17}O, ^{18}O and ^{19}O. For ^{18}O, which corresponds to

$(d_{5/2})^2$ configuration the experimental data (Fig. 3.7) yields the following equations:

$$a - \frac{35}{2}b + 6c = 0 \,, \, a - \frac{23}{2}b = 1.98 \text{ and } a + \frac{5}{2}b = 3.55.$$

Solving we get $a = 3.27$, $b = 0.112$ and $c = -0.22$.

Hence, $H_{eff} = 3.27 + 0.224 \sum_{i<k} \vec{j}_i \cdot \vec{j}_k - 0.22 \sum_{i<k} q_{ik}$.

Using above interaction and Eq. (3.59b), we try to predict the energy levels

of ^{19}O: $E_{5/2} = 3a - \frac{35}{2}b + 4c$, $E_{9/2} = 3a - \frac{3}{2}b$, $E_{3/2} = 3a - \frac{45}{2}b$.

Substituting values of a, b and c we get low lying states of ^{19}O (Fig. 3.7).

Exercise 3.9. Using the seniority quantum number v, find the ground state energy of a system of n particles consisting of zero coupled pairs and un-paired particles representing respectively even-even ($v = 0$) and odd A ($v = 1$) nuclei.

The zc pairs interact with each other and also with the core. Let a be the interaction energy between two zc pairs and let b be the interaction energy between a pair and a particle. Then $2b$ is the interaction energy between two zc pairs. The total energy

$$E_I^{n,v} = E_I^{v,v} + \frac{n-v}{2}a + \frac{v(n-v)b}{2} + \frac{1}{2}\frac{n-v}{2}\left\{\frac{n-v}{2} - 1\right\}(2b)$$

$$= E_I^{v,v} + \frac{n-v}{2}a + \frac{n-v}{4}(n+v-2)b.$$

In the ground state, we expect largest number of zc and hence the seniority is smallest. For even n, $v = 0$ and for odd n, $v = 1$, $v^2 = v$. It follows from the above that

$$E_{gs}^{n,v} = \frac{n-v}{2}\left[a - \frac{b}{2}\right] + \frac{n(n-1)b}{4}.$$

Exercise 3.10. Find a relationship between the quasi-spin quantum number and the seniority quantum number.

If $| j^k I \, v \rangle$ represents a state of k particles of the total angular momentum I and seniority v, which is number of unpaired particle in the state then

$$s^2 | j^k I \, v \rangle = \{s_+ s_- + s_z(s_z - 1)\} | j^k I \, v \rangle$$

$$= \left(\frac{v - \Omega}{2}\right)\left(\frac{v - \Omega}{2} - 1\right) | j^k I \, v \rangle$$

since $s_+ s_- \mid j^k I \ v\rangle = 0$ and $s_+ s_-$ being the number operator for zero coupled pairs. Again

$$s^2 \mid j^k I \ v\rangle = s(s+1) \mid j^k I \ v\rangle$$

Equating both sides we get

$$s(s+1) = \left(\frac{\Omega - v}{2}\right)\left(\frac{\Omega - v}{2} + 1\right), \text{ where } \Omega = (2j+1)/2.$$

Hence, $\qquad\qquad s \equiv \dfrac{\Omega - v}{2}.$ $\qquad\qquad\qquad\qquad\qquad\qquad$ (3.60)

3.4 LARGE BASIS SHELL MODEL

We saw that in the case of oxygen isotopes, to produce a few low lying excited states, the energy matrix could be diagonalized by hand. But the dimension of the matrix increases with increasing number of particles outside the closed core and the increasing number of levels over which particles are to be distributed. For example, to study the spectroscopy of $^{62}_{28}Ni_{34}$ we must consider 6 neutrons in the shell from $N = 28$ to $N = 50$. In order to decide what single particle levels are important and what their energies are, we must know something about the spectrum of single particle nucleus $^{57}_{28}Ni_{29}$. ^{56}Ni is taken as doubly magic core because it consists of 28 neutrons and 28 protons. The energy levels of ^{57}Ni observed in the pick up reaction $^{58}Ni(p,d)$, ^{57}Ni indicates that we must distribute 6 neutrons over the $2p_{1/2}$, $1f_{5/2}$ and the $2p_{1/2}$ states. We see that the number of ways of constructing $J = 2$ state is 33. Hence we are required to diagonalize a 33×33 matrix. The dimension of the matrix becomes 308×308 if we include just one more level, i.e. $1g_{9/2}$. Therefore, the problem of finding the eigen values and eigen functions of the Hamiltonian is reduced to a laborious numerical diagonalization of the Hamiltonian with the aid of computers. Apart from calculation difficulties, this approach does not help us to understand essential physics. *Pairing force* theory developed in the next section provides us an elegant method of treating the residual interaction by means of which it is considerably easier to study properties of nuclear many-body system.

A. Theory of Paring Force

We have already witnessed that the shell model achieved considerable success in explaining many ground and excited state properties of nuclei. The systematic variation of some of these properties with mass number are also reasonably reproduced. However, there are certain features of nuclear properties which are not explained by the shell model. For example, (1) existence of an energy gap in the spectra of even-even nuclei and absence of such gap in the spectra of odd-A nuclei, and (2) the density of one particle levels.

Bohr, Mottelson and Pines[7] drew attention to the similarity of the excitation spectra of nuclei to those of superconducting metallic states. The existence of energy gap in superconducting states and hence the phenomena of superconductivity in metals at low temperature were explained by Bardeen Cooper and Schrieffer in terms of their "pairing correlation" theory, also known as the BCS theory. Drawing analogy from the BCS theory, the pairing of nucleon can be described by quantum numbers of independent particle model. For example, if one nucleon is described by $n\ell jm$, then paired nucleon is described by $n, \ell, j, -m$. A definite energy is needed to separate the pair of nucleons since the pairing interaction is an attractive interaction as evidenced by the binding energy data and zero spin of ground state of even-even nuclei. This leads to appearance of an energy gap between the ground state and the first excited state in even-even nuclei in which all the nucleons are paired in its ground state so that the total angular momentum of the nucleus is zero. To produce the first 2+ state one has to break a pair and hence requiring the extra energy.

Bohr, Mottelson and Pines[7] adopted BCS method to calculate pairing energy in the nucleus and made the following additional assumptions:

(1) The pairing force has a short range.
(2) The pair correlations are expected in the neighbourhood of Fermi surface.
(3) The pairing energy is significant for energy spectra of intermediate and heavy nuclei for which separation of single particle energies are sufficiently small.

To write BCS wave function we use the following notations for creation and anhilation operators.

$$a_v^\dagger \equiv a_{jm}^\dagger \quad a_{\bar{v}}^\dagger = (-)^{j-m} a_{j-m}^\dagger;$$

and
$$a_{v'} \equiv a_{j'm'} \quad a_{\bar{v}'} = (-)^{j'-m'} a_{j'-m'},$$

where $v, v' > 0$, i.e. they assume only positive values. v and $\bar{v}$ now onwards will represent respectively $|jm\rangle$ and $|j-m\rangle$. For convenience we suppress the j subscript.

In the extended single particle shell model description, we encountered several states of $I = 0$ and we call the g.s., the state of seniority $v = 0$. Higher states are $v = 2$. For even-even nuclei all nucleons are paired up to $I = 0$. Therefore, the many-body wave functions in pairing Coupling scheme is given by (3.56c) or (3.57b).

Let us now consider the operator $n_v = a_v^\dagger a_v$, which yields zero acting on a state v unoccupied and unity when the orbit n is occupied. Thus n_v measures the occupancy of the single particle orbit n.

Bardeen, Cooper and Schrieffer (BCS) showed that a many body wave function can be constructed if we assume that the number of states is large and occupancies of different paired state are uncorrelated.

$$\psi_{BCS} = \Pi(U_v + V_v a_v^\dagger a_{\bar{v}}^\dagger)|0\rangle, \tag{3.61}$$

where the product 'Π' runs over $n/2$ states. Here V_v^2 is the occupation probability that the paired state $v\bar{v}$ is occupied state and U_v^2 is the probability that the paired state $v\bar{v}$ is empty. Since ψ is normalized, we have the condition: $U_v^2 + V_v^2 = 1$.

Henceforth, ψ refers to the BCS wave function unless otherwise stated. It is also clear from Eq. (3.61) that the BCS wave function represents an admixture of states with an even number of particles of one kind. Thus, BCS wave function is described by a real parameter V_v and U_v.

The operators a_v and $a_v^\dagger$ satisfy the anticommutation relation, $a_v a_{v'}^\dagger + a_{v'} a_v^\dagger = \delta_{vv'}$.

An important feature of the BCS wave function is that it is not an eigen state of number operator

$$\hat{n} = \sum_{v>0} (a_v^\dagger a_v + a_{\bar{v}}^\dagger a_{\bar{v}}). \tag{3.62}$$

In other words, $\hat{n}\psi \neq n\psi$, where n is the number of particles and ψ does not describe a state with fixed number of nucleons, rather it is an admixture of states with different number of nucleons. However, the nucleus is a system of definite number of nucleons, and the approximate wave function ($v = 0$) does not belong to a state of fixed number of nucleons.

In order to correct approximately the short coming of the BCS wave function one imposes the condition

$$n = \langle\psi|\hat{n}|\psi\rangle = \sum_{v>0} \langle\psi|(a_v a_v^\dagger + a_{\bar{v}}^\dagger a_{\bar{v}})|\psi\rangle. \tag{3.63a}$$

Substituting the BCS wave function (3.61) we get

$$n = 2\sum_{v>0} V_v^2. \tag{3.63b}$$

The fluctuation in the number of particles

$$(\Delta n)^2 = \langle\psi|(\hat{n} - n)^2|\psi\rangle = 4\sum_{v>0} U_v^2 V_v^2. \tag{3.64}$$

We can also easily pick up that part of ψ which belongs to a fixed number of particles n and this is given by

$$P_n \, \psi = \prod_{v>0} \frac{1}{(n/2)!} \left(\frac{V_v}{U_v} \, a_v^\dagger \, a_{\bar{v}} \right)^{n/2} |0\rangle.$$

This is the same as the wave function $(s^\dagger)^{n/2} |0\rangle$ *(see* Eq. 3.56c) if we take

$$\frac{V_v}{U_v} = 1.$$

It is this feature of the BCS wave function that saves one time and effort. That is only a selected set of seniority zero wave functions is taken for the ground state instead of all possible combinations as is done in the shell model calculation. For example, ^{116}Sn represented by a doubly magic core of $Z = N = 50$ plus 16 particles in 5 orbits $(g_{7/2}, d_{5/2}, h_{11/2}, d_{3/2}$ and $s_{1/2})$ have a total 110 possible seniority zero wave functions. Thus, an ordinary shell model calculation necessitates diagonalizing a 110×110 matrix in order to solve for the 110 constants C_α^n (Eq. 3.45). The BCS wave function, on the other hand, treats the problem as though there are only 5 independent constants, viz. the ratios V/U for $j = 1$ to 5.

The Hamiltonian of the system of interacting nucleons can be written in the form:

$$H = \sum_v \epsilon_v \, a_v^\dagger \, a_{\bar{v}} - G \sum_{vv'} a_v^\dagger \, a_{\bar{v}} \, a_{\bar{v}'} \, a_{v'}, \qquad (3.65)$$

where G is the strength of the residual pair interaction, which is defined as $\langle v' \, \bar{v}' | V | \bar{v} \, v \rangle = -G$ and ϵ_v is the energy of an individual nucleon in the self consistent field. In the above it is assumed that the transition matrix is independent of v and v'. The Hamiltonian (3.65) describes a system with a well defined number n of particles.

B. Ground and Excited States

We now calculate the expectation value of the residual interaction responsible for pairing using BCS wave function, ψ_{BCS}. The required expectation value

$$\langle \psi_{BCS} | H | \psi_{BCS} \rangle = \epsilon_0 \; (v = 0)$$

$$= 2 \sum_{v>0} \epsilon_v V_v^2 - G \left[\sum_{v>0} U_v V_v^2 + \sum_{v>0} V_v^2 \right] \qquad (3.66)$$

In order to find the lowest-energy state, we vary U_v and V_v subject to the condition that expectation value of number operator $\langle \psi | \hat{n} | \psi \rangle = n$, the number of particles being considered in the problem. This means that we must introduce a Lagrange multiplier λ and define an auxiliary Hamiltonian

$$H' = H - \lambda \hat{n} \qquad (3.67)$$

We use ψ as a trial wave function in a variational equation

$$\delta \langle \psi \mid H' \mid \psi \rangle = \delta \langle \psi \mid H - \lambda \hat{n} \mid \psi \rangle = 0.$$

It is seen that λ is nothing but the chemical potential. For variation of n, the change in the energy is

$$\delta \langle H' \rangle = \delta \langle H \rangle - \lambda \delta \langle \hat{n} \rangle.$$

Setting $\delta \langle H' \rangle = 0$, we get $\lambda = \dfrac{\partial \epsilon_0}{\partial n}$.

We also find that the term $\sum V_\nu^4$ is small and should be included in the single particle energy ϵ_ν. Employing Eqs. (3.66) and (3.67), we get

$$\langle \psi \mid H' \mid \psi \rangle = \sum_{\nu>0} 2(\epsilon_\nu - \lambda) V_\nu^2 - G \left(\sum_{\nu>0} U_\nu V_\nu \right)^2 - G \sum_{\nu>0} V_\nu^4.$$

We obtain by imposing the condition $\dfrac{\partial}{\partial V_\nu} \langle \psi \mid H' \mid \psi \rangle = 0$, the following

relation:

$$4(\epsilon_\nu - \lambda) V_\nu - 2G \left\{ \sum_{\mu>0} U_\mu V_\mu \left(U_\nu + V_\nu \frac{\partial U_\nu}{\partial V_\nu} \right) \right\} - 4G V_\nu^3 = 0.$$

Noting from the normalization condition $U_\nu^2 + V_\nu^2 = 1$, that

$$V_\nu \left(\frac{\partial U_\nu}{\partial V_\nu} \right) = - \frac{V_\nu^2}{U_\nu}$$

we get $\quad 4(\epsilon_\nu - \lambda) V_\nu - 2G \left\{ \sum_{\mu>0} U_\mu V_\mu \right\} \left\{ U_\nu - \dfrac{V_\nu^2}{U_\nu} \right\} - 4G V_\nu^3 = 0,$

or $\qquad\qquad 2(\epsilon_\nu' - \lambda) U_\nu V_\nu = \Delta(U_\nu^2 - V_\nu^2), \qquad (3.68a)$

where $\qquad\qquad \epsilon_\nu' = \epsilon_\nu - GV_\nu^2$ and $\Delta = G \sum_{\mu>0} U_\mu V_\mu. \qquad (3.68b)$

Note that ϵ_ν' is the renormalized single particle energy in the orbit ν which includes the contribution $-GV_\nu^2$ of the pairing interaction of the one-body field. By squaring (3.68a) and making use of the relation

$$U_v^4 + V_v^4 = 1 - 2U_v^2 V_v^2, \text{ we obtain:}$$

$$4(\epsilon_v' - \lambda)^2 \ U_v^2 V_v^2 = \Delta^2 (U_v^2 - V_v^2)^2 = \Delta^2 (1 - 4 \ U_v^2 - V_v^2),$$

Hence,

$$U_v^2 V_v^2 = \frac{1}{4}\left[\frac{\Delta^2}{(\epsilon_v' - \lambda)^2 + \Delta^2}\right] = \frac{1}{4}\left[1 - \frac{(\epsilon_v' - \lambda)^2}{(\epsilon_v' - \lambda)^2 + \Delta^2}\right].$$

Using again the normalization condition, we get

$$U_v^2(1 - U_v^2) = \frac{1}{4}\left[1 - \frac{(\epsilon_v' - \lambda)^2}{(\epsilon_v' - \lambda)^2 + \Delta^2}\right].$$

$$U_v^4 - U_v^2 + \frac{1}{4}\left[1 - \frac{(\epsilon_v' - \lambda)^2}{(\epsilon_v' - \lambda)^2 + \Delta^2}\right] = 0. \tag{3.69}$$

Solving the quadratic equation (3.69) we get

$$U_v^2 = \frac{1}{2}\left[1 + \left\{1 + \left(1 - \frac{(\epsilon_v' - \lambda)^2}{(\epsilon_v' - \lambda)^2 + \Delta^2}\right)\right\}^{1/2}\right]$$

$$= \frac{1}{2}\left[1 + \frac{(\epsilon_v' - \lambda)}{\sqrt{(\epsilon_v' - \lambda)^2 + \Delta^2}}\right], \tag{3.70a}$$

and

$$V_v^2 = \frac{1}{2}\left[1 - \frac{(\epsilon_v' - \lambda)}{\sqrt{(\epsilon_v' - \lambda)^2 + \Delta^2}}\right]. \tag{3.70b}$$

The above equations can be written in a neat form

$$U_v^2 = \frac{1}{2}\left[1 + \frac{\epsilon_v' - \lambda}{E_v}\right] \text{ and } V_v^2 = \frac{1}{2}\left[1 - \frac{\epsilon_v' - \lambda}{E_v}\right], \tag{3.71}$$

where

$$E_v = \left\{(\epsilon_v' - \lambda)^2 + \Delta^2\right\}^{1/2}. \tag{3.72}$$

Our next aim is to obtain an expression for energy gap Δ and strength of the pairing interaction G and also the average number of particles n.

Substituting U_v and V_v from (3.71) and (3.72) in (3.68b), we get

$$\Delta = G\sum_{v>0}\left[\frac{1}{4}\left\{1 - \frac{(\epsilon_v' - \lambda)^2}{(\epsilon_v' - \lambda)^2 + \Delta^2}\right\}\right]^{1/2} = \frac{G}{2}\sum_v \frac{1}{\epsilon_v}, \tag{3.73}$$

which always has the trivial solution $\Delta = 0$. The non-trivial solution is obtained by solving

$$\frac{G}{2} = \sum_{v>0} \frac{1}{E_v} = \sum_{v>0} \frac{1}{\left\{ (\epsilon'_v - \lambda)^2 + \Delta^2 \right\}^{1/2}}. \tag{3.74}$$

Similarly,

$$n = 2 \sum_{v>0} V_v^2 = \sum_{v>0} \left[1 - \frac{\epsilon'_v - \lambda}{\left\{ (\epsilon'_v - \lambda)^2 + \Delta^2 \right\}^{1/2}} \right], \tag{3.75}$$

which determines the value of the unknown quantity Δ, when n is specified. Eqs. (3.73)-(3.74) are usually called the gap equations. The modification of the particle distribution due to pairing interaction can be seen by plotting V_v^2 against ϵ'_v (Fig. 3.8). We see that the pairing interaction scatters pairs of particles from originally filled levels to high lying levels giving rise to diffuse Fermi surface. Eqs. (3.61) and (3.65), yields the g.s. energy

$$\langle \psi \mid H \mid \psi \rangle = \sum_{v>0} 2 \epsilon_v V_v^2 - \frac{\Delta^2}{G} - G \sum_v V_v^4, \tag{3.76}$$

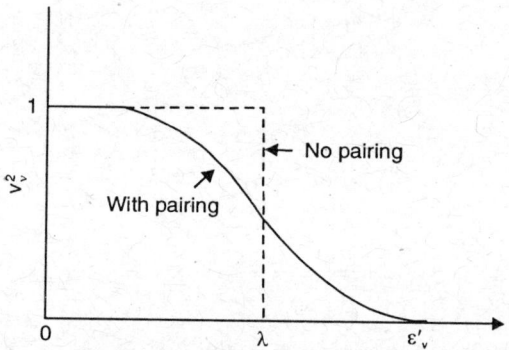

Fig. 3.8: Variation of V_v^2 against ϵ'_v

which reduces for $G \to 0$ and vanishing $\Delta (= 0)$ to a sum over occupied single particle energies. The constraint on the BCS wave function ψ_{BCS} is that

$2 \sum_{v>0} V_v^2$ is even. Again it follow from (3.71-3.72) in the limit of $G \to 0$, i.e.

$\epsilon_{v'} = \epsilon_v$ and $\Delta = 0, U_v$ and V_v become 0 and 1 respectively if $\epsilon_v < \lambda$ and 1

and 0 if $\epsilon_v > \lambda$. In other words all states upto $\epsilon_v = \lambda$ are completely occupied and all states $\epsilon_v > \lambda$ are vacant. Here λ is the Fermi level. The presence of pairing force smears the Fermi level, the nucleon state v such that

$|\epsilon_v - \lambda| \sim \Delta$ are completely full or completely empty, but states with $|\epsilon_v - \lambda| \sim \Delta$ are partially occupied. λ is determined from (3.63b). Of course this equation has to be solved simultaneously with the gap equation (3.74).

The effect of pairing correlation is to smooth out the sharp Fermi surface so that instead of being 100% occupied or unoccupied, the various particle states have certain probability of being occupied.

Finally let us discuss excited states of such a system. We saw earlier that the presence of pairing force smears the Fermi level, i.e. the pairs of particles in the interacting system do not stay all the time below Fermi level; they may be found on levels above Fermi level. Therefore, in an alternative approach to pairing force problem, the Hamiltonian H is written in terms of hypothetical particle known as "quasi-particle". A quasi-particle in $|jm\rangle$ state is a mixture of nucleon state $|jm\rangle$ and a hole state $|j-m\rangle$. In terms of the particle creation and annihilation operators, the quasi-particle creation and annihilation operators are defined by

$$\alpha^\dagger_{jm} = U_j \, a^\dagger_{jm} - (-)^{j-m} \, V_j \, a_{j-m} \tag{3.77a}$$

$$\alpha_{jm} = U_j \, a_{jm} - (-)^{j-m} \, V_j \, a^\dagger_{j-m} \tag{3.77b}$$

where $a^\dagger_{jm} (a_{jm})$ are particle creation (annihilation) operators and $a^\dagger_{j-m} (a_{j-m})$ are hole creation (annihilation) operators.

$U_j = 0 \, V_j = 1$ for $\epsilon_j < \lambda$ and $U_j = 1 \, V_j = 0$ for $\epsilon_j > \lambda$, where λ is Fermi-level. If we destroy a particle state in the Fermi sea (set of occupied states) and thereby create a hole in that state, this becomes equivalent to creating a quasi-particle. In other words, quasi-particle is either a hole in state $|\phi_0\rangle$ or a particle above the occupied state in $|\phi_0\rangle$.

It has already been mentioned that a definite energy is needed to separate the pair of nucleons (pair interaction is attractive interaction, and this leads to the appearance of energy gap between the ground state and the first excited state in even-even nuclei in which all nucleons are paired and the total angular momentum of the nucleus is thus equal to zero). Therefore, the ground state of even-even nuclei contain no quasi-particle, i.e. BCS ground state is a quasi-particle vacuum. This implies $\alpha_{jm} |BCS\rangle = 0$. The energy E_j of a quasi-particle is found by calculating the expectation value of the Hamiltonian in the state with one quasi-particle

$$E_j = \sqrt{(\epsilon_j - \lambda)^2 + \Delta^2} \,. \tag{3.78}$$

Eq. (3.78) gives $E_j \geq \Delta$. Therefore, a two quasi-particle state $(j_1 \, j_2)$ lies at an energy $E_{j_1} + E_{j_2} \leq 2\Delta$ above the vacuum state and that is why Δ is known as energy gap. 2Δ gives the energy required to break a zero coupled pair and Δ may be estimated from the nucleon separation energy.

Since BCS wave function corresponds to quasi-particle vacuum, i.e. $|BCS\rangle \equiv |0\rangle_{qp}$ and represents an even-even nucleus, we can associate with it a zero point energy ϵ_0. One quasi-particle corresponds to a nucleus with energy $(\epsilon_0 + \epsilon_1)$, i.e.

$$\alpha^\dagger_{v_1} |0\rangle_{qp} = (\epsilon_0 + \epsilon_1)\, \alpha^\dagger_{v_1} |0\rangle_{qp}. \tag{3.79}$$

The two quasi-particle state corresponds to low lying states of an even-even nucleus. They have excitation energy

$$\alpha^\dagger_{v_1} \alpha^\dagger_{v_2} |0\rangle_{qp} = (\epsilon_0 + \epsilon_1 + \epsilon_2)\alpha^\dagger_{v_1} \alpha^\dagger_{v_2} |0\rangle_{qp}. \tag{3.80}$$

3.5 MANY-BODY THEORY OF NUCLEUS

Some theorists consider shell model wave functions just as a convenient basis for the many-body theory of the nucleus. The interaction used in this approach is the one between free nucleons. If the nuclear many-body problem could have been solved to a good accuracy, the resulting wave functions would have been the real wave functions of the nucleus. There is still no solid theoretical argument which shows that the shell model is a good approximation for nuclear states. In order to calculate energies of nuclear states of simple shell model configuration of valence nucleons, the residual or effective interaction between them must be known. Such an interaction can not be interaction between free nucleons. The latter is very strong at short distances due to the presence of an infinite repulsive core leading to short range correlations in the real wave function of the nucleus. No such correlations are present in simple shell model wave functions.

A. The Hartree Fock Method

In the shell model we have seen that many of the ground state properties can be described by making the simple assumption of independent particle motion in an average field subject to the Pauli exclusion principle. Now we shall examine how the average one-body field, or the "mean field" is generated through a given two-nucleon effective interaction. The procedure adopted to achieve this is called "Hartree Fock" (HF) method, which is well known from atomic physics. It essentially employs 'variational principle' for finding the lowest eigen value of the Hamiltonian H representing a many nucleon system [Eq. 3.35] with a wave function which is a single determinant of one particle states [Eq. 3.40].

Our aim is to find out best possible independent particle trial function for determining the ground state energy E_0. That is, if we start with an arbitrary set of single particle functions ϕ_i, the HF method[12] consists in constructing many particle wave function ψ in terms of ϕ_i and minimizing the energy $E = \int \psi^* H \psi \, d\tau$, with respect to all possible variations of single particle functions ϕ_i. The variational principle has the merit that it always gives $E \geq E_0$.

In this procedure[12] the single particle Hamiltonian h is used to generate single particle states ϕ_i

$$h\phi_i = \epsilon_i \phi_i \quad \text{or} \quad h \, | \, i \rangle = \epsilon_i | \, i \rangle. \tag{3.81}$$

h is either guessed or calculated in terms of two-body effective interaction. Now,

$$E_{HF} = \langle \psi | \, H \, | \psi \rangle = \sum_{i=1}^{A} \langle i \, | \, t \, | \, i \rangle + \frac{1}{2} \sum_{i,j} \langle ij \, | \, V_{ij} \, | \, ij \rangle, \tag{3.82}$$

where E_{HF} is called *HF* energy; $\langle i \, | \, t \, | \, i \rangle$ is the kinetic energy of the single particle in the orbit 'i' and $\langle ij \, | \, V \, | \, ij \rangle = -\langle ij \, | \, V \, | \, ji \rangle$ is an antisymmetrized matrix element of two-body interaction V. A practical method to solve *HF* equation (3.82) is by matrix method in which it is advantageous to expand single particle states on a basis $| \, \lambda \rangle$ of known wave functions:

$$| \, i \rangle = \sum_{\lambda} C_{\lambda}^{i} | \, \lambda \rangle \tag{3.83}$$

The basis $| \, \lambda \rangle$ may be shell model states $| \, n\ell j m_j \rangle$ in a harmonic oscillator. Any basis may be used to expand the single particle orbit, provided one is able to calculate matrix element of the interaction V_{ij} with the basis states.

The set of single particle states, both filled or empty is assumed to form a orthonormal set of wave functions:

$$\sum_{\lambda} C_{\lambda}^{i*} C_{\lambda}^{i'} = \delta_{ii'} \quad \text{and} \quad \sum_{i} C_{\lambda}^{i*} C_{\lambda'}^{i} = \delta_{\lambda \lambda'}. \tag{3.84}$$

The wave function ψ is determined by the coefficients C_{λ}^{i} which become variational parameters. The energy will be stationary with normalization condition (Eq. 3.84) when

$$\frac{\partial}{\partial C_{\lambda}^{i*}} \left[\langle \psi | H \, | \psi \rangle - \epsilon_i \sum_{\lambda} C_{\lambda}^{i*} C_{\lambda}^{i} \right] = 0. \tag{3.85}$$

where ϵ_i is introduced as the Lagrange multiplier. With the help of expansion (3.83) of the orbits, the energy $\langle \psi | H \, | \psi \rangle$ in (3.85) can be expressed in terms of known matrix element of t and V in the known basis $|\lambda\rangle$. The derivative (3.85) may be calculated directly and one obtains

$$\frac{\partial}{\partial C_\lambda^{i*}}\left[\sum_{\lambda\lambda'}\langle\lambda|t|\lambda'\rangle C_\lambda^{i*}C_\lambda^i + \sum_{i=1}^{A}\sum_{\lambda_1\lambda_2}\langle\lambda\lambda_1|V|\lambda'\lambda_2\rangle C_\lambda^{i*}C_{\lambda_1}^{i*}C_{\lambda'}^i C_{\lambda_2}^{i*}\right.$$

$$\left.-\epsilon_i\sum_\lambda C_\lambda^{i*}C_\lambda^i\right]=0$$

or $\left[\sum_{\lambda'}\langle\lambda|t|\lambda'\rangle + \sum_{i=1}^{A}\sum_{\lambda_1\lambda_2}C_{\lambda_1}^{i*}\langle\lambda\lambda_1|V|\lambda'\lambda_2\rangle C_{\lambda_2}^i\right]C_{\lambda'}^i = \epsilon_i\,C_\lambda^i.$ (3.86a)

Eq. (3.86a) have the form of an eigen value problem

$$\sum_{\lambda'}\langle\lambda|h|\lambda'\rangle C_{\lambda'}^i = \epsilon_i\,C_\lambda^i,\qquad(3.86b)$$

where h is a Hamiltonian given by its matrix elements

$$\langle\lambda|h|\lambda'\rangle = \langle\lambda|t|\lambda'\rangle + \sum_{i=1}^{A}\sum_{\lambda_1\lambda_2}\langle\lambda\lambda_1|V|\lambda'\lambda_2\rangle C_{\lambda_1}^i\,C_{\lambda_2}^i\qquad(3.87)$$

and h is called HF Hamiltonian. Eq. (3.82) is the same as Eq. (3.87) expressed in the basis λ. Eqs. (3.86b) and (3.87) are the HF equations for the orbit i. They may be solved by well known iteration procedure stated below:

(1) An initial set of coefficients C_λ^i is guessed and also we guess which orbits are occupied.

(2) With this set of coefficients C_λ^i, the matrix elements of (3.87) of the HF Hamiltonian are calculated.

(3) Eq. (3.86a) is solved by diagonalizing the Hamiltonian matrix (3.87). A new set of coefficients C_λ^i is obtained, and A orbits are chosen to be filled by nucleons (often the ones with lowest energy). Then one returns to step 2. This process is repeated until successive diagonalizations produce the same set of coefficients C_λ^i. When this happens, the single particle potential is consistent with two-body potential that goes into (3.35). The accuracy of the method will depend upon the truncation of the sum in the expansion (3.83).

The deformed orbitals in a major shell are generated in a self consistent manner, and an intrinsic determinantal state then can be constructed from these orbitals. After determining which shape of the nucleus is energetically most favourable, the results are compared with those of a Nilsson type calculation. To find the nuclear spectra, the various states of total angular momentum I are then projected from such an intrinsic state.

B. Nuclear Matter

In the degenerate Fermi gas model of the nucleus, discussed in section (1.2B), there is some validity in the free motion of nucleons because the Pauli principle prevents the strong N-N force from causing virtual scattering of two nucleons into the occupied states. However, for nucleons whose energy is near the Fermi energy, the long range collision will transfer small virtual momentum and put nucleons into unoccupied states. These are the only scattering allowed by the Pauli principle and most of the scattering effects of attractive potential is eliminated. The object of the present section is to find an appropriate equation to describe the relative motion of two nucleons in "nuclear matter" and to examine the profound effects of other nucleons on the scattering of two nucleons in question.

The theoretical investigation of the properties of "nuclear matter" as deduced from the nature of nucleon – nucleon potential was developed by Brueckner[8] and Bethe and Goldstone[9] and subsequently by Brueckner and Gammel[10] and by Gomes et al.[11].

Analysis of the scattering of two nucleons in nuclear matter yields an interesting result. Their motions are essentially those of free particles except when they are close together by certain distance, called the "healing distance", which is about 1.0 fm. But the average internucleon distance in nuclear matter is around 1.77 fm. Consequently, effect of two particle correlations is not large. However, it is not negligible either. The small ratio of the healing distance to the average internucleon distance also implies that if two nucleons are closer than the healing distance, a third nucleon will very rarely be simultaneously within the healing distance of either of the first two. Naturally the question arise 'why is the healing distance so short? For the collision of two nucleons in vacuum, the wave functions never heal, for the scattered wave is phase shifted with respect to free particle motion and the actual wave function does not approach to the free particle wave function even at infinity. A record of the scattering event persists in this phase shift even into the asymptotic region. On the other hand, scattering in nuclear matter is distinguished from this situation by the fact that most, if not all, of the scattering states are already occupied and consequently the Pauli principle operates to prevent or strongly inhibit the scattering, so that, following their encounter, the two interacting particles will in general revert to their initial states. Although during encounter the wave function can be strongly perturbed, which can be viewed as a "wound" that becomes negligible as particles separate for distances $|\vec{r}_1 - \vec{r}_2|$ larger than the "healing distance", which is roughly equal to $\lambda_F = \dfrac{1}{k_F}$. The average separation of two nucleon in nuclear matter in accordance to (1.35a) is

$$d = \rho^{-1/3} = \left(\frac{3}{2}\, \pi^2\right)^{1/3} \lambda_F$$

or

$$\frac{d}{\lambda_F} = \left(\frac{3}{2}\, \pi^2\right)^{1/3} = 2.45 \, .$$

Thus, if one of the partners of a pair of nucleon singled out for consideration, makes a close encounter with yet with a third nucleon, the original pair wave function completely heals at separation $\approx d$ (= 1.77 fm).

Here, the nuclear matter is conceived as an idealized degenerate Fermi gas of uniform density, containing equal number of neutrons and protons (A/2 in each case) with Coulomb force turned off. A is assumed to be so large that the surface effects can be neglected. Often, it simplifies our thinking as well as computation to consider such hypothetical system of uniform infinite nuclear matter. Nuclei can not actually be very large because of repulsive Coulomb interactions among the protons cause large nuclei to fall apart. For this reason, in the hypothetical nuclear matter, Coulomb interactions are neglected. In infinite uniform nuclear matter, the properties are the same in all points of space, i.e. there is no physically distinguished point from which to determine the position. In other words, single particle potential is independent of position r in ordinary space. However, the energy of the particle may depend upon its momentum. A momentum dependent potential can be considered non-local in ordinary space (*see* Exercise 5.16). This means that energy of a particle at point r depends not only on r but also on wave function at other points $\vec{r}'$ near $\vec{r}$. More precisely the energy operator is not diagonal in coordinate space. The S. Eq. describing the relative motion of two nucleons in nuclear matter then reads as

$$-\frac{\hbar^2}{2m} \nabla_r^2\, \psi(r) + \int d\vec{r}'\, V(\vec{r}, \vec{r}')\psi(\vec{r}') = E\psi(r), \qquad (3.88)$$

where $V(\vec{r}, \vec{r}')$ is the actual potential felt by a nucleon at r, influenced by the presence of another nucleon at r'. The momentum dependence of non-local potential $V(\vec{r}, \vec{r}')$ can be seen by assuming the wave function $\psi(r)$ to be plane waves of constant momentum p, that is, $\psi = \exp\left[i(\vec{p}/\hbar)\cdot\vec{r}\right]$. Substituting ψ in (3.88), we get

$$E = \frac{p^2}{2m} + \int d\vec{r}'\, V(\vec{r}, \vec{r}')\exp\left(\frac{i\vec{p}\cdot(\vec{r}' - \vec{r})}{\hbar}\right) = \frac{p^2}{2m} + V(p), \qquad (3.89)$$

where $V(p)$ is the Fourier transform of $V(\vec{r}, \vec{r}')$. This demonstrates the equivalence between non-local and momentum dependent potential. The total energy of the nucleus in the ground state would be

$$E = \sum_{p<p_F} \frac{p^2}{2m} + \frac{1}{2} \sum_{p<p_F} V(p), \text{where } p_F \text{ is the Fermi momentum.} \qquad (3.90)$$

So the average energy per nucleon in accordance to Eq. (1.36) is

$$E_{av} = \frac{3}{5}\left(\frac{p_F^2}{2m}\right) + \frac{1}{2A} \sum_{p<p_F} V(p),$$

For momentum upto p_F and for some distance beyond, it appears to be a good approximation to treat the momentum dependence of $V(p)$ quadratic:

$$V(p) = -V_0 + bp^2 \text{ so that,}$$

$$E = \frac{p^2}{2m} - V_0 + bp^2 = \frac{p^2}{2}\left(\frac{1}{m} + 2b\right) - V_0 = \frac{p^2}{2m^*} - V_0, \qquad (3.91)$$

where $\dfrac{1}{m^*} = \left(\dfrac{1}{m} + 2b\right)$ is called the effective mass.

The above relation indicates that the energy has the same momentum dependence as a free particle of mass m^*. The approximation in which only terms upto order of p^2 is considered is called the "effective mass" approximation. In summary, we can say that a nucleon in nuclear matter has an effective mass which is smaller than in free space, its energy and momentum are related by (3.91) with $m^*/m = 0.7$.

In order that the ideal Fermi gas represent a real nucleus, the depth of the potential well should be so chosen that the energy required to detach an individual nucleon coincides with the experimental value of nucleon separation energy (~ 8 MeV). Since the least strongly bound nucleons are those with energies close to E_F, the depth of the potential well V_0 resulting from nuclear forces, is the sum of T_F given by (1.35b) and nucleon separation energy, i.e. (39 + 8) = 47 MeV, which is commensurate with the average interaction of the shell model. In order to improve upon this simple model we must specify the character of nucleon-nucleon potential operative between a pair of nucleon immersed in the Fermi sea of nucleons.

Let us consider for simplicity an attractive Wigner type central two-body potential which is same for all states and consider only the direct term and all pairs of nucleons $\left(\approx \dfrac{A(A-1)}{2} \approx A^2\right)$. Then

$$\langle V \rangle = \frac{1}{2} \rho \, A \int V_w(r_{ij}) \, d^3 r_{ij} = A \rho \tilde{V}_w \qquad (3.92)$$

where $\tilde{V}_w$ is half of the volume integral of Wigner interaction and is independent of A. The average kinetic energy per nucleon (1.36), when added to $\dfrac{\langle V \rangle}{A}$ yields

$$E_{av} = C_1 \rho^{2/3} - C_2 \rho, \, (3.93)$$

where C_1 and C_2 are positive constants independent of A. For minimum energy:

$$\frac{\partial E_{AV}}{\partial \rho} = \frac{C_1}{\rho^{1/3}} - C_2 \quad \text{or} \quad \frac{C_1}{\rho^{1/3}} - C_2 = 0. \qquad (3.94)$$

The above condition (3.94) is satisfied when $\rho \to \infty$, i.e. the system must collapse. In order to prevent collapse, the two nucleon force is assumed to have a repulsive core in the potential. This assumption is also necessary to explain saturation property of nuclear forces. We also saw the evidence of repulsive core from the analysis of high-energy N-N scattering data (Ch. 2).

In the present chapter we have described how the nuclear shell model has been very successful in explaining systematics of the ground state properties and low lying spectrum of the light nuclei. Nuclear matter theory, which validates shell model was also briefly described at the end. However, observed low lying spectrum of heavy nuclei have dramatically simple structure in which energy levels can be grouped into bands with strong electric quadrupole transitions between the bands. This is indicative of nuclear collective motion. A simpler description of motion of the nucleus as a whole in terms of few variables will be presented in the next chapter.

Exercise 3.11. Calculate the average energy per nucleon in the effective mass approximation:

$$E_{av} = \int_0^{p_F} \frac{p^2 \, dp \; p^2/2m + b/2p^2 - V_0/2}{\int_0^{p_F} p^2 \, dp}$$

$$= \frac{3}{10} \frac{p_F^2}{m} + \frac{3}{10} p_F^2 \, b - \frac{V_0}{2}$$

$$= \frac{3}{10} p_F^2 \left(\frac{1}{m} + b \right) - \frac{V_0}{2} = \frac{3}{10} \frac{p_F^2}{2m} \left(1 + \frac{m}{m*} \right) - \frac{V_0}{2}.$$

Exercise 3.12. Generalize (3.89) to the case in which the wave functions are not plane waves.

The wave function $\psi(r)$ can be written as a superposition of plane waves, namely, $\psi(\vec{r}) = \int \phi(p) e^{i\vec{p}\cdot\vec{r}/h} d\vec{p}$. The potential term in Eq. (3.88) becomes

$$\int d\vec{r}' V(\vec{r}, \vec{r}' - \vec{r}) \psi(\vec{r}')$$

$$= \iint d\vec{r}' \phi(\vec{p}) V(\vec{r}, \vec{r}' - \vec{r}) e^{i\vec{p}\cdot\vec{r}'/h} d\vec{p} = \int e^{i(\vec{p}\cdot\vec{r}/h)} d\vec{p} \, \phi(\vec{p}) V(\vec{r}, \vec{p})$$

$$= V(\vec{r}, -i\hbar\nabla_r) \int \phi(\vec{p}) \int e^{i(\vec{p}\cdot\vec{r}/h)} dp = V(\vec{r}, p_{op}) \psi(r),$$

where $V(\vec{r}, \vec{p})$ is the Fourier transform of $V(\vec{r}, \vec{r}' - \vec{r})$ and $P_{op} = -i\hbar\nabla_r$.

REFERENCES

1. Robert Oppenheimer on success of the shell model (1964). Quoted by J.D. Jensen in Nobel Lectures in Physics (1963-70). Elsevier, Amsterdam page 48.
2. W.M. Elasser, J. Phys. Et. Radium **4**, 549 (1939).
3. A. Bohr and B.R. Mottelson, Nuclear Structure, Vol. I and II, Benjamin, New York (1969).
4. M.G. Mayer, Phys. Rev. **75**, 1969 (1949); Ibid. **78**, 16, 20 (1950); O. Hasul, J.H.D. Jensen and H.E. Suess, Phys. Rev. **75**, 1776 (1949).
5. P.J. Brussaard and P.W.M. Glaudemans, Shell-Model Applications in Nuclear Spectroscopy North Holland (1977).
6. Igal Talmi and Amos De Shalit, Nuclear Shell Theory, Academic Press (1963).
7. A. Bohr, B.R. Mottelson and D. Pines, Phys. Rev. **110**, 936 (1958).
8. K.A. Brueckner, Phys. Rev. **96**, 508 (1954); Phys. Rev. **97**, 1353 (1955).
9. H.A. Bethe and J. Goldstone, Proc. Roy. Soc. (London) **A238**, 551 (1956).
10. K.A. Brueckner and J.L. Gammel, Phys. Rev. **109**, 1023 (1958).
11. L.C. Gomes, J.D. Walecka and V.F. Weisskopf, Ann. Phys. **3**, 241 (1958).
12. George Ripka, Advances in Nuclear Physics, Vol. 1, Plenum Press, New York (1968).

SUGGESTED BOOKS FOR FURTHER READING

13. Structure of the Nucleus, M.A. Preston and R.K. Bhaduri, Addision-Wesley, MA (1975).
14. Theory of Nuclear Structure, M.K. Pal East West Press, New Delhi (1982).
15. Nuclear Shell Model, Kres L.G. Heide, Springer-Verlag Berlin (1990).
16. Nuclear Models, W. Grenier and J.A. Maruhn, Springer (1996).
17. Fifty years of Shell Model. The quest for the effective interaction Igal Talmi, Advances in Nuclear Physics Ed. J.W. Negale and E.W. Vogt Vol. **27** (2003) Plenum Publisher.
18. Shell Model, E. Caurier et al. Rev. Mod. Phys. **77**, 427-488 (2005).
19. Unified Theory of Nuclear Models and Forces, G.E. Brown, North Holland (1971).
20. Theory of Nuclear Shell Model, R.D. Lawson Clarendon Press, Oxford 1980.

4

Theory of Nuclear Collective Motion

"Rotation is a universal phenomenon"

– E.T. Whittaker[1]

4.1 DEFORMABLE LIQUID DROP

Classical systems like the Earth, the Sun and a liquid drop would be spherical if they were not rotating. The intimate link between shape of a macroscopic object to its rotary capabilities has been a subject of study for a long time. Newton in the 17th century argued that rotation will make shape of the Earth slightly oblate with the symmetry axis coinciding rotation axis. Lord Rayleigh[2] investigated the stability and oscillations of electrically charged liquid drop in 1877. But unlike the stars or planets the nucleus is first of all governed by the laws of quantum mechanics and analogy with astronomical objects should not be pushed too far.

It took quite sometime to realize that at low excitation energies most of nuclei have non-spherical equilibrium shape even when they are not rotating. The experimental observations which give some immediate indications of non-spherical shape, are the large quadrupole moments observed in nuclei which are far away from the magic nuclei. The other observations which gave indication of nuclear collective motion (motion of a nucleus as a whole in which all nucleons participate) are: (1) the large transition probability between collective levels, (2) rotational and vibrational spectra at the low energy excitation and (3) giant resonances at the high energy excitation.

At high spin, nucleus shows coexistence of properties determined by the individual nucleon motion and also collective participation of nucleons in the motion of a nucleus as a whole. The shell model has been developed with the concept of an average force field in a nucleus. By the collective model, we understand as those models of the nucleus, in which the average field deviates from the spherical symmetry. There are two approaches to study collective behaviour of nuclei. The first one, is the macroscopic or the phenomenological approach in which the deformation is introduced as a semi-emperical parameter, without raising the basic question how the intrinsic deformation inside the nucleus arises. The second one, is the microscopic approach in which deformed orbitals in a major shell are generated in a self-consistent manner and an intrinsic determinantal state is then constructed from these orbitals. After determining which shape of the nucleus is energetically most favourable, the results are compared with those of the phenomenological models. In other words, in a microscopic model one finds self-consistent solution of many-body Hamiltonian leading to the ground state many body wave function in a deformed field. We will mostly discuss here the phenomenological models.

Bohr and Wheeler[3] extended the liquid drop model (Sec. 1.2) by assuming that the liquid drop is deformable about the spherical equilibrium shape, for the explanation of nuclear fission. Such a deformed nuclear surface can be represented with respect to the space fixed coordinate system as:

$$r\,(\theta,\,\varphi) = r_0 \left[1 + \sum_{\lambda\mu} \alpha_{\lambda\mu} \cdot Y_\lambda^\mu(\theta,\varphi) \right], \tag{4.1}$$

where r_0 is the radius of the spherical drop (Fig. 4.1), $\alpha_{\lambda\mu}$ are deformation parameters and Y_λ^μ are the spherical harmonics. We can rewrite (4.1) as

$$r(\theta,\varphi) = r_0 \left[1 + \eta \right], \text{ where } \eta = \sum_{\lambda\mu} \alpha_{\lambda\mu} \cdot Y_\lambda^\mu. \tag{4.2}$$

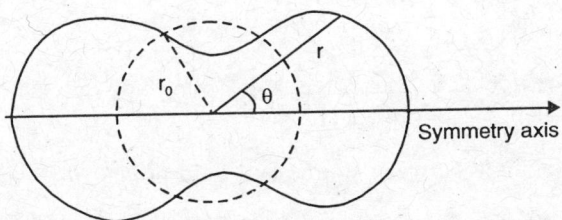

Fig. 4.1: Co-ordinates of axially symmetric deformation. The dotted circle is the spherical nucleus of radius r_0

The nuclear volume $V = \int d\omega \int_0^{r_0(1+\eta)} r^2\, dr = \dfrac{1}{3} r_0^3 \int (1+\eta)^3\, d\omega,$

where $d\omega$ is an element of solid angle. The constancy of nuclear volume demands

$$\frac{4\pi r_0^3}{3} = \frac{4\pi r_0^3}{3} + r_0 \int d\omega \, \eta (1+\eta)$$

or
$$\int \eta (1+\eta) \, d\omega \approx 0, \qquad \text{for } \eta \ll 1.$$

Integrating over 4π solid angle, we get $\sqrt{4\pi} \, \alpha_{00} + \sum_{\lambda\mu} |\alpha_{\lambda\mu}|^2 = 0$.

Following Eder[28], we estimate the self Coulomb energy:

$$E_c = \frac{1}{2} \iint \frac{\rho(s) \, \rho(s')}{|\vec{s} - \vec{s}'|} \, d^3s \, d^3s'. \tag{4.3}$$

For a uniform charge distribution $\rho = \dfrac{3Ze}{4\pi \, r_0^3}$, we have

$$E_c = E_c^0 + \frac{1}{2} \rho^2 \int d\omega \int d\omega' \int_0^r s^2 ds \int_0^r \frac{s'^2 \, ds'}{|\vec{s} - \vec{s}'|}$$

$$= E_c^0 + \frac{1}{2} \rho^2 \int d\omega \int d\omega' \int_{r_0}^r s^2 ds \left\{ 2 \int_0^{r_0} + \int_{r_0}^r \right\} \frac{s'^2 \, ds'}{|\vec{s} - \vec{s}'|}, \tag{4.4}$$

where $E_c^0 = \dfrac{3}{5} \dfrac{Z^2 e^2}{r_0} = $ Coulomb energy for spherical nucleus Eq. (1.31).

$$|\vec{s} - \vec{s}'|^{-1} = \sum_{\lambda=0}^{\infty} \frac{s'^\lambda}{s^{\lambda+1}} P_\lambda(\cos \chi)$$

$$= 4\pi \sum_{\lambda=0}^{\infty} \sum_{\mu=-\lambda}^{+\lambda} \frac{s'^\lambda}{(2\lambda+1) s^{\lambda+1}} Y_\lambda^{\mu^*}(\theta', \phi') Y_\lambda^\mu(\theta, \phi), \tag{4.5a}$$

where χ is the angle between source and field points, $d\omega = \sin\theta \, d\theta \, d\phi$,

$d\omega' = \sin\theta' \, d\theta' \, d\phi'$ and $\cos\chi = \cos\theta \cos\theta' + \sin\theta \sin\theta' \cos(\phi - \phi')$. (4.5b)

To evaluate (4.4) we employ the addition theorem of spherical harmonics. The approximate value of the first integral (say I_1) in (4.4) is evaluated by using (4.2) and (4.5a):

$$I_1 = \rho^2 \, 4\pi \int d\omega' \int_{r_0}^r s \, ds \left\{ \int_0^s s'^2 \, ds' + s \int_s^{r_0} s' \, ds' \right\}$$

$$= \frac{2\pi}{3}\rho^2 \int d\omega \int_{r_0}^{r} ds\,(3r_0^2 s^2 - s^4)$$

$$= \frac{2\pi}{3}\rho^2 \int d\omega \left[r_0^2 r^3 - \frac{r^5}{5} - \frac{4}{5}r_o^5 \right] = \frac{2\pi}{3}\rho^2 r_0^5 \int d\omega \left[2\eta(\eta+1) - \eta^2 \right]$$

$$= \frac{-2\pi}{3}\rho^2 r_0^5 \sum_{\lambda\mu} |\alpha_{\lambda\mu}|^2, \text{ since } \int \eta(\eta+1)\,d\omega = 0 \text{ and } \eta \ll 1 \text{ so that}$$

only the second order term in η is retained in the above calculation.

The second integral (say I_2) in (4.4) is evaluated as follows:

$$I_2 = \frac{1}{2}\rho^2 \int d\omega \int d\omega' \int_{r_0}^{r} s^2 ds \int_{r_0}^{r} \frac{s'^2\,ds'}{|\vec{s} - \vec{s}\,'|}$$

$$= \frac{1}{2}\rho^2 r_0^5 \int d\omega \int d\omega' \sum_{\lambda\mu} \eta\eta' P_\lambda(\cos\chi)$$

$$= 2\pi\,\rho^2 r_0^5 \sum_{\lambda\mu} |\alpha_{\lambda\mu}|^2/(2\lambda+1).$$

Hence,
$$E_c = E_c^0 + 2\pi\rho^2 r_0^5 \left[\sum_{\lambda\mu} |\alpha_{\lambda\mu}|^2 \left\{ \frac{1}{2\lambda+1} - \frac{1}{3} \right\} \right]$$

$$= E_c^0 - \frac{4\pi}{3}\rho^2 r_0^5 \sum_{\lambda\mu} \frac{\lambda-1}{2\lambda+1} |\alpha_{\lambda\mu}|^2,$$

or
$$E_c = E_c^0 \left[1 - \frac{5}{4\pi} \sum_{\lambda\mu} |\alpha_{\lambda\mu}|^2 \left\{ \frac{\lambda-1}{2\lambda+1} \right\} \right]. \tag{4.6}$$

It follows from (4.6) that as the deformation increases, the average distance between protons increases and hence the Coulomb energy decreases.

The surface energy $E_s = \gamma \int dS$, where S is the surface area and γ is the nuclear surface tension. The element of deformed surface

$$dS = r_0^2 d\omega \left[1 + 2\eta + \eta^2 + \frac{1}{2}\left(\frac{\partial\eta}{\partial\theta} \right)^2 + \frac{1}{2\sin^2\theta}\left(\frac{\partial\eta}{\partial\varphi} \right)^2 \right],$$

so that
$$E_s = \gamma \int r_0^2 d\omega + 2\gamma\, r_0 \int \eta\,(\eta+1)d\omega$$

$$+\frac{1}{2}\,\gamma\,r_0^2\int d\omega\left\{\left(\frac{\partial\eta}{\partial\theta}\right)^2+\frac{1}{\sin^2\theta}\left(\frac{\partial\eta}{\partial\varphi}\right)^2-2\eta^2\right\}.$$

Carrying out the angular integration we get

$$E_s=E_s^o+\frac{1}{2}\gamma r_0^2\left[\sum_{\lambda\mu}|\,\alpha_{\lambda\mu}\,|^2\{\lambda(\lambda+1)-2\}\right]$$

$$=E_s^o+\frac{1}{2}\,\gamma\,r_0^2\sum_{\lambda\mu}(\lambda-1)(\lambda+1)\,|\,\alpha_{\lambda\mu}\,|^2\,. \tag{4.7}$$

We see that the surface energy increases with the deformation. The total deformation potential energy,

$$V_{pot}(\alpha_{\lambda\mu})=(E_s+E_c)-(E_s^o+E_c^o)$$

$$=\frac{1}{2}\sum_{\lambda\mu}|\,\alpha_{\lambda\mu}\,|^2\left[\gamma\,r_0^2\,(\lambda+2)\,(\lambda-1)-\frac{3}{2\pi}\,\frac{Z^2e^2}{r_0}\,\frac{(\lambda-1)}{2\lambda+1}\right]$$

$$=\frac{1}{2}\sum_{\lambda\mu}C_\lambda\,|\,\alpha_{\lambda\mu}\,|^2\,, \tag{4.8a}$$

where the "stiffness" parameter $C_\lambda=\dfrac{\lambda-1}{4\pi}\left[(\lambda+2)\,E_s^o-\dfrac{10E_c^o}{2\lambda+1}\right].$ (4.8b)

The surface and Coulomb terms depend differently on the shape of the nucleus and there is a competition between them as shape changes. For a given volume, the surface energy increases with deformation, whereas the Coulomb energy decreases. In the expansion (4.8a), $\lambda=0$ term is negligible for small deformation and $\lambda=1$ term is absent because it corresponds to the displacement of the centre and not any real deformation. Hence, the lowest order deformation possible in a nucleus is $\lambda=2$ deformation, called the quadrupole deformation.

A. Nuclear Fission

An interesting application of the liquid drop concept is to the phenomenon of nuclear fission. In the fission process a relatively heavy nucleus splits into two large fragments with energy release of nearly 200 MeV. Fission can be either spontaneous or reaction induced. Characteristically, it shows that for low internal excitation of the fissioning nucleus, the two fragments most probably have unequal masses, a situation known as the asymmetric fission. In what follows, we will discuss the theory of nuclear fission within the framework of deformable and charged liquid drop picture.

For an axially symmetric quadrupole deformation the potential energy

$$V_{pot}(\alpha_{20}) = \frac{2}{5}\left[E_s^o - \frac{1}{2}E_c^o \right]\alpha_2^2, \tag{4.9}$$

where $\alpha_2 = (5/4\pi)^{1/2}\,\alpha_{20}$. If $V_{pot}(\alpha_2)$ is positive, it means that there is an effective energy barrier for deformation and fission, while a negative or zero $V_{pot}(\alpha_2)$ indicates that deformation is energetically favoured. Therefore, at the point of fission $(1/2)E_c^o \geq E_s^o$ or $(1/2)\,a_c\left(Z^2/A^{1/3}\right) \geq a_s\,A^{2/3}$, which defines a critical value of Z^2/A from (1.39b) as

$$\left(\frac{Z^2}{A}\right)_c \geq \frac{2a_s}{a_c} \geq \frac{2a_2}{a_3} \geq 50. \tag{4.10}$$

The Eq. (4.9) is redefined in terms of the fissility parameter

$$x = (Z^2/A)/(Z^2/A)_c = \frac{E_c^0}{2E_s^0} \text{ as } V_{pot}(\alpha_2) = \frac{2}{5}\,E_s^0(1-x)\,\alpha_2^2, \tag{4.11}$$

which is valid for small deformation α_2. Nuclei with $x \geq 1$, (which is the case for $Z > 114$ and $A > 258$) can not exist and those for which $x < 1$ fission can not take place according to the classical liquid drop model (Fig. 4.2a). However, for nuclei with $x < 1$ there is always quantum mechanical tunneling effects similar to alpha decay and in principle fission can take place. For example, ^{238}U corresponds to $(Z^2/A) = 35.56$ and $x = 0.71$; the classical model disallows fission of ^{238}U but due to the quantum mechanical tunneling, fission takes place. It is not observed because the spontaneous fission life time for ^{238}U is 8×10^{15} years. Thus, on the energetic consideration alone spontaneous fission is possible but the presence of fission barrier keeps these nuclei stable. The spontaneous fission is actually a negligible mode of decay compared to α-particle decay until $A \geq 250$. An example of a case where the spontaneous fission actually becomes the main decay channel is $^{254}_{98}Cf$ for which $x = 0.76$ and the fission life time » 60.5 days. It is found experimentally that the spontaneous fission life time decreases with increasing Z^2/A. For $^{258}_{100}Fm$, the half life time is 380 m sec. For $x > 1$, nuclei have no barrier and fission immediately due to the Coulomb repulsion. The fission process is therefore basically governed by the characteristic features of the fission barrier in the map of the potential energy versus deformation for a fissioning nucleus (Fig. 4.2a).

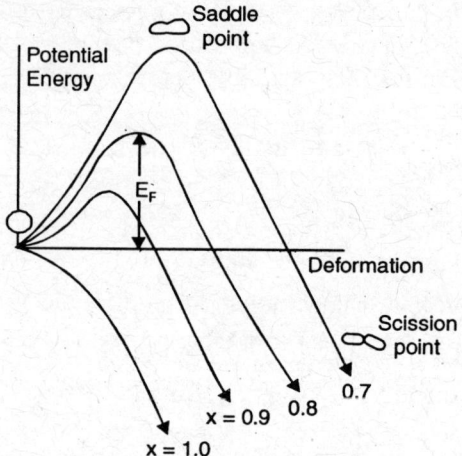

Fig. 4.2: (*a*) Potential barrier for nuclear fission (see text)

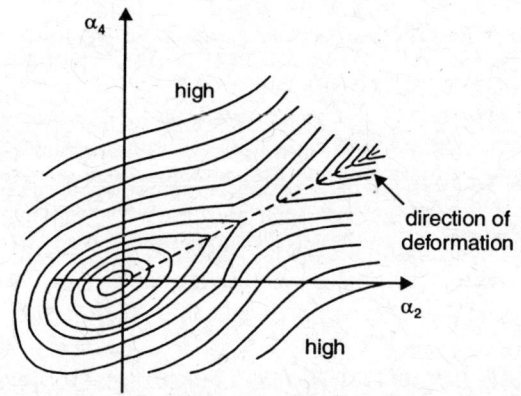

Fig. 4.2: (*b*) Potential energy contour of a nucleus as a function of α_2 and α_4

As mentioned earlier, (4.11) is valid for small deformation α_2. When we consider higher order term in α_2 and if $V_{pot}(\alpha_2)$ continues to increase indefinitely as α_2 increases, then α_2 is not a natural parameter to use for large deformation. Bóhr and Wheeler[3] calculated the potential energy to fourth order in α_2 and to second order in α_4 assuming axial symmetry. α_4 is called the necking parameter.

If formation and separation of two daughter nuclei occur in a completely symmetrical manner in a continuous fission process, then $r(\pi - \theta) = r(\theta)$. This means that terms of even α will appear in the expression (4.1), which will then read as

$$r = r_0\left[1 + \alpha_2 P_2(\cos\theta) + \alpha_4 P_4(\cos\theta) + \cdots\cdots\right] \qquad (4.12)$$

and potential energy as a function of deformation α becomes

$$V_{pot}(\alpha) = V_{pot}(\alpha_2, \alpha_4) = \frac{2}{5} E_s^o (1-x) F(\alpha_2, \alpha_4), \qquad (4.13)$$

where $F(\alpha_2, \alpha_4)$ is a polynomial in α_2 and α_4 obtained by Bohr and Wheeler[3] (see Eder[28]). If we plot the potential energy $V_{pot}(\alpha)$ as a function of deformation parameter, for different x we get what is shown in Fig. 4.2(*a*). Fig. 4.2(*b*) shows a schematic diagram of the potential energy contour of a nucleus as a function of α_2 and α_4 calculated from the liquid drop model. We can determine the maximum value of the potential energy $V_{pot}(\alpha_2, \alpha_4)$ to see how much work is needed for nuclear fission to occur. We first keep α_2 fixed and find the α_4 value for which $V_{pot}(\alpha_2, \alpha_4)$ is maximum. Therefore, we set $\dfrac{\partial V_{pot}(\alpha_2, \alpha_4)}{\partial \alpha_4} = 0$, which yields for heavy nuclei $(x \approx 1)$: $\alpha_4 = \dfrac{27}{85}\alpha_2^2$. The path in the (α_2, α_4) plane that maintains this relation is called deformation axis (Fig. 4.2*b*).

Spontaneous and Induced Fissions: We have already mentioned earlier that the spontaneous fission can be explained also in terms of quantum mechanical tunneling through the fission barrier. While describing the dynamics of an oscillating drop we noted that the liquid drop model predicts a single hump around the saddle point. If we parametrize this path by a single parameter α, which is a measure of deformation of the nucleus along this path, and denote the value of α at the saddle point by α_s, then the potential energy of the charged liquid drop representing the nucleus may be written as

$$V(\alpha) = V(\alpha_s) - C(\alpha - \alpha_s)^2 + \cdots\cdots, \qquad (4.14)$$

where C is a measure of the curvature of the barrier and $V(\alpha_s) = E_f$ is the height of the barrier. The Hamiltonian for this nucleus near the saddle point is

$$H(\alpha') = \frac{1}{2} B(\dot{\alpha}')^2 - C(\alpha')^2 + E_f, \qquad (4.15)$$

where $\alpha' = \alpha - \alpha_s$ and the kinetic energy of the nucleus around the saddle point is $(1/2) B(\dot{\alpha}')^2$, with B being the associated inertial parameter near the saddle point. If we substitute $C = (1/2) B\omega_f^2$, where ω_f^2 has the dimensions of frequency, then (4.15) will resemble the Hamiltonian of an inverted harmonic oscillator. In order for fission to take place, the nucleus as a whole has to tunnel through this barrier. To calculate the transmission coefficient, we employ the WKB approximation (Sec. 1.2B) as we did earlier for the explanation of alpha decay. For an inverted harmonic-oscillator barrier, the WKB result is exact[4]. If an energy $E(< E_f)$ has been imparted to the nucleus, then the transmission coefficient is given by

$$T = \frac{1}{1 + e^{2S}} \approx e^{-2S}, \tag{4.16}$$

where $2S = 2\pi(E_f - E)/\hbar\omega_f$ and $\hbar\omega_f$ is of the order of 1 MeV for nuclei with $A \approx 240$. The vibrating nucleus is making a certain number of strikes n on the barrier per second. This number is equated to the frequency of vibrations. Setting $\hbar\omega_f \approx 1$ MeV, we can estimate n as follows:

$$2\pi\nu_f = \frac{1 \text{ MeV} \times c}{\hbar c} = \frac{1 \text{ MeV} \times 3 \times 10^{10} \text{ cm sec}^{-1}}{197 \text{ MeV} \times 10^{-13} \text{ cm}}. \tag{4.17}$$

Hence, $n = 10^{20.38}$ sec^{-1}. The half life for spontaneous fission $t_{1/2} = \dfrac{\log 2}{n \cdot T}$

$\cong 10^{-20.38} T^{-1}$ sec, where T is given by Eq. (4.16). For $E = E_f$, $T = 1/2$.

The threshold energy for fission may be supplied if one uses a flow of low energy neutrons to induce neutron capture reaction. These push the nucleus into an excited state above the fission barrier, as a result the nucleus splits up. This process is known as induced nuclear fission. Neutron capture by nuclei with an odd neutron number release not just some binding energy but also some pairing energy. This small extra contribution to the energy balance makes a decisive difference to nuclear fission properties. For example, neutron capture by ^{238}U, will release 4.9 MeV binding energy, which is below threshold energy of 5.5 MeV for nuclear fission of ^{239}U. Neutron capture by ^{238}U can therefore only lead to nuclear fission if the neutron possesses at least 0.6 MeV energy (fast neutrons). On the top of this neutron capture cross section is proportional to $1/v$, where v is the velocity of neutron, and so it is very small. On the other hand, neutron capture in ^{235}U releases 6.4 MeV and the fission barrier of ^{236}U is just 5.5 MeV. Thus, fission may be induced in ^{235}U by slow neutrons.

B. Deformed Potential Energy Landscape and Fission Isomers

A group of Russian workers around 1962 reported spontaneous fission activities in ^{240}Am, ^{242}Am isotopes with half lives of order of milliseconds. This would mean that there are excited states in these nuclei, which instead of decaying to the ground states by gamma emission, tunnel through the fission barriers and exhibit fission decay. Normal isomers are explained in terms of their high spins which hinder gamma emission and give them a relatively long half life (Ch. 3 - Sec. 3.2). Fission isomers can not be explained as normal spin isomers because spins of these levels differ by only 2 to 3 units of angular momentum from those of the corresponding ground states. Some of the known isomers with their each half life in nanoseconds given inside bracket are:

^{243}Cm (37), ^{240}Pu (4), ^{236}U (105), ^{238}U (195)

An explanation for the existence of fission isomers emerged from the major progress in our understanding of potential energy landscape of a fissioning nucleus through the pioneering work of Mayer and Swaitecki[5] and Strutinsky[6]. They developed a macroscopic - microscopic approach to calculate potential energy of a deformed nucleus taking into account the energy correction introduced by shell effects. This work revealed that the hitherto considered single-humped fission barrier was indeed a double humped fission barrier for a range of fissioning nuclei in actinide region.

Nuclear shell effects: The BW semi-empirical mass formula is good enough to reproduce the main trend in the variation of binding energy of nuclei with mass number A and charge Z. The BW formula predicted binding energy is generally in error by only several MeV in a quantity that goes up to 2000 MeV. The variation of error from one nucleus to the next is very much smaller than the error itself. There is a systematic shell effect in the deviations. This can be seen easily. We plot the difference $M(N,Z)_{expt} - M(N,Z)_{LD}$ versus the mass number A in Fig. 4.3, where $M(N,Z)_{LD}$ is the spherical liquid drop value given by BW formula. The quantity $M(N,Z)_{expt} - M(N,Z)_{LD}$ is called the experimental shell effects. It is seen that deviations are particularly large for the magic nuclei. Similar large discrepencies show up when we plot it against proton number Z. For closed shell nuclei $M(N,Z)_{LD}$ is higher than $M(N,Z)_{expt}$. This discrepency is not quite surprising because BW formula does not take into account shell effects and assumes a smooth continuous distribution of levels, whereas in the real nucleus, the single particle levels are discrete with large energy gap between major shells (Fig. 4.4b). For mid shell nuclei, the shell correction is most negative for deformed shape. Liquid drop model favours spherical shape.

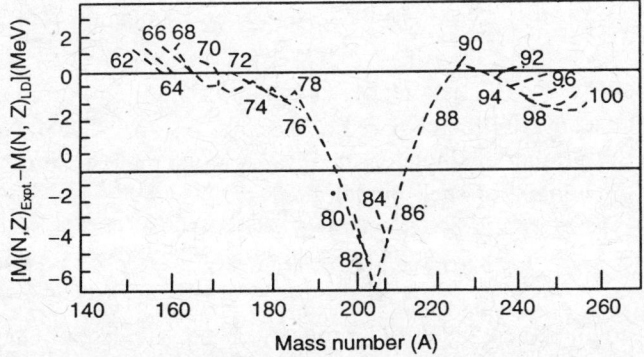

Fig. 4.3: $[M(N,Z)_{expt} - M(N,Z)_{LD}]$ versus A, showing manifestation of the shell effects

The total potential energy of a nucleus as a function mass number A or (N or Z) and deformation α can be written as

$$E = E_{\text{Macroscopic}} + \Delta E_{\text{Microscopic}} \qquad (4.18)$$

The contribution of microscopic energy comes from the shell correction energy and pairing correction energy. The macroscopic part is calculated by the liquid drop (LD) model. The small shell structure energy deviations are then added to the smooth energy bulk contribution calculated by the liquid drop model. The shell structure contribution to the liquid drop model dependence is oscillatory with respect to the nuclear deformation and leads to a secondary minimum in the fission barrier of actinide region. With the existence of doubled humped barrier, the explanation of fission isomers became straight forward. The salient features of Strutinsky's method[6] are given below:

We have already calculated the liquid drop energy (Eq. 4.8) using the parametrization in (Eq. 4.1). The single particle potential can be generated such that the overall geometrical shape follows that of the nucleus (e.g., Nilsson potential – Sec. 4.3). We may also assume a single particle potential of Yukawa form and fold it with the nuclear density, which is taken to be constant inside the shape. This gives the folded Yukawa potential.

If ϵ_i's are the eigen values of a properly chosen (realistic) shell model one-body potential for a nucleus of mass number A, then

$$E(A) = \sum_{i=1}^{A} \epsilon_i, \qquad (4.19)$$

Eq. (4.19) takes into account the quantum nature of the levels and their fluctuations due to the shell effects. Thus, the shell model energy given by (4.19) can be divided into a smooth part $\overline{E}(A)$ and an oscillating part, which represents the shell correction ΔE_{SC}. Therefore,

$$E(A) = \overline{E}(A) + \Delta E_{SC}. \qquad (4.20)$$

The shell correction energy results from the non-homogeneity of the single particle levels.

The smooth part $\overline{E}(A)$ is already contained in the liquid drop energy. In order to calculate the oscillating part one uses the concept of level density. Let $g(\epsilon) \, d\epsilon$ be the number of single particle levels in the interval between ϵ and $\epsilon + d\epsilon$. In the context of shell model

$$g(\epsilon) = \sum_{i=1}^{\infty} \delta(\epsilon - \epsilon_i), \qquad (4.21)$$

where ϵ_i's are the eigen values of a properly chosen single particle potential and the sum over i is taken over all states taking account of the degeneracy of the states. The total number of particles and shell model energy are then given by

$$A = \int\limits_{-\infty}^{\lambda} g(\epsilon) d\epsilon \quad \text{and} \quad E(A) = \int\limits_{-\infty}^{\lambda} \epsilon g(\epsilon) \, d\epsilon, \qquad (4.22)$$

where λ is the Fermi energy that can be arbitrarily chosen between the last filled orbit and the first unfilled orbit.

The level density $g(\epsilon)$ has a discrete character which we smooth out as follows:

$$g(\epsilon) = \overline{g}(\epsilon) + \delta g(\epsilon) = \overline{g}(\epsilon) + g_{osc}(\epsilon), \qquad (4.23)$$

The total number of particles $A = \int\limits_{-\infty}^{\overline{\lambda}} \overline{g}(\epsilon) d\epsilon,$ \qquad (4.24)

where $\overline{\lambda}$ is the average Fermi energy usually different from λ, and the lower limit signifies that each level in principle can also spread into negative ranges of ϵ (even when the bottom of the level is at $\epsilon = 0$). The lower limit can be the energy of the lowest level rather than $-\infty$. Just as $E(A)$ is the energy associated with $g(\epsilon)$, the energy associated with $\overline{g}(\epsilon)$ is

$$\overline{E}(A) = \int\limits_{-\infty}^{\overline{\lambda}} \epsilon \overline{g}(\epsilon) d\epsilon. \qquad (4.25)$$

The shell correction energy can be obtained by using (4.25) with the knowledge of the smoothened level density $\overline{g}(\epsilon)$.

So, once we know how to make separation (4.20), we will be able to estimate ΔE_{SC}. The ways of doing the separation, as done by Strutinsky, is to spread the single particle energy over same width and smooth the individual delta function occurring in (4.21) in such a manner that these overlap with each other to yield a smooth $\overline{g}(\epsilon)$. Therefore, we express

$$\delta(\epsilon - \epsilon_i) = \frac{1}{\gamma} \delta\left(\frac{\epsilon - \epsilon_i}{\gamma} \right),$$

which is valid for any γ and we identify γ as a smoothing distance, which is of the same order as shell spacings.

We expand $\delta(x)$ in terms of harmonic oscillator wave functions:

$$\delta(x) = e^{-x^2} \sum_m A_m H_m(x), \qquad (4.26)$$

where $x = \dfrac{\epsilon - \epsilon_i}{\gamma}$ and $H_m(x)$ is a generalized Laguerre polynomial of order m. Since $\delta(x)$ is an even function of x, $A_m = 0$ for odd m and

$$\int_{-\infty}^{\infty} \delta(x) H_n(x)\,dx = \int_{-\infty}^{\infty} e^{-x^2} \sum_m A_m H_m(x) H_n(x)\,dx.$$

The γ indicates width of the Gaussian function appearing above.

Now

$$H_n(0) = (-)^{n/2}\, n\,! \Big/ \left(\frac{n}{2}\right)! = \sum_{m=\text{even}} A_m\, \delta_{mn}\, 2^n\, n!\sqrt{\pi}$$

Hence, $\quad A_m = \dfrac{1}{\sqrt{\pi}}\dfrac{(-)^{m/2}}{2^m \left(\dfrac{m}{2}\right)!}$ and $\quad \delta(x) = e^{-x^2} \displaystyle\sum_m \dfrac{1}{\sqrt{\pi}}\dfrac{(-)^{m/2}}{2^m \left(\dfrac{m}{2}\right)!} H_m(x).$

Therefore,

$$\overline{g}(\epsilon) = \frac{1}{\sqrt{\pi}}\frac{1}{\gamma} \sum_{i=\text{all levels}} e^{-(\epsilon-\epsilon_i)^2/\gamma^2} \sum_{m=\text{even}} \frac{(-)^{m/2}}{2^m \left(\dfrac{m}{2}\right)!} H_m\left(\frac{\epsilon - \epsilon_i}{\gamma}\right). \quad (4.27)$$

We can see from (4.27) that each separate contribution to the density is spread out. Next we choose γ such that $\gamma = \hbar\omega = (41/A^{1/3})\,\text{MeV}$ and divide the series (4.27) into two parts

$$\sum_{m=\text{even}} (\ ) = \sum_{m=0}^{p} (\ ) + \sum_{\substack{m=p+2 \\ p=\text{even}}}^{\infty} (\ ). \qquad (4.28)$$

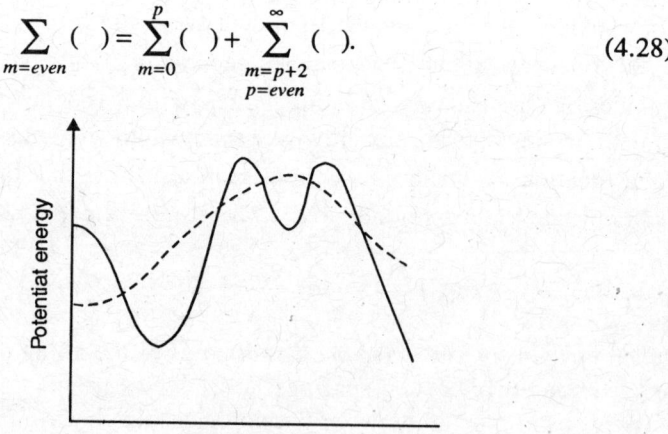

Fig. 4.4: (*a*) The liquid drop potential energy after shell correction, showing double humped barrier, i.e. more than one minimum. The dotted curve (liquid drop) and solid curve (after shell correction)

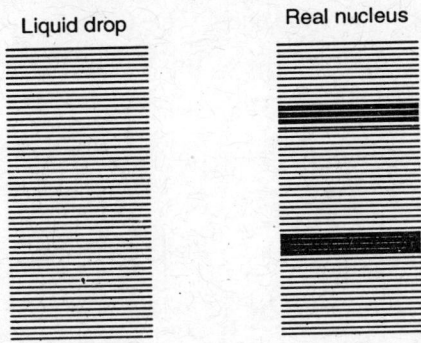

Fig. 4.4: (b) Density of levels

The first part of (4.28) gives the average result and the second part, the fluctuations. If we take p to be too large, we get too small a shell correction whereas if we take p too small, we get too crude a result for the average energy, which amounts to a gross shell correction. It turns out that $p = 6$ is a suitable choice which gives an accuracy around 0.5 MeV. By doing the summation in (4.28) over $m = 0, 2, 4, 6$, we get $\bar{g}(\epsilon)$. The average Fermi energy $\bar{\lambda}$ is determined by employing the condition (4.24). Knowing $\bar{\lambda}$ and using (4.25) we get the average energy $\bar{E}$. The difference $E(A) - \bar{E}$ directly gives the shell correction energy E_{SC}.

We find from the above analysis that the variation of the single particle energy with deformation gives a shell correction which modulates the liquid drop potential energy as shown in Fig. 4.4(a). The minima in this curve correspond to the dumpings of the single particle levels. We also see that the shell correction shifts the first minimum in the total potential energy to a non-zero deformation and hence confirm the existence of nuclei which are deformed in their ground state. It also modulates the barrier heights. It gives better values for the ground state energies showing that the deviation from the liquid drop values are well reproduced by E_{SC}. The calculated ground state deformations are in good agreement with the measured values, obtained from the intrinsic quadrupole moments. We see from Fig. (4.4a) that the shell correction to liquid drop energy leads to a *secondary minimum* in the fission barrier of actinide nuclei. From the present analysis we find that the shell correction drives the nucleus towards deformed shapes and pairing correction gives maximum stability for spherical shapes. Equilibrium shape depends on the relative importance of different terms. With the existence of double hummed barrier, the explanation of fission isomers became straight forward.

Exercise 4.1. Employing the BW mass formula calculate the energy released in the spontaneous fission of $^{238}_{92}U$.

For the symmetric fission, the energy released

$$Q = 2\left[(BE)_{Z/2}^{A/2} - (BE)_Z^A\right]$$

$$= a_2(1-\sqrt{2})A^{2/3} + a_3\left(1 - \frac{1}{3\sqrt{4}}\right)\frac{Z^2}{A}\cdot A^{2/3}$$

$$= A^{2/3}\left[-4.36 + 0.266\frac{Z^2}{A}\right] \text{ MeV} = 190 \text{ MeV}.$$

Exercise 4.2. A nucleus with $x \geq 1$ fissions into two equal fragments. Estimate the fission time.

The equal fragments touching each other have Coulomb energy

$$E_C = \left(\frac{Z}{2}\right)^2 \frac{e^2}{2r_o}.$$

If the fragments start at rest, a typical observed relative kinetic energy is half of this value and the corresponding velocity of the fragment, $v = \left(Z^2 e^2 / mAr_o^2\right)^{1/2}$. The time it takes to move a distance r_o is then a measure of fission time T_f so that

$$T_f = \frac{r_o}{v} = \frac{A}{Z}\left(\frac{ma_o^3}{e^2}\right) \approx 2 \times 10^{-22} \text{ sec}.$$

Exercise 4.3. The nucleus $_{92}^{238}U$ can undergo spontaneous fission into two equal fragments and it can also undergo α-decay. Show that the barrier penetrating probability for fission is much lower than that for alpha decay.

We employ WKB expression for the transition probability (Eq. 1.51).

$$T_{\ell=0}^{WKB} \cong e^{-2S}$$

where $S = \frac{\sqrt{2\mu E}}{\hbar} \int_{r_o}^{r_B}\left(\frac{r_B}{r} - 1\right)^{1/2} dr$, such that $r_B = \frac{Z_1 Z_2 e^2}{E}$. $Z_1 e$ and $Z_2 e$ are the two fission fragments and r_o is the range of nuclear potential ($r_o \ll r_B$). We substitute $(r_B / r) = u$ such that $\frac{du}{dr} = -\frac{r_B}{r^2}$ and get

$$\int_{r_o}^{r_B}\left(\frac{r_B}{r} - 1\right)^{1/2} dr = \int_1^{r_B/r_o} \frac{r_B}{u^2}\sqrt{u-1}\, du$$

$$= r_B \left| -\frac{\sqrt{u-1}}{u} + \tan^{-1} \sqrt{u-1} \right|_1^{r_B/r_0}$$

$$\approx r_B \left[\frac{\pi}{2} - \left(\frac{r_0}{r_B} \right)^{1/2} \right] \qquad \text{since } \frac{r_B}{r_0} \gg 1.$$

Therefore, $S = \dfrac{1}{\hbar} \sqrt{\dfrac{2\mu}{E}}\, Z_1 Z_2 e^2 \left[\dfrac{\pi}{2} - \left(\dfrac{r_0}{r_B} \right)^{1/2} \right].$

For symmetric fission $\mu_f = 60$, $E = E_f = 200$ MeV, $Z_1 = Z_2 = 46$ and for α-emission $\mu_\alpha = 4$, $E = E_\alpha = 4$ MeV and $Z_1 = 2$; $Z_2 = 90$ so that,

$$\frac{S \text{ (Fission)}}{S\ (\alpha - \text{emission})} \cong 6.$$

Hence, barrier penetrability for fission is much lower than that for α-decay.

Exercise 4.4. The single particle density of states is given by

$$g(\epsilon) = \sum_{N=0}^{\infty} \delta \left(\epsilon - \left(N + \frac{1}{2} \right) \hbar\omega \right),$$

show that the shell correction to the energy is zero.

Since, on the average, there is one state in the energy interval $\hbar\omega$, the smoothed density of states should be simply

$$g(\epsilon) = \frac{1}{\hbar\omega} \quad \text{for } \epsilon \geq 0,$$

$$= 0 \quad \text{for } \epsilon < 0$$

Employing the relation (4.24), we get

$$A = \int_{-\infty}^{\bar\lambda} \bar{g}(\epsilon)\, d\epsilon = \int_0^{\bar\lambda} \frac{1}{\hbar\omega} d\epsilon \quad \text{or} \quad \bar\lambda = A\hbar\omega.$$

Hence, $\qquad \bar{E}(A) = \displaystyle\int_0^{\bar\lambda} \epsilon\, \bar{g}(\epsilon) d\epsilon = \dfrac{A^2 \hbar\omega}{2}.$

Now, $\qquad E(A) = \displaystyle\sum_{i=1}^{A} \epsilon_i = \left[\dfrac{1}{2} + \dfrac{3}{2} + \cdots\cdots \dfrac{A-1}{2} \right] \hbar\omega = \dfrac{A^2}{2} \hbar\omega.$

Therefore, $E_{SC} = E(A) - \bar{E}(A) = \dfrac{A^2 \hbar \omega}{2} - \dfrac{A^2 \hbar \omega}{2} = 0.$

4.2 COLLECTIVE BEHAVIOUR OF NUCLEI

Bohr and Mottelson[7], in order to describe nuclear excited states not only considered nucleus as charged, incompressible and deformable liquid drop but assumed that it can possess irrotational and non-viscous motions. Under these assumptions the nucleus can undergo deformation oscillation at constant density in which surface tension of the nucleus acts as a restoring force. Since the excited nuclear states have definite angular momentum, it is justified to express nuclear surface by the relation (4.1). Any collective motion is expressed by letting $\alpha_{\lambda\mu}$ varying with time.

The kinetic energy T of the system, which originates from the hydrodynamic streaming flow of nuclear fluid of velocity v and constant density ρ is

$$T = \frac{1}{2} \rho \int v^2 d\tau = \frac{1}{2} \sum_{\lambda} B_{\lambda} \, |\dot{\alpha}_{\lambda\mu}(t)|^2, \qquad (4.29)$$

where B_{λ} is analogous to an inertial parameter and its hydrodynamic value is

obtained from the flow property of the nuclear fluid as $B_{\lambda} = \dfrac{\rho \, r_0^5}{\lambda}$ (Exercise 4.5).

In the irrotational fluid motion of the nucleus, where the density does not change and the circulation of the velocity field (ϕ) is everywhere zero, we have $\nabla^2 \phi = 0$. Such flows are often called surface-wave flows and they are characteristic of viscous-free fluid. This is different from the rigid body motion.

A. Collective Vibration and Rotation

To derive collective model Hamiltonian, we first obtain a classical form of the Hamiltonian following the hydrodynamic model, which can finally be quantized. The energy for charged and deformable vibrating liquid drop under harmonic approximation is then

$$E = \frac{1}{2} \sum_{\lambda\mu} B_{\lambda} \, |\dot{\alpha}_{\lambda\mu}|^2 + \frac{1}{2} \sum_{\lambda\mu} C_{\lambda} \, |\alpha_{\lambda\mu}|^2. \qquad (4.30)$$

Since there is no viscosity, the potential is conservative. The amplitude $\alpha_{\lambda\mu}$ performs harmonic oscillations with frequency $\omega_{\lambda} = \sqrt{C_{\lambda}/B_{\lambda}}$ and corresponding classical equation of motion is $\ddot{\alpha}_{\lambda\mu}(t) + \omega_{\lambda}^2 \alpha_{\lambda\mu}(t) = 0$.

In order to quantize collective oscillations of the liquid drop and introduce the concept of phonon excitation, we first define momentum conjugate to the variable $\alpha_{\lambda\mu}$:

$$\pi_{\lambda\mu} = \frac{dT}{d\dot\alpha_{\lambda\mu}} = B_\lambda\,\dot\alpha_{\lambda\mu}^\dagger,$$

and the Hamiltonian $\quad H = \sum_{\lambda\mu}\left[\frac{|\pi_{\lambda\mu}|^2}{2B_\lambda} + \frac{1}{2}C_\lambda|\alpha_{\lambda\mu}|^2\right],$ $\qquad\qquad$ (4.31)

where $\alpha_{\lambda\mu}$ and $\pi_{\lambda\mu}$ satisfy the commutation relation $\left[\alpha_{\lambda\mu}\,\pi_{\lambda\mu}\right] = i\hbar.$

$\lambda = 0$ would give rise to monopole breathing mode oscillation and it is ruled out by the assumption of incompressibility. It is important to know whether nuclei are compressible. If they are readily so, a common form of excitation of the nucleus will be density oscillations or the so called breathing mode in which the nucleus will alternately expand and contract without the change of shape. Excited states corresponding to density changes are not observed, at least in the low lying regions of nuclear spectra. For a medium mass nucleus, the minimum excitation energy is ≈ 15 MeV. $\lambda = 1$ mode or dipole mode corresponds to the protons moving one way with the neutrons moving the opposite way in the centre of mass (c.m.) frame. These oscillations lead to giant dipole resonance and they have excitation energy ≈ 20 MeV and hence excluded from consideration. Therefore, the low lying excited states can originate only from $\lambda = 2$ oscillations.

We express $\alpha_{2\mu}$ and $\dot\alpha_{2\mu}$ in terms of quadrupole phonon creation and annihilation operators $b_\mu^\dagger$ and b_μ respectively as follows :

$$\alpha_{2\mu} = \left(\frac{\hbar}{2B_2\omega}\right)^{1/2}\left[b_\mu + (-)\,b_{-\mu}^\dagger\right] \text{ and }$$

$$\dot\alpha_{2\mu} = i\omega\left(\frac{\hbar}{2B_2\omega}\right)^{1/2}\left[b_\mu + (-)\,b_{-\mu}^\dagger\right] \qquad\qquad (4.32)$$

where $\omega = \sqrt{C_2/B_2}$ is the frequency of the quadrupole phonon. b_μ and $b_\mu^\dagger$ satisfy the following commutation relations: $[b_\mu, b_\nu] = [b_\mu^\dagger, b_\nu^\dagger] = 0$ and $[b_\mu, b_\nu^\dagger] = \delta_{\mu\nu}.$

Also $b_\mu |n_\mu\rangle = \sqrt{n_\mu} \ |n_\mu - 1\rangle$ and $b_\mu^\dagger |n_\mu\rangle = \sqrt{n_\mu + 1} \ |n_\mu + 1\rangle$ (4.33)

where n_μ is the number of oscillations in 2μ mode. Expressing H in terms of b_μ and $b_\mu^\dagger$ we get

$$H = \frac{1}{2} \hbar\omega \sum_\mu (b_\mu^\dagger \ b_\mu + b_\mu \ b_\mu^\dagger)$$ (4.34)

and $$E = \frac{1}{2} \hbar\omega \langle n_\mu | \sum_\mu (b_\mu^\dagger \ b_\mu + b_\mu \ b_\mu^\dagger) |n_\mu\rangle = \left(N + \frac{5}{2}\right) \hbar\omega,$$ (4.35)

where $N = \sum n_\mu$ is the number of phonons and $((5/2) \hbar\omega)$ is the zero point energy. N takes integral values $0, 1, 2, \ldots$. For example, $N = 0$ and $N = 1$ correspond respectively to ground and first excited states. It can be shown for the irrotational motion that a phonon of type $(\lambda\mu)$ carries angular momentum λ with Z component μ and parity $(-)^\lambda$.

Since each quadrupole phonon has 2 units of angular momentum, two phonon excitation will have total $I = 0, 2$ and 4. Only even values are allowed because of symmetry properties of two phonons (Bosons). For example, $|IM\rangle_{12} = (-)^{2+2-I} |IM\rangle_{21}$ for two-phonon state of total angular momentum

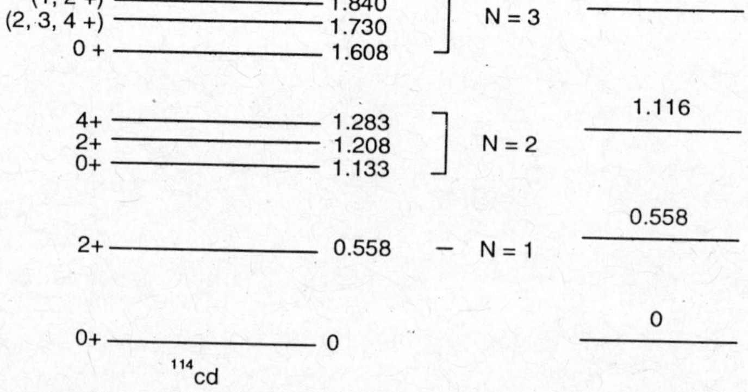

Fig. 4.5: Vibrational spectra of ^{114}Cd. Levels are identified with respect to oscillator quanta N

quantum number I with Z-component M and hence, I is even for symmetric function.

Equally spaced energy levels arising due to the quadrupole vibrations are shown in Fig. 4.5. For $N = 2$, there is a triplet degenerate even states with spin 0^+, 2^+, 4^+ and for $N = 3$ there are 5 degenerate even states 0^+, 2^+, 3^+, 4^+, 6^+. For illustration we show in Fig. 4.5, a scheme of the observed levels of ^{114}Cd and identify the various phonon states from $N = 0$ to $N = 4$.

Permanent Deformation and Collective Rotation: Nuclei in the rare earth $(150 \leq A \leq 190)$ and actinide $(A > 226)$ regions show the existence of non-zero equilibrium deformation (permanent deformation) and certain symmetries. This is evidenced by the large quadrupole moment and enhanced $E2$ transition probability compared to the single particle estimates. Explanation of such large quadrupole moment was first given by Rainwater[8]. He represented a strongly deformed nucleus as an ellipsoid of rotation with uniform charge density. In general, for any point on the surface we have

$$\frac{x^2}{a^2} + \frac{y^2}{b^2} + \frac{z^2}{c^2} = 1, \tag{4.36}$$

where a, b and c are semi axes of the ellipsoid. (Fig. 4.6).

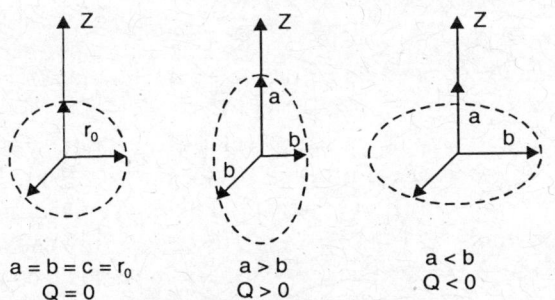

$$
\begin{array}{ccc}
a = b = c = r_0 & a > b & a < b \\
Q = 0 & Q > 0 & Q < 0
\end{array}
$$

Fig. 4.6: Semi-axes of spheroid

The equation of the deformed ellipsoid, symmetrical about Z-axis (also called spheroid), can be represented by

$$\frac{x^2}{b^2} + \frac{y^2}{b^2} + \frac{z^2}{a^2} = 1, \tag{4.37}$$

where a is the semi-major axis parallel to the Z-axis and b is the other two axes of same length, perpendicular to the Z-axis. The volume of the prolate spheroid formed by the rotation of ellipse about its semi major axis $= (4\pi/3)\, ab^2$. Since the charge density is uniform, we can write (*see* Eq. 1.20)

$$Q = \frac{3Z}{4\pi ab^2} \int (3z^2 - r^2)d\tau. \tag{4.38}$$

We choose the new variables, $x = \rho\cos\phi$, $y = \rho\sin\phi$ and $z = z$, so that

$$\frac{\rho^2}{b^2} + \frac{z^2}{a^2} = 1 \text{ or } \rho = b\sqrt{1 - z^2/a^2}. \tag{4.39}$$

Hence, $Q = \dfrac{3Z}{4\pi ab^2} \displaystyle\int_{-a}^{+a} dz \int_{0}^{b\sqrt{1-z^2/a^2}} (2z^2 - \rho^2)\, 2\pi\rho \, d\rho$

$$= \frac{3Z}{ab^2}\left[\int_{0}^{a} z^2 dz\{b^2(1 - z^2/a^2)\} - \int_{0}^{a} \frac{b^4}{4}(1 - z^2/a^2))^2\, dz \right]$$

$$= \frac{3Z}{ab^2}\left[\frac{2a^3 b^2}{15} - \frac{2}{15}ab^2 \right] = \frac{2Z}{5}(a^2 - b^2)$$

$$= \frac{8Z}{5}\frac{a-b}{a+b} r_0^2 \text{ , where } r_0 = \frac{a+b}{2}. \tag{4.40}$$

Similarly, an oblate spheroid is formed by the rotation of an ellipse about its minor axis.

Employing (4.40) and the measured value of quadrupole moment, we can estimate the magnitude of deformation. As an example, let us consider ^{175}Lu nucleus for which $Z = 71$, $r_0 = 7$ fm and the observed quadrupole moment = (5.1 ± 0.3) barn. It follows from Eq. (4.40) that $a = 7.7$ fm and $b = 6.5$ fm. The deformation parameter δ, can be introduced from the geometry of the system such that

$$\delta = \frac{a-b}{r_0} = \frac{2(a-b)}{a+b} \approx 0.17,$$

which indicates degree of deviation from the spherical shape. When the semi-major axis has length a and other two minor axes have equal length b, then one can define the deformation parameter as $\delta = \dfrac{3(a-b)}{3a-b}$. For $a = 2b$, $\delta = 3/5 = 0.6$, the nucleus is superdeformed. For $a = 3b$, $\delta = 3/4 = 0.75$, the nucleus is hyper deformed. $a = b$ represents the spherical nucleus.

In addition to the large values of quadrupole moments, nuclei in the rare earth and actinide regions show the following characteristic feature for the low lying states: $E(I) \propto I(I+1)$, where I is the spin of the nuclear excited state.

The above I (I+1) dependence of energy levels is a characteristic of rigid rotator motion. In order to treat the rotation of a permanently deformed nucleus, Bohr[7] suggested that nuclear excitation of such a nucleus should be considered not in the space fixed coordinate system but in the body fixed coordinate system, coinciding with the principal axis of the nucleus. We can express deformation parameter (a_{2v}) with respect to a coordinate system fixed in body in terms of deformation parameters ($\alpha_{2\mu}$) with respect to a coordinate system fixed in space.

Let us represent the deformed nuclear surface in body fixed system (denoted by X', Y', Z' or equivalently 1, 2, 3) by a length vector as follows :

$$r(\theta',\phi') = r_0\left[1 + \sum_v a_{2v}\, Y_2^v(\theta',\phi')\right]. \tag{4.41}$$

In order to specify the orientation of a body in three dimensional space, we need three angular variables and in order to specify the state of motion of the body, we need three quantum numbers. The orientation of a body in three dimensional space will involve three angular variables, since three parameters are needed to specify complete rotation. We choose three Euler angles θ_1, θ_2, θ_3 for this purpose and they are defined (Fig. 4.7) as follows : A rotation by $\theta_1(0 \le \theta_1 \le 2\pi)$ about OZ taking axis OY to position OY_1; next rotation by θ_2 $(0 \le \theta_2 \le \pi)$ about OY_1 taking axis OZ to position OZ'. Finally, a rotation through $\theta_3 (0 \le \theta_3 \le 2\pi)$ about OZ' resulting coordinate frame S to go to S'.

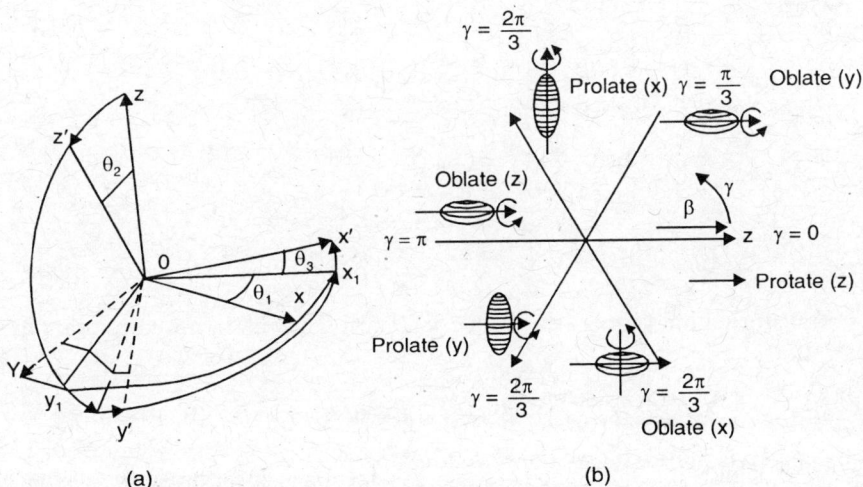

Fig. 4.7: (*a*) Euler angles defining rotations in the three dimensional space and (*b*) Nuclear shapes with respect to β and γ

The complete rotation described by the three Euler anlges θ_1, θ_2 and θ_3, is then

$$R(\theta_1\theta_2\theta_3) = e^{-\theta_3 L_z} \, e^{-i\theta_2 L_{Y_1}} \, e^{-i\theta_1 L_z} . \tag{4.42a}$$

An operator $\hat{O}$ in S-system transforms to $R \ \hat{O} \ R^\dagger$ in S', where R is the rotation operator which takes S to S'. Using this property we get

$$R\left(\theta_1 \, \theta_2 \, \theta_3\right) = e^{-i\theta_1 L_z} \, e^{-i\theta_2 L_Y} \, e^{-i\theta_3 L_z} . \tag{4.42b}$$

This factorization of rotation operator enables us to express all three rotations in the same coordinate frame S.

This is applicable to total angular momentum J where L is replaced by J while $Y_L^M (\theta, \varphi)$ by $| JM \rangle$. The application of rotation operator $R(\theta_1 \, \theta_2 \, \theta_3)$ to a state $| JK \rangle$, of well defined total angular momentum J and its Z-component K can only change K and not J so that

$$| JK \rangle' = R(\theta_1\theta_2\theta_3)| JK \rangle = \sum_K D_{MK}^{J*} (\theta_1\theta_2\theta_3)| JM \rangle,$$

where $| JK \rangle'$ and $| JM \rangle$ refer to rotated frame (S') and space fixed frame (S) respectively. Hence, the matrix element of rotation operator $R(\theta_1 \, \theta_2 \, \theta_3)$, i.e. D-function is

$$D_{MK}^{J*} (\theta_1\theta_2\theta_3) = \langle JM | \, e^{-i\theta_1 J_z} \, e^{-i\theta_2 J_Y} \, e^{-i\theta_3 J_z} \, |JK \rangle. \tag{4.42c}$$

The D_{MK}^{J} satisfy the following conditions:

$$\int_0^{2\pi} d\theta_1 \int_0^{\pi} \sin \theta_2 \, d\theta_2 \int_0^{2\pi} d\theta_3 \, D_{MK}^{J*} (\theta_1\theta_2\theta_3) D_{M'K'}^{J'} (\theta_1\theta_2\theta_3)$$

$$= \frac{8\pi^2}{2J+1} \delta_{MM'} \, \delta_{KK'} \, \delta_{JJ'}, \tag{4.42d}$$

$$\sum_M D_{MK'}^{J*} \, D_{MK}^{J} = \delta_{KK'} \quad \text{and} \quad \sum_K D_{M'K}^{J*} \, D_{MK}^{J} = \delta_{MM'}. \tag{4.42e}$$

Transformation properties of spherical harmonics under finite rotation are as follows:

$$Y_2^\mu (\theta', \phi') = \sum_\nu D_{\mu\nu}^{2*} (\theta_k) \, Y_2^\nu (\theta, \phi) \quad \text{or} \quad Y_2^\nu (\theta, \phi) = \sum_\mu D_{\mu\nu}^{2} (\theta_k) \, Y_2^\mu (\theta', \phi'),$$

where θ_k stands for θ_1, θ_2 and θ_3. Since distance is a scalar invariant, we have

$$\sum_\nu a_{2\nu} Y_2^\nu (\theta', \phi') = \sum_\mu \alpha_{2\mu} Y_2^\mu (\theta, \phi). \tag{4.43a}$$

Multiplying both sides of (4.43a) by $Y_2^{\nu'*}(\theta',\phi')$ and integrating we get

$$\int \sum_\nu a_{2\nu} Y_2^\nu(\theta',\phi')\, Y_2^{\nu'*}(\theta',\phi')\, d\Omega'$$

$$= \int \sum_\mu a_{2\mu} \sum_{\nu'} D_{\mu\nu'}^2(\theta_k)\, Y_2^{\nu'}(\theta',\phi')\, Y_2^{\nu'*}(\theta',\phi')\, d\Omega'.$$

Hence, $\quad a_{2\nu} = \sum_\mu \alpha_{2\mu}\, D_{\mu\nu}^2(\theta_k) \quad$ or $\quad \alpha_{2\mu} = \sum_\mu D_{\mu\nu}^{2*}(\theta_k)\, a_{2\nu}.$ \hfill (4.43b)

Symmetry properties of deformation: We choose the principal axis of the nucleus as body fixed axis. This choice enables us to examine certain symmetry properties of deformed nucleus. Since the body fixed axis coincides with the symmetry axis, we have "reflection symmetry". On reflection through the plane $Z' = 0$ $(\theta',\phi') \rightarrow (\pi - \theta',\phi')$, the spherical harmonics $Y_2^0, Y_2^{\pm2}$ remain unchanged, while the functions $Y_2^{\pm1}$ change sign. Since $r(\theta', \phi')$ given by Eq. (4.41) should remain unchanged by this transformation, it follows $a_{21} = a_{2-1} = 0$.

On reflection through the plane $X' = 0$, $(\theta', \phi') \rightarrow (\theta', \pi - \phi')$, the spherical harmonics $Y_2^{\pm2}$ go over into $Y_2^{\mp2}$ and again invariance of $r(\theta', \phi')$ demands $a_{22} = a_{2-2}$. Thus, in the body fixed axis, the deformation of the nucleus can be characterized by only two real parameters a_{20} and a_{22}. It is convenient to introduce two new parameters β and γ and express a_{20} and a_{22} in terms of them:

$$a_{20} = \beta\cos\gamma \quad \text{and} \quad a_{22} = a_{2-2} = \frac{1}{\sqrt{2}}\beta\sin\gamma. \hfill (4.44a)$$

The significance of β is seen as follows:

$$\text{Deformation} = \sum_\mu |\alpha_{2\mu}|^2 = \sum_\mu \sum_\nu a_{2\nu}^* D_{\mu\nu}^{2*}(\theta_k) \sum_{\nu'} D_{\mu\nu'}^2(\theta_k) a_{2\nu'}$$

$$= \sum_{\nu\nu'} a_{2\nu}^* a_{2\nu'} \sum_\mu D_{\mu\nu}^{2*}(\theta_k) D_{\mu\nu'}^2(\theta_k) = \sum_{\nu\,\nu'} a_{2\nu}^* a_{2\nu'} \delta_{\nu\nu'} = \sum_\nu |a_{2\nu}|^2$$

$$= \beta^2\cos^2\gamma + \frac{1}{2}\beta^2\sin^2\gamma + \frac{1}{2}\beta^2\sin^2\gamma = \beta^2.$$

In the above we have used (4.42e) and (4.44a). Thus, the parameter β is the measure of size of deformation. Let us write down the length of body fixed axes which are principal axes for the nuclear surface:

$$r(\theta',\phi') = r_0\left[1 + \beta\cos\gamma\, Y_2^0(\theta', \phi') + \frac{\beta}{\sqrt{2}}\sin\gamma\left\{Y_2^2(\theta',\phi') + Y_2^{-2}(\theta',\phi')\right\}\right]$$

$$A_1 = r(\theta' = \pi/2; \varphi' = 0) = r_0 \left[1 - \sqrt{\frac{5}{16\pi}}\, \beta\, (\cos\gamma - \sqrt{3}\, \sin\gamma) \right],$$

$$A_2 = r(\theta' = \pi/2; \phi' = \pi/2) = r_0 \left[1 - \sqrt{\frac{5}{16\pi}}\, \beta\, (\cos\gamma + \sqrt{3}\, \sin\gamma) \right], \text{ and}$$

$$A_3 = r(\theta' = 0; \phi') = r_0 \left[1 + \sqrt{\frac{5}{4\pi}}\, \beta \cos\gamma \right].$$

Generally, $A_k = r_0 \left[1 + \sqrt{\dfrac{5}{4\pi}}\, \beta\cos\left(\gamma - \dfrac{2\pi k}{3} \right) \right]$,

where $k = 1, 2$ and 3 refer respectively to body fixed X', Y', Z' axes. For $\gamma = 0$, $2\pi/3$, $4\pi/3$, 2π, the nucleus has prolate shape (elongated about 3rd axis). For $\gamma = \pi/3$, π, $2\pi/3$, the nucleus has oblate shape. Hence, γ determines the shape of deformation (*see* Fig. 4.7b).

Their significance can be visualized if we look at the stretching or shrinking of radii of nuclear ellipsoid along the principal axes :

$$\delta\, r_k = A_k - r_0 = \sqrt{\frac{5}{4\pi}}\, \beta\cos\left(\gamma - \frac{2\pi k}{3} \right). \qquad (4.44b)$$

We can write the classical collective Hamiltonian using (4.30) including rotation in the body fixed frame

$$H = \frac{1}{2}B(\dot{\beta}^2 + \beta^2\dot{\gamma}^2) + \frac{1}{2}\sum_{k=1}^{3} \mathscr{I}_k\, \omega_k^2 + \frac{1}{2}C\beta^2 = H_v + T_r, \qquad (4.45)$$

where $H_v = \dfrac{1}{2}B(\dot{\beta}^2 + \beta^2\dot{\gamma}^2) + \dfrac{1}{2}C\beta^2$ represents the vibrational energy, of

which the first term represents beta vibration and the second term gamma vibration. H_v remains unaffected by rotation due to length invariance in rotation.

$T_r = \dfrac{1}{2}\sum_{k=1}^{3} J_k^2/\omega_k^2 = \dfrac{1}{2}\sum_{k=1}^{3} J_k^2/\mathscr{I}_k$, is the rotational kinetic energy (*see* Exercise

4.9) where J_k is the rotational angular momentum of the core, $\mathscr{I}_k$ the moment of inertia with respect to the space fixed axis and ω_k is the angular velocity of the principal axis with respect to space fixed axis. Vast majority of nuclei have well deformed non-spherical equilibrium shape with equilibrium deformation $\beta = \beta_0$ and $\gamma = \gamma_0$. In a more general way Bohr's classical collective Hamiltonian of a nucleus rotating and vibrating about the equilibrium shape is

$$H = T_\beta + T_\gamma + \frac{1}{2} \sum_{k=1}^{3} (J_k^2 / \mathscr{S}_k) + V(\beta, \gamma),$$

where $V(\beta, \gamma) \approx V_0 + \frac{1}{2} C_\beta (\beta - \beta_0) + \frac{1}{2} C_\gamma (\gamma - \gamma_0)^2 \mathscr{S}_k (\beta_0, \gamma_0),$ \hfill (4.46)

where V_0 is the intrinsic equilibrium energy. The rotation and vibration are coupled because moment of inertia $\mathscr{S}_k$ are function of β and γ (Exercise 4.9).

The Rotational Spectra of Even-Even Nuclei: The quantized Hamiltonian for rotation is obtained from Eq. (4.45). If we represent a deformed even-even nucleus by such a rotator, the nuclear spin or the total angular momentum will arise solely due to rotational motion $(\vec{I} = \vec{J})$. For a nucleus which is symmetric about the third axis $\mathscr{S}_1 = \mathscr{S}_2 = 3B\beta^2 = \mathscr{S} \neq \mathscr{S}_3,$

$$H = \frac{\hbar^2}{2\mathscr{S}} (I^2 - I_3^2) + \frac{\hbar^2 I_3^2}{2\mathscr{S}_3}. \hfill (4.47a)$$

In order to choose commuting operators for description of rotational motion, we consider I^2, I_z (in space fixed system) and I_3 (in body fixed system), which commutes with I^2, I_z. We represent simultaneous eigenfunctions of I^2, I_z and I_3 by $|IMK\rangle$ so that

$$I^2 |IMK\rangle = I(I+1) |IMK\rangle, \; I_z |IMK\rangle = M |IMK\rangle \; \text{and} \; I_3 |IMK\rangle = K |IMK\rangle.$$

For a nucleus with axial symmetry, there is an additional quantum number associated with I_3 with eigen value K (Fig. 4.8a).

Hence, $$E = \frac{\hbar^2}{2\mathscr{S}} [I(I+1)] + \frac{\hbar^2}{2} \left(\frac{1}{\mathscr{S}_3} - \frac{1}{\mathscr{S}} \right) K^2. \hfill (4.47b)$$

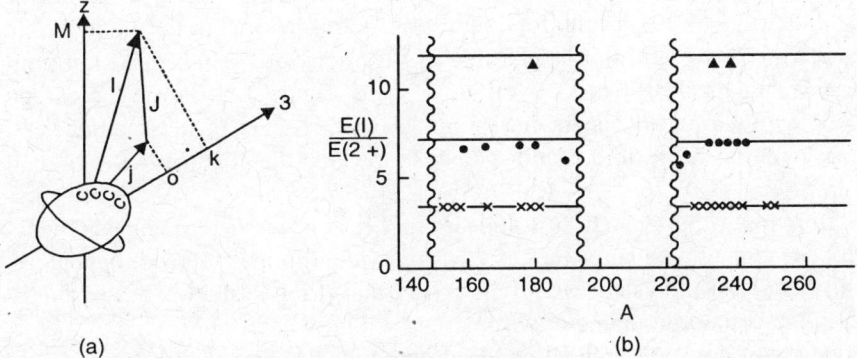

(a)   (b)

Fig. 4.8: (*a*) The vector diagram of rotor-particle coupling and (*b*) Characteristic of rigid rotator spectra of even-even nucleus. The triangle, dot and cross denote respectively, $E(8^+)/E(2^+)$, $E(6^+)/E(2^+)$ and $E(4^+)/E(2^+)$

The normalized wave function (AppendixD)

$$| IMK \rangle = \sqrt{\frac{2I+1}{8\pi^2}} \left[(D^I_{MK} \ \theta_1\theta_2\theta_3) \right]. \tag{4.48a}$$

If the nucleus is an ellipsoid of revolution, it is transformed into itself on reflection through principal plane perpendicular to the symmetry axis. The reflection operation through a plane $Z' = 0$ is equivalent for an axially symmetric nucleus to a rotation R_2 (π)through an angle of 180° about 2nd axis.

$$R_2(\pi) D^I_{MK}(\theta_k) = \sum_{K'} D^{I*}_{KK'}(0,\pi,0) \, D^I_{MK'} = (-)^{I-K} D^I_{M,-K}$$

The rotational invariance of wave function implies

$$| IMK \rangle = \frac{1}{\sqrt{2}} \left(\frac{2I+1}{8\pi^2} \right)^{1/2} \left[D^I_{M,K} + (-)^{I-K} D^I_{M,-K} \right]. \tag{4.48b}$$

Most nuclei in their ground state have an axial symmetry ($\gamma \doteq 0$). In this case $\mathscr{I}_1 = \mathscr{I}_2 = \mathscr{I}$ and $I_3 = 0$ or $K = 0$. For $K = 0$ only even values of I are allowed so that

$$| IMK \rangle = \sqrt{\frac{2I+1}{8\pi^2}} D^I_{MO}, \text{ where } I = 0, 2, 4, \ldots\ldots \tag{4.49}$$

Hence, $E(I\pi) = \dfrac{\hbar^2}{2\mathscr{I}} I(I+1).$

So that, $\dfrac{E(4+)}{E(2+)} = \dfrac{10}{3}, \dfrac{E(6+)}{E(2+)} = 7$ and $\dfrac{E(8+)}{E(2+)} = 12,$

which is the characteristic of rigid rotator spectra (Fig. 4.8*b*). It should be noted that Eq. (4.49) is valid for quadrupole deformation. For the excited state bands (β or γ vibration), it is possible to have non-zero values of K. For non-zero K, I takes all values K, $K + 1$, $K + 2$, etc. Figure 4.8b shows the energy ratio of rotational levels for heavy nuclei with respect to mass number A. The deviation from rigid rotator values is towards the edge of the rotational region.

To explain this deviation it appears that it is necessary to consider what happens to energy eigen values as the rigidity condition is relaxed. In analogy with the classical system we might expect that if the system is not perfectly rigid, the equatorial diameter will increase slightly as it rotates and as a consequence of constant volume condition the polar diameter will decrease. So we introduce another correction term known as the centrifugal stretching term to (4.49), which then reads as

$$E(I) = aI(I+1) + bI^2(I+1)^2, \tag{4.50}$$

where $a = \hbar^2 / 2\mathcal{J}_{rig}$ and b is an adjustable parameter.

Rotator Particle Coupling – Spectra of Odd A Nuclei

We shall now extend the collective rotational model discussed in the previous section to odd-A nuclei. We assume that the odd nucleon of angular momentum 'j' is coupled to even-even core angular momentum $\vec{J}$ to give total angular momentum $\vec{I}(= \vec{J} + \vec{j})$. The last $\vec{j}$ is not necessarily for only one particle but for all particles not included in core. The Hamiltonian of the nucleus undergoing rotational and intrinsic particle motion is the sum of the particle Hamiltonian $H_p(r')$, the rotator Hamiltonian H_r and the rotator-particle interaction H_{int} as follows:

$$H = H_p(r') + H_r + H_{\text{int}}. \tag{4.51}$$

Because of isotropy of space I is conserved. This does not mean j is conserved. Therefore, when we consider intrinsic motion of the particles relative to the orientation of the nucleus, there is no rotational invariance.

In the "adiabatic" approximation (no interaction between particle and rotational motion), we neglect H_{int} so that $H = H_r + H_p$. However, the neglect of H_{int} does not necessarily mean complete separation between single particle and rotational motions. For axially symmetric nucleus

$$H_r = \frac{\hbar^2}{2\mathcal{J}}(\vec{I} - \vec{j})^2 + \left(\frac{\hbar^2}{2\mathcal{J}_3} - \frac{\hbar^2}{2\mathcal{J}}\right)(I_3 - j_3)^2 = \frac{\hbar^2}{2\mathcal{J}}(\vec{I} - \vec{j})^2, \tag{4.52}$$

since for an axially symmetric nucleus I_3 and j_3 are equal due to the requirement of rotational invariance. Now,

$$\vec{I} \cdot \vec{j} = I_2 j_2 + I_3 j_3 \text{ and } I_+ j_- = 2(I_1 j_1 + I_2 j_2).$$

Hence, $2I \cdot j = (I_+ j_- + I_- j_+) + 2I_3 j_3$, so that

$$H_r = \frac{\hbar^2}{2\mathcal{J}} I^2 + \frac{\hbar^2}{2\mathcal{J}} j^2 - \frac{\hbar^2}{2\mathcal{J}}[(I_+ j_- + I_- j_+) + 2I_3 j_3] \tag{4.53}$$

Hence, $H = H_0 + H_r^0 + H_c$, $\tag{4.54}$

where $\quad H_0 = -\dfrac{\hbar^2}{2m}\nabla^2 + V(r') + \dfrac{\hbar^2}{2\mathcal{J}} j^2, \quad H_r^0 = \dfrac{\hbar^2}{2\mathcal{J}}\left[I^2 - 2I_3 j_3\right]$

and $\quad H_c = -\dfrac{\hbar^2}{2\mathcal{J}}[(I_+ j_- + I_- j_+)] = H_{RPC}. \tag{4.55}$

H_c represents rotation-particle coupling and sometime it is denoted by H_{RPC}. It physically arises because we are looking at the system in a rotating frame and thus a "coriolis force" arises whose effect is given by H_{RPC}.

We observe that j_3 is not a constant of motion, since $V(r')$ is non-spherical. However, I is a constant of motion. Therefore, we choose I^2, I_z, I_3 and j_3 as commuting operators for the description of the coupled system. We represent the simultaneous eigen functions of I^2, I_z, I_3 and j_3 by $|IMK\,\Omega\rangle$, so that $I^2\,|IMK\,\Omega\rangle = I(I+1)\,|IMK\Omega\rangle$, $I_z|IMK\,\Omega\rangle = M\,|IMK\,\Omega\rangle$, $|IMK\,\Omega\rangle = K\,|IMK\,\Omega\rangle$ and $j_3|IMK\,\Omega\rangle = \Omega\,|IMK\,\Omega\rangle$. Explicit form of $|IMK\,\Omega\rangle$ without symmetry requirement is

$$|IMK\,\Omega\rangle = \sqrt{\frac{2I+1}{8\pi^2}}\,D_{MK}^I(\theta_k)\chi_\Omega(r').\tag{4.56}$$

The above factorization of the rotator wave function (D_{MK}^I) and particle wave function (χ_Ω) is possible in adiabatic or no coupling approximation and the total angular momentum and rotator angular momentum are both labelled by I. Since the particle excitation energy (~ 8 MeV) is much higher than the energy of the ground state rotational band, namely, 2+, 4+ (~ 2 MeV – 4 MeV), the adiabatic approximation is justifiable. But large single particle energy is not a sufficient condition for the validity of the wave function. Since the nucleus does not rotate about the symmetry axis $I_3\,|IMK\,\Omega\rangle = (I_3 - j_3)\,|IMK\,\Omega\rangle = 0$. Hence, $K = \Omega$. To find the transformation properties of χ_Ω under rotation of π about 2nd axis, we expand χ as

$$\chi_\Omega = \sum_j C_{j\Omega}\,\chi_{j\Omega}, \quad \text{where } \chi_{j\Omega} = \sum_{\Omega'}D_{\Omega'\Omega}^j(\theta_k)\,\chi'_{j\Omega'},\tag{4.57a}$$

where $\chi'_{j\Omega'}$ refers to the space fixed system. Hence,

$$R_2\,\chi_{j\Omega} = \sum_{\Omega'}\chi_{j\Omega}\,R_2\,D_{\Omega'\Omega}^j(\theta_k)$$

$$= (-)^{j-\Omega}\sum_{\Omega'}D_{\Omega'\Omega}^j(\theta_k)\,\chi'_{j\Omega} = (-)^{j-\Omega}\,\chi_{-\Omega},\tag{4.57b}$$

Since $\Omega = K$ for axially symmetric nucleus, $R_2\,\chi_K = (-)^{j-K}\,\chi_{-K}$ is required for the reflection symmetry. The normalized wave function for $K = \Omega \neq 0$ satisfying symmetry requirement is then

$$|IMK\,\Omega\rangle = \left(\frac{2I+1}{16\pi^2}\right)^{1/2}\left[D_{M,K}^I(\theta_k)\,\chi_K + (-)^{I+j-2K}\,D_{M,-K}^I\,\chi_{-K}\right]$$

$$= \left(\frac{2I+1}{16\pi^2} \right)^{1/2} \left[D^I_{M,K}(\theta_k) \chi_K + (-)^{I-j} D^I_{M,-K} \chi_{-K} \right]. \qquad (4.58)$$

The above state has a definite parity. The parity of $|IMK\rangle$ is the same as the parity of χ_K (or fixed K) but different values of I form a rotational band.

Using the Hamiltonian (4.54) and the wave function (4.58) we get the following expression for the rotational spectrum of odd-A deformed nuclei:

$$E_I = \langle IMK|H_0|IMK\rangle + \langle IMK|H^0_r|IMK \; \Omega\rangle + \langle IMK|H_c|IMK\rangle$$

$$= E_0 + \frac{\hbar^2}{2\mathscr{I}} \left[I(I+1) - 2K^2 + a\left(I + \frac{1}{2}\right)(-)^{I+1/2} \delta_{K,1/2} \right], \qquad (4.59a)$$

where 'a' is called the decoupling parameter, which decouples particle motion and rotational motion and E_0 is the single particle energy of the odd nucleus. The explicit form of a is

$$a = \sum_j |C_{j\frac{1}{2}}|^2 (-)^{1/2-j} \left(j + \frac{1}{2} \right). \qquad (4.59b)$$

The decoupling parameter can have a very strong effect on the appearance

of $K = \frac{1}{2}$ rotational band. If it is too small, the band will display the normal

$I(I+1)$ spacings. If a is large and negative, the band will have an anomalous behaviour. For illustration (Fig. 4.9) we compare energy levels of $^{169}_{69}Tm_{100}$ obtained with $a = -0.7575$ and $a = 0$. Generally speaking, inclusion of coriolis term has two effects: (1) The spectrum deviates from $I(I+1)$ sequence. (2) The core angular momentum K ceases to be good quantum number, which causes the mixing of K-bands.

The matrix elements of H_C, arising due to the presence of $\vec{I} \cdot \vec{j}$ term, lead to so called band mixing (coriolis mixing). This is the only term, which is non-diagonal in K. H_C has non-diagonal elements between states with $\Delta K = 1$ and diagonal elements only for $K = 1/2$. We consider specifically mixing of only two bands. The required matrix element is

$$\langle IMK| \; \vec{I} \cdot \vec{j} \; |IMK+1\rangle = \frac{2I+1}{16\pi^2} \left\langle \left(D^{I*}_{M,K}\chi_K + (-)^{I-j} D^I_{M,-K} \chi_{-K} \right) \right.$$

$$\left. |\vec{I} \cdot \vec{j}| \left(D^I_{M,K+1} \chi_{K+1} + (-)^{I-j} D^I_{M,-K-1} \chi_{-K-1} \right) \right\rangle$$

A careful application of raising and lowering operators and integration of the product of D-functions yield:

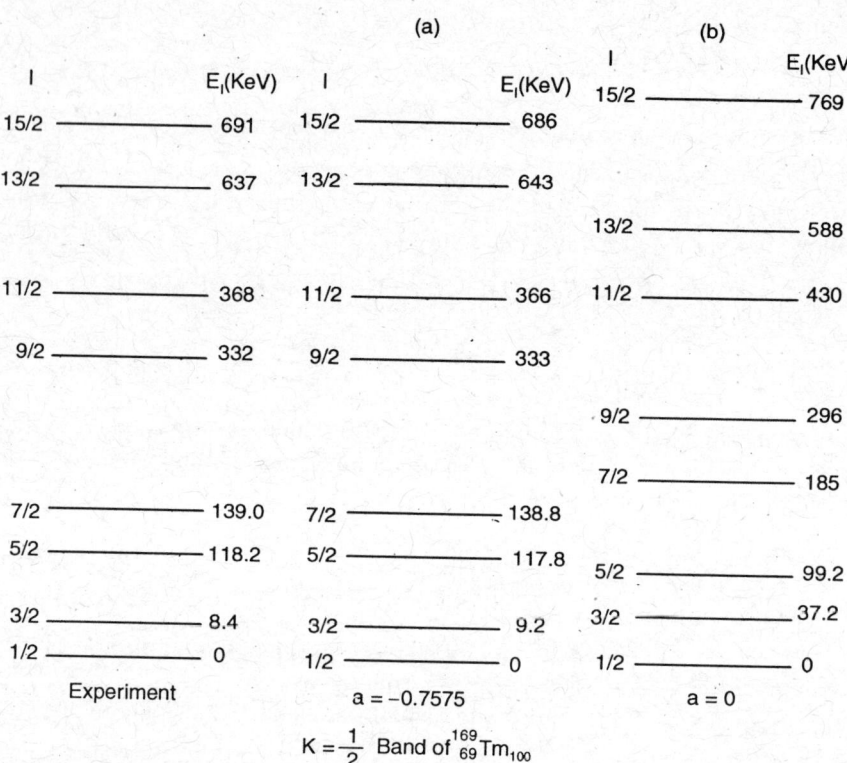

Fig. 4.9: Comparison of the energy levels of $^{169}_{69}\text{Tm}_{100}$ in the rotor-particle model with experiment

$$\langle IMK \,|\, \vec{I} \cdot \vec{j} \,|\, IMK+1\rangle = \frac{1}{\sqrt{2}}\{(I+K+1)\,(I-K)\}^{1/2}\,\langle \chi_k \,|\, j_- \,|\, \chi_{k+1}\rangle \quad (4.60)$$

B. Electromagnetic Moments

A rotating liquid drop that classically has a mechanical angular momentum, should have a magnetic moment due to superposition of the individual circular currents that are produced by rotation of charge mass elements. In this primitive picture, the gyromagnetic ratio of the rotator is given by $g_R = \dfrac{Z}{A}$.

The magnetic dipole operator for a coupled system, consisting of a rotator and a single nucleon (odd-A deformed nucleus) representing the intrinsic motion, is $\vec{\mu} = g_R \, \vec{J} + g_\Omega \, \vec{j}$, where g_Ω is the gyromagnetic ratio associated with single nucleon motion in well defined angular momentum state 'j'.

Now, $\quad g_\Omega \, \vec{j} = g_K \, \vec{j} = g_j \, \vec{j} = g_s \, \vec{s} + g_\ell \, \vec{\ell} = g_s \, \vec{s} + g_\ell (\vec{j} - \vec{s})$

Hence, $\vec{\mu} = g_R \vec{I} + (g_\ell - g_R)\vec{j} + (g_s - g_\ell)\vec{s} = g_R \vec{I} + \vec{G}$,

where $\vec{G} = (g_\ell - g_R)\vec{j} + (g_s - g_\ell)\vec{s}$, $G_Z = (g_\ell - g_R)j_Z + (g_s - g_\ell)s_Z$ and

$\vec{I} = \vec{J} + \vec{j}$. Substituting $j_Z = \ell_Z + s_Z$, we get

$$G_Z = g_\ell \ell_Z + g_s s_Z - g_R j_Z = g_j j_Z - g_R j_Z.$$

The observed magnetic moment

$$\mu = \langle IIK \mid \mu_Z \mid IIK \rangle = \langle IIK \mid g_R I_Z + G_Z \mid IIK \rangle. \tag{4.61}$$

Employing the wave function given by (4.58) we get

$$\mu = g_R I + \frac{g_K - g_R}{I + 1} K^2 \text{ for } K \neq \frac{1}{2} \tag{4.62}$$

In Table (4.2), we compare rotator model predicted g.s. magnetic moments for $^{187}_{75}$Re and $^{111}_{48}$Cd$_{63}$ with Schmidt's and experimental values.

Table 4.2: Rotational model predicted g.s. magnetic moments

Nucleus	I_{gs}	$\mu_{Schmidt}$ (nm)	$\mu_{Rotator}$ (nm)	$\mu_{Experiment}$ (nm)
$^{187}_{75}$Re$_{112}$	5/2	4.79	3.71	3.20
$^{111}_{48}$Cd$_{63}$	5/2	−1.91	−1.08	−0.71

The spectroscopic quadrupole moment Q is given by expectation value of the quadrupole moment operator $e\hat{Q}(r)$ in the space fixed system

$$Q = \langle \psi^\tau_{IMK} \mid e\hat{Q}(r) \mid \psi^\tau_{IMK} \rangle_{M=I}. \tag{4.63}$$

Here, τ represents quantum numbers describing the intrinsic motion and $M = I$ indicates that the expectation value is calculated in a state in which the total angular momentum has its maximum projection value.

The relation between the intrinsic quadrupole moment and spectroscopic quadrupole moment is obtained by transforming the quadrupole moment operator $e\hat{Q}(r)$ relative to S into quadrupole moment operator relative to S' fixed in the nucleus. Therefore,

$$Q = \langle e\hat{Q}(r) \rangle_{MI} = \langle IMK \mid \frac{16\pi}{5} \int \rho(r') \, r'^2 \, Y_2^0(\theta') \, d^3r' \, D_{00}^2(\theta_k) \mid IMK \rangle$$

$$= \frac{2I + 1}{16\pi^2} Q_0 \int D_{MK}^{I*}(\theta_k) D_{00}^2(\theta_k) D_{MK}(\theta_k) \, d\Omega_k, \tag{4.64}$$

where Q_0 is the intrinsic quadrupole moment expressed in barn. Employing the properties of the D-functions we get

$$Q = Q_0 [\langle I2\, I0 | II \rangle \langle I2\, K0 | IK \rangle = Q_0 \left[\frac{3K^2 - I(I+1)}{(I+1)\,(2I+3)} \right]. \quad (4.65)$$

If I is sufficiently large and $I(I+1) > 3K^2$, the spectroscopic quadrupole moment will be opposite in sign to the intrinsic quadrupole moment. For the ground state rotational band $I = K$

$$Q = \frac{I(2I-1)}{(I+1)\,(2I+3)} Q_0. \quad (4.66)$$

Gamma Transitions Between Collective Levels: Since the solution of the Schrodinger equation possesses the stationary property, a close correspondence between model eigen values and observed nuclear level structure is not particularly a good test of the model. The test of the model comes from various gamma transitions between the excited states. Strongly enhanced electric quadrupole transitions ($E2$) are considered to be signature of nuclear deformation.

We begin with the electromagnetic transition properties of vibrational nuclei. In our model of oscillating liquid drop, the charge density is constant upto a radius $r\,(\theta, \phi)$ with value $\rho = \dfrac{3Ze}{4\pi r_0^3}$ and is zero outside. The electric 2^λ-pole operator has the form:

$$\hat{Q}_{\lambda\mu} = \int r^\lambda Y_\lambda^\mu (\theta, \phi) \rho(r) d^3r$$

$$= \frac{3Ze}{4\pi r_0^3} \int \int_0^{r(\theta,\phi)} r^{\lambda+2}\, Y_\lambda^\mu (\theta, \phi)\, dr\, d\Omega, \quad (4.67)$$

for a uniform charge density. Employing (4.1) for the nuclear surface we get

$$\hat{Q}_{\lambda\mu} = \frac{3Ze}{4\pi r_0^3} r_0^{\lambda+3} \int \frac{(1+\eta^\dagger)^{\lambda+3}}{\lambda+3} Y_\lambda^\mu (\theta, \varphi)\, d\Omega,$$

since r is real. Expanding and retaining only the first order term in η, we get

$$\hat{Q}_{\lambda\mu} = \frac{3Ze}{4\pi} r_0^\lambda \int \frac{(\lambda+3)}{\lambda+3} \eta^\dagger Y_\lambda^\mu (\theta, \varphi)\, d\Omega = \frac{3Ze}{4\pi} r_0^\lambda\, \alpha_{\lambda\mu}^\dagger.$$

We now use the second quantized expression for $\alpha_{\lambda\mu}^\dagger$ (4.32) and obtain the second quantized form of 2^λ pole operator as

$$\hat{Q}_{\lambda\mu} = \left(\frac{\hbar}{2B_\lambda\omega}\right)^{1/2} \frac{3Ze}{4\pi} r_0^\lambda \left[b_{\lambda\mu} + (-)^\mu b_{\lambda\mu}^\dagger\right]. \tag{4.69}$$

We note that $\hat{Q}_{\lambda\mu}$ contains only a single creation and an annihilation operator for a phonon of type $\lambda\mu$. Therefore, $\hat{Q}_{\lambda\mu}$ connects states differing by only one phonon of multipole λ. The reduced matrix element from one phonon state ($n_\lambda = 1$) to no phonon state ($n_\lambda = 0$) for the electric transition is

$$B(E\lambda, n_\lambda = 1 \rightarrow n_\lambda = 0)$$

$$= \left(\frac{3Ze}{4\pi} r_0^\lambda\right)^2 \left(\frac{\hbar}{2B_\lambda\omega}\right) |\langle n_\lambda = 1 | b_{\lambda\mu} + (-)^\mu b_{\lambda\mu}^\dagger | n_\lambda = 0 \rangle|^2$$

$$= \left(\frac{3Ze}{4\pi} r_0^\lambda\right)^2 \left(\frac{\hbar}{2\sqrt{B_\lambda C_\lambda}}\right) \text{ since } \omega = \sqrt{\frac{C_\lambda}{B_\lambda}}. \tag{4.70a}$$

Furthermore, by using the properties of $b_{\lambda\mu}$ we can obtain the following sum rule:

$$\sum_f B(E\lambda, n_\lambda I_i \rightarrow (n_\lambda - 1)I_f) = n_\lambda B(E, n_\lambda = 1 \rightarrow n_\lambda = 0). \tag{4.70b}$$

For excitation solely due to vibration, the nuclear spin I equals the angular momentum of the phonon ($I = J$). Here, the summation on f is over all states with given I_f and $n_\lambda = 1$ phonon. For example, we consider a particular state $n_\lambda = 2$ which gives $I = 0, 2, 4$ and $n_\lambda = 1$ giving $I = 2$. Employing (4.70b) we get,

$$B(E2; 2^2 \rightarrow 2^1) = 2B(E2, 2^1 \rightarrow 0) \tag{4.70c}$$

and
$$B(E2; 4^1 \rightarrow 2^1) = 2B(E2, 2^1 \rightarrow 0), \tag{4.70d}$$

where 2^2 denotes the second 2+ excited state. It follows from the above discussion that

$$\frac{B(E2; 2^2 \rightarrow 2^1)}{B(E2, 2^1 \rightarrow 0)} = 2 \quad \text{and} \quad \frac{B(E2; 4^1 \rightarrow 2^1)}{B(E2, 2^1 \rightarrow 0)} = 2. \tag{4.70e}$$

Experimental results are in reasonable agreement with the theoretical ratios defined in (4.70e).

Again, $\langle n_\lambda | Q_{\lambda\mu} | n_\lambda \rangle \equiv \langle n_\lambda | b_{\lambda\mu} + (-)^\mu b_{\lambda\mu}^\dagger | n_\lambda \rangle \left(\frac{3Ze}{4\pi} r_0^\lambda\right)\left(\frac{\hbar}{2\sqrt{B_\lambda C_\lambda}}\right)^{1/2}$

and applying this result, we find the following outcome: (1) $\langle n_\lambda = 2| Q_{\lambda\mu} | n_\lambda = 0 \rangle = 0$, i.e. there should be no cross over $E2$ transition $2^2 + \rightarrow 0+$. (2) $\langle n_\lambda = 1| Q_{\lambda\mu} | n_\lambda = 1 \rangle = 0$, i.e. the quadrupole moment of the first excited state in vibrational model is zero and (3) there is a strong $E2$ transition from (2^1+) to $(0+)$ g.s.

E2 Transition Between Rotational Levels: We employ (1.75b) and (1.76b) respectively to calculate electric and magnetic transition probabilities. The reduced transition probability $B(E2; I_iK \rightarrow I_fK)$ for an electric quadrupole transition between two members of the same rotational band with quantum number K, is related to the intrinsic quadrupole moment Q_0. To see this, we first calculate expectation value of Q_{20} using the rotational wave function:

$$\langle I_f M_f K| Q_{20} | I_i M_i K \rangle$$

$$= \frac{[(2I_f +1)(2I_i +1)]^{1/2}}{8\pi^2} \sum_v \langle \chi_K| Q'_{2v} |x_K \rangle \int D^{I_f}_{M_f K} D^2_{0v} D^{I_i}_{M_i K} \, d\Omega.$$

The above relation is obtained by transforming Q_{20} to body fixed frame. Employing the properties of the D-functions and the C coefficients <1>, we get

$$\langle I_f M_f K| Q_{20} | I_i M_i K \rangle$$

$$= \frac{[(2I_f +1)(2I_i +1)]^{1/2}}{8\pi^2} \cdot \frac{8\pi^2}{(2I_f +1)} \langle I_i 2M_i 0 | I_f M_f \rangle$$

$$\times \langle I_i 2Kv | I_f K \rangle \, \delta_{M_i M_f} \sum_v \langle x_K | Q'_{2v} | x_K \rangle \, \delta_{v0}$$

$$= \frac{(2I_i +1)}{(2I_f +1)} \left(\langle I_f 2M_f 0 | I_f M_f \rangle \right)^2 \left(\langle I_i 2K \, 0 | I_f K \rangle \right)^2 e^2 \left(\frac{5}{16\pi} \right) Q_0^2 .$$

Hence, $B(E2; I_iK \rightarrow I_fK) = (2I_i +1)^{-1} \sum_{M_i M_f} |\langle I_f M_f K | Q_{20} | I_i M_i K \rangle |^2$

$$= \sum_{M_f} \frac{(2I_i +1)}{(2I_f +1)} \left(\langle I_i 2M_f 0 | I_f M_f \rangle \right)^2 \left(\langle I_i 2K \, 0 | I_f K \rangle \right)^2 e^2 \left(\frac{5}{16\pi} \right) Q_0^2$$

$$= \frac{5e^2}{16\pi} Q_0^2 \left[\langle I_i \, 2K \, 0 | I_f K \rangle \right]^2 , \tag{4.71a}$$

since $\sum_{M_f} \left(\langle I_i 2M_f 0 | I_f M_f \rangle \right)^2 = (2I_f +1)$.

For the ground state band ($K = 0$) and the even-even nuclei, we have

$$B(E2; I0 \to I-2, 0) = \frac{5e^2}{16\pi} Q_0^2 \left[\langle I2\ 00 \mid I-20\rangle\right]^2$$

$$= \frac{5e^2 Q_0^2}{16\pi} \frac{2I(I-1)}{2I(2I-1)\ (2I+1)} \tag{4.71b}$$

Hence, $\dfrac{B(E2;\ 4 \to 2)}{B(E2,\ 2 \to 0)} = \dfrac{10}{7}.$ \hfill (4.71c)

However, according to the vibrational model the above ratio is 2 (4.70e). The cross-section for the Coulomb excitation is directly proportional to upward reduced E2 gamma transition probability $B(E2\ \uparrow)$. Experimentally, Q_0 is usually determined from the E2 Coulomb excitation:

$$B(E2;\ I,K \to I+2,K)$$

$$= \frac{5}{16\pi} e^2 Q_0^2 \left[\langle I2\ 00 \mid I+20\rangle\right]^2 = \frac{5}{16\pi} e^2 Q_0^2 \frac{3}{2} \cdot \frac{(I+1)\ (I+2)}{(2I+1)\ (2I+1)}. \tag{4.72a}$$

For $I = 0$, $\quad B(E2;\ 0 \to 2) = \dfrac{5}{16\pi} e^2 Q_0^2.$ \hfill (4.72b)

Under certain assumptions, the experimental value of the intrinsic quadrupole moment Q_0 (referred to the body fixed axes) may provide a value of the deformation parameter β.

Following (1.20) and (4.41), and assuming a uniform charge distribution (4.4) we get

$$Q_0 = \int \sqrt{\frac{16\pi}{5}} r^2\ Y_2^0(\theta,\phi)\ \rho\ r^2\ dr\ d\Omega$$

$$= \sqrt{\frac{16\pi}{5}} \frac{3Z}{4\pi r_0^3} \int Y_2^0 \frac{r^5}{5}\ d\Omega = \frac{3Zr_0^2}{5\sqrt{5\pi}} \int Y_2^0 (1+\beta Y_2^0)^5\ d\Omega$$

$$= \frac{3Zr_0^2}{5\sqrt{5\pi}} \int Y_2^0 (1+5\beta Y_2^0 + 10\beta^2 Y_2^0 Y_2^0)\ d\Omega$$

$$= \frac{3Zr_0^2}{\sqrt{5\pi}} \beta \left(1+\frac{4}{7}\sqrt{\frac{5}{4\pi}}\beta\right), \tag{4.72c}$$

where we have used (2.53b) for the integration of the product of three spherical harmonics. The deformation β, estimated from the observed Q_0 for some rare earth nuclei are shown below :

$^{152}_{}Sm$ $\qquad\qquad\qquad Q_0 = 5.7$ barn $\qquad\qquad \beta = 0.33$

$^{160}_{64}Gd$ $\qquad\qquad\qquad Q_0 = 10.0$ barn $\qquad\qquad \beta = 0.46$

$^{178}_{72}Hf$ $\qquad\qquad\qquad Q_0 = 8.1$ barn $\qquad\qquad \beta = 0.31$

As further tests of the rotational model, there are various intensity rules. For example, for E2 transitions between the levels of ground state rotational band ($I_0 = K$ where $I_0 = g.s.$ spin) the ratio

$$\frac{B(E2; I_0 \to I_0 + 2)}{B(E2, I_0 \to I_0 + 1)} = \frac{(\langle I_0\ 2K\ 0\ |\ I_0 + 2K \rangle)^2}{(\langle I_0\ 2K\ 0\ |\ I_0 + 1K \rangle)^2} = \frac{2(I_0 + 1)}{I_0(2I_0 + 3)}. \qquad (4.72d)$$

The above ratios obtained by the Coulomb excitation measurements for odd A nuclei are compared with experimental values in Table 4.3. By and large the agreement between theory (4.72d) and experiment is reasonably good.

Table 4.3: $B(E2)$ ratios for odd A nuclei

Nucleus	I_0	$B(E2; I_0 \to I_0 + 2) / B(E2, I_0 \to I_0 + 1)$	
		Experiment	Theory
$^{157}_{64}Cd$	5/2	0.28	0.35
$^{173}_{71}Lu$	7/2	0.23	0.26
$^{179}_{92}Hf$	9/2	0.22	0.20
$^{235}_{92}U$	7/2	0.16	0.26
$^{237}_{92}Np$	5/2	0.44	0.35

Exercise 4.5. Show that the inertial parameter $B_\lambda = \rho\, r_0^5 / \lambda$ for nuclear fluid motion, where ρ is the nuclear density, r_0 is the radius of the spherical drop and λ is the order of deformation from spherical equilibrium shape.

The kinetic energy

$$T = \frac{1}{2}\rho \int v^2\, d\tau,$$

where v is velocity of the nuclear fluid. The incompressibility, non-viscous and irrotational character of the nuclear fluid enable us to express $v = \text{grad } \phi$, and $T = \frac{1}{2}\rho \int (grad\ \phi)^2\, d\tau = \frac{1}{2}\rho \int div(\phi * grad\phi)\, d\tau$. According to (4.30),

$v = r_0 \sum_{\lambda\mu} \dot{\alpha}_{\lambda\mu} Y_\lambda^\mu$. The general solution of the Laplace equation $\nabla^2 \phi = 0$ is

$\phi = \sum_{\lambda\mu} n_{\lambda\mu} \dfrac{u_\lambda(r)}{r} Y_\lambda^\mu$, where $n_{\lambda\mu}$ is a constant and $u_\lambda(r)$ satisfies

$\dfrac{d^2 u_\lambda(r)}{dr^2} - \dfrac{\lambda(\lambda+1)}{r^2} u_\lambda(r) = 0$, so that $u_\lambda(r) = r^{\lambda+1}$ and $\phi = \sum_{\lambda\mu} n_{\lambda\mu} r^\lambda \, Y_\lambda^\mu$.

Hence, $T = \dfrac{\rho}{2} \int [\phi * grad \ \phi]_{r=r_0} r_0^2 d\Omega = \dfrac{\rho r_0^2}{2} \sum_{\lambda\mu} \lambda \, n_{\lambda\mu}^2 \, r_0^{2\lambda-1}$,

by the Gauss theorem. Now using $v = \text{grad } \phi$ and v obtained from (4.30), we get: $r_0 \dot{\alpha}_{\lambda\mu} = \lambda \, n_{\lambda\mu} \, r_0^{\lambda-1}$.

Therefore, $T = \dfrac{1}{2} \rho \, r_0^5 \sum_{\lambda\mu} \dfrac{|\dot{\alpha}_{\lambda\mu}|^2}{\lambda} = \dfrac{1}{2} \sum_{\lambda\mu} B_\lambda \, |\dot{\alpha}_{\lambda\mu}|^2$, where $B_\lambda = \dfrac{\rho \, r_0^5}{\lambda}$.

Exercise 4.6. Assuming that the mean field of the nucleus deviates slightly from the spherical symmetry, determine the correction to single particle energy, associated with the deformation.

We employ (4.41) to express the deformed potential as

$$V(r) = V(r_0) + r_0 \sum_\nu a_{2\nu} Y_2^\nu \left(\frac{\partial V(r)}{\partial r} \right)_{r=r_0}$$

Perturbation $\Delta V = V(r) - V(r_0) = r_0 \left[\sum_\nu a_{2\nu} Y_2^\nu \left(\frac{\partial V(r)}{\partial r} \right)_{r=r_0} \right]$

$$= -V_0 \, r_0 \, \delta(r - r_0) \sum_\nu a_{2\nu} Y_2^\nu \text{ for a square well potential.}$$

We employ perturbation theory to calculate the first order correction to energy, ΔE as follows: We use the single particle wave function as unperturbed wave function and get

$$\Delta E = -V_0 \, r_0 \int |u_{n\ell}(r)|^2 \, \delta(r - r_0) \, dr \times \sum_\nu \langle jm_j | Y_2^\nu | jm_j \rangle$$

$$= V_0 r_0 \, |u_{n\ell}(r_0)|^2 \frac{1}{4} \sqrt{\frac{5}{4\pi}} \sum_\nu \frac{3m_j^2 - j(j+1)}{j(j+1)} a_{2\nu} \delta_{\nu 0}$$

$$= c_j \left[3m_j^2 - j(j+1) \right] \beta \cos\gamma,$$

where $\quad c_j = \frac{1}{4}\sqrt{\frac{5}{4\pi}}\, V_0 r_0 \mid u_{n\ell}(r_0) \mid^2 / \, j(j+1).$

Exercise 4.7. Show that the constants a and b appearing in (4.50) have the

following values: $a = \dfrac{\hbar^2}{2\mathscr{I}_{rig}}$ and $b = -12a^3(\hbar\omega_\beta)^{-2}.$

We substitute the value of $(\beta - \beta_0)$ from Exercise 4.8 in (4.46) for $\gamma = 0$ and $V_0 = 0$. We get

$$\frac{1}{2}C(\beta - \beta_0) = \frac{1}{2}B\omega_\beta^2 \left[12\left(\frac{\hbar^2}{2\mathscr{I}_{rig}}\right)^2 \beta_0 \frac{J(J+1)}{(\hbar\omega_\beta)^2} \right]^2$$

$$= 12\left(\frac{\hbar^2}{2\,\mathscr{I}_{rig}}\right)^3 \frac{[J(J+1)]^2}{(\hbar\omega_\beta)^2} \quad \text{since } 3B\beta_0^2 = \mathscr{I}_{rig}.$$

Now, $\quad \dfrac{\hbar^2}{2\mathscr{I}}J(J+1) = \dfrac{\hbar^2}{2\mathscr{I}_{rig}}\left[1 - 24\left(\dfrac{\hbar^2}{2\mathscr{I}_{rig}}\right)^2 \dfrac{J(J+1)}{(\hbar\omega_\beta)^2}\right]J(J+1).$

Hence,

$$E_J(\beta) = \frac{\hbar^2}{2\mathscr{I}_{rig}}J(J+1) - 24\left(\frac{\hbar^2}{2\mathscr{I}_{rig}}\right)^3 \frac{J^2(J+1)^2}{(\hbar\omega_\beta)^2} + 12\left(\frac{\hbar^2}{2\mathscr{I}_{rig}}\right)^3 \frac{J^2(J+1)^2}{(\hbar\omega_\beta)^2}$$

$$= \frac{\hbar^2}{2\mathscr{I}_{rig}}J(J+1) - 12\left(\frac{\hbar^2}{2\mathscr{I}_{rig}}\right)^3 \frac{J^2(J+1)^2}{(\hbar\omega_\beta)^2}.$$

Comparing with equation (4.50) and identifying J as I, we get

$$a = \frac{\hbar^2}{2\mathscr{I}_{rig}} \text{ and } b = -12a^3(\hbar\omega_\beta)^{-2}.$$

Exercise 4.8. Show that in a hydrodynamic model, when the rotation and vibration about $\beta = \beta_0$ is considered together and rotation-vibration interaction is neglected, the stretched moment of inertia

$$\mathscr{I} = \mathscr{I}_{rig} \left[1 + 24 \left(\frac{\hbar^2}{2\mathscr{I}_{rig}} \right)^2 \frac{J(J+1)}{(\hbar\omega_\beta)^2} \right].$$

Using (4.46) we can write the total vibrational and rotational energy for small deformation as

$$E_J(\beta) = V_0 + \frac{1}{2} C(\beta - \beta_0)^2 + \frac{\hbar^2}{6B\beta^2} J(J+1).$$

Next we treat β as variational parameter and minimize $E_J(\beta)$ for a specific value of J, so that

$$\frac{\partial E_J(\beta)}{\partial \beta} = C(\beta - \beta_0) - \frac{\hbar^2}{3B\beta^3} J(J+1) = 0,$$

or
$$B\omega_\beta^2 (\beta - \beta_0) = \frac{\hbar^2}{3B\beta^3} J(J+1) \text{ since } C = B\omega_\beta^2.$$

Therefore,
$$\beta - \beta_0 = 12 \left(\frac{\hbar^2}{2\mathscr{I}} \right)^2 \beta \frac{J(J+1)}{(\hbar\omega_\beta)^2},$$

where $\mathscr{I} = 3B\beta^2$ is the hydrodynamic moment of inertia. When $\beta \to \beta_0$ then $\mathscr{I} \to \mathscr{I}_{rig}$, so that

$$\beta - \beta_0 = 12 \left(\frac{\hbar^2}{2\mathscr{I}_{rig}} \right)^2 \beta_0 \frac{J(J+1)}{(\hbar\omega_\beta)^2}$$

or
$$\beta = \beta_0 \left[1 + 12 \left(\frac{\hbar^2}{2\mathscr{I}_{rig}} \right)^2 \frac{J(J+1)}{(\hbar\omega_\beta)^2} \right].$$

In this approximation the stretched moment of inertia

$$\mathscr{I} \approx 3B\beta_0^2 \left[1 + 24 \left(\frac{\hbar^2}{2\mathscr{I}_{rig}} \right)^2 \frac{J(J+1)}{(\hbar\omega_\beta)^2} \right]$$

$$\approx \mathscr{I}_{rig} \left[1 + 24 \left(\frac{\hbar^2}{2\mathscr{I}_{rig}} \right)^2 \frac{J(J+1)}{(\hbar\omega_\beta)^2} \right].$$

Exercise 4.9. Compare the hydrodynamic value of the moment of inertia of an axially symmetric deformed nucleus with quadrupole deformation β, about an axis perpendicular to the symmetry axis with the moment of inertia of a rigid sphere of radius r_0 having the same density

$$(\mathscr{I})_{\text{hyd.}} = 3B\beta^2 = \frac{3 \cdot (\text{Am})}{\frac{4\pi}{3} r_0^3} \cdot \frac{r_0^5 \beta^2}{2} = \frac{9}{8\pi} \text{Am } r_0^2 \beta^2, \text{ since } B = \frac{\rho \, r_0^5}{2} \text{ for } \lambda = 2.$$

For a sphere, $(\mathscr{I})_{\text{rigid}} = \frac{2}{5} \text{Am } r_0^2$. Hence, $\mathscr{I}_{\text{hyd}} = \left(\frac{45}{16\pi} \beta^2\right) \mathscr{I}_{\text{rig}}$.

For the most deformed nuclei: $\beta^2 = \frac{1}{10}$ and $\frac{45}{16\pi} \approx 1$, so that $\mathscr{I}_{\text{hyd}} \approx \frac{1}{10} \mathscr{I}_{\text{rig}}$.

The hydrodynamic value is nearly one tenth of the rigid-body value, indicating thereby that the nuclear rotation actually caries little effective mass with it. The analysis of low lying states of deformed even-even nuclei yields

$$\mathscr{I}_{\text{expt}} = \frac{1}{12} \mathscr{I}_{\text{rig}} = 5 \, \mathscr{I}_{\text{hyd}}.$$

Hence, experiment shows that the moment of inertia is seriously underestimated in the hydrodynamic model.

Exercise 4.10. Derive rotational kinetic energy T_r given by (4.45) from (4.30).

$$T_r = \frac{1}{2} B \sum_\mu |\dot{\alpha}_{2\mu}|^2$$

$$= \frac{1}{2} B \sum_\mu |\frac{d}{dt} \sum_\nu D^{2*}_{\mu\nu} a_{2\nu}|^2 = \frac{1}{2} B \sum_\mu |\sum_\nu (\dot{a}_{2\nu} D^{2*}_{\mu\nu} + a_{2\nu} \dot{D}^2_{\mu\nu})|^2$$

$$= \frac{1}{2} B \left[\sum_{\mu\nu\nu'} \dot{a}_{2\nu} a^*_{2\nu'} D^2_{\mu\nu} D^{2*}_{\mu\nu'} + \sum_{\mu\nu\nu'} a_{2\nu} \dot{D}^2_{\mu\nu} a^*_{2\nu'} \dot{D}^{2*}_{\mu\nu'} \right]$$

+ cross terms which vanish.

The first term $= \frac{1}{2} B \sum_{\mu\nu\nu'} \dot{a}_{2\nu} \dot{a}^*_{2\nu'} D^2_{\mu\nu} D^{2*}_{\mu\nu'}$

$$= \frac{1}{2} B \sum_\mu |\dot{a}_{2\nu}|^2, \quad \text{since} \sum_\mu D^2_{\mu\nu} D^{2*}_{\mu\nu'} = \delta_{\nu\nu'}$$

$$= \frac{1}{2} B \left[\dot{a}_{20}^2 + 2\dot{a}_{22}^2 \right] = \frac{1}{2} B \left[\dot{\beta}^2 + \beta^2 \dot{\gamma}^2 \right], \text{ which follows from (4.44).}$$

The time derivative of the rotation matrix appearing in the second term is carried out by decomposing the rotation θ_k $(k = 1, 2, 3)$, which carries space fixed axes (X, Y, Z) into coincidence with body fixed frame $(1,2,3)$, where the angular momentum is represented by J_k $(k = 1,2,3)$ and $\omega_k = \dot{\theta}_k$, where θ_k are the Euler angles.

Now
$$\frac{\partial}{\partial t}(D_{MK}^{J*}) = \frac{\partial}{\partial t}\left\langle JM \mid e^{-i\theta \, \hat{n}\cdot\vec{J}} \mid JK \right\rangle \qquad \text{(from 4.42c)}$$

$$= -i \sum_{k=1}^{3} \omega_k \left\langle JM \mid J_k \mid \sum_N \mid JN \right\rangle \left\langle JN \mid e^{-i\theta_k \hat{n}\cdot\vec{J}} \mid JK \right\rangle,$$

since $\sum_N \mid JN \rangle \langle JN \mid = 1$.

Hence, the second term $= \dfrac{1}{2} B \sum_{k\nu\nu'} (-)^{\nu'} \left\langle 2, \nu \mid J_k^2 \mid 2, -\nu' \right\rangle a_{2\nu} \, a_{2\nu'} \, \omega_k^2$.

We define the effective moment of inertia as

$$\mathscr{I}_k = B \sum_{\nu\nu'} (-)^{\nu'} \left\langle 2, \nu \mid J_k^2 \mid 2, -\nu' \right\rangle a_{2\nu} \, a_{2\nu'},$$

where ν, ν' run only through the values 0 and 2 and $a_{22} = a_{2-2}$.

The matrix elements of J_k^2 are as follows:

$$\left\langle 2\nu \mid J_1^2 \mid 2\nu \right\rangle = \left\langle 2\nu \mid J_2^2 \mid 2\nu \right\rangle = \frac{1}{2}\left[2(2+1) - \nu^2\right].$$

$$\left\langle 2\nu + 2 \mid J_1^2 \mid 2\nu \right\rangle = -\left\langle 2\nu + 2 \mid J_2^2 \mid 2\nu \right\rangle = \frac{1}{4}\left\{(1+\nu)\,(2+\nu)\,(3-\nu)\,(4-\nu)\right\}^{1/2}$$

$$\left\langle 2\nu - 2 \mid J_1^2 \mid 2\nu \right\rangle = -\left\langle 2\nu - 2 \mid J_2^2 \mid 2\nu \right\rangle = \frac{1}{4}\left\{(1+\nu)\,(2+\nu)\,(3-\nu)\,(4-\nu)\right\}^{1/2}$$

$$\left\langle 2\nu \mid J_3^2 \mid 2\nu' \right\rangle = \nu^2 \, \delta_{\nu\nu'}.$$

Hence,
$$\mathscr{I}_1 = B \left(3a_{20}^2 + 2\sqrt{6}\, a_{20}\, a_{22} + 2a_{22}^2\right)$$

$$= 4B\beta^2 \sin^2\left(\gamma - \frac{2\pi}{3}\right) \text{ and } \mathscr{I}_3 = 8Ba_{22}^2 = 4B\beta^2 \sin^2\gamma.$$

Exercise 4.11. If μ_{gs} and μ_1 are the magnetic moments of the ground and first excited states of a rotational band $\left(K \neq \dfrac{1}{2}\right)$, show that

$$g_R = \frac{1}{2}\left[\frac{K+2}{K+1}\, \mu_1 - \mu_{gs} \right] \text{ and } g_K = \frac{1}{2K}\left[\frac{3K+2}{K+1}\, \mu_{gs} - \frac{K+2}{K+1}\mu_1 \right].$$

Employ the relation (4.62) for the ground state ($I = K$) to get

$$\mu_{gs} = g_R \frac{K}{K+1} + g_K \frac{K^2}{K+1}, \text{ which gives the relation}$$

$$(K+1)\,\mu_{gs} = K\,g_R + K^2 g_K.$$

Similarly, for the first excited state ($I = K+1$), we get

$$\mu_1 = g_R \frac{(K+1)\,(K+2) - K^2}{K+1+1} + g_K \frac{K^2}{K+1+1}$$

or $\qquad (K+2)\mu_1 = (3K+2)g_R + K^2\, g_K.$

Solving for μ_{gs} and μ_1, we get

$$g_R = \frac{1}{2}\left[\frac{K+2}{K+1}\, \mu_1 - \mu_{gs} \right] \text{ and } g_K = \frac{1}{2K}\left[\frac{3K+2}{K}\, \mu_{gs} - \frac{K+2}{K+1}\mu_1 \right].$$

4.3 NILSSON MODEL AND NEW REGIONS OF DEFORMATION

In a natural extension of shell model which allows some description of collective motion, we assume an overall average non-spherical potential for all nucleons and fill this well according to the Pauli principle. This extension is in some sense a 'Unification' of single particle model and collective model. This unification was first suggested by Nilsson[9] and hence the model is known as the Nilsson model. The anisotropic harmonic oscillator potential for an axially symmetric quadrupolar deformed nucleus is

$$V = \frac{1}{2}\,\mu\,\omega_0^2\, r'^2 \left[1 + \beta Y_2^0\right]^{-2} \cong \frac{1}{2}\mu\,\omega_0^2\, r'^2 \left(1 - 2\beta Y_2^0\right)$$

$$= \frac{1}{2}\,\mu\,\omega_0^2\left[r'^2 - \beta r'^2 \sqrt{\frac{5}{4\pi}}\,(3\cos^2\theta' - 1) \right]$$

$$= \frac{1}{2}\mu\,\omega_0^2\left[x_1^2 + x_2^2 + x_3^2 - \beta\sqrt{\frac{5}{4\pi}}\,(3x_3^2 - x_1^2 - x_2^2 - x_3^2) \right], \text{ since } r'\cos\theta' = x_3.$$

Finally,

$$V = \frac{1}{2}\,\mu\left[\omega_0^2\left(1 + \sqrt{\frac{5}{4\pi}}\,\beta\right)x_1^2 + \omega_0^2\left(1 + \sqrt{\frac{5}{4\pi}}\,\beta\right)x_2^2 + \omega_0^2\left(1 - 2\sqrt{\frac{5}{4\pi}}\,\beta\right)x_3^2 \right]$$

$$= \frac{1}{2}\, \mu \left[\omega_1^2 x_1^2 \;+\; \omega_2^2 x_2^2 \;+\; \omega_3^2 x_3^2 \right], \tag{4.73a}$$

where $\omega_1^2 = \omega_2^2 = \omega_0^2 \left(1 + \frac{2}{3}\, \delta\right)$ and $\omega_3^2 = \omega_0^2 \left(1 - \frac{4}{3}\, \delta\right)$ with $\delta = \frac{3}{2} \sqrt{\frac{5}{4\pi}}\, \beta.$

$$\tag{4.73b}$$

The relation (4.73b) indicates that the single particle oscillates faster in (1-2) plane than along 3rd symmetry axis if deformation δ is positive. Since frequencies are inversely proportional to corresponding nuclear radii, then

$$\omega_1 = \frac{\omega_0}{A_1}, \qquad \omega_2 = \frac{\omega_0}{A_2} \quad \text{and} \quad \omega_3 = \frac{\omega_0}{A_3}. \tag{4.74}$$

The constancy of nuclear volume demands $A_1 A_2 A_3 = \omega_1 \omega_2 \omega_3 = \omega_{00}^3$, with ω_{00} being a constant independent of δ (also γ).

Hence, $\omega_0(\delta) = \omega_{00} \left[1 - \frac{4}{3}\, \delta^2 - \frac{16}{27}\, \delta^3 \right]^{-1/6}$ where $\omega_{00} = \omega_0(\delta = 0).$ (4.75)

Finally, s.p. Hamiltonian takes the form:

$$H = -\frac{\hbar^2}{2\mu} \nabla'^2 + \frac{1}{2}\, \mu \left(\omega_1^2 x_1^2 + \omega_2^2 x_2^2 + \omega_3^2 x_3^2 \right) + C\vec{\ell} \cdot \vec{s} + D\ell^2, \tag{4.76}$$

where the constant C is the strength of the spin orbit coupling and D is a parameter that takes into account the deviation of oscillator potential from the realistic one. We introduce the dimensionless variable $\zeta_k = \alpha\, x_k$, and

$\rho^2 = \sum_k \zeta_k^2$, where $\alpha = \sqrt{\dfrac{\mu \omega_0(\delta)}{\hbar}}$. Therefore,

$\nabla^2 \equiv \alpha^2 \nabla'^2$ or $\dfrac{-\hbar^2}{2\mu} \nabla'^2 \equiv -\dfrac{1}{2}\, \hbar \omega_0(\delta) \nabla_\zeta^2$. The above scale transformation

eliminates coupling between major shells and we can write

$$H = H_{00} + H_\delta, \tag{4.77}$$

where $\quad H_{00} = \frac{1}{2}\, \hbar \omega_0(\delta) \left[-\nabla_\zeta^2 \;+\; \rho^2 \right]$

and $\quad H_\delta = -\frac{1}{3}\, \hbar \omega_0(\delta) \sqrt{\frac{16\pi}{5}}\, \rho^2 Y_2^0 + C\vec{\ell} \cdot \vec{s} + D\ell^2.$

For the actual calculation, we write H in the form:

$$H = H_{00} - \chi \hbar \omega_{00} \left[\frac{H_\delta}{\chi \hbar \omega_{00}} - 2\vec{\ell} \cdot \vec{s} - \frac{D}{\chi \hbar \omega_{00}} \ell^2 \right], \qquad (4.78)$$

where $\chi = \dfrac{C}{2\hbar\omega_{00}}$ and $\dfrac{H_\delta}{\chi \hbar \omega_{00}} = -\dfrac{1}{3\chi} \dfrac{\omega_0(\delta)}{\omega_{00}} \sqrt{\dfrac{16\pi}{5}} \rho^2 Y_2^0.$

Next we set $\eta = \dfrac{1}{3\chi} \dfrac{\omega_0(\delta)}{\omega_{00}}$ and $\mu = \dfrac{D}{\hbar \omega_{00}\chi}$ so that

$$H = H_{00} - \chi\hbar\omega_{00} \left[1 + \eta\sqrt{\frac{16\pi}{5}} \rho^2 Y_2^0 - 2\vec{\ell} \cdot \vec{s} - \mu\ell^2 \right]. \qquad (4.79)$$

Representation: The s.p. states for a spherically symmetric h.o. potential may be characterized by the quantum numbers (N, ℓ, jm_j) where $N = \{2(n-1) + \ell\}$ is the principal quantum number. For an anisotropic potential j, ℓ and N no longer remain good quantum numbers. For an axially symmetric deformed nucleus, only projection of j along Z'-axis, i.e. W remains good quantum number and, therefore, states are labelled by W. Hence, we may separate N into quantum numbers n_1, n_2, n_3, representing respectively quanta of excitation of 1, 2 and 3 components of oscillator motion such that

$$N = n_1 + n_2 + n_3.$$

The quantum numbers of components are not constants of motion but for large positive deformation and $\hbar\omega_1 - \hbar\omega_3 \geq 0$, we may treat n_3 as an approximately good quantum number. It is also called the asymptotic quantum number for the large deformation. We expand

$$\chi_\Omega = \sum_{\ell\lambda} a_{\ell\lambda}(\Omega) |N \ell \tfrac{1}{2}\lambda \sigma\rangle = \sum_{\ell\lambda} a_{\ell\lambda}(\Omega) |N \ell \lambda \sigma\rangle, \text{ where } \Omega = \lambda + \sigma.$$

$$(4.80a)$$

Since the intrinsic spin of single particle is implicit then by comparing Eq. (4.57a) with (4.80a), we get

$$C_{j\Omega} = \sum_\lambda a_{\ell\lambda} \langle \ell \tfrac{1}{2} \lambda \sigma \mid j \Omega\rangle, \qquad (4.80b)$$

where $\langle \ell \tfrac{1}{2} \lambda \sigma \mid j \Omega\rangle$ are the C-coefficients. Since the parity is a good quantum number, there is only one ℓ for each j.

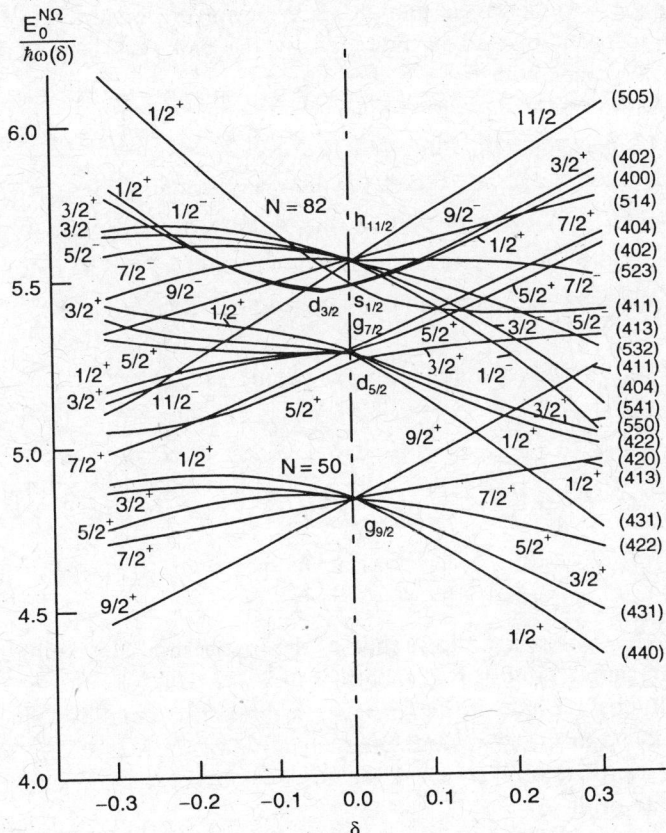

Fig. 4.10: The single particles levels (j^π) plotted against the deformation parameter δ (Nilsson diagram : Ref. 11)

The binding states of an individual nucleon in a strongly deformed nucleus are obtained by the exact diagonalization of the Nilsson Hamiltonian (4.79). The parameter η is related to the nuclear quadrupole deformation $\delta = 0.95\,\beta$. The parameters χ and μ determine the sequence and energy spread of levels and their values are chosen in such a way that for $\delta = 0$, the shell model levels are reproduced. Nilsson[11] tabulated the coefficients $a_{\ell\lambda}(\Omega)$ as a function of deformation parameter δ. The values of χ and μ and $\hbar\omega_{00}$ determined by fitting levels in the spherical shell model are roughly:

$$\hbar\omega_{00} = 0.05 \qquad \text{for } Z < 50$$
$$= 0.08 \qquad \text{for } Z < 20.$$

μ depends strongly on A; $\mu = 0$ for $8 < Z < 20$ and $8 < Z < 20$. For $50 < Z < 82$; $\mu = 0.55$ for $N = 4$ and $\mu = 0.50$ for $N = 5$.

The single particle levels plotted against η or δ are shown in Fig. 4.10. Each single particle level is indicated by the value of 'j' and the parity ($\pi = \pm 1$) together with the set of asymptotic quantum numbers. For large deformation n_3 becomes a good quantum number. $\delta = 0$ corresponds to the shell-model levels specified by $(N\ n\ \ell\ j)$.

In the Nilsson diagrams (Fig. 4.10) there are as many lines corresponding to each state of anisotropic potential as there are different projections $\Omega = |m_j|$ of the single particle angular momentum j, i.e. $(2j+1)/2$ lines. The energy of a nucleon in the quantum state (n_1, n_2, n_3) is given by

$$\epsilon_{n_1 n_2 n_3} = \left(n_1 + \frac{1}{2}\right)\hbar\omega_1 + \left(n_2 + \frac{1}{2}\right)\hbar\omega_2 + \left(n_3 + \frac{1}{2}\right)\hbar\omega_3$$

$$= \hbar\omega_0\left[(n_1 + n_2 + 1)\left(1 + \frac{1}{3}\delta\right) + \left(n_3 + \frac{1}{2}\right)\left(1 - \frac{2}{3}\delta\right)\right]$$

$$= \hbar\omega_0\left[N + \frac{3}{2} + \frac{\delta}{3}(N - 3n_3)\right], \tag{4.81}$$

with $N = n_1 + n_2 + n_3$. It follows that, as the major shell N is being filled by nucleons for large values of n_3 such that $3n_3 > N$, it is advantageous to put nucleons in the prolate orbits ($\delta > 0$), whereas for smaller values of n_3 ($3n_3 < N$) oblate orbits ($\delta < 0$) are energetically favoured. It also follows that in the lowest prolate orbit ($N = n_3$), the energy is lower than the energy of the lowest oblate orbit ($n_3 = 0$).

A. Calculation of Equilibrium Deformation

In order to calculate equilibrium deformation we rewrite non-isotropic oscillator (Eq. 4.73) in the form

$$V_\sigma = \frac{1}{2}m\omega_0^2\left[e^\sigma\left(x_1^2 + x_2^2\right) + e^{-2\sigma}\,x_3^2\right]. \tag{4.82}$$

The exponential factors lead to a volume conserving scale transformation, carrying the isotropic harmonic oscillator to the non-isotropic form. For example, the volume of the spheroid $\left(= \frac{4}{3}\pi(e^\sigma)^2(e^{-2\sigma})\right)$ is independent of σ. Here, e^σ and $e^{-2\sigma}$ are respectively major and minor axes. We write Eq. (4.82) in the form

$$V_\sigma = \frac{1}{2}m\omega_0^2\left[e^\sigma(x_1^2 + x_2^2) + e^{-2\sigma}\,x_3^2\right]$$

and identify $\sigma = (2/3)\delta$ by making a comparison with (4.73).

$$\in_{n_1 n_2 n_3} = \left(n_1 + \frac{1}{2} \right) \hbar\omega_1 + \left(n_2 + \frac{1}{2} \right) \hbar\omega_2 + \left(n_3 + \frac{1}{2} \right) \hbar\omega_3$$

$$= \hbar\omega_0 \left[(N - n_3 + 1) \, e^{\sigma/2} + \left(n_3 + \frac{1}{2} \right) e^{-\sigma} \right]. \qquad (4.83)$$

The equilibrium deformation is obtained by setting $\left(\dfrac{d\in}{d\sigma} \right)_{\sigma=\sigma_0} = 0$:

$$\frac{N - n_3 + 1}{2} e^{\sigma_0/2} - \left(n_3 + \frac{1}{2} \right) e^{-\sigma_0} = 0.$$

Hence, $\qquad \sigma_0 = \dfrac{2}{3} \log \dfrac{(2n_3 + 1)}{(N - n_3 + 1)}. \qquad (4.84)$

For k total particles in the oscillator shell

$$\sigma_0 = \frac{2}{3} \log \left[\sum_{i=1}^{k} (2n_3 + 1)_i \bigg/ \sum_{i=1}^{k} (N - n_3 + 1)_i \right]. \qquad (4.85)$$

In Fig. 4.11 we show with respect to A, the deformation parameter $\delta \left(= \dfrac{3}{2} \sigma_0 \right)$, calculated in the Nilsson model and that obtained from the experimental value of intrinsic quadrupole moment for rare earth nuclei.

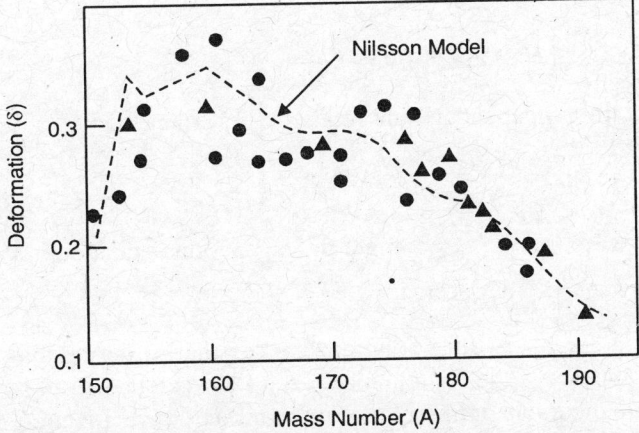

Fig. 4.11: Comparison of deformation parameter δ, calculated in the Nilsson model with that estimated from the measured quadrupole moment (4.72c)

Cranking Model: The treatment of motion of a system of particles in a non-spherical potential well provides us also a systematic basis for calculating moment of inertia of a nucleus. This treatment was first given by Inglis[10] and is known as the cranking model. Let us consider a nucleus that is rotating with angular velocity ω about a body fixed 1-axis (or X-axis) and has an axis of symmetry about 3-axis. The same nucleus could be described in a space fixed system, where without any loss of generality we can assume that the space fixed X-axis coincides with the body fixed 1-axis.

Suppose the Hamiltonian for a deformed well in space fixed system is H. Then the Hamiltonian in rotating frame $H(\theta) = e^{i\theta_1 J_1} H e^{-i\theta_1 J_1}$. Here, θ_1 is time dependent and is given by $\theta_1 = \omega t$. ω is very small and this is revealed by the experimental observation that the rotational energy is small compared to the single particle energy. The wave function referred to the space fixed system will satisfy the time dependent Schrodinger equation

$$H(\omega t)\psi = i\hbar \frac{d\psi}{dt}. \tag{4.86}$$

The wave function Φ in the rotating frame is related to ψ by transformation, $\Phi = e^{-i\theta_1 J_1}\psi$.

Hence, S. Eq. (4.86) takes the form:

$$H(\omega t)\, e^{i\theta_1 J_1}\Phi = \left[-\hbar\omega J_1 \Phi + i\hbar \frac{d\Phi}{dt}\right] e^{i\theta_1 J_1}.$$

Multiplying both sides of the above expression by $e^{-i\theta_1 J_1}$, we get

$$H(\omega t)\, \Phi = \left[-\hbar\omega J_1 \Phi + i\hbar \frac{d\Phi}{dt}\right]$$

$$\left[H + \hbar\omega \cdot J_1\right]\Phi = i\hbar \frac{d\phi}{dt}$$

where H is time independent and rotational invariant, i.e.

$$H = e^{-i\theta_1 J_1} H(\omega t)\, e^{i\theta_1 J_1}.$$

We can write $\Phi = e^{-\frac{i}{\hbar}\lambda t}\chi$ and get the eigen value equation

$$\left[H + \hbar\omega J_1\right]\chi = \lambda\chi, \tag{4.87}$$

where χ is a stationary state solution. We see that in the rotating frame an extra term appears in the Hamiltonian. Eq. (4.87) can be solved by using the perturbation theory in terms of the unperturbed wave function χ_n, which satisfies $H\chi_n = E_n \chi_n$.

The second term $\hbar\omega J_1$ in (4.87) is treated as perturbation and we get

$$\chi = \chi_n + \hbar\omega \frac{\langle n \mid J_1 \mid 0\rangle}{E_n - E_0} \chi_n + \cdots\cdots$$

$\langle \chi \mid \hbar\omega J_1 \mid \chi\rangle$

$$= \left\langle \left\{ \chi_0 + \hbar\omega \sum_{n\neq 0} \frac{\langle n\mid J_1\mid 0\rangle}{E_n - E_0} \chi_n \right\} \mid \hbar\omega J_1 \mid \times \left\{ \chi_0 + \hbar\omega \sum_{n\neq 0} \frac{\langle n\mid J_1\mid 0\rangle}{E_n - E_0} \right\} \right\rangle$$

$$= 2\hbar^2\omega^2 \sum_{n\neq 0} \mid \langle n\mid J_1\mid 0\rangle \mid^2 /(E_n - E_0). \tag{4.88}$$

The diagonal matrix element $\langle \chi_0 \mid J_1 \mid \chi_0 \rangle = 0$ and $\langle \chi_n \mid J_1 \mid \chi_n \rangle = 0$, since χ is a determinental wavefunction with definite value of angular momentum $J_3 = \Omega$. J_1 is a linear combination of the raising and lowering operators J_+ and J_-, which would alter Ω by ± 1. The energy of the system in stationary coordinate frame is

$$\langle \chi_0 \mid J_1 \mid \chi_0 \rangle = 0 \tag{4.89}$$

It is clear that the correction to the energy must be regarded as the energy of the rotation, and the moment of inertia is identified as twice the coefficient of ω^2. Thus, we obtain the well known cranking formula for moment of inertia:

$$\mathscr{I}_{cranking} = 2\hbar^2 \sum_{n\neq 0} \frac{\mid\langle n\mid J_1\mid 0\rangle\mid^2}{(E_n - E_0)}. \tag{4.90}$$

We shall show that the cranking formula for moment of inertia for a rotating anisotropic harmonic oscillator potential without the spin orbit force, yields the rigid body moment of inertia.

In the expression (4.90) we substitute $J_1 = \left(x_2 \dfrac{\partial}{\partial x_3} - x_3 \dfrac{\partial}{\partial x_2} \right)$.

The operator J_1 connects $\Delta n_1 = \pm 1$ and $\Delta n_3 = \pm 1$ corresponding to $\Delta N = 2$. Employing the expectation value of $x_2 \dfrac{\partial}{\partial x_3}$ etc. in anisotropic harmonic oscillator basis we get

$$\mathscr{I}_{cranking} = \frac{\hbar}{2\omega_2\omega_3} \left[\frac{(\omega_2 + \omega_3)^2}{(\omega_2 - \omega_3)} \left\{ \sum_{occ} \left(n_3 + \frac{1}{2} \right) - \sum_{occ} \left(n_2 + \frac{1}{2} \right) \right\} \right.$$

$$+ \frac{(\omega_2 - \omega_3)^2}{(\omega_2 + \omega_3)} \left\{ \sum_{occ} \left(n_3 + \frac{1}{2} \right) + \sum_{occ} \left(n_2 + \frac{1}{2} \right) \right\} \Bigg].$$

Thus, the moment of inertia depends on the number of occupied states (*occ*) in the self consistent field. The self consisting condition is

$$\mathscr{I}_{cranking} = \frac{\hbar}{2\omega_2\omega_3} \left[\frac{(\omega_2 + \omega_3)^2}{(\omega_2 - \omega_3)} \left\{ \sum_{occ} \left(n_3 + \frac{1}{2} \right) - \sum_{occ} \left(n_2 + \frac{1}{2} \right) \right\} \right.$$

or $\omega_1 N_1 = \omega_2 N_2 = \omega_3 N_3.$

Therefore, $\mathscr{I}_{cranking} = \frac{\hbar}{2\omega_2\omega_3} [(\omega_2 + \omega_3)(N_2 + N_3) + (\omega_2 - \omega_3)(N_3 - N_2)]$

$$= \hbar \left[\frac{N_3}{\omega_3} + \frac{N_2}{\omega_2} \right] = \hbar \left[\sum_{occ} \frac{\left(n_3 + \frac{1}{2} \right)}{\omega_3} + \sum_{occ} \frac{\left(n_2 + \frac{1}{2} \right)}{\omega_2} \right].$$

For harmonic oscillator

$$\sum_{occ} \langle x_2^2 \rangle = \frac{\hbar}{M\omega_2} \sum_{occ} \left(n_2 + \frac{1}{2} \right) \quad \text{and} \quad \sum_{occ} \langle x_3^2 \rangle = \frac{\hbar}{M\omega_3} \sum_{occ} \left(n_3 + \frac{1}{2} \right)$$

Hence, $\mathscr{I}_{cranking} = M \left[\sum_{occ} \langle x_2^2 + x_3^2 \rangle \right] = \mathscr{I}_{rigid}.$

B. Superdeformed Shapes

One of the most important developments in nuclear structure physics was the prediction and observation of super deformed (SD) shape at high angular momentum. These highly deformed shapes are stabilized by collective rotation unlike the fission isomers in heavy nuclei for which a minimum in potential exists even at spin zero. The first experimental observation[11] of superdeformation at high spin was in ^{152}Dy nucleus in which rotational transition indicated dynamic moment of inertia $\mathscr{I} = 85 \ \hbar^2 \text{MeV}^{-1}$ which for rigid rotation implies a deformation parameter $\beta_2 = 0.65$. The possible existence of superdeformed states characterized by substantial distortions in nuclear shapes was predicted already in the seventies. The term "superdeformed was probably first introduced by Cohen et. al.[12].

The nucleus considered as a drop of ideal liquid when rotated fast enough may suddenly loose its stability with respect to axially symmetric oblate shape in favour of strongly elongated prolate shape. Subsequently, SD bands in several nuclei in the neighbourhood of ^{132}Ce and ^{152}Dy, multiple SD bands in a single isotope, and new SD regions at masses $A \approx 190$ and $A = 140$ were observed.

The theories that have been successful in explaining superdeformation also predict other exotic shapes. In particular, rotational bands built upon hyperdeformed (HD) shapes with axis ratio around 3:1 are expected.

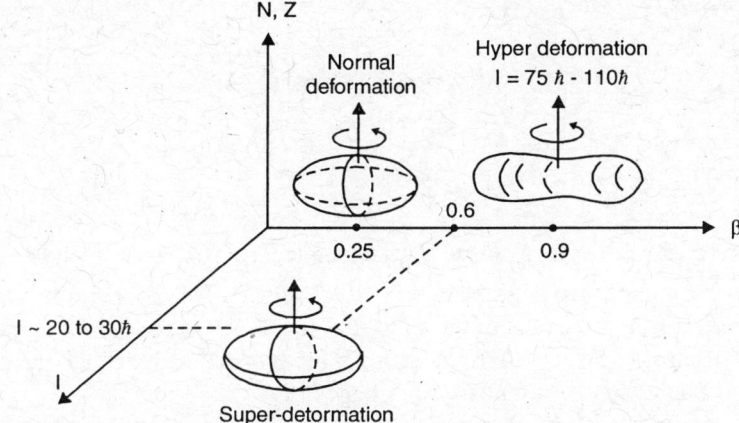

Fig. 4.12: The super and hyperdeformed shapes at the high spin

Recently Galindo-Uribari et al.[13] found evidence for hyper deformation (HD) at high angular momentum in ^{153}Dy. The rotational transitions in these nuclei corresponds to a dynamic moment of inertia $\mathscr{I} = 130\ \hbar^2 MeV^{-1}$, which suggests a hyperdeformed prolate shape with $\beta_2 = 0.9$. The behaviour of nuclei at the most extreme deformations and highest angular momenta is of great interest and is schematically shown in Fig. 4.12. Superdeformed or hyperdeformed shapes of nuclei at high spin are offering challenging new tests of nuclear models. The formation of different nuclear shapes is influenced essentially by the existence of the shell structure in the nucleus. It turns out that in certain areas in the deformation space and for certain numbers of neutrons or protons, the nuclear shell structure tends to favour the formation of systems that are more stable as compared to the adjacent nuclei and adjacent deformations.

Let us examine this effect by considering independent motion of nucleons in a potential of a non-rotating three dimensional anisotropic harmonic oscillator. The single particle energies of a nucleon in this potential well as before (Eq. 4.76) can be characterized in terms of Nilsson's deformation parameters $\delta \left(= \dfrac{3}{2} \sqrt{\dfrac{5}{4\pi}}\ \beta \right)$ and γ, three frequencies w_i, and three quantum numbers n_i $(i = 1,\ 2,\ 3)$ as follows:

$$\in_{n_1 n_2 n_3} = \hbar\omega_1 \left(n_1 + \frac{1}{2} \right) + \hbar\omega_2 \left(n_2 + \frac{1}{2} \right) + \hbar\omega_3 \left(n_3 + \frac{1}{2} \right), \qquad (4.91)$$

where
$$\omega_1 = \omega_0 (\delta, \gamma) \left\{ 1 + \frac{1}{3}\delta\cos\gamma + \frac{\delta}{\sqrt{3}}\sin\gamma \right\},$$

$$\omega_2 = \omega_0 (\delta, \gamma) \left\{ 1 + \frac{1}{3}\delta\cos\gamma - \frac{\delta}{\sqrt{3}}\sin\gamma \right\},$$

and
$$\omega_3 = \omega_0 (\delta, \gamma) \left\{ 1 - \frac{2}{3}\delta\cos\gamma \right\}.$$

Periodicity in Deformation: The dependence of ω_0 on δ and γ follows from the volume conservation condition (4.74-4.75). The Bohr and Mottelson condition for the existence of the shell structure leads to the requirement that the three harmonic oscillator frequencies ω_i are proportional to three (generally not very large) integer numbers such that

$$\omega_1 : \omega_2 : \omega_3 = a:b:c \qquad (4.92)$$

It is very easy to see that the single particle states with the same value of the quantum number N_{shell} defined as

$$N_{shell} = an_1 + bn_2 + cn_3 \qquad (4.93)$$

have the same energy. Thus, the quantum number N_{shell} determines a set of degenerate states, i.e. a shell for the deformation specified by Eq. (4.92). Obviously further symmetries of the system are able to increase degeneracy in the shell and therefore, to make the bunching of levels even more pronounced. Thus, in the case of spherical symmetry, $a = b = c = 1$, the harmonic oscillator (h.o.) become isotropic and the quantum number overlaps with main h.o. quantum number N:

$$N_{shell} = N = n_1 + n_2 + n_3 \qquad (4.94)$$

and the corresponding magic numbers define the closed shell with the number of nucleons 2, 8, 20, 40, 70, ...

In the case of axial symmetric deformation, there are some simplification, $a = b \neq c$, although the corresponding shell structure exhibits less degeneracy than in the isotropic case. Here we can distinguish two possible groups of shapes with different shell structure:

Prolate - $(a, b, c) = (2,2,1)$ (superdeformed) and $(a, b, c) = (3,3,1)$ (hyperdeformed).

Oblate - $(a, b, c) = (1,1,2)$ (superdeformed) and $(a, b, c) = (1,1,3)$ (hyperoblate).

The series of magic number corresponding to the closed (deformed) shells determined by the sets of numbers (a, b, c) are :

$$N_{shell} = 2, 4, 10, 16 ...(\text{superdeformed})$$

$$N_{shell} = 2, 4, 6, 12 ... (\text{hyperdeformed})$$

In Table 4.4 we summarise shell structure in the anisotropic harmonic oscillator.

Table 4.4: Shell Structure in the Deformation Space Table

	$a : b : c$	δ	γ	*Magic #s*
Spherical	1 : 1 : 1	0	0	2, 8, 20, 40, 70 ......
Super deformed	2 : 2 : 1	0.6	0	2, 4, 10, 16, 28, 40 ......
Hyper deformed	3 : 3 : 1	0.857	0	2, 4, 6, 12, 18, 24 ......
Super oblate	1 : 1 : 2	0.750	60°	2, 6, 14, 26, 44 ......
Hyper oblate	1 : 1 : 3	1.2	60°	2, 6, 12, 22, 36 ......
Super triaxial	4 : 3 : 2	0.577	30°	2, 4, 6, 10, 12, 18 ......
Hyper triaxial	3 : 2 : 1	0.839	30°	2, 4, 8, 14, 20 ......

So far we have discussed the possibilities for the formation of the exotic states within the framework of an ideal anisotropic harmonic oscillator. However, more realistic calculations employing modified h.o. (Nilsson), Saxon Wood, etc. have been made. The main failure of pure h.o. nuclear potential consists on the lack of possibility of including the strong $\vec{\ell} \cdot \vec{s}$ coupling in the nuclear Hamiltonian.

Exercise 4.12. Employing the Nilsson basis (4.80a) show that the decoupling parameter a is given by

$$a = (-)^{\ell} \sum_{\ell} (a_{\ell 0}^2 + 2\sqrt{\ell(\ell+1)}\ a_{\ell 0}\ a_{\ell 1}).$$

According (4.80b)

$$C_{j\frac{1}{2}} = \sum_{\ell} a_{\ell \lambda} \langle \ell \frac{1}{2} \lambda \sigma \mid j \frac{1}{2} \rangle$$

Since $\lambda + \sigma = \Omega = 1/2$ and $\sigma = \pm 1/2$, then $1 = 0$ and 1.
Therefore,

$$a = \sum_{j} (-)^{j+\frac{1}{2}} \left(j + \frac{1}{2} \right) \left[\sum_{\ell} \langle \ell \frac{1}{2} 0 \frac{1}{2} \mid j \frac{1}{2} \rangle a_{\ell 0} + \langle \ell \frac{1}{2} 1 - \frac{1}{2} \mid j \frac{1}{2} \rangle a_{\ell 1} \right]^2$$

$$= -\left[(-)^\ell \sum_\ell \left\{ \langle \ell \frac{1}{2} 0 \frac{1}{2} | \ell - \frac{1}{2} \frac{1}{2} \rangle a_{\ell 0} + \langle \ell \frac{1}{2} 1 - \frac{1}{2} | \ell - \frac{1}{2} \frac{1}{2} \rangle a_{\ell 1} \right\}^2 \right]$$

$$- \left[(-)^\ell \sum_\ell \left\{ \langle \ell \frac{1}{2} 0 \frac{1}{2} | \ell - \frac{1}{2} \frac{1}{2} \rangle a_{\ell 0} + \langle \ell \frac{1}{2} 1 - \frac{1}{2} | \ell - \frac{1}{2} \frac{1}{2} \rangle a_{\ell 1} \right\}^2 \right]$$

Substituting the values of the C-coefficients, we get

$$a = (-)^\ell \sum_\ell \left[(\ell+1) \left\{ \frac{\ell+1}{2\ell+1} a_{\ell 0}^2 + \frac{\ell+1}{2\ell+1} a_{\ell 1}^2 + \frac{2\sqrt{\ell(\ell+1)}}{2\ell+1} a_{\ell 0} a_{\ell 1} \right\} \right]$$

$$- \ell \left\{ \frac{\ell}{2\ell+1} a_{\ell 0}^2 + \frac{\ell+1}{2\ell+1} a_{\ell 1}^2 \right\} - \frac{2\sqrt{\ell(\ell+1)}}{2\ell+1} a_{\ell 0} a_{\ell 1} \right]$$

$$= (-)^\ell \sum_\ell \left[a_{\ell 0}^2 + 2\sqrt{\ell(\ell+1)}\, a_{\ell 0} a_{\ell 1} \right].$$

Exercise 4.13. The Nilsson wave for the level $\frac{1}{2} - [510]$ for $\eta = 4$ has the following form: $\psi = 1.000 | 550+ \rangle - 2.731 | 530+ \rangle + 2.44 | 510+ \rangle - 0.606 | 551- \rangle + 0.78 | 531- \rangle + 1.431 | 511- \rangle$.

Employing the above wave function calculate the decoupling parameter and hence compare the same as obtained from the energy levels of $^{183}_{74}W_{109}$ given below for $K = 1/2$ band :

E (KeV)	0	46	99	207
$I\pi$	$\frac{1}{2}-$	$\frac{3}{2}-$	$\frac{5}{2}-$	$\frac{7}{2}-$

Employing (4.59a) we get the ratio $\dfrac{E(3/2-)}{E(5/2-)}$ as

$$\frac{46}{99} = \left(\frac{13}{4} + 2a \right) \Big/ \left(\frac{33}{4} - 3a \right), \text{ which yields } (a)_{expt.} = 0.093.$$

The normalization of the given Nilsson wave function

$$N^2 = a_{50}^2 + a_{30}^2 + a_{10}^2 + a_{51}^2 + a_{31}^2 + a_{11}^2$$

$$= 1 + 7.453 + 5.954 + 0.367 + 0.608 + 2.035 = 17.42$$

Hence,

$$a = -\frac{1}{17.42}\left[a_{50}^2 + 2\sqrt{30}\ a_{50}\,a_{51} + a_{30}^2 + 2\sqrt{12}\,a_{30}\,a_{31} + a_{10}^2 + 2\sqrt{2}\ a_{10}\,a_{11}\right]$$

$$= -\frac{1}{17.42}\ [1 + 7.453 + 5.954 + 9.760 - 6.63 - 19.0] = 0.083$$

Finally, $a_{Nilsson} = 0.083$ and $a_{expt.} = 0.093$. The agreement is reasonably good.

4.4 BEHAVIOUR OF NUCLEI AT HIGH SPIN

There are two basic ways of generating high spin states in a nucleus: (1) collective (in phase) motions of the nucleons, for example, vibrations and rotations, (2) single-particle effects, for example, pair breaking and particle hole excitations. The individual spins of a few nucleons j_i generate the total nuclear spin.

The behaviour of nuclei at high spin indicates the interplay between the collective motion and single particle motion. It is now possible to know experimental excitation energy at high spin ($I = 30\text{-}66$) by means of heavy ion reactions. One observes strange behaviour of nuclei at high spin rotational states. However, any irregularity of rotational energy of a nucleus is hardly visible in $E(I)$ versus $I(I+1)$ plot. For example, a plot of energy versus spin for rotational nucleus ^{162}Er shows a smooth parabolic behaviour in accordance to $I(I+1)$ law with a small kink around $I = 14$. Actually deviations from the pure rotational law are observed which lower the excitation energies for higher angular momenta below the $I(I+1)$ law. A measure of the deviation from the pure rotator is the moment of inertia, which depends on the angular momentum.

The deviation from a pure rotator is normally represented in a plot of twice the moment of inertia as a function of the square of the rotational frequency. A pure rotator has a constant moment of inertia and yields therefore a line parallel to the abscissa. In the Variable Moment of Inertia (VMI) model, we express[13]

$$\mathscr{I} = \mathscr{I}_a + \mathscr{I}_b\,\omega^2$$

and get a straight line increasing with the square of the rotational frequency. The parameters $\mathscr{I}_a$ and $\mathscr{I}_b$ are known as the Harris parameters[13].

But when one plots moment of inertia $\mathscr{I}$ versus square of rotational frequency $\hbar\omega$, this becomes a large effect (Fig. 4.13). Moment of inertia ($\mathscr{I}$) tends to increase slowly at first, possibly due to an effective centrifugal stretching effect and finally at more or less critical angular momentum $\mathscr{I}_c(\approx 14)$ breaks into a back bending curve.

This varying nature nuclear moment of inertia as a function of rotational frequency (or spin) sheds light on the mechanism of the build-up of angular momentum in nuclei: Is it the total angular momentum of the nucleus due to pure collective rotation, or is it due to the many-particle angular momentum alignment along the rotation axis, or is it due to a mixture of the two effects, i.e. partly collective and partly particle aligned angular momentum.

A. Nuclear Fast Rotation

Backbending phenomenon was predicted first on a theoretical ground and was observed much later in even-even rotational nuclei.[14,15] A measure of the deviation from pure rotator is the moment of inertia, which depends on angular momentum:

$$\frac{\hbar^2}{2\mathscr{I}_I} = \left[\frac{\partial E_I}{\partial I(I+1)}\right]_{I=const.} \tag{4.95}$$

But, it is not possible to calculate the derivate of (4.95) numerically or analytically for one angular momentum alone since energy is given for even I and not as a continuous function of I. Therefore, we use slope between I and $I-2$:

$$E_{I-2} = \frac{\hbar^2}{2\mathscr{I}_{(I-2)}}[(I-2)(I-1)].$$

Hence, $\qquad \partial E_I = \Delta E_{I,I-2} = E_I - E_{I-2} = \frac{\hbar^2}{2\mathscr{I}_I}2(2I-1),$

where $\mathscr{I}_I$ denotes the mean value of moment of inertia between I and I-2. We rewrite (4.95) as

$$\frac{2\mathscr{I}_I}{\hbar^2} = \frac{2(2I-1)}{\Delta E_{I,I-2}}. \tag{4.96a}$$

The rotational frequency $\hbar\omega = \frac{\hbar^2}{\mathscr{I}_I}\sqrt{I(I+1)}$ since $\mathscr{I}_I\omega = \hbar\sqrt{(I+1)I}$.

Taking an average of the difference of $I(I+1)$ between I and $I-2$, i.e.

$$\left[I(I+1)\right]_{AV} = \frac{I(I+1)+(I-2)(I-1)}{2} = I^2 - I + 1$$

$$\partial I(I+1) = I^2 + I - (I-2)(I-1) = 2(2I-1), \text{ we get}$$

$$(\hbar\omega)^2 = \frac{I^2 - I + 1}{(2I-1)^2}(\Delta^2 E_{I,I-2}). \tag{4.96b}$$

The relations (4.96a) and (4.96b) enable us to obtain $(2\mathscr{I}/\hbar^2)$ versus $(\hbar\omega)^2$ plot directly from the experimental spectra. Table (4.5) and Fig. (4.13) give results of such an analysis for ^{162}Er showing backbending.

The basis of understanding backbending phenomenon is the single particle shell model with the inclusion of pairing correlation and the Bohr-Mottelson's model for collective motion.

The nucleus possesses stable quadrupole deformation. There is a rotation symmetry in the X-Y plane as well as reflection symmetry with respect to this plane. Because of pair correlation nucleus possesses a stable energy gap Δ. The Cooper pairs are formed from pair $(j\Omega, j\text{-}\Omega)$ where j is the single particle angular momentum and Ω is the projection on the symmetry axis Z. Strictly so for I = 0, ω = 0. As soon as nucleus begins to rotate around the X-axis in this picture centrifugal and Coriolis force will cause a break down of these symmetries.

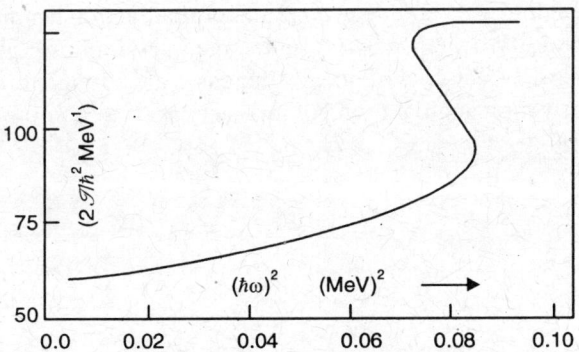

Fig. 4.13: Back bending phenomenon

Table 4.5: Shows $2\mathscr{I}/\hbar^2$ vs $(\hbar\omega)^2$ values for ^{162}Er

$2\mathscr{I}/\hbar^2$ $(MeV)^{-1}$	$(\hbar\omega)^2$ $(MeV)^2$
126.6	0.0950
126.3	0.0769
120.9	0.0748
93.1	0.0844
81.8	0.0794
75.1	0.0645
69.8	0.0468
65.3	0.0291
61.7	0.0137
58.8	0.0033

Centrifugal stretching along the direction of Z-axis will increase the moment of inertia. However, the effect is too small to account for the observed variation of moment of inertia. The Coriolis force breaks the pair and an alignment of single particle momenta with respect to rotation axis takes place. The effect of alignment, called Coriolis Anti-Pairing (CAP) is to decrease the pair correlation. Thus, the energy gap is reduced and moment of inertia increases.

We see that the first back bend occurs at critical angular momentum $I_c = 14$. Other best studied back bending rare earth nuclei are ^{158}Er (Lee et al.[14]) or ^{164}Er (Johnson et al.[15]) and ^{156}Er (Reich et al.[16]). In Fig. 4.14 we show the $2\mathscr{I}/\hbar^2$ versus $(\hbar\omega)^2$ plot for ^{158}Er. The first backbending is much like that of ^{162}Er but a second backbending occurs at frequency $(\hbar\omega)^2 = 0.18$ (around $I_c = 28\hbar$). This second backbend has now been seen in many rare earth nuclei and it has become a subject of theoretical study.

In Fig. 4.14 (right) we give the energy levels of ^{156}Er for an exercise to prepare backbending plot and determine the critical angular momentum.

Band Crossing: We shall now present some quantitative explanations of backbending. As the rotational frequency increases the coriolis force increases and at sufficiently high angular momentum one expects a coherent breakdown

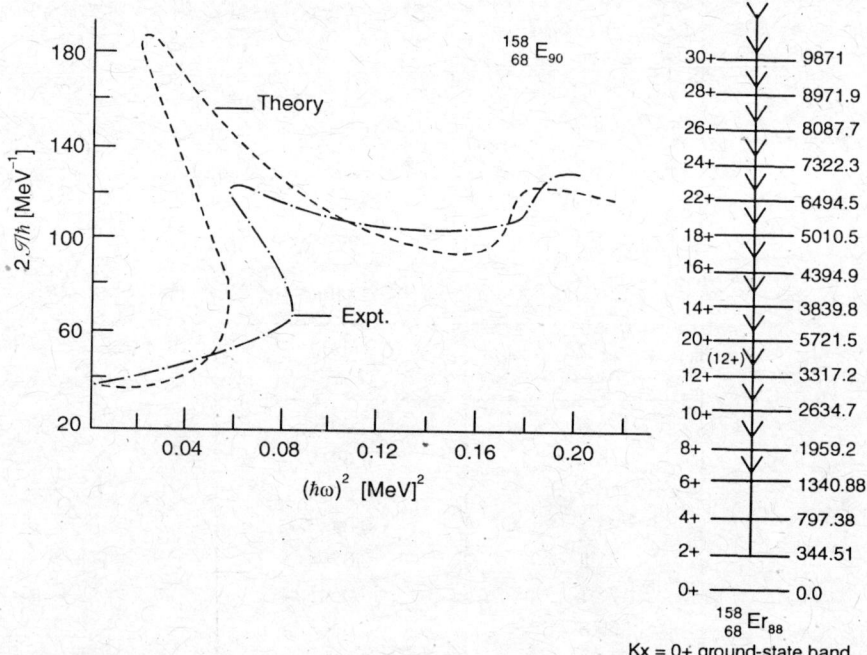

Fig. 4.14: $2\mathscr{I}/\hbar^2$ versus $(\hbar\omega)^2$ curve showing the first and second back bends (left), Rotational levels of ^{156}Er (Reich et al.[16]) (right)

of pairing correlation analogous to Meisner effect. Thus, backbending results from the intersection of paired and unpaired bands. It indicates that the nucleus is undergoing through a normal to superfluid type of "phase transition".[17] For unpaired band, the moment of inertia of a nucleus attains the rigid body value $\mathscr{I}_{rig}$. For paired band let us suppose that the moment of inertia is $\mathscr{I}$, which is much smaller than rigid-body value (experiments show that $\mathscr{I}_{rig} \approx 2\mathscr{I}$). The effect of pairing is to decrease the moment of inertia and hence increase the rotational energy.

The pairing correlation energy in the ground state is roughly Δ^2 / d, where Δ is the pairing gap energy and 'd' is the average level spacings. If the band crossing is at $I = I_c$ (Fig. 4.15) then equating the rotational energy of two bands at I_C, we get

$$\frac{\Delta^2}{d} + \frac{\hbar^2}{2\mathscr{I}_{rig}} I_c(I_c + 1) = \frac{\hbar^2}{2\mathscr{I}} I_c(I_c + 1).$$

Substituting $\mathscr{I}_{rig} \approx 2I$ we get $\dfrac{\Delta^2}{d} = \dfrac{\hbar^2}{2\mathscr{I}_{rig}} \left[I_c(I_c + 1) \right].$ \hfill (4.97)

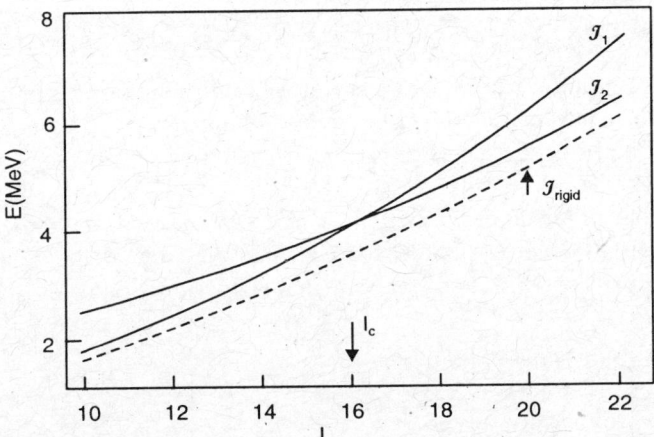

Fig. 4.15: Band crossing of paired and unpaired bands with moment of inertia $\mathscr{I}_1$ and $\mathscr{I}_2$ respectively. The rigid body moment of inertia is shown by the dotted curve

Hence, we can estimate I_c from (4.97). For illustration, let us consider a nucleus of mass number $A = 150$ and $\beta = 0.3$ (*see* Ex. 4.9). Analysis of low lying levels of deformed nuclei shows $\mathscr{I}_{rig} \approx 2\mathscr{I}$.

Now $\qquad \dfrac{\Delta^2}{d} = \dfrac{(11.4)^2}{A} \cdot \dfrac{A}{63} \approx \dfrac{130}{63}$ MeV,

so that
$$\frac{130}{63} \text{ MeV} = \frac{5}{4} \frac{(197)^2 \text{ MeV}^2 \text{ fm}^2}{938 \text{ MeV } 1.44 \text{ fm}^2} I_c(I_c + 1).$$

Hence, $[I_c]_{theory} \approx 15$, which is close to $[I_c]_{expt} \approx 16$, observed for the rare earth nuclei.

We have seen earlier that the addition of a centrifugal stretching term (4.50) leads to better fit to the ground state rotational levels of deformed even-even nuclei. Mantri and Sood[18] showed that by expressing E_I cubic in I, one can provide an estimate of the critical angular momentum I_c (Exercise 4.14).

The quantitative understanding of a second anomaly of the moment of inertia which has been detected in 1977 and at the same time explained by Faessler and Ploszajczak[19] and others[20-21] as the alignment of a high 'ℓ' neutron/proton pair. A first systematic study of the variation of the interaction between the ground state band and the aligned band was performed by Bengtsson, Hamamoto and Mottelson[22].

Nature of the second back bend results from the following consideration: The Coriolis force is not equally strong for all pairs. For single particle states with good angular momentum j and angular momentum projection to the symmetry axis Ω, the Coriolis strength is determined by the matrix element.

$$\langle j\Omega \pm 1 \mid j_x \mid j\Omega \rangle = \frac{1}{2}\sqrt{j(j+1) - \Omega(\Omega \pm 1)} \tag{4.98}$$

The largest single-particle angular momentum in the rare-earth nuclei has $i_{13/2}$ level for neutrons. The lowest angular momentum projection Ω is found in the first half of the rare earth nuclei. If in a rare earth nucleus a Nilsson level is near the Fermi surface which consists mainly of $i_{13/2}$ and has also small angular momentum projection Ω, one can expect that the pair of two nucleons in this level is broken apart at a lower angular momentum than others. Stephens and Simmon[20] assume that always only one pair in $i_{13/2}$ level

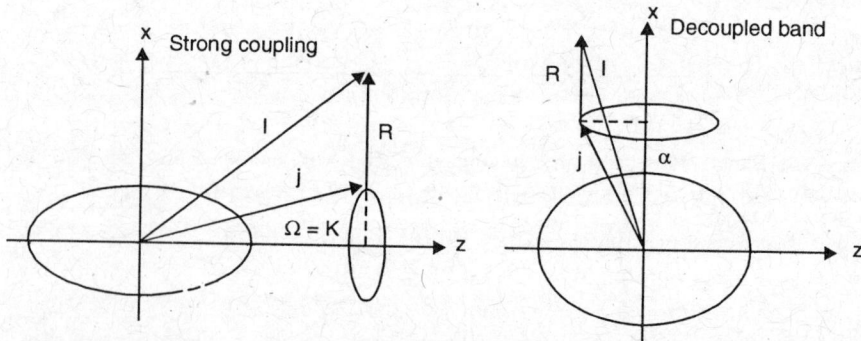

Fig. 4.16: Nucleon spin aligned along deformation axis (left), Nucleon spin aligned along rotation axes (right)

can be broken apart. Thus, for high j orbitals, the strong Coriolis force allows another type of coupling scheme where the nucleon spin is aligned along the rotation direction instead of following the direction of the deformation axis (Fig. 4.16).

The second band is therefore a two quasi particle band which starts at the energy of 2Δ above the ground state and has a larger moment of inertia due to pair blocking, and therefore smaller slope than the ground state rotational band.

B. Shell Effects at High Spin

The motion of a particle in an axially symmetric deformed nucleus generates a spectrum as shown in the Nilsson diagram (Fig. 4.10). A deformed nucleus has another degree of freedom, the rotation of the system around an axis perpendicular to the symmetry axis. In the completely adiabatic situation the nucleonic orbits follow the motion of the whole nucleus; but as the rotational frequency is increased the particles feel coriolis force and more symmetries are lost, for example, time reversal invariance. The wave function is different depending whether the particle is moving in the direction or counter to the direction of rotation and an additional term $-\omega\, j_x$ is added to the independent nuclear Hamiltonian in the rotating coordinate system:

$$h' = h_0 - \omega\, j_x. \tag{4.99}$$

We assume symmetry about 3-axis. Eq. (4.99) describes the dynamical coupling of intrinsic nuclear degrees of freedom with those characterizing nuclear rotation and h' is called "cranking" one particle Hamiltonian or "Routhian"[19-21]. The second term on the right hand side of (4.99) appears in the formula when the time dependent Schrodinger equation with the rotation potential is transformed to the body fixed frame of reference. Alternatively, we may treat h' as given by (4.99) as an auxiliary Hamiltonian which comes into consideration when nuclear energy is minimized with a constraint expressing the constant value of angular momentum, which in this formalism is identified with its own component on the rotation axis. The quantity w which is the angular frequency about the rotation axis is identified as the Lagrange multiplier.

The rotation directly influences the single particle energies and hence the shell effects at large I are quite different than for no rotation. Diagonalizing (4.99) we get $\langle \chi_i' \mid h' \mid \chi_i' \rangle = e_i'\, \chi_i'$, the Routhians, i.e. the energies in the rotating system (e_i') and the single particle eigen functions (χ_i') in the rotating frame. The energies in the laboratory frame are therefore, $e_i = \langle \chi_i' \mid h_0 \mid \chi_i' \rangle$ and the total angular momentum $I = \langle \chi_i' \mid j_x \mid \chi_i' \rangle$. Therefore, the total single particle energy is obtained as follows:

$$\sum_i \langle \chi_i' | h' | \chi_i' \rangle = \sum_i \langle \chi_i' | h_0 | \chi_i' \rangle - \hbar\omega \sum_i \langle \chi_i' | j_x | \chi_i \rangle$$

or
$$E_{sp} = \sum_i e_i' + \hbar \, \omega \, I \tag{4.100}$$

In the spirit of the Strutinsky approach[6], one calculates a smoothed out energy and an average angular momentum to obtain $\bar{E} = \sum_i \bar{e}_i' + \hbar \, \omega \, \bar{I}$.

The shell correction energy $E_{shell}(I) = E_{sp}(I) - \bar{E}(\bar{I})$.

In the above description, which is based on a dynamical coupling of single particle degrees of freedom to the nuclear rotation, we have neglected the existence of short range attractive pairing forces producing superfluid correlations. As is well known, these forces are very important in the region of low angular momenta. The coupling of the intrinsic nuclear degrees of freedom to these characterising nuclear rotation is given again by the cranking Hamiltonian with pairing interaction added to it. In this phenomenological approach the Hamiltonian is:

$$h' = h_0 + h_{pair} - \lambda N - \omega \, j_x \tag{4.101}$$

The system governed by (4.101) in fact describes the interplay and competition between three major tendencies in nuclei detailed in the following :

(1) Single particle potential attempting to locate the nucleons on the lowest possible orbits

(2) The pairing force that tends to locate pairs of nucleons in the state of zero angular momentum.

(3) The combined effect of the centrifugal and coriolis force that tends to align nucleon angular momentum with the rotation axis and to increase the moment of inertia. In addition nuclear rotation may change the form of single particle potential itself and/or shape of the nucleus. In (4.101), λ is the Lagrange multiplier defined earlier (Chapter 3) and $\hat{N}$ is the number operator such that $N = \langle \tilde{O} | \hat{N} | \tilde{O} \rangle$ where $| \tilde{O} \rangle$ denotes the ground state. This term is included in the Hamiltonian to conserve particle number N.

The trajectory of the single particle energy as a function of ω then describes the effect of Coriolis and centrifugal forces. Experimentally observed rotational 'bands' base on a single particle state, corresponding to one of these curves (Routhians) in the rotating frame. These calculated Routhians depend on the different deformation parameters, pairing strength and deformed single particle orbitals. Thus, an experimental study and comparison with theory can tell us about how the deformation and pairing correlation are changing as a function of the rotational frequency ω.

The Symmetry Properties of Cranked Hamiltonian: The coriolis term in the Cranking Hamiltonian h', which is proportional to j_x violates the assumed axial symmetry and time reversal invariance of h_0, but h' remains invariant with respect to space reflection (parity) and with respect to a rotation $R_x(\pi)$ of $180°$ about the X-axis. Thus, the eigen values e' and eigen states $|\chi'\rangle$ of the Cranking Hamiltonian can be labelled by the parity of the state π and by the eigen value of the rotation operator $R_x(\pi)$, given by

$$R_x(\pi)\,|\chi\rangle = e^{-i\pi\alpha}\,|\chi\rangle = r_1|\chi\rangle = e^{-i\pi j_x}\,|\chi\rangle$$

The quantum number r_1 is known as the signature. Usually, we call the signature exponent α, the signature quantum number. Since rotation of 2π leaves the wave function with a negative sign we have:

$$e^{-2\pi i\alpha}\,|\chi\rangle = (\cos 2\pi\alpha - i\sin 2\pi\alpha)|\chi\rangle = -|\chi\rangle \text{ for } \alpha = \pm\frac{1}{2}$$

Signature and parity remain the only good quantum numbers at high spins because of the K-mixing due to coriolis term. The square of the R_x operator acting on χ_k (*see* 4.57) gives

$$R_x^2\,\chi_k = \sum_j (-1)^{2j}\,C_j\,\chi_{jk}.$$

Since $2j$ is an odd integer for odd A nuclei, and is an even integer for even A nuclei, then $R_x^2\,\chi_k = (-)^A\,\chi_k$. For odd A nuclei we get $\alpha = \frac{1}{2}$ or $-\frac{1}{2}$.

This implies that because of 4π symmetry of the system, states having a total angular momentum differing by $2\hbar$ have the same signature. That is for odd-A nuclei,

$$\alpha = \frac{1}{2}: \qquad I = \frac{1}{2}, \frac{5}{2}, \frac{9}{2}, \frac{13}{2}, \dots\dots \qquad \text{and}$$

$$\alpha = -\frac{1}{2}: \qquad I = \frac{3}{2}, \frac{7}{2}, \frac{11}{2}, \dots\dots$$

and for even nuclei,

$$\alpha = 0 \quad: \qquad I = 0, 2, 4, \dots\dots \qquad \text{and}$$

$$\alpha = 1 \quad: \qquad I = 1, 3, 5, \dots\dots$$

A given $K = \frac{1}{2}$ band splits into two chains corresponding to $\alpha = \pm\frac{1}{2}$.

Giant Resonances and Nuclear Deformation: Giant Resonances are fundamental collective excitations of the entire nucleus, that provide the information about the basic properties of nucleus. Predominantly, one studies nuclear systems at high excitations and high angular momenta. Giant monopole resonance, a compressional mode without change of shape is of special importance because its energy is related to the incompressibility of a finite nucleus (*see* Eq. 6.61). The mode that is most easily studied is the Giant Dipole Resonance (GDR). Fundamental questions that are being addressed by such studies are:

(1) How does the nuclear shape affects GDR photonuclear cross sections?

(2) How does the restoring force of collective mode change with excitation energy?

The GDR γ-strength is given in terms of the classical electric dipole (E1) sum rule and for the spherical nucleus, three degenerate giant dipole modes exist. But in the case of ellipsoid, these three degenerate modes split into two, one mode oscillating along the major axis of the ellipsoid and two degenerate modes oscillating along minor axes. Therefore, the single resonance peak in the case of spherical nucleus splits into two peaks for the ellipsoidal nucleus (Fig. 4.17).

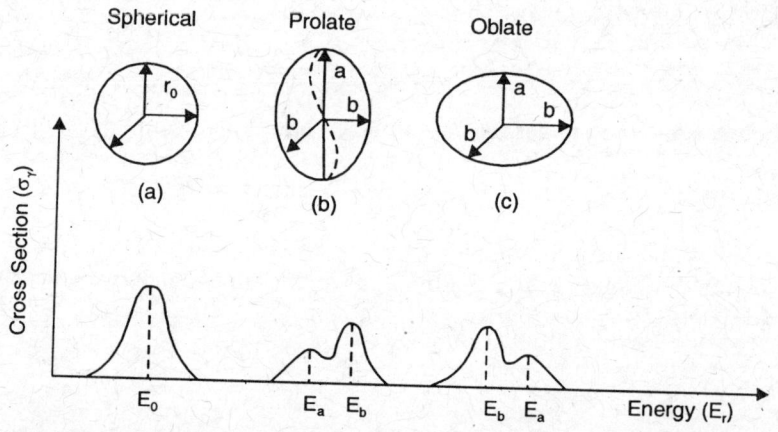

Fig. 4.17: Splitting of giant resonance peak for prolate and oblate spheroids

Since giant resonances are standing waves in the nucleus, one expects wavelengths $\lambda_i \propto r_i$ where r_i are the radii in the three ellipsoidal axis. Hence, $\omega_i \propto r_i$, i.e. the energies $E_i \propto r_i$. Danos and Okamoto[23] showed that

$$\frac{E_b}{E_a} = 0.911 \frac{a}{b} \quad \text{or} \quad \frac{E_b - E_a}{E_a} = 0.911 \left(\frac{a - b}{b} \right)$$

Therefore, if γ = 0 then (4.44b) yields

$$\frac{a}{b} = \frac{r_i}{r_2} = \frac{1 + \sqrt{\frac{5}{4\pi}}\beta}{1 - \frac{1}{2}\sqrt{\frac{5}{4\pi}}\beta} = 1 + \frac{3}{2}\sqrt{\frac{5}{4\pi}}\beta, \qquad (4.102)$$

where β is nuclear deformation for an axially symmetric deformed nucleus.

Thus, in the case of deformed nuclei, photon absorption cross section exhibits two peaks in the giant resonance region (*see* Fig. 4.17), the separa-

tion of the two peaks is proportional to β: $\beta \cong \frac{2}{3}\sqrt{\frac{4\pi}{5}}\left(\frac{E_b}{E_a} - 1\right)$.

The above relation assumes that GDR is a pure volume mode. The above formalism is extended to triaxial shapes by Chakraborty et al[24]. The energy of the ground state GDR is very well described by the droplet model:

$$E_{GDR} = \left\{ \frac{Mr_0^3}{8J\hbar^2} A^{2/3} + \frac{3Mr_0^2}{8Q} A^{1/3} \right\}^{-1/2},$$

where the first term is a surface term and the second is a volume term, each with its force constant $J = 36.8$ MeV and $Q = 17$ MeV.

Exercise 4.14. Expressing the excitation energy E_I as cubic polynomial in I in centrifugal stretching model, such as $E_I = aI + bI^2 + cI^3$, estimate the value of I_c connecting with the backbending phenomenon.

Now $$\frac{\partial E_I}{\partial I(I+1)} = \frac{a + 2bI + 3cI^2}{2I + 1}$$

or $$2I_I = (2I+1)/(a + 2bI + 3cI^2)$$

and $$\omega = \frac{\partial E_I}{\partial \sqrt{I(I+1)}} = \frac{2\sqrt{I(I+1)}\,(a + 2bI + 3cI^2)}{(2I+1)}.$$

Hence, $$\omega^2(I) = (a + 2bI + 3cI^2)^2 \times \frac{I^2 + I}{I^2 + I + \frac{1}{4}} \approx (a + 2bI + 3cI^2)^2.$$

Therefore, $\frac{\partial \omega(I)}{\partial I} = 2b + 6cI$. At the point of inflexion $\frac{\partial \omega(I)}{\partial I} = 0$, so that

$$2b + 6cI_c = 0.$$

Hence, $I_c = \dfrac{b}{3|c|}$, because the analysis of experimental ground state rotational band of even-even nuclei by a cubic polynomial gives the best fit when c is negative [Ref. 17].

4.5 ALGEBRAIC APPROACH TO COLLECTIVE MOTION

It is now quite well understood from the earlier discussion (Sec. 4.3-4.4) that many collective nuclear properties are dominated by the quadrupole degrees of freedom. Five quadrupole shape variables were used to parametrize the surface of deformed liquid drop to describe rotation and vibration properties of the nucleus. This approach, originally developed by Bohr and Mottelson, is now called the "Geometrical Model". More recently Arima and Iachello[25] have introduced an alternative "Algebraic Model", called the "Interacting Boson Model (IBM)", in which collective nuclear properties in even-even nuclei are described in terms of a system of N interacting bosons made out of nucleon pairs with angular momentum $L = 0$ and $L = 2$. The $L = 0$ bosons are called monopole or s-bosons and $L = 2$ bosons are called quadrupole or d-bosons. These are the most strongly bound pair states inside the nucleus. Therefore, in this algebraic approach one employs six boson variables (instead of the five shape variables in the BM approach). These bosons interact via one boson and two boson forces.

The $L = 0$ and $L = 2$ pairs of protons and neutrons are formed as

$$s^\dagger = \frac{1}{2}\sum_j \alpha_j \left[a_j^\dagger \, a_j^\dagger \right]_0^{(0)} \text{ and } d_m^\dagger = P\sum_{j_1 j_2} \beta_{j_1 j_2} \left[a_{j_1}^\dagger \, a_{j_2}^\dagger \right]_m^{(2)},$$

where P projects out seniority eigen states.

A subspace constructed from these pairs is called sd-subspace. The validity of this truncation is examined and is found good for low spin. For example, if both protons and neutrons occupy degenerate $^0g_{7/2}$, $^1d_{5/2}$, $^1d_{3/2}$ and $^2s_{1/2}$ shell, one can calculate the probability of finding nucleons in sd subspace which is shown in Fig. 4.18. One sees that the probabilities are large upto 4+ state revealing that the truncation into sd subspace works well both for spherical and deformed nuclei.

IBM in a way is the quantized version of BM model and its advantage lies in the introduction of some algebraic equations with the use of boson operators. These equations are easier to solve than the differential equations. Another advantage of IBM is the recognition of several 'dynamical symmetries" in the spectra of medium mass and heavy nuclei, which enables one to apply group theory to study the physical properties. The mathematical apparatus of group theory, which is essential for studying symmetries in nuclear and particle physics is briefly described in Appendix D1.

A. Interacting Boson Model

In the original formulation of IBM (now called IBM-1) no distinction is made between neutron and proton, and low lying collective quadrupole states of even-even nuclei are generated in terms of a system of N interacting bosons. The five components of d-bosons and the s-boson span a six dimensional space with group structure $U(6)$. The many-body states are classified according to the totally symmetric irreducible representation N of $U(6)$, where N is the total number of bosons, which is conserved. In order to calculate energies, one also needs appropriate operators to express the Hamiltonian H which preserves boson number (N) and angular momentum (L).

We express H in second quantized form in terms of creation $s^{\dagger}$, $d_m^{\dagger}$ and annihilation s, d_m operators, respectively, for s and d-bosons. The boson creation and annihilation operators satisfy the following commutation relations:

$[s, s^{\dagger}] = 1$, $[s, s] = [s^{\dagger}, s^{\dagger}] = 0$, $[d_m, d_n^{\dagger}] = \delta_{mn}$ and $[s, d_m^{\dagger}] = [s^{\dagger}, d_m]$

$= [s, d_m] = [s^{\dagger}, d_m^{\dagger}] = 0$, where m or $n = 0$, ± 1, ± 2. Also $\tilde{d}_m = (-)^m d_{-m}$ annihilates the time reverse of the state created by d_m and $\tilde{s} = s$.

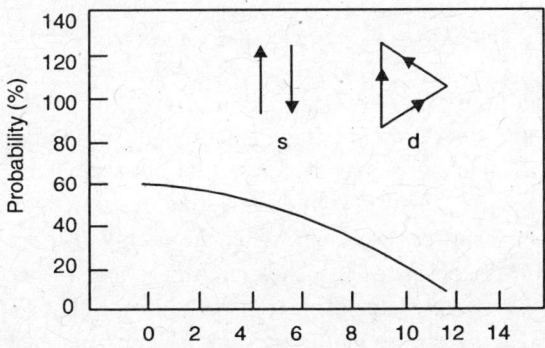

Fig. 4.18: Probabilities of shell model eigen states in the s-d subspace

The IBM has group structure $U(6)$ with 36 generators since $U(n)$ has n^2 generators. Bilinear expression can be formed with one creation ($s^{\dagger}$ or $d_m^{\dagger}$) and one annihilation (s or d_m) operators and they may be organized into 8 spherical tensor operators. We denote the 36 generators of $U(6)$ by G_i ($i = 1......36$) and they are given in Table 4.6, specifying L and m. Eight spherical tensor operators are:

$$s^{\dagger}s^{(0)}, d^{\dagger}d^{(L)} \ L = 0, 1, 2, 3, 4, s^{\dagger}d^{(2)} \text{ and } d^{\dagger}d^{(2)} \qquad (4.103)$$

The generators of $U(6)$ are expressed in the form

$$T_{M_3}^{L_3} = \left[U^{L_1} \times V^{L_2}\right]_{M_3}^{(L_3)} = \sum_{M_1 M_2} \langle L_1 L_2 M_1 M_2 \mid L_3 M_3 \rangle \, U_{M_1}^{L_1} V_{M_2}^{L_2}. \tag{4.104}$$

Table 4.6: Thirty six generators of IBM

Type	Number
$s, \ s^\dagger$	1
$[d^\dagger \times \tilde{d}]_m^{(L)}$; $L = 0, 1, 2, 3, 4$	5+5
$m = -L \dots 0 \dots L$	1+3+5+7+9

Therefore, the generators G_i that satisfy the relation $[G_i, G_j] = \sum_k \alpha_{ij}^k \, G_k$, form a Lie algebra of group $U(6)$ and one says that IBA Hamiltonian expressed in terms of $G_i = 1 \cdots\cdots 36$ has the group structure of $U(6)$. The Lie algebra is also called the spectrum generating algebra. The s-boson number operator $\hat{n}_s = s^\dagger s$ is such that $\hat{n}_s \mid n_s \rangle = n_s \mid n_s \rangle$, $s \mid 0 \rangle = 0$, $s^\dagger \mid 0 \rangle = 1$. Similarly $d^\dagger \mid 0 \rangle = 1$ and $\tilde{d} \mid 0 \rangle = 0$. The boson number operator $\hat{N}$ defined by

$$\hat{N} = s^\dagger s^{(0)} + \sum_m d_m^\dagger \, d_m = s^\dagger s^{(0)} + \sqrt{5} \, d^\dagger d^{(0)} \tag{4.105}$$

commutes with all the generators defined in (4.103). It is therefore a constant multiple of the unit operator, and may be excluded from the set of generators and the states of N bosons may be classified according to the irreducible representation of $SU(6)$. Thus, N would not affect the excitation spectrum and the single number N suffices as a label. When the energy difference $\epsilon = \epsilon_d - \epsilon_s$ between s- and d-boson states vanishes, and there are no two body interactions between bosons, all states of the representation (N) of $SU(6)$ are degenerate. In general the Hamiltonian H_{IBA} may be constructed from the scalar operators in (4.103) together with the two body interactions obtained by contraction of the tensor operators with $L \geq 1$. The number of bosons N corresponds to the number of valence nucleon pairs counted from nearest single particle major shell, and holes replace particles if more than half of the shell is full.

Example: N = half of the total number of valence nucleons; if there are n_s number of s-bosons and n_d number of d-bosons, then $N = n_s + n_d$ $= s^\dagger s + \sum_m d_m^\dagger d_m$. Since the total number of bosons in a given nucleus is fixed, there is a restriction on the total number of basic states available as well as on the maximum angular momentum L_{max} that can be constructed ($L_{max} = 2N$). The total angular momentum L is a good quantum number. The valence number counting is done relative to the nearest closed shell, i.e. bosons are counted as particles if the neutron and proton number is before mid shell and as holes if the neutron and proton number is after mid shell, e.g., $^{196}_{78}Pt_{118}$ has 4 proton holes (relative to $Z = 82$) and 8 neutron holes (relative to $N = 126$). Hence the total number of boson $N = n_\pi + n_v = 2 + 4 = 6$. Similarly $^{128}_{54}X_{74}$ has four valence protons (relative to $Z = 50$) and 8 valence neutron holes (relative to $N = 82$) so that the total number of bosons. $N = n_\pi + n_v = 2 + 4 = 6$. Here π and v label the proton and the neutron respectively. Once the correlated nucleon pairs are formed, then one can not distinguish between neutron bosons and proton bosons, i.e. they behave as identical bosons.

B. Spectrum Generating Algebra

Group Structure of IBM: As mentioned earlier, s and d bosons have 6 degrees of freedom and hence they can be taken as basis states of the $U(6)$ group. In other words, six dimensional space provides a basis for representation of $U(6)$. Hence, IBM is also called $U(6)$ model. Therefore, we should consider all possible subgroups of $U(6)$ where generators of a subgroup are subset of the set of 36 generators of group $U(6)$ and which are also closed under commutation. Since the nuclear states have definite spin, a restriction is imposed such that each subgroup chain must contain rotation group $O(3)$ as one of the subgroups. With this restriction only three chains are possible, each giving rise to a characteristic spectrum. They are:

I.	$U(6)$	$U(5)$	$O(5)$	$O(3)$	(4.87a)
II.	$U(6)$	$SU(3)$	$O(3)$		(4.87b)
III.	$U(6)$	$O(6)$	$O(5)$	$O(3)$	(4.87c)

The chain I ($U(5)$) corresponds to vibrational, the chain II ($SU(3)$) corresponds to rotational and the chain III ($O(6)$) corresponds to γ-soft limits, thus unifying various collective aspects of nuclei. In nuclear physics we do not consider the reduction $O(3) \supset O(2)$ because nuclear states are not split with respect to magnetic quantum number M unless a magnetic field is applied.

The Casimir operator plays important role in providing the representation of groups. Corresponding to each chain, we can express the Hamiltonian in terms of the Casimir operators of those subgroups. If the Hamiltonian of the system can be written as a sum of the Casimir operators, then the problem is an eigen value problem and eigen values are just the sum of linear combination of eigen values of Casimir operators of the group. The system is said to possess 'dynamic symmetry' if the Hamiltonian can be written only in terms of the Casimir operators. Casimir operators are diagonal in the representation provided by the corresponding group. Three possible chains yield three dynamic symmetrics (*see* Appendix D1).

The possible values of various quantum numbers are as follows. Total number of bosons N is fixed for a given nucleus. In a given nucleus, all bosons can be of d-type so that $N = n_d$ or one s-type boson and $(N-1)$ d-type boson, i.e. $n_d = N - 1, \cdots\cdots$.

Therefore, $N = n_d$ can take values from 0 to N. It may have that none of the d-bosons are coupled to $L = 0$ so that seniority quantum number of uncoupled $L = 0$ particles $= v = n_d$. When one pair of d-bosons is coupled to $L = 0$ then $v = n_d - 2$ and so on. Therefore, $v = n_d, n_d - 2, n_d - 4, \ldots\ldots$ 0 or 1 depending upon whether there are even number of d type boson or odd.

Now we shall demonstrate how in special cases analytic solutions of the eigen value problems can be found. An example is the lowest order Hamiltonian, which can be written in the form

$$H_0 = \epsilon_s \, \hat{n}_s + \epsilon_d \, \hat{n}_d = \epsilon_s \, \hat{N} + \epsilon \hat{n}_d, \quad \text{where} \quad \epsilon = \epsilon_d - \epsilon_s. \tag{4.103}$$

$H_0 = \epsilon \, \hat{n}_d$ with spectrum $E_{nd} = \epsilon n_d$, where n_d is the number of d-bosons. For $N = 3$, $n_d = 0, 1, 2, 3$ and the corresponding L values are respectively 0, 2, (0,2,4), (0,2,3,4,6). The spectrum is schematically shown in Fig. 4.19. (4.87) applies when the energy difference ϵ is much larger than all surviving n_d. Therefore, corresponding to each chain, we can express a Hamiltonian in terms of the Casimir operator C of those subgroups.

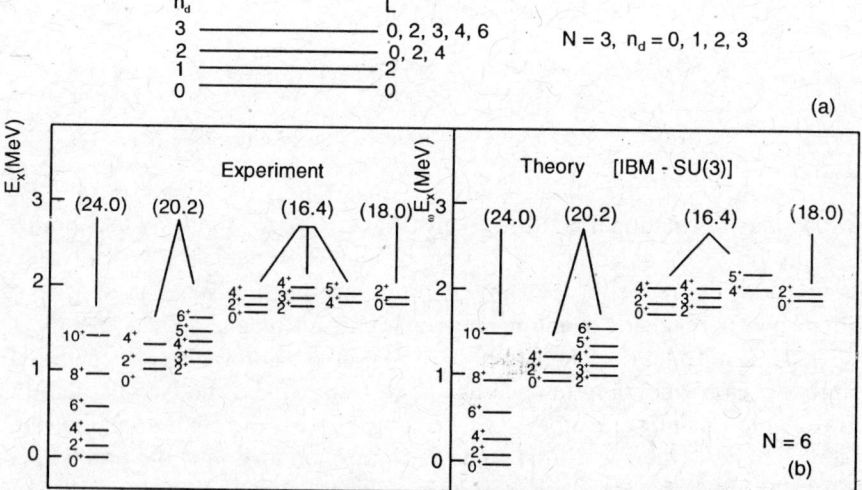

Fig. 4.19: IBM predicted rotational levels (*see* text)

Chain I

$$H_{IBA}(I) = \in C_1(U(5)) + \eta \, C_2(U(5)) + \delta \, C_2(O(5)) + \gamma \, C_2(O(3)),$$

where $C_1(U(5))$ is a linear Casimir operator of $U(5)$ and others are quadratic Casimir operators. $\in$, η, δ, γ are constants.

The basis states can be labelled by the quantum numbers $|N, \, n_d, \, v, \, L\rangle$ which characterizes the irreducible representation of $U(6)$, $U(5)$, $O(5)$ and $O(3)$ respectively. Symbolically,

$$\left| \begin{array}{cccc} U(6) & U(5) & O(5) & O(3) \\ N & n_d & v & L \\ & (d^+\tilde{d})_m^{0-4} & (d^+\tilde{d})^{1,3} & (d^+\tilde{d})_m^1 \end{array} \right\rangle.$$

Here n_d represents the number of d-bosons, v the boson seniority, i.e. the number of zero pairs of d-bosons and L the angular momentum. Since the decomposition of representation $O(5) \supset O(3)$ is not unique, an extra lebel n_s, which can be related to the number of d-boson triplet coupled to zero angular momentum has to be introduced.

Chain II

Chain II admits two Casimir operators $C_2\ (SU(3))$ and $C_2\ (O(3))$. (*see* Appendix D1). $SU(3) \to O(3)$ generates rotational bands. Symbolically,

$$\left| \begin{array}{ccc} SU(3) \to O(3) \to O(2) \\ \lambda\mu & \kappa \, L & M \end{array} \right\rangle,$$

where in a given N, λ and μ the multiple occurrence of L are distinguished by κ label.

For a given value of N, λ and μ, there may be more than one state, of same angular momentum quantum number L. Therefore, an additional quantum number is needed to distinguish the states. Hence, the complete set of basis states for the chain II is $|N, (\lambda,\mu), \kappa, LM\rangle$. The Hamiltonian in this case when expressed in terms of quadratic Casimir operator is

$$H_{IBA}(II) = \alpha \, C_2(SU(3)) + \beta \, C_2(O(3)), \text{ where } \alpha \text{ and } \beta \text{ are constants.}$$

The expression for the eigen values of $C_2\ (SU(3))$ and $C_2\ (O(3))$ in the representation $\|N(\lambda\mu), \kappa, LM\rangle$ characterizing $SU(3) \to O(3) \to O(2)$ reduction can be obtained by using the fact that the Casimir operator of the $SU(3)$ group commutes with the generators A_{ij} of the group, i.e. $[A_{ij}, C_2] = 0$,

where $C_2(SU(3)) = \dfrac{3}{2}\sum\limits_{\alpha\beta} A_{\alpha\beta} \, B_{\beta\alpha}.$

To show this we employ one of the basic properties of the unitary group of N dimension, $U(N) : N^2$ Cartesian generators A_{ij}, $i, j = 1...N$ of $U(N)$ satisfy lie algebra: $[A_{ij}, A_{\alpha\beta}] = A_{i\beta}\, \delta_{j\alpha} - A_{j\alpha}\, \delta_{i\beta}$.

Employing the above relation we get

$$[A_{ij}, C_2 SU(3)] = \left[A_{ij}, \sum_{\alpha\beta} A_{\alpha\beta}\, A_{\beta\alpha} \right]$$

$$= \sum_{\alpha\beta} \{ [A_{ij} A_{\alpha\beta}] A_{\beta\alpha} + A_{\alpha\beta} [A_{ij} A_{\beta\alpha}] \}$$

$$= \sum_{\alpha\beta} \{ A_{i\beta} \delta_{j\alpha} A_{\beta\alpha} - A_{\alpha j} \delta_{i\beta} A_{\beta\alpha} + A_{\alpha\beta} A_{i\alpha} \delta_{j\beta} - A_{\alpha\beta} \delta_{\beta j} A_{\beta j} \}$$

$$= \left[\sum_{\beta} A_{i\beta} A_{\beta j} - \sum_{\alpha} A_{\alpha j} A_{i\alpha} + \sum_{\alpha} A_{\alpha j} A_{i\alpha} - \sum_{\beta} A_{i\beta} A_{\beta j} \right] = 0.$$

The matrix element of $C_2 (SU(3))$ is diagonal in the $SU(3)$ irreducible representation $|\lambda\mu\rangle$. The explicit expression for the Casimir operator of $SU(3)$ group $C_2 (SU(3))$ in spherical form is

$$\frac{2}{3} C_2 (SU(3)) = \frac{1}{6} Q \cdot Q + \frac{1}{2} \hat{L} \cdot \hat{L}.$$

Now

$$\langle \lambda\mu | C_2 (SU(3) | \lambda\mu \rangle = \frac{3}{2} \left\{ \frac{1}{6} (2\lambda + \mu)^2 + (2\lambda + \mu) + 2 \cdot \frac{\mu}{2} \left(\frac{\mu}{2} + 1 \right) \right\}$$

$$= (\lambda^2 + \mu^2 + \lambda\mu + 3\lambda + 3\mu).$$

Hence, $E_{II}(N, (\lambda\mu), L) = \alpha\ (\lambda^2 + \mu^2 + \lambda\mu + 3\lambda + 3\mu) + \beta\ L(L+1)$ (4.104)

The eigen values of the operators $C_2 (SU(3))$ and $C_2 (O(3))$ in the representation $|N, (\lambda, \mu), \kappa, LM\rangle$ are $\frac{2}{3} (\lambda^2 + \mu^2 + \lambda\mu + 3\lambda + 3\mu)$ and $2L(L+1)$. Therefore, the expectation values of the Hamiltonian in the above representation is

$$E_{II}(N, (\lambda, \mu), L) = \frac{2}{3} \alpha \{ \lambda^2 + \mu^2 + \lambda\mu + 3\lambda + 3\mu \} + 2\beta L(L+1).$$

The allowed values (λ, μ) contained in $[N]$ are given by
$$[N] = (2N, O) \oplus (2N - 4, 2) \oplus (2N - 8, 4) \oplus \cdots\cdots \oplus (O, N)$$
$$\oplus \cdots\cdots \oplus (2N - 6, 0) \oplus (2N - 10, 2) + \cdots\cdots \quad (N = \text{even}).$$

The spectrum is characterized by a series of bands, in which the energy spacing is proportional to $L(L+1)$, as in the rigid rotor model. The ground state band has $(\lambda, \mu) = (2N,0)$ and the excited bands with $(\lambda, \mu) = (2N - 4,2)$, $K = 0$ and 2 can be associated with b and g band respectively. A classical limit of the Hamiltonian shows that for $N \to \infty$ the $SU(3)$ limit corresponds to the axially symmetric rotor with prolate deformation.

Now we can summarise the underlying steps to determine collective levels of even-even nuclei within the framework of IBM as follows:

(*i*) We, construct a basis by looking for all subgroups G' of the largest group G;

(*ii*) analyze the quantum number that are associated with the various subgroups G;

(*iii*) construct the various Casimir operators for the groups G';

(*iv*) diagonalize a given Hamiltonian.

For a Hamiltonian that is written in terms of the invariant operators of a given group chain only:

$$H = \alpha C(G) + \beta C(G') + \alpha C(G'') + ...$$

The energy is given by

$$E = \alpha\langle C(G)\rangle + \beta\langle C(G')\rangle + \alpha\langle C(G'')\rangle + ...$$

For odd A nuclei, we couple single fermion degrees of freedom to the collective boson degrees of freedom of even-even core nucleus. Therefore, we start with a set of N bosons (s- and d-bosons) and a odd nucleon occupying single particle orbits with $j = j_1, j_2, \cdots$ which describe the single particle freedom (for details see Ref. 25, 33).

Exercise 4.15. Show that the total boson number operator commutes with any generator of O (6) group.

The total boson number operator, $\hat{N} = s^{\dagger}s + d^{\dagger}d$. Now

$$[\hat{N}, s^{\dagger}d] \,|\, n_s,n_d\rangle = (s^{\dagger}s + d^{\dagger}d)\, s^{\dagger}d \,\,|n_s,n_d\rangle - s^{\dagger}d(s^{\dagger}s + d^{\dagger}d)\, |n_s,n_d\rangle$$

$$= s^{\dagger}ss^{\dagger}\sqrt{n_d} \,\,|n_s,n_d - 1\rangle + d^{\dagger}\, ds^{\dagger}\sqrt{n_d} \,\,|n_s,n_d - 1\rangle$$

$$- s^{\dagger}\, ds^{\dagger}\sqrt{n_s} \,\,|n_s - 1, n_d\rangle - s^{\dagger}\, dd^{\dagger}\sqrt{n_d} \,\,|n_s,n_d - 1\rangle$$

$$= s^{\dagger}s\sqrt{n_d(n_s + 1)} \,\,|n_s + 1,n_d - 1\rangle$$

$$+ d^{\dagger}d\sqrt{n_d(n_s + 1)} \,\,|n_s + 1, \, n_d - 1\rangle$$

$$- s^{\dagger}d \; n_s \,|n_s,n_d\rangle - s^{\dagger}d \; n_d \,|n_s, \, n_d\rangle$$

$$= (n_s + 1)\sqrt{n_d(n_s + 1)} \mid n_s + 1,\, n_d - 1\rangle$$

$$+ (n_d - 1)\sqrt{n_d(n_s + 1)} \mid n_s + 1,\, n_d - 1\rangle$$

$$- n_s\sqrt{n_d(n_s + 1)} \mid n_s + 1,\, n_d - 1\rangle$$

$$- n_d\sqrt{n_d(n_s + 1)} \mid n_s + 1,\, n_d - 1\rangle = 0.$$

Exercise 4.16. Show that generators of U(6) given in Table 4.6 form a lie algebra.

$$[d^\dagger d,\, s^\dagger s] \mid n_s, n_d\rangle = (d^\dagger d\ \ s^\dagger s) \mid n_s, n_d\rangle - s^\dagger s d^\dagger d \mid n_s, n_d\rangle$$

$$= d^\dagger d\, n_s \mid n_s, n_d\rangle + s^\dagger s\, n_d \mid n_s, n_d\rangle$$

$$= (n_s n_d - n_d n_s) \mid n_s, n_d\rangle = 0 \text{ or } [d^\dagger d,\, s^\dagger s] = 0.$$

Again,

$$[d^\dagger s,\, s^\dagger s] \mid n_s, n_d\rangle = (d^\dagger s s^\dagger\ s) \mid n_s, n_d\rangle - s^\dagger s d^\dagger s \mid n_s, n_d\rangle$$

$$= n_s\sqrt{n_s(n_d + 1)} \mid n_s - 1,\, n_d + 1\rangle$$

$$- \sqrt{n_s(n_d + 1)}\, (s s^\dagger - 1) \mid n_s - 1,\, n_d + 1\rangle$$

$$= \sqrt{n_s(n_d + 1)} \mid n_s - 1,\, n_d + 1\rangle$$

$$= d^\dagger s \mid n_s, n_d\rangle \text{ or } [d^\dagger s,\, s^\dagger s] = d^\dagger s.$$

REFERENCES

1. Quoted by S. Chandrasekhar in his book "Ellipsoidal Figures of Equilibrium, Dover, New York 1987.
2. J.W.S. Rayleigh, Theory of Sound, Vol. II, Macmillan, New York **364** (1877).
3. N. Bohr and J. Wheeler, Phys. Rev. **56**, 426 (1939).
4. D. Hill and J. Wheeler, Phys. Rev. **89**, 1102 (1953).
5. W.D. Myers and W.J. Swiatecki, Nucl. Phys. **81**, 1 (1966).
6. V.M. Strutinski, Nucl. Phys. **495**, 420 (1967), Nucl. Phys. **A122**, 1 (1968).
7. A. Bohr, Dan. Mat. Fys. Medd. 26, No. 14 (1952).
   A. Bohr and B.R. Mottelson, Dan. Mat. Fys. Medd **27**, No. 16 (1953).
8. J. Rainwater, Phys. Rev. **79**, 432 (1950).
9. S.G. Nilsson, Dan. Mat. - Fys. Medd. **29**, No. 16 (1955).
10. D. Inglis, Phys. Rev. **96**, 1059 (1954); **103**, 1786 (1956).
11. P.J. Twin et al., Phys. Rev. Lett. **57**, 811 (1986).
12. S. Cohen, S. Plasil and W.J. Swiatecki, Ann. Phys. N.Y. **82**, 557 (1974).
13. S.M. Harris, Phys. Rev. **B138**, 509 (1965).
    A. Galendo Uribari et al., Phys. Rev. Lett. 71, 231 (1993).

14. I.Y. Lee et al., Phys. Rev. Lett. **38**, 1454 (1977).
15. A. Johnson, H. Ryde and S.A. Hjorth, Nucl. Phys. **A179**, 753 (1972); Phys. Lett. **B34**, 605 (1971).
    F.S. Stephens, Rev. of Mod. Phys. **47**, 43 (1975).
16. P.W. Reich, Nuclear Data Sheets **99**, 753 (2003).
17. B.R. Mottelson and J.G. Valantin, Phys. Rev. Lett. **5**, 511 (1960).
18. A.N. Mantri and P.C. Sood, Phys. Rev. **C9**, 2076 (1974).
19. A. Faessler and M. Ploszajezak, Phys. Lett. **76B**, 1 (1978).
20. F.S. Stephens and R.S. Simon, Nucl. Phys. **A183**, 257 (1972).
    T. Mathur and S.N. Mukherjee, Phys. Rev. **C44**, 909 (1991).
21. A.K. Jain and P.C. Sood, J. Phys. **G4**, 81 (1978).
22. R. Bengtsson, I. Hamamoto and B. Mottelson, Phys. Lett. **73B**, 259 (1978), At. And Nucl. Data Tables, **35**, 15 (1986).
23. M. Danos, Nucl. Phys. **5**, 23 (1958); K. Okamoto, Phys. Rev. **110**, 143 (1958).
24. D.R. Chakraborty et al., Phys. Rev. **C36**, 1886 (1987); Phys. Rev. **C37**, 1437 (1988).
25. A. Arima and F. Iachello, Ann. Phys. (N.Y.) **99**, 253 (1976); Ann. Phys. (N.Y.), **111**, 201 (1978), Phys. Lett. **66B**, 205 (1977), Ann. Phys. (N.Y.) **123**, 436 (1979), Phys. Lett. **B57**, 39 (1975), Phys. Rev. Lett. **35**, 1069 (75).

SUGGESTED BOOKS FOR FURTHER READING

26. Structure of the Nucleus, M.A. Preston and R.K. Bhaduri, Addison Wesley Publishing Co., Reading, Mass, 1975.
27. Nuclear Theory Vol. I, J.M. Eisenberg and W. Greiner, North Holland, Amsterdam IIIrd Enlarged Ed. 1987.
28. Nuclear Forces, Gernot Eder, The M.I.T. Press, Cambridge, Massachusetts, 1988.
29. Physics of Rotating Nuclei Ed. S.N. Mukherjee and Y.R. Waghmare Eastern Wiley, 1995.
30. Nuclear Fast Rotation Z. Szymanski, Clarendon Press, Oxford, 1983.
31. Nuclear Structure W.F. Hornyack, Academic Press N.Y. 1975.
32. Nuclear Shells and Shapes S.G. Nilsson and I. Ragnarsson Cambridge University Press, N.Y. 1995.
33. Interacting Boson Models of Nuclear Structure D. Bonatsos, Clarendon Press, 1988.

5

Nuclear Reaction Dynamics

"The language of dynamics presents a remarkable consistency and completeness. An unambiguous formulation can be given to each legitimate problem."

– Ilya Prigogine[1]

A new dimension was introduced into the nuclear studies by the invention of charged particle accelerators by Cockcroft and Walton[2] and Lawrence[3]. Descriptions of modes of interaction of low energy light projectiles with nuclei were accomplished. The phenomena of resonances in nuclear reactions were observed. The resonance reaction showed the characteristic of sharp increase in the cross section for certain energies of the impinging projectile. In 1936 Bohr[4] proposed the 'Compound Nucleus' (CN) theory, based on application of equilibrium thermodynamics to nuclei. It was highly successful in explaining various features of the resonance behaviour.

In the analysis of low energy deuteron induced reactions, it was observed experimentally that the (d, p) reactions were more frequent than the (d, n) reactions and angular distribution of the emergent particle p or n is peaked in the forward direction. These observations contradicted the predictions of the CN theory. Oppenheimer and Phillips[9] explained the reaction by stating that the reaction takes place by a 'Direct Reaction (DR)' mechanism without the formation of compound nucleus.

5.1 REACTION MECHANISM AND CROSS SECTIONS

Nuclear reactions are thus classified into the "Compound Nuclear (CN) reactions" and 'Direct Reactions' (DR) with the main difference in the time taken by the process in each case. CN occurs slowly whereas DR occurs rapidly.

To understand the CN reaction mechanism, let us consider a neutron of 1 MeV energy (wavelength $\lambda = 28.6$ fm) entering a nucleus. It gains kinetic

278

energy of about 50 MeV due to the presence of nuclear potential and its wavelength inside will be approximately 4 fm. The sudden change of wavelength at the nuclear surface results in a large probability of reflection when the neutron tries to escape after having crossed the nucleus. A standing wave in the nucleus is then built up so that sufficient time for the interaction with the nucleus is available.

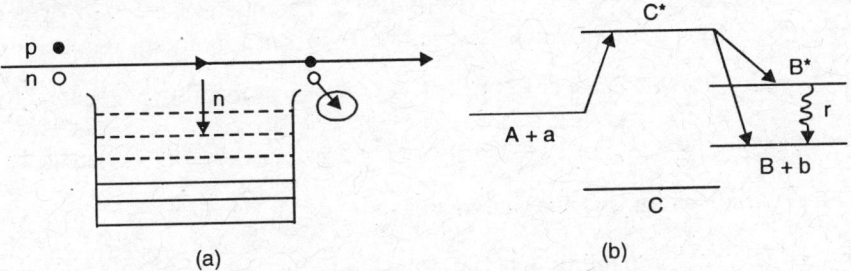

Fig. 5.1: (*a*) Direct reaction mechanism for (*d, p*) reaction and (*b*) Formation and decay of compound nucleus with the emission of gamma ray in the reaction $A + a \rightarrow C^* \rightarrow B + b$

During this time, the first struck nucleon successively strikes others, and so on, until the randomness introduced by many collisions involving many nucleons causes loss of memory of what type of particle brought in the energy that is now shared through the nucleus by the multiple collision process. Eventually, however, in the random course of events, one or more nucleons will acquire sufficient energy to overcome nuclear attraction to come out as reaction products. The low energy neutrons are most probable decay mode since unlike charged particles they see no Coulomb barrier.

That the CN process is more probable at the low energy can also be seen by comparing the mean free path (*L*) for nucleon-nucleon (N-N) collision with the nuclear diameter. In moving distance *dx*, the average number of collisions encountered by a nucleon = $\rho \sigma dx$. Hence, the distance travelled between successive collision: $L = 1/\rho\sigma$, where ρ = nucleon density in the nucleus = 0.14 fm^{-3} and σ is the N-N scattering cross section. Employing the experimental value of $\sigma = 30$ mb for nucleon c.m. energy around 40 MeV, we get the mean free path

$$L \approx \frac{1}{\rho\sigma} \approx 2.38 \text{ fm} , \qquad (5.1)$$

which is small compared to the diameter of a medium mass nucleus (≈ 12 fm), for multiple collisions to take place.

When an incident particle enters a nucleus to form a compound nucleus, a quantum state of the system must be formed. Since states have different energies, one could think that a compound nucleus be formed only if the incident particle just had the right energy to form these states. If that were the case, the cross section for any nuclear reaction would look like spikes. However, this is not quite true because energies of CN states have finite width. If Γ is the mean width of the compound nucleus, then the mean life time τ of the compound nucleus is given by the uncertainty relation, $\Gamma\tau = \hbar$. For the level width ~ 1.0 eV, the CN life time $\tau \approx 1 \times 10^{-15}$ sec. The CN is not a stationary state and it decays sooner in one way or another. This is shown in Fig. (5.1b) for reaction of the type $a + A \rightarrow C^* \rightarrow B + b$. The rate at which CN decays to a given final state i, which is specified by a set a quantum numbers, is

$1/\tau_i = \Gamma_i / \hbar$. Also the decay constant $\lambda_i = 1/\tau_i$ = probability per second for

the decay i. The total rate of decay of CN, $1/\tau = \sum_i \Gamma_i / \hbar$, where $\Gamma_i (= \hbar \lambda_i)$

is the partial width and the summation is over all decay modes such that

$\Gamma = \sum_i \Gamma_i$. The probability that CN decays into the final state i,

$$P_c(i) = \Gamma_i / \Gamma. \tag{5.2}$$

CN can decay in many ways and the dominant modes, which the energy conservation laws allow, are emission of neutrons, protons, α-particles, γ-rays and light nuclei. The neutron induced reaction of CN decays are: (n,n), (n,p), (n,a), (n,g) etc., such that

$$\Gamma_R = \Gamma_n + \Gamma_p + \Gamma_\alpha + \Gamma_\gamma \quad \text{and} \quad \Gamma = \Gamma_n + \Gamma_R. \tag{5.3}$$

The decay rate $\Gamma_i / \hbar$ for a particular decay "i" can be factorized as

$$\Gamma_i / \hbar = F_i (1/\tau_i) \, T_i, \tag{5.4}$$

where F_i is the preformation factor, i.e. the probability of finding the particle to be emitted within the nucleus, $1/\tau_i$ is the knocking rate, i.e. the frequency with which the particle hits the nuclear boundary and T_i is the transmission coefficient, i.e. the probability that a particle which hits the wall will penetrate and escape. In the case of nucleon emission the preformation factor $F_i = 1$, because nucleons are always present in the nucleus.

At the opposite extreme are the direct reactions (DR) which occur quickly. The reaction is completed in a time of the order of transit time τ of the

projectile across the target. Therefore, $\tau = \dfrac{2R}{v}$. As mentioned earlier, one MeV

nucleon will acquire a kinetic energy = 50 MeV due to the average potential, as it enters the nucleus and will acquire a velocity $v = 0.3$ c at that energy, so that

$$\tau = \frac{2 \times 1.2 A^{1/3} \times 10^{-13} \text{ cm}}{0.3 \times 3 \times 10^{10} \text{ cm sec}^{-1}} \approx 8 \times 10^{-23} \text{ sec for } A = 27. \tag{5.5}$$

This time is much shorter than the time for CN formation to take place (~ 10^{-15} sec). A further consequence is that if a reaction happens quickly it will vary only slowly if the bombarding energy is changed over intervals of order of several MeV. The best known example of DR is the deuteron stripping reaction, in which a deuteron incident on a target is stripped off one of its nucleon and the other nucleon is set free (Fig. 5.1a). In this type of reactions the final state is formed directly from the initial state without going through any intermediate state. The reaction is characterized as surface process. The collision may occur with the minimum rearrangement of the constituent nucleons and in the process only few degrees of freedom are involved.

Although, the direct particles are emitted immediately and the compound-nucleus particles some time later, the time lapse is still very small compared to the resolution of the most refined detecting systems. The particles can not therefore be distinguished experimentally, and so the cross-sections calculated for the two processes have to be combined before comparison is made with the experimental data. In Fig. 5.2 we show some typical reactions induced by protons.

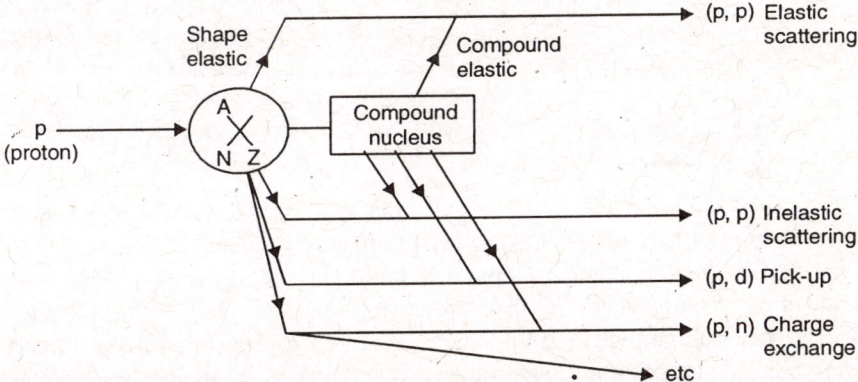

Fig. 5.2: Some typical reactions induced by protons via direct and compound nucleus mechanisms (Adopted from Hodgson Ref. 45)

If the projectile retains its full energy and leaves the target nucleus in the ground state, then the process is called direct or shape elastic scattering whereas if it has less than the full energy, it is compound inelastic scattering. If in the CN process the emitted particle is the same as the projectile and has its full energy, then it is called compound elastic scattering. Other reactions shown in Fig. 5.2 are the direct pickup reaction and charge exchange reaction. All these reactions enable several different residual nuclei to be formed from the target and the measurable features of the reactions tell us about the states of the residual nucleus. Therefore, the nuclear reaction dynamics also brings nuclear structure information.

Energy Spectrum: It is difficult to say at what energy a given reaction will proceed according to DR or CN reaction mechanism. At low energies, formation of CN is more likely whereas DR will prevail at higher energies. In a (d, p) reaction if the detector is fixed at a particular angle and the energy of the scattered proton is plotted, a spectrum schematically illustrated in Fig. 5.3 is typically obtained. Energy distribution of protons is found to consist of a number of discrete peaks at higher energies which become closer together and merge into a continuous distribution at lower energies. Examination of angular distribution of protons at low energies when averaged over a suitable energy interval, reveals that the continuous distribution is symmetrical about 90° and usually almost isotropic. These protons are therefore due to CN process.

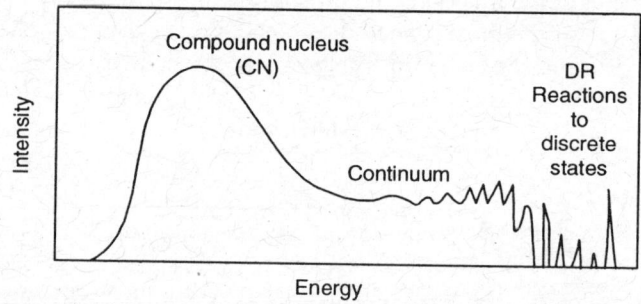

Fig. 5.3: A schematic diagram of the energy spectrum of emergent protons in a (d, p) reaction.

Reaction Channels: Let us consider the two-body reaction of the type $A(a,b)B$. We define a channel as a possible pair of product nucleus and outgoing particle each in a definite quantum state. The specification of a channel includes, the nature of the particle, the energy, the spin and orbital motion. The channel containing the initial states of A and a – is called the entrance channel or incident channel. The states of B and b form the exit channel or the final channel. In the above reaction we have assumed that only two particle channels are important. We are thus neglecting electromagnetic channels and three-particle channels such as $(p,2p)$.

We first consider energy and momentum conservations. Kinematics of the reaction is shown in Fig. 5.4(a) with respect to the laboratory system. All particles are treated non-relativistically. The Q-value of the ground state reaction is the difference of final and initial kinetic energies, i.e. $Q = E_B + E_b - E_a$. If the residual nucleus is in an excited state with excitation energy E_B^* then the Q-value is $Q - E_B^*$.

Conservation of linear momentum in the direction of incident particle gives

$$(2M_a E_a)^{1/2} = (2M_b E_b)^{1/2} \cos\theta + (2M_B E_B)^{1/2} \cos\varphi. \qquad (5.6)$$

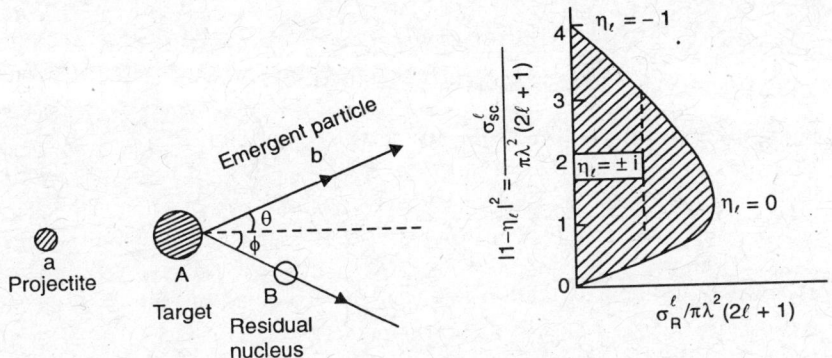

Fig. 5.4: (*a*) Reaction kinematics of the reaction $A(a,b)B$ and (*b*) Limits of scattering cross-section for given reaction cross-section

The linear momentum in a direction normal to the plane containing θ is zero, so that

$$(2M_b E_b)^{1/2} \sin\theta - (2M_B E_B)^{1/2} \sin\phi = 0. \tag{5.7}$$

Since in an experiment generally the angular distribution of the light particle is measured, we eliminate E_B and φ and get:

$$\sin^2\varphi = \frac{M_b E_b}{M_B E_B} \sin^2\theta \tag{5.8a}$$

$$\cos^2\varphi = \frac{M_a E_a}{M_B E_B} + \frac{M_b E_b}{M_B E_B} \cos^2\theta - \frac{2\sqrt{M_a M_b E_a E_b}}{M_B E_B} \cos\theta. \tag{5.8b}$$

Adding (5.8a) and (5.8b) and simplifying, we get

$$\left(1 + \frac{M_b}{M_B}\right)E_b - \frac{2(M_a M_b E_a)^{1/2}}{M_B} \cos\theta \sqrt{E_b} + \left\{\left(\frac{M_a}{M_B} - 1\right)E_a - Q\right\} = 0.$$

This is a quadratic equation in E_b, known as the Q-equation whose solution is

$$\sqrt{E_b} = v \pm (v^2 + w)^{1/2}, \tag{5.9}$$

where $\quad v = \dfrac{(M_a M_b E_a)^{1/2}}{M_B + M_b} \cos\theta \quad$ and $\quad w = \dfrac{Q\, M_B + (M_B - M_a)E_a}{M_b + M_B}.$

The smallest value of bombarding energy $(E_a)_{th}$ for which the reaction can take place (threshold energy) is obtained for the minimum positive value of E_b, i.e. for $v^2 + w = 0$.

Hence, $E_a = \dfrac{-Q(M_B + M_b)}{(M_b + M_B - M_a) - (M_a M_b \sin^2\theta)/M_B}$.

E_a is minimum for $q = 0$. Using $M_a + M_A = M_B + M_b + Q/c^2$, we get

$$(E_a)_{\text{th}} = \frac{-Q(M_a + M_A - Q/c^2)}{M_A - Q/c^2} = -Q\frac{M_a + M_A}{M_A},$$

by employing the approximation $M_A \gg Q/c^2$. We can rewrite (5.9) as

$$E_b = w + 2u^2 \cos^2\theta\left[1 \pm \left\{1 + \frac{w}{u^2 \cos^2\theta}\right\}^{1/2}\right], \tag{5.10}$$

where $u = (M_a M_b E_a)^{1/2}/(M_B + M_b)$.

Let ϵ_a and ϵ_b be the kinetic energies of the projectile (a) and emergent particle (b) in the c.m system.

$$\epsilon_a = \frac{1}{2}\mu\,[\vec{v}_a - \vec{v}_A]^2 = E_a\frac{M_A}{M_a + M_A} \text{ since } \vec{v}_A = 0 \text{ and } \mu = \frac{M_a M_A}{M_a + M_A}.$$

Therefore, $\qquad \epsilon_b = \epsilon_a + Q = E_a\dfrac{M_A}{M_a + M_A} + Q$

or $\qquad E_b\dfrac{M_B}{M_B + M_b} = E_a\dfrac{M_A}{M_a + M_A} + Q.$

For a (d,p) reaction with target of mass A: $E_b = \dfrac{A}{A+1}E_a + \dfrac{A+2}{A+1}Q$.

The conservation of angular momentum implies that the vector sum of the spins of the particles in the entrance channel ($\vec{S}_{aA}$) and their relative angular momentum ($\vec{\ell}_{aA}$), which is often called the entrance channel spin must be equal to the spin of the compound state $(\vec{I}_c)$ and to the vector sum of the spins in the exit channel $(\vec{S}_{bB})$ and their relative angular momentum ($\vec{\ell}_{bB}$):

$$\vec{S}_{aA}\left\{=\left(\vec{I}_A + \vec{i}_a\right)\right\} + \vec{\ell}_{aA} = \vec{I}_c = \vec{S}_{bB}\left\{=\left(\vec{I}_B + \vec{i}_b\right)\right\} + \vec{\ell}_{bB}, \tag{5.11}$$

where $\vec{I}_A$, $\vec{i}_a$, $\vec{I}_B$ and $\vec{i}_b$ are respectively spins of target, projectile, residual nucleus and emergent particle. The conservation of parity by the strong and electromagnetic interactions demands that the parity of the entrance channel be equal to the parity of the compound state and to the parity of the exit channel:

$$\pi_A + \pi_a = \pi_c = \pi_B + \pi_b.$$

A. Route to Cross Section

The partial wave analysis (*see* Appendix B2) provides an easy route to evaluate reaction cross sections. The reaction cross-section is equal to the number of particles taken out of the beam per unit incident particle, i.e.

$$\sigma_R^\ell = \frac{\text{Number of particles taken out of beam per second}}{\text{Number of incident particles per unit area per second}}.$$

The numerator is the net inward radial flux N_R over a large sphere of radius r_0. N_R is equal to the integral of the probability current density over the surface of the sphere r_0, to be calculated by taking the complete wave function $\psi(\vec{r})$. The particles which enter this sphere never come out again, that is, they do not appear in the entrance channel again. Now,

$$N_R = \frac{i\hbar}{2\mu} \int \left(\psi^* \frac{\partial \psi}{\partial r} - \psi \frac{\partial \psi^*}{\partial r} \right)_{r=r_0} r^2 \, d\Omega,$$

the negative sign has been omitted since inward flux is of interest here. Employing the partial wave expansion, we get

$$\left(\psi^* \frac{\partial \psi}{\partial r} - \psi \frac{\partial \psi^*}{\partial r} \right)_{r=r_0} = \sum_{\ell=0}^{\infty} \frac{(2\ell+1)^2}{4k^2} |P_\ell(\cos\theta)|^2 \frac{(-2ik)}{r_0^2} \left[1 - |\eta_\ell|^2 \right]$$

or

$$N_R = \frac{k\hbar}{\mu} \frac{1}{4k^2} \int \sum_{\ell=0}^{\infty} \frac{(2\ell+1)^2}{r_0^2} |P_\ell(\cos\theta)|^2 r_0^2 \, d\Omega \left[1 - |\eta_\ell|^2 \right].$$

$$= v \frac{\pi}{k^2} \sum_{\ell=0}^{\infty} (2\ell+1) \left[1 - |\eta_\ell|^2 \right].$$

Hence, $\quad \sigma_R = \dfrac{N_R}{N_{inc}} = \pi\lambda^2 \sum_{\ell=0}^{\infty} (2\ell+1) \left[1 - |\eta_\ell|^2 \right] = \sum_{\ell=0}^{\infty} \sigma_R^\ell, \qquad (5.12)$

where $\sigma_R^\ell = \pi\lambda^2 (2\ell+1) \left[1 - |\eta_\ell|^2 \right]$. The partial scattering cross section

$$\sigma_{SC}^\ell = \pi\lambda^2 (2\ell+1) |1 - \eta_\ell|^2.$$

Since, $|\eta_\ell|^2 > 0$, the maximum reaction cross section $(\eta_\ell = 0)$: $(\sigma_R^\ell)_{max} = \pi\lambda^2 (2\ell+1)$. Since, σ_R^ℓ can not be negative, it must be that $|\eta_\ell|^2 < 1$. The maximum value of the scattering cross section $(\eta_\ell = -1)$,

$$(\sigma_{sc}^{\ell})_{max} = 4\pi\lambdabar^2 (2\ell + 1) = (4\sigma_R^{\ell})_{max}.$$

The scattering cross section takes on its largest value which corresponds to $\eta_\ell = -1$ with $\sigma_R^\ell = 0$. If σ_R^ℓ takes on its maximum possible value which corresponds to $\eta_\ell = 0$, then the scattering cross section $(\sigma_{sc}^\ell)_{max} = \pi\lambdabar^2$ $(2\ell + 1)$. Scattering and reaction cross sections are both then of equal size. $\sigma_{sc}^\ell = 0$ for $\eta_\ell = +1$ with $\sigma_R^\ell = 0$. The total partial cross section,

$$\sigma_{tot}^\ell = \sigma_R^\ell + \sigma_{sc}^\ell = \pi\lambdabar^2 (2\ell + 1)\left[2 - \eta_\ell - \eta_\ell^*\right]. \qquad (5.13)$$

The scattering and reaction cross-sections can be represented diagramatically as shown in Fig. 5.4(b) which indicates that there can not be any reaction without some scattering but there can be scattering without reaction.

Compound Nucleus Cross Sections: According to the compound nucleus theory proposed by Neils Bohr[4], the nuclear reaction takes place in two steps: (i) the projectile together with the target nucleus forms the compound nucleus; (ii) the compound nucleus then decays to the final products. It is assumed that the mode of decay of CN is independent of mode of its formation, except for the requirement of various conservation laws. In other words, CN looses its memory as how it was formed at the time of decay. The above assumption is called the "independent hypothesis". In the reaction of type $A(a,b)$, if $\sigma_c(a)$ is the cross section for the formation of CN, where $C \equiv (A + a)^*$ and $P_c(b)$ is the probability of decay of CN with emission of particle b leaving the final nucleus B independent of how C was formed then:

$$\sigma(a,b) = \sigma_{CN}(a)\, P_c(b) \qquad (5.14a)$$

$$\sigma_{CN}(a) = \frac{\pi(2\ell + 1)}{k_a^2}\, T_a \text{ and } P_c(b) = \frac{\Gamma_b}{\Gamma}, \qquad (5.14b)$$

where, $\dfrac{1}{k_a} = \lambdabar_a$ is the reduced wave length, Γ_b and Γ are the partial and total widths and T_a is the transmission coefficient. The reciprocity theorem, which states that a time reversed reaction has the same cross section as the reaction itself, implies: $T_a P_b = T_b P_a$. In other words, the reactions $A + a \rightleftharpoons B + b$ are invariant under time reversal. This is true for all channels, so that

$$\frac{P_a}{T_a} = \frac{P_b}{T_b} = \xi \text{ (say)},$$

where ξ is a constant. This relation tells us that the probability of a reaction in a particular channel is proportional to the transmission coefficient in that channel. Since P values are probabilities, $\sum_a P_a = 1 = \sum_b P_b$ so that $\xi \sum_b T_b = 1$ and hence, $P_b = T_b / \sum_a T_a$, where T_a is the transmission coefficients in all the channels. Eq. 5.14 gives

$$\sigma_{CN}(a,b) = \frac{\pi(2\ell+1)}{k_a^2} \, T_a T_b / \Sigma T_a.$$

Since the compound nucleus can decay back into the incident channel, these summations are over the incident channel as all as well the outgoing channels.

Finally, $\qquad \sigma_{CN}(a,b) = \dfrac{\pi(2\ell+1)}{k_a^2} \, T_a T_b \Big/ \left(\sum_i T_i \right).$ $\qquad$ (5.15)

Thus, σ_{CN} is the product of the transmission coefficients in the incident and outgoing channels divided by the sum of transmission coefficients in all the channels. The amount of particles particularly coming out in the exit channel 'b' obviously depends on how much particles are transmitted in the entrance channel 'a'.

Equation (5.15) is called the Hauser Feshbach[4] formula for spinless particles, which may not be sufficiently accurate for general use, principally because the process of formation and decay are not completely independent of each other for overlapping resonances. This may be brought out more clearly by derivation of the formula for σ_{CN} by averaging over all resonances (Exercise 5.8).

Experimental Verification: Let us consider the reaction of the type $a + A \Leftrightarrow B + b$. According to the reciprocity theorem (Appendix B2), $k_a^2 \sigma(a, b) = k_b^2 \sigma(b, a)$. Bohr's independent hypothesis yields

$$\sigma(a, b) = \sigma_{CN}(a)\frac{\Gamma_b}{\Gamma} \quad \text{and} \quad \sigma(b, a) = \sigma_{CN}(b)\frac{\Gamma_a}{\Gamma} \text{ according to (5.14).}$$

Hence, $k_a^2 \sigma_{CN}(a)\dfrac{\Gamma_b}{\Gamma} = k_b^2 \sigma_{CN}(b)\dfrac{\Gamma_a}{\Gamma}$ or $\dfrac{k_a^2 \sigma_{CN}(a)}{\Gamma_a} = \dfrac{k_b^2 \sigma_{CN}(b)}{\Gamma_b} = f \text{(say)}$

Hence, $P_c(a) = \dfrac{\Gamma_a}{\Gamma} = \dfrac{k_a^2 \, \sigma_{CN}(a)}{\Gamma f} = \dfrac{k_a^2 \, \sigma_{CN}(a)}{k_a^2 \sigma_{CN}(a) + k_b^2 \, \sigma_{CN}(b) + \cdots}.$ $\qquad$ (5.16)

The validity of Bohr's hypothesis of CN reaction was verified by Ghoshal[6] in the following experiment. Alpha and proton beams were used to strike respectively $^{60}_{28}Ni$ and $^{63}_{29}Cu$ targets to produce the same excitation in $^{64}Zn^*$.

Since the excitations produced through the two processes are the same, the probability of decay of the compound nucleus $P_c(i)$ in the channel 'i' is the same, for it depends upon excitation produced in CN and not upon mode of formation. Therefore, at any one energy of the compound nucleus, the Bohr's assumption leads us to expect that the ratio of the cross sections for the (a, n), the $(a, 2n)$ and the (a, pn) are independent of the incident particle α. Therefore, for $a = p$ or α we have

$$\sigma(p,n):\sigma(p,2n):\sigma(p,pn) = \sigma(\alpha,n):\sigma(\alpha,2n):\sigma(\alpha,pn)$$

The above relations were confirmed in Ghosal's experiment[6] within the accuracy of 10%. It may be remarked that this is not a completely convincing test because alpha particle brings in more angular momentum than the proton and therefore, spectrum of states excited in the compound nucleus is not the same in two cases.

The compound nucleus theory described above can be used to calculate differential cross section provided transmission coefficient T in the incident and exist channels and in all the competing channels are known. These transmission coefficients can be calculated by employing a model for the nucleus. The transmission coefficient is defined as that fraction of the incoming current which is transmitted. For particles incident on a nucleus represented by a potential, this definition implies that particles do not appear again neither by being reflected nor by penetrating the nucleus. Thus, the fraction in question is precisely the part which in the optical model language has been absorbed. Hence, the transmission coefficient of the ℓth partial wave is $T_\ell = 1-|\eta_\ell|^2$. For particle transmission, the incident energy E must equal to atleast the effective potential $V_{eff}(r)$, which the particle has to surmount (Fig. 5.5):

$$E = V_{eff}(r) = V_N(r) + V_{Coul}(r) + V_\ell(r) = V_c(r) + \frac{\hbar^2 \ell(\ell+1)}{2\mu r^2}. \quad (5.17)$$

In (5.17), $V_c(r) = V_N(r) + V_{Coul}(r)$, $V_{Coul}(r) = Z_P\, Z_T\, e^2/r$. $V_N(r)$, the nuclear potential may have an Saxon Woods form. In the last term, called the centrifugal potential; ℓ is the angular momentum and μ the reduced mass of the particle.

There are various ways of calculating the nuclear potential $V_N(r)$ by folding either the nucleon-nucleus or the nucleon-nucleon potentials over the respective density distribution of nuclei. In the single folding potential method, the interaction potential is determined by folding the density of one nucleus $\rho_1(\vec{r})$ with experimentally obtained nucleon-nucleus optical potential $U(r)$ of the other, i.e.

$$V_N(r) = \int d^3r' \rho_1(r)\, U(r - r').$$

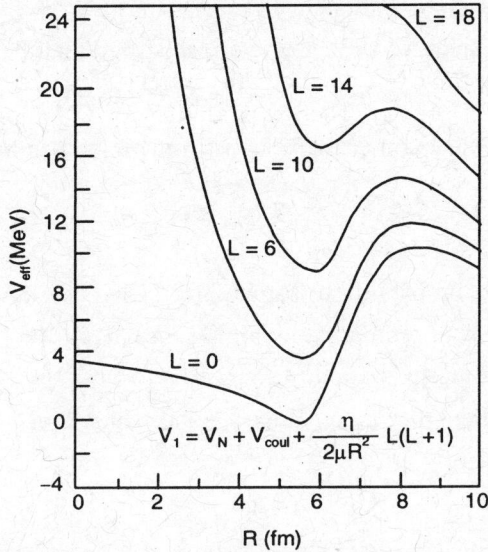

Fig. 5.5: Schematic diagram of the effective potential barrier for various values of angular momentum

The effective potential $V_{eff}(r)$ as a function of ion-ion separation distance R are displayed in Fig. 5.5 for several values of angular momentum ℓ. The potential energy diagram exhibits a pocket which gradually vanishes at higher ℓ values. Now if the projectile has enough energy to overcome the barrier, then it is assumed to be trapped inside the pocket leading to *CN* reaction or fusion. The cross section for *CN* formation

$$\sigma_R = \frac{\pi}{k^2} \sum_{\ell=0}^{\infty} (2\ell+1) T_\ell,$$

where T_ℓ is calculated assuming a particular model for a nucleus. In the sharp cut off model, one assumes $T_\ell = 1$ for $\ell \leq \ell_{max}$ and $T_\ell = 0$ for $\ell > \ell_{max}$. Therefore,

$$\sigma_R = \frac{\pi}{k^2} \sum_{\ell=0}^{\ell_{max}} (2\ell+1) = \frac{\pi}{k^2} (\ell_{max} + 1)^2.$$

If the target and projectile are both spherical with $r_0 = a_0 \, A_i^{1/3}; i = T, P$ then $\hbar \ell_{max} = a_0 (A_T^{1/3} + A_P^{1/3}) \sqrt{2\mu(E - V_{eff})}$, where ℓ_{max} is the maximum value of the partial waves which contribute to CN cross section and A_T and A_p are respectively the mass number of target and projectile.

The limitation to CN formation can be very severe for very heavy projectiles. Dynamical and viscosity effects are so strong that the complete fusion and subsequently the CN formation are entirely inhibited. This will be discussed in the next chapter. The cross section for CN formation are found smaller than total reaction cross sections in many occasions. Further,

$$\sigma_{CN} = \frac{\pi}{k^2} \sum_{\ell=0}^{\infty} (2\ell+1) T_\ell \, P_\ell^{CN},$$

where P_ℓ^{CN} is the probability of formation of CN once the particle is transmitted through the barrier. For rather high values of ℓ, $T_\ell = 1$. We make again the sharp cut off approximation $P_\ell^{CN} = 1(0)$ for $\ell < \ell_c (> \ell_c)$, where ℓ_c is called the critical angular momentum beyond which no CN formation takes place. Therefore, for $\ell_c \gg 1$

$$\sigma_{CN} = \frac{\pi}{k^2} \sum_{\ell=0}^{\ell_c} (2\ell+1) = \frac{\pi}{k^2} (\ell_c+1)^2 \approx \pi \lambda^2 \, \ell_c^2. \tag{5.18}$$

Employing (5.17) we get $E \cong V_c(r) + \dfrac{\hbar^2 \ell_c^2}{2\mu r_c^2}$, where r_c = critical distance (strong absorption radius). Hence,

$$\sigma_{CN} = \pi \left(\frac{\hbar^2}{2\mu E} \right) [E - V_c(r)] \frac{2\mu r_c^2}{\hbar^2} = \pi r_c^2 \left[1 - \frac{V_c(r)}{E} \right]. \tag{5.19}$$

We can derive (5.18) from the relation $\ell = kb$ as well.

$$\sigma_{CN} = \int d\sigma_{CN} = \int 2\pi b \, db = \frac{2\pi}{k^2} \int_0^{\ell_c} \ell d\ell = \pi \lambda^2 \ell_c.$$

The cross section for *CN* formation given above is also called cross section for "complete fusion". It is convenient to define a factor by which *CN* cross sections are reduced by the non-elastic direct reactions. Thus, the total cross section

$$\sigma_{CN} - \sigma_{DR} = R\sigma_{CN},$$

so that the reduction factor $R = 1 - \dfrac{\sigma_{DR}}{\sigma_{CN}}$.

Cross Section from Nuclear Interactions: We again consider the reaction of the type $A(a,b) B$ and employ the standard perturbation theory to calculate cross section in terms of nuclear interaction involved in the reaction process:

$$\sigma_{fi} = \frac{W_{fi} V_a}{1 \cdot v_{aA}},$$ (5.20)

where V_a is the normalization volume for the incident particles and v_{aA} is the projectile-target relative velocity, or the number of particles incident per unit area per sec. Here, the initial channel, $i \equiv (A + a)$ and the final channel, $f \equiv B + b$. The transition probability

$$W_{fi} = \frac{2\pi}{\hbar} \left| \int \psi_f^* H_{fi} \psi_i \, d\tau \right|^2 \frac{dn}{d\epsilon},$$ (5.21)

where H_{fi} is the interaction Hamiltonian and $dn / d\epsilon$ is the number of final states in the energy interval ϵ and $\epsilon + d\epsilon$ of the total energy. In case of degeneracy, each sublevel is to be counted separately. If b and B exist as free particles after the reaction, the number of states of the total system is equal to the number of possible states for one of the particles, since the conservation of energy and momentum completely determines the motion of the second particle. Generally, one studies energy distribution of lighter particles. At present we ignore the particle spin. In the phase space of particle b, the number of states having momenta of magnitudes $p_b + dp_b$ is

$$dn = \frac{4\pi p_b^2 \, dp_b \, V_b}{(2\pi\hbar)^3}.$$

We set normalization volume $V_a = V_b = 1$. In the c.m. system, the energy conservation law yields:

$$\epsilon = \epsilon_b + \epsilon_B = \left(p_b^2 c^2 + M_b^2 c^4 \right)^{1/2} + \left(p_B^2 c^2 + M_B^2 c^4 \right)^{1/2}.$$

$$\frac{d\epsilon}{dp_b} = \frac{p_b c^2}{\epsilon_b} + \frac{c^2}{2\epsilon_B} \frac{dp_B^2}{dp_b} \text{ and } p_b^2 = p_B^2 \text{ i.e., } 2p_b = \frac{dp_B^2}{dp_b} \text{ since } \vec{p}_b = -\vec{p}_B.$$

$$\frac{d\epsilon}{dp_b} = \frac{p_b c^2}{\epsilon_b} + \frac{p_b c^2}{\epsilon_B} = M_b v_b c^2 \left(\frac{1}{\epsilon_B} + \frac{1}{\epsilon_b} \right)$$

$$= v_b \left(\frac{\epsilon_b}{\epsilon_B} \right) + v_b = v_b + v_B = v_{bB}. \quad \text{Hence } \frac{dn}{d\epsilon} = \frac{4\pi p_b^2}{(2\pi\hbar)^3} \frac{1}{v_{bB}}.$$

So that,

$$W_{fi} = \frac{2\pi}{\hbar} \left| \int \psi_f^* H_{fi} \psi_i \, d\tau \right|^2 \frac{4\pi p_b^2}{(2\pi\hbar)^3} \frac{1}{v_{bB}}$$

and
$$\sigma_{fi} = \frac{p_b^2}{\pi \, \hbar^4} \frac{1}{v_{aA} \, v_{bB}} \left| T_{fi} \right|^2 \quad \text{with } T_{fi} = \int \psi_f^* \, H_{fi} \, \psi_i \, d\tau \quad (5.22a)$$

and
$$\frac{p_b^2}{v_{aA} \, v_{bB}} = \frac{p_b \, \mu_b \, v_{bB}}{\dfrac{p_a}{\mu_a} \, v_{bB}} = \mu_a \, \mu_b \, (k_b / k_a).$$

Hence,
$$\frac{d\sigma_{fi}}{d\Omega} = \frac{\mu_a \, \mu_b}{(2\pi \, \hbar^2)^2} \frac{k_b}{k_a} \left| T_{fi} \right|^2 . \qquad (5.22b)$$

The above formalism is quite general in the sense that almost any cross section can be obtained in terms of transition matrix element (Eq. 5.22a) and this may in turn be obtained by solving S. Eqs. for (Aa) and (Bb) systems represented by a complex potential, usually called an optical potential.

B. Optical Potentials

We have already mentioned that the character of nuclear reactions will depend upon bombarding energy. At low energy, nuclei can only scatter elastically. As the energy is raised, inelastic and reaction channels open. Thus, some flux is removed from the elastic channel. The requirement for the removal of flux can be understood by treating the reaction dynamics quantum mechanically. For example, in the reaction $A(a,b) B$, let the Hamiltonian for the projectile and target respectively be H_a and H_A (a or A also denotes respective mass number). The corresponding S. Eqs. are

$$(H_a - \epsilon_a) \psi_a = 0 \text{ and } (H_A - \epsilon_A) \psi_A = 0,$$

where ψ_a and ψ_A are respectively the wave functions for projectile and target and ϵ_a and ϵ_A are the respective eigen values. The relative coordinate between two nuclei is

$$\vec{r}_\alpha = \frac{1}{a} \sum_{i=1}^{a} \vec{r}_i - \frac{1}{A} \sum_{i=a+1}^{A+a} \vec{r}_i$$

and the corresponding kinetic energy operator $T_\alpha = -\dfrac{\hbar^2}{2\mu_\alpha} \nabla_\alpha^2$.

Two nuclei interact with each other through the two nucleon interaction V_{ij} and the potential energy, $V_\alpha = \sum_{i \in a, j \in A} V_{ij}$. The total Hamiltonian, excluding the c.m. energy is

$$H = H_a + H_A + T_\alpha + V_\alpha,$$

where $(H_a + H_A)$ is the intrinsic Hamiltonian of the projectile and target system which can also be written as $H_\alpha = H_a + H_A$. If we denote the intrinsic state of two nuclei by $\psi_\alpha = \psi_a + \psi_A$ then

$$(H_\alpha - \epsilon_\alpha)\,\psi_\alpha = 0, \text{ where } \epsilon_\alpha = \epsilon_a + \epsilon_A.$$

Finally, $T_\alpha + V_\alpha$ refers to only relative motion. The interaction between two nuclei may cause apart from the elastic scattering, the excitation of target or projectile or both. It may result transfer of one or few nucleons from a to A or vise versa. We introduce another potential $U(r_\alpha)$ that depends upon the relative distance of projectile and target and represents the average interaction between all the nucleons of projectile and target. S. Eq., would be

$$(H_\alpha + T_\alpha + U_\alpha - E)\,\psi = 0. \tag{5.23}$$

If we write the wave function $\psi = \psi(\vec{\xi})\phi(\vec{r}_\alpha)$ then S. Eq. (5.23) is separable into nuclear coordinates in $H_\alpha(\xi)$ and relative coordinates contained in U_α so that

$$(T_\alpha + U_\alpha - E_\alpha)\,\phi(\vec{r}_\alpha) = 0. \tag{5.24a}$$

$$(T_\alpha + U_a^* - E_\alpha)\,\varphi^*(\vec{r}_\alpha) = 0. \tag{5.24b}$$

We multiply (5.24a) by $\phi_\alpha^*(r_\alpha)$ and (5.24b) by $\phi_\alpha(r_\alpha)$ and subtract to get

$$\frac{-\hbar^2}{2\mu_\alpha}\left[\varphi_\alpha^*(r_\alpha)\,\nabla^2\,\varphi_\alpha(r_\alpha) - \varphi_\alpha(r_\alpha)\,\nabla^2\,\varphi_\alpha^*(r_\alpha)\right]$$

$$= (U_\alpha^* - U_\alpha)\varphi_\alpha^*(r_\alpha)\,\varphi_\alpha(r_\alpha).$$

Using the definition of current (Appendix B2), we get

$$-i\hbar\,\vec{\nabla}\cdot\vec{j} = (U_\alpha^* - U_\alpha)\,\varphi_\alpha^*(r_\alpha)\,\varphi_\alpha(r_\alpha) = -2i\,\text{Im}\,U_\alpha\,\rho_\alpha.$$

or $\qquad \hbar\,\vec{\nabla}\cdot\vec{j} = 2\rho_\alpha\,\text{Im}\,U_\alpha \qquad$ where $\rho_\alpha = \varphi_\alpha^*\,\varphi_\alpha$.

Integrating this over a volume containing the scattering centre, we find that if U_α is real there is no change of flux, whereas, if $\text{Im}\,U_\alpha$ is negative there is a loss. Therefore, U describes accurately elastic cross section when nuclear reaction can also occur only if it is complex with a negative imaginary part. The imaginary parts of the potential accounts for non-elastic processes.

In optics, a complex refractive index was introduced for the description of absorption of light. Hence, nuclear potential U_α which has also an imaginary part is called "the optical potential". Parameterization of the optical potential is done consistently with our knowledge of distribution of nucleons as revealed by the electron scattering experiment. Accordingly the optical potential has the following Saxon-Woods form factor:

$$f(r) = \frac{1}{1 + \exp\{(r - r_o)/a\}}$$

so that

$$U(r) = \frac{-V_0}{1 + \exp\{(r - r_v)/a_v\}} + \frac{-iW_0}{1 + \exp\{(r - r_w)/a_w\}}$$

$$= V(r) + iW(r), \tag{5.25}$$

which is called volume type.

Nucleons have spin 1/2, and even if the incident beam is unpolarised we find that the scattered beam is polarized. This polarization can be calculated from the optical potential if we add an additional spin dependent term of the form:

$$V_{so}(r) = -(V_{so} + iW_{so})\frac{1}{r}h(r)\,\vec{\ell}\cdot\vec{s},$$

where

$$h(r) = \frac{df_s(r)}{dr} \quad \text{with } f_s(r) = \frac{1}{1 + \exp\{(r - r_{so})/a\}}.$$

Usually, the nuclear radius r_0 is expressed in the form $r_v = r_{0v}\,A^{1/3}$ and $a_v(a_w)$ is independent of A. The schematic behaviour of the real and imaginary parts of optical potential as a function of energy is shown in Fig. (5.6a). The depth of the potential, $V_0(W_0)$ is different for neutrons or protons (*see* Ex. 3.6). The radius and diffuseness parameters in general are: $r_v = r_w = 1.25$ fm and $a_v = a_w = 0.65$ fm.

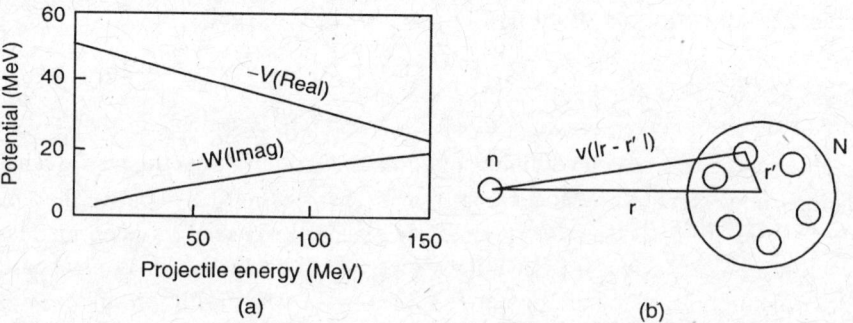

Fig. 5.6: (a) Variation of the real and the imaginary part of optical potential with respect to the incident energy (b) Non Local Potential $U(r)$

In practice, data can not accurately determine all these parameters at each energy. It is thus more meaningful to characterize the potential (Eq. 5.25) by its global properties. We may characterize it by radial moments defined as follows:

$$[r^n]_V = \frac{4\pi}{A} \int_0^\infty r^n V(r)\, dr \quad \text{and} \quad [r^n]_W = \frac{4\pi}{A} \int_0^\infty r^n W(r)\, dr.$$

The two aspects of optical potential are: (i) its phenomenology, i.e. the choice of parameters in codifying experimental data and (ii) derivation of optical potential from a basic theory. For the best fit individual nuclei, all parameters can be varied. The fact that $|W_0|$ increases with energy corresponds to the opening of more reaction channels that drain flux from the elastic channel. The absorption is on the surface; more specifically the imaginary part of the potential has the radial dependence which is surface peaked, i.e. of the radial derivative form

$$g(r) = -4a\, \frac{df(r)}{dr} = 4 \exp\{(r-r_0)/a\}/[1 + \exp\{(r-r_0)/a\}]^2, \qquad (5.26)$$

where the factor $4a$ is introduced to ensure $g\,(r_0) = 1$.

It is found empirically that at low energies the surface peaked potential dominates and as the energy increases, the absorption progressively spreads through the nuclear volume. Usually the imaginary potential is written as

$$W(r) = W_{surf}(r) + W_{vol}(r).$$

The optimum fit chosen is the one which corresponds with the set of parameters that gives the minimum value of χ^2. Phenomenological local nucleon-nucleus optical potentials have been obtained by various authors by fitting elastic scattering data for variety of target nuclei and for a wide range of bombarding energy.

Nucleon-nucleus optical potential calculated from nucleon-nucleon (N-N) potential is found to be non-local. Mathematically this means that the term $U(r)\,\psi(r)$ in (5.23) is replaced by $\int U(r,r')\,\psi(r')\, dr'$ where $U(r,r')$ is non-local potential and integration is over all space.

The optical potential between a nucleon and a nucleus can be calculated in several ways of increasing sophistication. The simplest one is to average the nucleon-nucleon interaction $v(|r-r'|)$ over the nuclear density $\rho\,(r)$ to give (Fig. 5.6b):

$$U(r) = \int v(|r-r'|)\,\rho(r')\, dr'.$$

$U(r)$ gives the real part of the potential and is only approximate because the interaction between the free nucleons is not the same as the interaction between a free nucleon and nucleons inside the nucleus. It does however indicate that the potential has a radius and surface diffuseness similar to that of the nucleus itself. Thus, the real part of the optical potential due to all the nucleons in the nucleus and the incident particle is approximately given by

$$U(r) = \sum_i v(|\vec{r} - \vec{r_i}|) \approx \int \rho(r')\, v(|\vec{r} - r'|)\, dr'.$$

Since N-N interaction has short range we can approximate it by a delta function

$$U(r) = \int \rho(r')\, V_0\, \delta(r - r')\, dr' = V_0\, \rho(r).$$

Thus, to first approximation we expect the optical potential to have a radial variation that follows the nuclear density quite closely (*see* Fig. 1.5), and for this reason, the Woods-Saxon form is particularly convenient. Feshbach, Porter and Weisskopf [7] assumed that the real and imaginary parts of the optical potential have the same form.

Exercise 5.1. What is the average energy of neutrons in the forward direction produced in the laboratory by a Van de-Graaff generator through the reaction $^3H(d,n)$ for the incident deuteron energy $E_d = 0.4$ MeV?

From (5.10) we get for $\theta = 0$, $E_n = w + 2u^2 \left[1 \pm \left(1 + \dfrac{w}{u^2} \right)^{1/2} \right]$

where $u = \dfrac{\sqrt{2}}{5}\, \sqrt{E_d}$ and $w = \dfrac{2}{5} E_d + \dfrac{4}{5} Q$

and $Q = BE(^3H) + BE(^2H) - BE(^4He)$. Therefore,

$(E_n)_{max} = 15.3$ MeV , $(E_n)_{min} = 12.83$ MeV and $\bar{E}_n = 14.06$ MeV.

5.2 RESONANCES

The appearance of sudden increase in the output for a certain given input is generally known as phenomenon of resonance in physics. Typical examples in nuclear reactions are the well known peaks seen in the scattering of thermal neutrons from nuclei. The game of "bump hunting along with fixing its parameter" is of crucial importance both from the point of view of particle spectroscopy and nuclear reactions. Resonances/metastables/quasi-stationaries occur in collision processes and they are characterized (apart from internal quantum numbers) by a set of labels E, Γ, $\ell\hbar$, π, where E is the real energy, Γ the width, $\ell\hbar$ the angular momentum and π the parity $(=(-)^\ell)$. In order to understand the dynamical origin of resonance, let us consider a two-body non-relativistic collision. In the ℓth partial wave the effective potential (V_{eff})

encountered by the projectile is given by (5.17). The given nuclear potential (V_N) plus Coulomb potential (V_{Coul}), i.e. $(V_N(r) + V_{Coul}(r))$ interferes with centrifugal term in such a way that $V_{eff}(r)$ develops a pocket. Figure 5.5 shows how $V_{eff}(r)$ changes as ℓ increases. The incident wave of energy E enters this pocket and gets reflected from its walls several times producing a metastable state and finally leaks out through the barrier. These metastable state or quasi-stationary states are called "pocket resonances", prevented from decaying by the Coulomb-plus centrifugal barrier. Since the shape of $V_{eff}(r)$ is so important that the corresponding resonance is called "shape resonance".

S-Wave Neutron Scattering: To obtain a more quantitative estimate of nuclear reaction cross section in the CN model[27], we consider the collision of low energy neutrons with a nucleus. The cross section σ_{CN} of the formation of CN is larger for neutrons than for charged particles because neutrons do not encounter a Coulomb barrier. For the low energy neutrons only $\ell = 0$ partial waves are involved. Hence, neutrons are not subjected to any centrifugal barrier as well. We assume that the target nucleus has a well defined spherical surface with radius r_0. The wave function of $\ell = 0$ neutrons outside nuclear surface is (Appendix B2-Eq12)

$$\frac{u_0}{r} = \frac{1}{2ikr}\left[\eta_0 e^{ikr} - e^{-ikr}\right], \tag{5.27}$$

where, $k\hbar = \sqrt{2\mu E}$; E = Energy of the incident particle in c.m.

Let u_i denotes the wave function inside the nucleus, which is to be determined by solving the radial S. Eq. with nuclear interaction. The outside and inside wave function must match at the nuclear boundary $r = r_0$, which requires that we set

$$u_i = u_0 \text{ and } \frac{du_i}{dr} = \frac{du_0}{dr} \text{ at } r = r_0. \tag{5.28}$$

We define a dimensionless quantity f as follows:

$$f = \left(\frac{r du_0}{u_0 dr}\right)_{r=r_0} = ikr_0 \frac{\eta_0 e^{ikr_0} + e^{-ikr_0}}{\eta_0 e^{ikr_0} - e^{-ikr_0}}$$

or

$$\eta_0 = \frac{f + ikr_0}{f - ikr_0} e^{-2ikr_0}. \tag{5.29}$$

In general f is a complex number and we set $f = kr_0(g + ih)$. (5.30)

So that,

$$\eta_0 = \frac{g + i(1+h)}{g + i(h-1)} e^{-2ikr_0}, \quad \{1 - |\eta_0|^2\} = \frac{-4h}{g^2 + (h-1)^2}.$$

Therefore, (5.12) yields $\sigma_R^0 = 4\pi\lambda^2 \dfrac{-h}{g^2 + (h-1)^2}.$ (5.31a)

Again $|1 - \eta_0|^2 = \left| \dfrac{f(1 - e^{-2ikr_0}) - ikr_0(1 + e^{-2ikr_0})}{f - ikr_0} \right|^2$

$= 4 \left| \dfrac{f \sin kr_0 - kr_0 \cos kr_0}{f - ikr_0} \right|^2$

$= 4 \left[\dfrac{\{g^2 + (h-1)^2\}\sin^2 kr_0 + 2h \sin^2 kr_0 - \sin^2 kr_0}{g^2 + (h-1)^2} + \cos^2 kr_0 - 2g \sin kr_0 \cos kr_0 \right].$

Therefore,

$$\sigma_{sc}^0 = 4\pi\lambda^2 \left[\sin^2 kr_0 + \frac{2\sin kr_0\{(h-1)\sin kr_0 - g \cos kr\}}{g^2 + (h-1)^2} + \frac{1}{g^2 + (h-1)^2} \right].$$

(5.31b)

We recognise at once that $h \le 0$, otherwise the reaction cross section (Eq. 5.31a) would be negative. When $h = 0$, f is real, there is no reaction. Therefore, for nuclear reaction to take place, f or η_0 must be complex. If the wave function vanishes at $r = r_0$, i.e. $[u(r)]_{r=r_0} = 0$, then $f = \infty$, hence, only the first term in the expression of σ_{sc}^0 will contribute. This case corresponds to the scattering by a hard sphere. The wave function does not penetrate inside the nucleus.

The hard sphere scattering cross section $= 4\pi\lambda^2 \sin^2 kr_0$. For small kr_0 and small f / kr_0, the first two terms in σ_{sc}^0 can be neglected in comparison to the last. The last term has a resonance character with a steep maximum for small f, the resonant value corresponds to $f = 0$. The reaction σ_R^0 has a resonant maximum at the same energy. Both cross sections remain finite at the resonance. In order to evaluate nuclear reaction cross sections in further detail, we obtain the quantity f from the behaviour of wave function inside the compound nucleus

$$\psi_i = \frac{u_i}{r} = A\frac{e^{iKr}}{r} + B\frac{e^{-iKr}}{r}.$$

(5.32)

K depends on the nuclear potential; for a square well potential,

$$K\hbar = \sqrt{2\mu(E + V_0)}.$$

A. Theory of Compound Nucleus Resonances

In order to understand the phenomena of CN resonance in detail, we again try to find out the behaviour of the wave function inside the compound nucleus. If the incident bombarding energy is low so that no channel other than the entrance channel is open, then the incident particle is not absorbed but only scattered. If nuclear reactions occur at all, this assumption is only fulfilled approximately.

Therefore, if the particle returns through the incident channel then $|A| \doteq |B|$ in (5.32). However, the incident and outgoing waves need not be in the same phase. We write

$$A = -Be^{2i\phi}, \; u_i = -Be^{2i\phi}\, e^{iKr} + Be^{-iKr}$$

and

$$u_i = -B\, e^{i\phi}\left[e^{i(Kr+\phi)} - e^{-i(Kr-\phi)}\right] = C\,\sin\,(Kr + \phi),$$

where

$$C = -2iBe^{i\phi}.$$

Then the dimensionless quantity

$$f = \left(\frac{r}{u_i}\,\frac{du_i}{dr}\right)\Bigg|_{r \to r_0} = Kr_0\cot\,(Kr_0 + \phi). \tag{5.33a}$$

We see that the logarithmic derivative f is real and according to (5.30)

$$g = \frac{K}{k}\cot(Kr_0 + \phi) \text{ and } h = 0. \tag{5.33b}$$

Then the cross sections (5.31b) for resonant processes for the simple case of s-wave neutron scattering,

$$\sigma_{sc}^0 = 4\pi\lambda^2\left[\sin^2 kr_0 - 2\,\sin kr_0\,\frac{\left\{\sin kr_0 + \frac{K}{k}\cot(Kr_0 + \phi)\cos kr_0\right\}}{1 + \frac{K^2}{k^2}\cot^2(Kr_0 + \phi)}\right.$$

$$\left. + \frac{1}{1 + \frac{K^2}{k^2}\cot^2(Kr_0 + \phi)}\right]. \tag{5.34}$$

Obviously, the reaction cross section $\sigma_R^0 = 0$, as we anticipated for the case $|A| = |B|$. If the denominator is small, resonances (sudden increase in σ) appear. This is to be expected whenever $f = 0$ that is,

$$\cot(Kr_0 + \phi) = 0 = \cot\{(2n + 1)\,\pi/2\}, \text{ where } n = 0,1,2,\ldots..$$

Setting the above condition in (5.34) we get $(\sigma_{sc}^0)_{res} = 4\pi\lambda^2\left[1 - \sin^2 kr_0\right]$.

For $k \to 0$; $\sin kr_0 \to kr_0$, so that $(\sigma_{sc}^0)_{res} = 4\pi\lambda^2\left[1 - k^2 r_0^2\right] \approx 4\pi r_0^2$. Hence, at resonance the scattering cross section reaches maximum possible values. Far from resonances, $\cot(kr_0 + \phi)$ are large, then we have $\sigma_{sc}^0 = 4\pi\lambda^2 \sin^2 kr_0$, which is the same as the cross section for scattering from hard sphere and for small incident energy. We again get $\sigma_{sc}^0 = 4\pi r_0^2$. Hence, outside the resonance and for small incident energies, the scattering cross section is independent of energy and equal to four times the geometrical cross section.

Behaviour Near an Isolated Resonance: In order to determine the behaviour of scattering cross section in the vicinity of an isolated resonance, we expand the logarithmic derivative $f(E)$ in a Taylor's series about the resonance energy E_R:

$$f(E) = f(E_R) + \left(\frac{\partial f}{\partial E}\right)_{E=E_R}(E - E_R) + \cdots$$

We denote $\left(\dfrac{\partial f}{\partial E}\right)_{E=E_R} = -\dfrac{2kr_0}{\Gamma}$ and neglect higher order terms so that

$f(E) = -\dfrac{2kr_0}{\Gamma}(E - E_R)$. Since at resonance $E = E_R$, $f(E_R) = 0$, $g = -\dfrac{2}{\Gamma}$

$(E - E_R)$ and $h = 0$ then (5.31b) gives

$$(\sigma_{sc}^0)_{res} = 4\pi\lambda^2 \frac{1}{\frac{4}{\Gamma^2}(E - E_R)^2 + 1} = 4\pi\lambda^2 \frac{\frac{\Gamma^2}{4}}{(E - E_R)^2 + \frac{\Gamma^2}{4}}. \qquad (5.35)$$

The cross section has the form of a resonance line (Lorentz form). The cross section becomes half of its maximum value at $(E - E_R) = \Gamma/2$. Thus, Γ is the full width at half maximum. The cross section is written in a more familiar form $\sigma_{sc}^0 = 4\pi\lambda^2 \dfrac{1}{1 + x^2}$, where $x = \dfrac{2(E - E_R)}{\Gamma}$ and σ_{sc}^0 is the resonance spectrum of damped oscillations. The interference term

$$(\sigma_{sc}^0)_{\text{interference}} = 4\pi\lambda^2\left[-2\sin^2 kr_0 \frac{1 - x\cot kr_0}{1 + x^2}\right], \qquad (5.36)$$

where $g = -x$. The interference contributes a negative amount when

$$(1 - x \cot kr_0) > 0 \text{ or } \tan kr_0 > x \text{ or } E < E_R + \frac{\Gamma}{2} \tan kr_0 \text{ (left of the maxima)}$$

and a positive amount when $1 - x \cot kr_0 < 0$ or $E < E_R + \frac{\Gamma}{2} \tan kr_0$ (right of

the maxima). The resonance term and the interference term superimpose so that the scattering cross section as a function of incident energy has a form as shown in Fig. (5.7). The solid line curve shows the cross section with interference term. The dotted curve shows the cross sections without interference term. The dashed curves gives the geometrical cross section and σ_{sc}^0 values for $k \to 0$. The single resonance shown has $E_R = 1$ KeV, $\Gamma = 0.02$ KeV, $r_0 = 5.0$ fm and cross sections are shown in the logarithmic scale. The hard sphere (hs) or geometrical cross section for the above case is:

$$(\sigma_{sc}^0)_{geom.} = (\sigma_{sc}^0)_{hs} = 4 \times 3.141 \times 25 \times 10^{-26} = 3.141 \text{ barn}$$

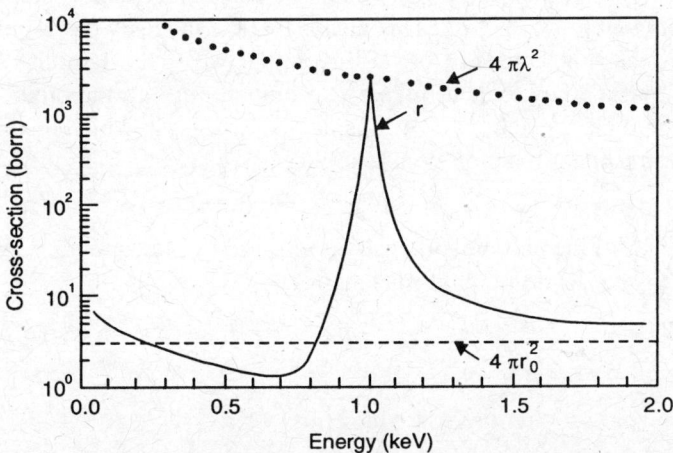

Fig. 5.7: S-wave neutron scattering cross-section as a function of incident neutron energy showing resonance behaviour (solid line). The dashed curve indicates hard sphere scattering ($4\pi r_0^2$) and the maximum cross section at the resonance ($4\pi\lambdabar^2$) is shown on the dotted curve

The foregoing considerations give us the possibility of describing cross sections in terms of line width (instead of scattering amplitude). The line width is related to the life time τ of the excited state of CN by $\Gamma \cdot \tau = \hbar$. For

the resonance in question $\tau = \frac{\hbar}{\Gamma} \approx 7.0 \times 10^{-16} \approx 4.8 \times 10^{-16}$ sec. The half life of the excited state

$$T_{1/2} = \tau \log_e^2 = 0.69 \times 7 \times 10^{-16} = 4.8 \times 10^{-16} \text{ sec.}$$

The maximum value of the resonance scattering cross section at the resonance energy $E = E_R = 1$ KeV is estimated as follows:

$$\lambdabar^2 = \frac{\hbar^2}{m} \frac{1}{2E} \approx 2 \times 10^4 \text{ fm}^2.$$

Hence,
$$(\sigma_{sc}^0)_{res}^{max} = 4\pi\lambdabar^2 = 2.65 \times 10^3 \text{ barn.}$$

B. Breit Wigner Formula

Resonant Scattering with Reaction: We generalize our consideration to include also the possibility that the incoming particle stays inside the nucleus and initiates a nuclear reaction. Then the form of the wave function $u_i(r)$ inside the nucleus is given by Eq. (5.32). The limit case $|A| = 0$ leads to continuum theory (complete absorption or black nucleus). $|A| = |B|$ leads to resonant scattering that is the process in which the incoming particle leaves the same channel by which it came in. Now we shall consider the intermediate case in which $0 < |A| < |B|$. This means that the intensity of outgoing wave is weaker than the intensity of the incoming wave (*see* Exercise 5.3). We expect a part of the incident intensity to branch out into the other channels because of nuclear reactions. We consider the process in which resonant scattering and resonant reaction both take place. We set

$$A = -Be^{2i\varphi}e^{-2\epsilon} \quad \text{for } \epsilon > 0.$$

The decay of the outgoing intensity is ensured by the factor $e^{-2\epsilon}$. The wave function inside the compound nucleus:

$$u_i = -B\left[e^{2i\varphi}e^{-2\epsilon}e^{iKr} - e^{-iKr}\right] = -Be^{i\varphi}e^{-\epsilon}\left[2i\sin(Kr + \varphi + i\epsilon)\right] \text{ for } r > r_0$$

$$(5.37a)$$

Hence,
$$f = \frac{\cot(Kr_0 + \varphi)\cos(i\epsilon) - \sin(i\epsilon)}{\cos(i\epsilon) + \cot(Kr_0 + \varphi)\sin(i\varphi)}$$

according to (5.33a) and (5.37a). The value of f now depends on two variables φ and ϵ. Now consider the case $\epsilon \ll 1$ (small probability for reactions):

$$f = Kr_0 \frac{\cot(Kr_0 + \varphi) - i\epsilon}{1 + i\epsilon\cot(Kr_0 + \varphi)}.$$

$$(5.37b)$$

We multiply numerator and denominator of (5.37b) by $(1 - i\epsilon\cot(Kr_0 + \varphi))$ and get

$$f = Kr_0 \frac{(1 - \epsilon^2)\cot(Kr_0 + \varphi) - i\epsilon[1 + \cot^2(Kr_0 + \varphi)]}{1 + \epsilon^2\cot^2(Kr_0 + \varphi)}.$$

In zero order approximation (that is, for $\in = 0$) we again have the case of pure scattering, and resonances in the cross section occur at $\cot(Kr_0 + \phi) = 0$. In the first order approximation ($\in \ll 1$), we can expand f in the vicinity of resonance as

$$f = [f(E)]_{E=E_R} + \left(\frac{\partial f}{\partial E}\right)_{\substack{E=E_R \\ \in=0}} (E - E_R) + \left(\frac{\partial f}{\partial \in}\right)_{\substack{\in=0 \\ E=E_R}} \in + \cdots, \qquad (5.38a)$$

since f is a function of both E and $\in$. Once again $[f(E)]_{E=E_R} = 0$ and

$$\left(\frac{\partial f}{\partial \in}\right)_{\substack{\in=0 \\ E=E_R}} = -iKr_0. \text{ Hence, } f = \left(\frac{\partial f}{\partial E}\right)_{\substack{E=E_R \\ \in=0}} (E - E_R) - iKr_0 \in.$$

We define two real quantities by $\left(\frac{\partial f}{\partial E}\right)_{\substack{E=E_R \\ \in=0}} = -\frac{2kr_0}{\Gamma_{sc}}$ and $\in = \frac{k}{K}\frac{\Gamma_R}{\Gamma_{sc}}$.

Hence, $f = -\dfrac{2kr_0}{\Gamma_{sc}}(E - E_R) - ikr_0\dfrac{\Gamma_R}{\Gamma_{sc}}$, $g = -\dfrac{2}{\Gamma_{sc}}(E - E_R)$ and $h = -\dfrac{\Gamma_R}{\Gamma_{sc}}$.

$$(5.38b)$$

Let us set $\Gamma_{tot} = \Gamma_R + \Gamma_{sc}$ so that

$$(\sigma_{sc}^0)_{res} = 4\pi\lambda^2 \frac{\Gamma_{sc}^2}{\Gamma_{tot}^2} \frac{1}{1 + \left\{\dfrac{2}{\Gamma_{tot}}(E - E_R)\right\}^2}, \qquad (5.39)$$

$$(\sigma_R^0)_{res} = 4\pi\,\lambda^2 \frac{\Gamma_R\Gamma_{sc}}{\Gamma_{tot}^2} \frac{1}{1 + \left\{\dfrac{2}{\Gamma_{tot}}(E - E_R)\right\}^2} = \frac{\Gamma_R}{\Gamma_{sc}}(\sigma_{sc}^0)_{res}. \qquad (5.40)$$

The resonance energy E_R corresponds to the excited state of the compound nucleus and the quantity Γ refers to this state and is related to the life time τ of the excited state (for $\Gamma = 1$ eV, $\tau \sim 6 \times 10^{-16}$ sec). The resonance levels that can be reached either by a particular compound elastic scattering or a particular compound nuclear reaction are only those that correspond to the compound nucleus excited states that can be formed without violating the conservation laws. If $(\sigma_{tot}^0)_{res}^{max}$ is taken as the cross section of formation of the compound nucleus then $\Gamma_{sc} / \Gamma_{tot}$ is the probability that the compound nucleus decays again into the original particles, and Γ_R / Γ_{tot} is the probability that reaction would go into channel different than the entrance channel. If $\Gamma_R \ll \Gamma_{sc}$ so that $\Gamma_{sc} \approx \Gamma_{tot}$ then $(\sigma_{sc}^0)_{res}$ reaches its maximum value $(4\pi\lambda^2)$. If $\Gamma_{sc} = \Gamma_R = \dfrac{1}{2}\Gamma_{tot}$ then the maximum value of s-wave reaction cross section at resonance $(\sigma_R^0)_{res}^{max} = \pi\lambda^2$.

The above cross section formulae (5.39-5.40) must be corrected for including the spin of the incident particle I_a, spin of the target I_A and the spin of the compound nucleus I_C. The total number of possible relative orientation of spins of projectile target system is $(2I_a + 1)(2I_A + 1)$. For the case $\ell = 0$, the resultant spin I_C of the compound nucleus with m-degeneracy is $(2I_C + 1)$ i.e. I_C takes the values: $(I_A + I_a), (I_A + I_a - 1), \cdots\cdots |I_A - I_a|$.

Hence, $$g = (2I_C + 1)/(2I_A + 1)(2I_a + 1). \tag{5.41}$$

Hence, inclusion of spin yields the following expression for total $\ell = 0$ cross section

$$\sigma_{tot}^0 = \sigma_{sc}^0 + \sigma_R^0 = 4\pi\lambda^2 g \frac{\Gamma_{sc}^2}{\Gamma_{tot}^2} \frac{1}{1 + \left\{ \dfrac{2}{\Gamma_{tot}}(E - E_R) \right\}^2}$$

$$+ 4\pi\lambda^2 g \frac{\Gamma_R / \Gamma_{sc}}{\Gamma_{tot}^2} \frac{1}{1 + \left\{ \dfrac{2}{\Gamma_{tot}}(E - E_R) \right\}^2}. \tag{5.42}$$

We apply (5.42) to a case when the projectile is a low energy neutron $(E \le 0.1 \text{ MeV})$, ℓ in this energy range naturally will be zero. We consider two possible exit channels (n, n), the neutron elastic scattering and (n, γ), the neutron capture reaction in which the compound nucleus decays with the emission of gamma ray. Here, $I_a = 1/2$ and parity $= +$ and $I_C = I_A = 1/2$ with the same parity of the target nucleus.

Hence, $$g = \frac{1}{2}\left(1 \pm \frac{1}{2I_A + 1} \right) \quad \text{for } I_C = I_A \pm \frac{1}{2}.$$

The resulting formulas, known as the Breit Wigner resonance formulas[8] are

$$\sigma(n, n) = \pi\lambda^2 \frac{\Gamma_{sc}^2}{\Gamma_{tot}^2} \frac{\Gamma_n^2}{(E - E_R)^2 + \dfrac{1}{4}\Gamma_{tot}^2} \frac{1}{2}\left(1 \pm \frac{1}{2I_A + 1} \right) \tag{5.43a}$$

$$\sigma(n,\gamma) = \pi \lambda^2 \frac{\Gamma_n \, \Gamma_\gamma}{(E - E_R)^2 + \frac{1}{4}\Gamma_{tot}^2} \frac{1}{2}\left(1 \pm \frac{1}{2I_A + 1}\right) \qquad (5.43b)$$

$$\sigma_{tot} = \pi \lambda^2 \frac{\Gamma_n \, \Gamma_{tot}}{(E - E_R)^2 + \frac{1}{4}\Gamma_{tot}^2} \frac{1}{2}\left(1 \pm \frac{1}{2I_A + 1}\right). \qquad (5.43c)$$

For application, we choose ^{115}In as target, which has two stable isotopes with abundance ratio $^{113}In / ^{115}In$ = 4.2/95.8. The capture of neutron by ^{115}In, produces $^{116}In^*$ in an excited state at excitation energy E_x = 1.457 eV + S_n where S_n is the neutron separation energy. $^{116}In^*$ decays with the re-emission of neutron (neutron resonance scattering) or by gamma emission. In γ-decay, which precedes through intermediate states, either the ground state or the first excited state of ^{116}In is reached. The ground state (1 +) decays with half life of 13 sec into the ground state (0+) of ^{116}Sn with emission of β^-. The long lived excited state (5+) because of a large difference of angular momentum does not decay by gamma emission but by β^- emission with a measurable half life of 54 min into first excited state (2+) of ^{116}Sn at an excitation energy of 1.27 MeV. In Table 5.1 we give some parameters for the slow neutron induced reaction on ^{115}In and ^{113}Cd. In Fig. 5.8, σ_{tot} vs neutron energy is shown for ^{115}In.

Table 5.1: Resonance Parameters

Nucleus	^{115}In	^{113}Cd
Γ_γ	0.072 eV	0.113 eV
Γ_n	0.003 eV	0.00065 eV
E_R	1.457 eV	0.178
Gs spin	9/2	1/2
$(\sigma_{tot})_{res.}^{max}$	39400 barn	7900 barn

R-Matrix Theory: The desire to relate the observed resonance behaviour of nuclear scattering and reaction cross sections to the underlying nuclear dynamics and the proper physical interpretation of resonance parameter are the main problems of CN reaction theory. R-matrix theory, originally developed by Wigner and Eisenbud[9], provides a powerful and convenient tool for handling these problems. Here we consider *s*-wave neutron scattering.

In this theory the nuclear wave function is expanded over a finite region of space in terms of some suitably chosen set of basis functions. A variety of

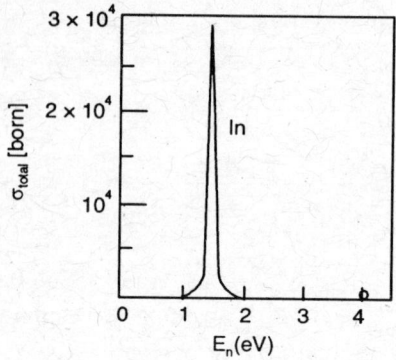

Fig. 5.8: Neutron capture cross section versus energy

rigorous formalism has been developed corresponding to various possible choices of these wave functions. Although, these approaches differ considerably in mathematical details and in ease of application to specific cases, the evidence to date indicates that all methods are capable of yielding equivalent results when correctly applied.

We begin with the simple case of potential scattering of a spinless particle in which possibility of reaction is excluded. Given a suitable Hamiltonian H for two interacting nuclei, it is desired to find a wave function ψ which satisfies

$$\frac{d^2\psi}{dr^2} + \frac{2\mu}{\hbar^2}\left[E - V_C(r) - \frac{\ell(\ell+1)\hbar^2}{2\mu r^2} \right]\psi = 0, \qquad (5.46)$$

subject to the appropriate boundary conditions. The philosophy of the R-matrix method is to divide into two parts the configuration space (one dimensional for radial motion) of colliding nuclei. One part is the internal (or interaction) region in which interaction between colliding nuclei is important. Another part is the external (or asymptotic) region in which both nuclei are separated by a distance which is larger than the nuclear force range. The distinction between internal and external regions is important because in the external region we can solve exactly the radial Schrodinger equation. The internal region is characterized by a parameter 'a_c' called R-matrix channel radius. The eigen states of the nuclear Hamiltonian in the interior region are denoted by Φ_λ with energy E_λ. These states are required to satisfy the boundary condition

$$r\frac{d\Phi_\lambda}{dr} + b\,\Phi_\lambda = 0 \text{ at } r = a_c \quad \text{or} \quad -b = \left[\frac{rd\Phi_\lambda}{dr}\Big/\Phi_\lambda \right]_{r=a_c}. \qquad (5.47)$$

The constant b is called the boundary condition parameter and it is a real number. The true nuclear wave function ψ for compound system is not stationary. Since Φ_λ form a complete set, it is possible to expand ψ in terms of Φ_λ's as $\psi = \sum_\lambda A_\lambda \Phi_\lambda$, where the expansion coefficients

$$A_\lambda = \int_0^{a_c} \Phi_\lambda^\dagger \psi \, dr.$$

Φ_λ and ψ satisfy:

$$-\frac{\hbar^2}{2\mu} \frac{d^2\psi}{dr^2} + V_C \psi = E\psi \qquad \text{for all } r, \qquad (5.48)$$

$$-\frac{\hbar^2}{2\mu} \frac{d^2\Phi_\lambda^*}{dr^2} + V_C \Phi_\lambda^* = E_\lambda \Phi_\lambda^* \qquad \text{for } r \leq a_c. \qquad (5.49)$$

Multiplying (5.48) by Φ_λ^* and (5.49) by ψ, subtracting and integrating, we get

$$\frac{\hbar^2}{2\mu}\left[\psi \frac{d\Phi_\lambda^*}{dr} - \Phi_\lambda^* \frac{d\psi}{dr}\right] = (E - E_\lambda) \int_0^{a_c} \psi \, \Phi_\lambda^* \, dr . \qquad (5.50)$$

Employing the boundary condition (Eq. 5.47) we get

$$\frac{\hbar^2}{2\mu}\left[\psi(a_c)\left(-\frac{b}{a_c}\right)\Phi_\lambda^* - \Phi_\lambda^* \frac{d\psi}{dr}\bigg|_{r=a_c}\right] = A_\lambda(E - E_\lambda).$$

Hence, $\quad A_\lambda = \frac{\hbar^2}{2\mu a_c}(E_\lambda - E)^{-1} \Phi_\lambda^*(a_c)[b\psi(a_c) + a_c \psi'(a_c)],$

where $\psi'(a_c) = \left[\dfrac{d\psi}{dr}\right]_{r=a_c}$

For V_C real, $\Phi_\lambda(a_c)$ is real so that $\Phi_\lambda^*(a_c) \equiv \Phi_\lambda(a_c)$. Therefore,

$$\psi(a_c) = \frac{\hbar^2}{2\mu a_c}\sum_\lambda \frac{\Phi_\lambda(a_c)\,\Phi_\lambda(a_c)}{(E_\lambda - E)}[b\psi(a_c) + a_c\psi'(a_c)]$$

$$= R[b\psi(a_c) + a_c\,\psi'(a_c)].$$

$$R = \frac{\hbar^2}{2\mu a_c}\sum_\lambda \Phi_\lambda(a_c)\Phi_\lambda(a_c)/(E_\lambda - E) = \sum_\lambda \frac{\gamma_\lambda^2}{(E_\lambda - E)} \qquad (5.51)$$

This is the fundamental form of R function which persists throughout its

generalization to a matrix and $\gamma_\lambda = \left(\dfrac{\hbar^2}{2\mu a_c}\right)^{1/2} \Phi_\lambda(a_c)$.

Finally, $\qquad R\left[b + a_c \dfrac{\psi'(a_c)}{\psi(a_c)}\right] = 1$,

which implies $\qquad a_c \dfrac{\psi'(a_c)}{\psi(a_c)} = \dfrac{1}{R} - b = \dfrac{(1-bR)}{R}$. $\qquad$ (5.52)

The left hand side of Eq. (5.52) is just the logarithmic derivative of the true wave function at the boundary $r = a_c$ and hence one can determine S-matrix element for $\ell = 0$ neutron scattering in terms of R-function. Employing (5.52) and following the discussions for cross section for s-wave neutron scattering from the condition at the nuclear surface given earlier (5.29), we can write (for $r_0 \equiv a_c$),

$$f = a_c \psi'_E(a_c) / \psi_E(a_c) = \dfrac{1-bR}{R}, \text{ which gives}$$

$$\eta_0 = \dfrac{f + ika_c}{f - ika_c} \exp(-2ika_c)$$

$$= \left[1 + \dfrac{2ika_c R}{1 - (bR + ika_c R)}\right] \exp(-2ika_c). \qquad (5.53)$$

Finally, we assume that E is near to a particular E_λ. We denote this value by $E_\alpha (\lambda \neq \alpha)$ so that for an isolated resonance

$$R = \dfrac{\hbar^2}{2\mu a_c} \dfrac{\Phi_\alpha^2(a_c)}{(E_\alpha - E)} = \dfrac{\Gamma_\alpha}{2ka_c} \dfrac{1}{(E_\alpha - E)}, \qquad (5.54)$$

where $\Gamma_\alpha = \dfrac{\hbar^2 k}{\mu} \Phi_\alpha^2(a_c)$. Defining $\Delta_\alpha = \dfrac{-b\Gamma_\alpha}{2ka_c}$, we get $R = -\dfrac{\Delta_\alpha}{b(E_\alpha - E)}$.

Hence, $\eta_0 = \left[1 + \dfrac{i\Gamma_\alpha}{(E_\alpha - E + \Delta_\alpha) - \dfrac{\Gamma_\alpha}{2}}\right] \exp(-2ika_c)$.

Finally, $\sigma_{sc}^0 = \dfrac{\pi}{k^2} |1 - \eta_0|^2 = \dfrac{\pi}{k^2} \left| e^{2ika_c} - 1 + \dfrac{\Gamma_\alpha}{E - (E_\alpha + \Delta_\alpha) + \dfrac{1}{2} i\Gamma_\alpha}\right|^2$.

$$(5.55)$$

From the above analysis we see that the procedure of imposing boundary conditions at the channel radius leads to isolated *s*-wave resonances of the Breit-Wigner form. If the constant $b \neq 0$, the position of the maximum in the cross section is shifted to $(E_\alpha + \Delta_\alpha)$. We should stress once more that upto this point our analysis is confined to single channel (elastic scattering) and single level (isolated resonance) (*see* Jackson Ref. 32).

In general, a nucleus can decay through many channels. When the formalism is extended to take this into account, the *R*-function becomes a matrix. In order to illustrate the numerical success of the multilevel ($\ell = 0, 1, 2$) and single channel (elastic scattering) *R*-matrix theory, we show the analysis of Das and Mukherjee[10] in Fig. 5.9.

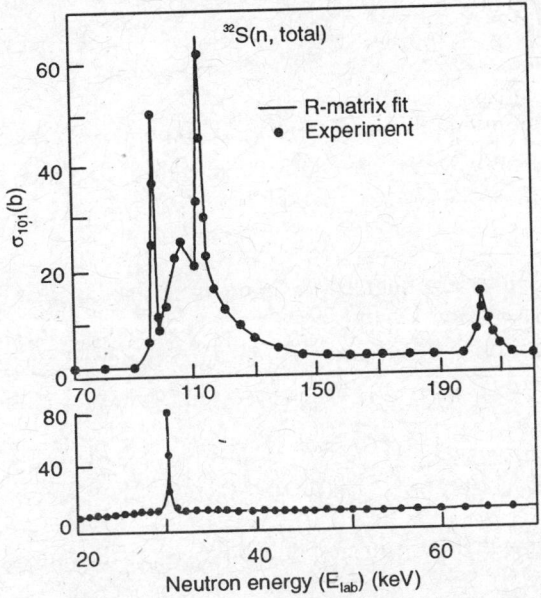

Fig. 5.9: R-matrix fit of the total elastic n + ^{32}S scattering. Theory: Das and Mukherjee[10]. Experiment: Halperin et al. (Ref. 11)

Exercise 5.2. Employing the Breit-Wigner formula and data given in Table 5.1, calculate the spin of the compound nucleus for ^{115}ln target.

At resonance $E = E_R$. Therefore, $(\sigma_{tot})_{res}^{max} = 4\pi \left(\dfrac{\hbar}{2\mu E} \right) \dfrac{\Gamma_R}{\Gamma_{tot}} g.$

Substituting σ and Γ values from Table 5.1 we get $g \cong 0.55 = \dfrac{11}{20}$ for ^{115}ln target.

Now, $$g = \frac{1}{2}\frac{2I_C + 1}{2I_A + 1} = \frac{1}{2}\frac{2I_C + 1}{2 \times \frac{9}{2} + 1} \text{ so that } I_C = 5.$$

Exercise 5.3. What is the effect on the outgoing amplitude (Eq. 5.32) due to the introduction of a complex potential?

Let us represent the complex potential by $V(r) = -V_0(1 + i\xi)$.

$$K = \left[\frac{2\mu}{\hbar^2}\{E + V_0(1 + i\xi)\}\right]^{1/2} = \sqrt{\frac{2\mu V_0}{\hbar^2}}\left\{\frac{E}{V_0} + (1 + i\xi)\right\}^{1/2}$$

$$\approx k_0\left(1 + \frac{1}{2}i\xi\right), \text{ where } \xi << 1 \quad \text{and} \quad \frac{E}{V_0} << 1 \text{ and } k_0 = \sqrt{\frac{2\mu V_0}{\hbar^2}}.$$

Hence, the outgoing solution (*see* Eq. 5.32), $u_i = -B\, e^{ik_0 r}\, e^{-\frac{1}{2}k_0\xi}$, which obviously represents wave exponentially attenuated and the attenuation depends upon the strength of the imaginary potential. The term $iV_0\xi$ accounts for the absorption.

Exercise 5.4. A spin zero nucleus has a neutron resonance at 80 eV and no other resonances nearby. For the 80 eV resonance, $\Gamma_n = 5$ eV, $\Gamma_\gamma = 1$ eV, $\Gamma_\alpha = 3$ eV, and all other Γ's are negligible. Find the cross section for (n, γ).

We employ (5.43), set $g = 1$, $\Gamma_{tot} = 9$ eV, and get

$$\sigma(n,\gamma) = \frac{3.141(197)^2 \text{ MeV}^2 fm^2}{2\,(939)\,\text{MeV}\,(80 \times 10^{-6})\,\text{MeV}}\frac{5 \times 1}{81/4} \cong 2000 \text{ barn}.$$

Exercise 5.5. Show that in neutron induced processes the integrated values of $\sigma(n, n)$ and $\sigma(n, \gamma)$ over a narrow isolated resonance are:

$$\int_{res} \sigma(n,n)\, dE = \pi^2\lambda^2 g\frac{\Gamma_n}{1 + \Gamma_\gamma/\Gamma_n} \text{ and } \int_{res} \sigma(n,\gamma)\, dE = \pi^2\lambda^2 g\frac{\Gamma_\gamma}{1 + \Gamma_\gamma/\Gamma_n}.$$

We employ (5.43) and take λ^2 outside integration because λ varies slowly with E.

$$\int_{res} \sigma(n,n)\, dE = 4\pi\lambda^2 g\int\frac{\Gamma_n^2}{\Gamma_t^2}\frac{dE}{1 + \left\{\frac{2}{\Gamma_{tot}}(E - E_R)\right\}^2}$$

$$= 4\pi\lambda^2 g \frac{\Gamma_n^2}{\Gamma_t^2} \int \frac{dx}{1+x^2/\in^2}, \text{ where } x = E - E_R \text{ and } \frac{1}{2}\Gamma_{tot} = \in.$$

or $\int\limits_{res} \sigma(n,n)\, dE = 4\pi\lambda^2 g \frac{\Gamma_n^2}{\Gamma_t^2} \in \left[\tan^{-1}\frac{x}{\in}\right]_0^\infty$ for $E > E_R$ and $\in$ very small

$$= 4\pi\lambda^2 g \cdot \frac{\Gamma_n^2}{\Gamma_t^2} \in \cdot \frac{\pi}{2} = \pi^2 \lambda^2 g \frac{\Gamma_n}{1+\Gamma_\gamma/\Gamma_n}.$$

In the same way we can prove the second relation.

When the areas under the experimental resonance curves (σ versus E) are determined, their ratio will give Γ_n/Γ_γ, and their sum will give $g\Gamma_n$. Hence, Γ_n can be inferred if g is known.

Exercise 5.6. The cross section σ is maximum for $\delta = \pi/2$ when we cross resonance energy $E = E_R$. Expanding $\delta(E)$ in the vicinity of resonance, derive the Breit-Wigner resonance formula.

$$\delta(E) = \delta(E_R) + \left(\frac{\partial\delta}{\partial E}\right)_{E=E_R} (E - E_R) + \cdots\cdots$$

$$\delta(E) \cong \pi/2 + \left(\frac{\partial\delta}{\partial E}\right)_{E=E_R} (E - E_R) \quad \text{since } \delta(E_R) = \pi/2.$$

If $\frac{\partial\delta}{\partial E}$ is small, that is, change in the cross section is slow, then resonance is broad. If $\frac{\partial\delta}{\partial E}$ is steep then the resonance is sharp. Hence, it is natural to connect inverse of slope with the width of the resonance $\Gamma = 2\left(\frac{\partial\delta}{\partial E}\right)^{-1}$. The factor 2 is put for convenience so that we obtain the standard Lorentz form:

$$\delta(E) = \pi/2 - \frac{2(E_R - E)}{\Gamma}.$$

Hence, $\qquad \tan\delta(E) = \cot\frac{2(E_R - E)}{\Gamma} = \dfrac{1}{\tan\dfrac{(E_R - E)}{\Gamma/2}}.$

Expanding $\tan \dfrac{(E_R - E)}{\Gamma/2}$ and retaining only the first order term in $(E_R - E)$ we get

$$\delta = \tan^{-1} \frac{\Gamma/2}{(E_R - E)} \quad \text{and} \quad \sin^2 \delta = \frac{\Gamma^2/4}{(\Gamma^2/4) + (E_R - E)^2}.$$

$$\sigma(n,n) = \frac{4\pi}{k^2} g \frac{\Gamma^2/4}{(\Gamma^2/4) + (E_R - E)^2} = \frac{\pi}{k^2} g \frac{\Gamma^2}{(E_R - E)^2 + \Gamma^2/4}.$$

This is again the well known Breit-Wigner resonance formula.

Exercise 5.8. Calculate the average cross section over one Breit-Wigner resonance for the compound nucleus formation for s-wave incident neutron, assuming that the levels do not overlap and the averaging interval equals the level separation, D.

Employing (5.30) and (5.38b) we can write

$$\eta_0^{(E)} = \left(1 - \frac{i\,\Gamma_{sc}}{E - E_R + \dfrac{i}{2}\Gamma_{tot}} \right) e^{-2ikr_0}.$$

If k varies slowly with energy, we obtain

$$\langle \eta_0 \rangle = \frac{1}{D} \int_{E_R - D/2}^{E_R + D/2} \eta_0(E)\, dE = \frac{1}{D} e^{-2ikr_0} \left(D - i\Gamma_{sc} \log \frac{D + i\Gamma_{tot}}{-D + i\Gamma_{tot}} \right).$$

For $\Gamma_{tot} \ll D$, we have by substituting $\log(-1) = -i\pi$, $\langle \eta_0 \rangle = \left(1 - \dfrac{\pi\,\Gamma_{sc}}{D} \right)$.

Hence, $\langle \sigma_{CN} \rangle \approx 2\pi^2 \lambda^2 \dfrac{\Gamma_{sc}}{D}$, neglecting the second order term in $\dfrac{\Gamma_{sc}}{D}$.

$\dfrac{\Gamma_{sc}}{D}$ is called the strength function, which is the measure of the average width of the levels per unit energy. We see that $\langle \sigma_{CN} \rangle$ depends only on Γ_{sc} and D but not available exit channels, which is in agreement with the Bohr's independent hypothesis.

Exercise 5.9. Employing the data in Table 5.1, calculate the neutron capture cross section for thermal neutrons $E_{th} = 0.0253$ eV for ^{115}In and ^{113}Cd targets.

Employing BW formula, we get

$$\sigma(n,\gamma)_{th} = (\sigma_{tot})_{th} \frac{\Gamma_\gamma(E_{th})}{\Gamma_{tot}(E_{th})}.$$

Now

$$\frac{(\sigma_{tot})_{th}}{(\sigma_{tot})_{E=E_R}} = \frac{E_R}{E_{th}} \frac{\Gamma_n(E_{th})}{\Gamma_n(E_{th})} \frac{1}{1+\left\{\dfrac{2}{\Gamma_{tot}(E_{th})}(E_{th}-E_R)\right\}^2}.$$

Hence,

$$\sigma(n,\gamma)_{th} = \frac{\Gamma_\gamma(E_{th})}{\Gamma_{tot}(E_{th})} \cdot \frac{E_R}{E_{th}} \frac{\Gamma_n(E_{th})}{\Gamma_n(E_R)} \frac{1}{\left(1+\left\{\dfrac{2}{\Gamma_{tot}(E_{th})}(E_{th}-E_R)\right\}^2\right)}.$$

Substituting $\dfrac{\Gamma_n(E_{th})}{\Gamma_n(E_R)} = \sqrt{\dfrac{E_{th}}{E_R}}$, we get

$$\sigma(n,\gamma)_{th} = (\sigma_{tot})_{E=E_R} \sqrt{\frac{E_R}{E_{th}}} \frac{1}{1+\left\{\dfrac{2}{\Gamma_{tot}(E_{th})}(E_{th}-E_R)\right\}^2} \frac{\Gamma_\gamma(E_{th})}{\Gamma_{tot}(E_{th})}.$$

For ^{115}In:

$$\sigma(n,\gamma)_{th} = 39,400 \text{ barn} \sqrt{\frac{1.457}{0.0253}} \frac{1}{1+\left(\dfrac{2.8634}{0.075}\right)^2} \frac{0.072}{0.075} = 197 \text{ barn}.$$

For ^{113}Cd:

$$\sigma(n,\gamma)_{th} = 7900 \text{ barn} \sqrt{\frac{0.178}{0.0253}} \frac{1}{1+\left(\dfrac{0.3054}{0.11365}\right)^2} \frac{0.11300}{0.11365} = 2549 \text{ barn}.$$

5.3 STATISTICAL THEORY OF COMPOUND NUCLEUS REACTIONS

Examination of any compilation of energy levels of light and heavy nuclei shows the following features. In light nuclei at fairly low excitation energy, levels are generally well separated, that is, spacings between the levels are much greater than the widths of the levels ($\Gamma/D \ll 1$). Therefore, in such cases of well separated CN levels, the concept of individual nucleon level remains meaningful and CN reactions can be described by theory of resonance reactions, the Breit-Wigner resonance theory, for example.

In heavy nuclei the levels are more closely spaced ($\Gamma/D \gg 1$) and if energy of excitation is sufficiently high ($E \approx 12\text{-}20$ MeV) overlapping levels will be excited. The density of levels grows sharply and division of nucleons into shells looses its significance. The number of levels excited is too great for them to be treated individually. In such cases statistical approach may provide reasonable description of CN reactions. The traditional presentation of statistical mechanics describes macroscopic large system (the number of states, $N \to \infty$). For small system ($N \sim 100\text{-}200$) the standard results of statistical mechanics have to be applied carefully.

Statistical concepts are used in two stages: First, for calculating nuclear level density within the framework of degenerate Fermi gas model. According to this model (1.2B), a nucleus in its ground state is a system of Fermions in its lowest energy state, which corresponds to zero temperature. An excitation energy of 1 MeV corresponds to a nucleon gas temperature $\approx 1.16 \times 10^{10}$ °K. For a given entrance channel, the excitation energy E is considered as heat energy distributed over many degrees of freedom of the excited nucleus. Heating of CN causes evaporation of nucleons.

Second, the energy distribution of evaporated nucleons are calculated by employing the Fermi model predicted level density and a statistical assumption that all possibilities of decay are equally likely. This statistical assumption when combined with conservation laws and principle of detailed balance, leads to statistical model for evaluating the average cross section.

A. Nuclear Level Density

To calculate nuclear level density, one needs a model for the nucleus. The rapid increase of level density $\rho(E)$ with excitation energy suggests an exponential energy dependence. In statistical thermodynamic Fermi model, the probability of occupation of a state of excitation energy ϵ is proportional to $\exp(-\epsilon/T)$, where T is the temperature. When one speaks of nucleus having an excitation energy E, one actually describes a situation in which levels near E are occupied and E is the average energy of this very narrow occupied band. Thus,

$$E = \frac{\int \epsilon \rho(\epsilon)\, e^{-\epsilon/T}\, d\epsilon}{\int \rho(\epsilon)\, e^{-\epsilon/T}\, d\epsilon}. \qquad (5.56)$$

To begin with, we assume energy as the only characterization for CN. Statistical thermodynamics relates $\rho(E)$ with entropy $S(E)$ as $S(E) = \log \rho(E)$ and defines nuclear temperature T as

$$\frac{dS(E)}{dE} = \frac{1}{T}. \tag{5.57}$$

In statistical thermodynamics, the probability distribution is proportional to $e^{S(E)}$. In the case of a nucleus, which is a finite system this role is played by the level density $\rho(E)$ which is in a sense nuclear entropy. The corresponding nuclear temperature T, then in analogy is defined as

$$\frac{1}{T} = \frac{\partial \log \rho(E)}{\partial E}.$$

For a Fermi gas, the functional relation between E and T must be consistent with the Nerst heat theorem, which states that thermal heat capacity $C_V = (dE / dT)_{T=0} = 0$. Therefore, the first term in the expansion of E must be proportional to T^2, i.e.

$$E = aT^2, \tag{5.58}$$

where a is constant, called level density parameter. The Fermi gas model yields

$$a = \frac{\pi^2 A}{4E_f}, \tag{5.59}$$

where E_f is the average Fermi energy (Eq. 1.36) for the neutron and proton gas.

Substituting $E_f = 24$ MeV, we get $a = \frac{A}{10}$ MeV^{-1} Employing (5.58) we get

$$\frac{dS(E)}{dE} = \sqrt{\frac{a}{E}} \quad \text{or} \quad S(E) = \log \rho = \int \sqrt{\frac{a}{E}} \, dE$$

or $$\rho(E) = C e^{2\sqrt{aE}}, \tag{5.60}$$

where, C is a constant. Many experimental results have been analyzed in terms of the above expression. The values of a and C, needed to provide approximate agreement with experiment, were earlier obtained by Blatt and Weisskopf[12] for low energy data. The spacings between the levels

$$D(E) = \frac{1}{\rho(E)}.$$ The value of C decreases with increasing A.

At higher excitation energies, there is a tendency for 'a' to approach the Fermi gas value. Since, nucleus is not a Fermi gas, the constants are fixed empirically. A more correct expression replacing (5.58) is

$$E = aT^2 - T.$$

One may expect many other shortcomings in the above results, due to the neglect of collective degrees of freedom and shell structure. For even-even nuclei, C is divided by 5 and for odd-odd nuclei, C is multiplied by 2. An estimate level density $\rho\,(E)$ for $E = 8$ MeV in the mass range $A \approx 115$ yields $\rho\,(8\ \text{MeV}) \approx 17.8 \times 10^4$. Sometime E is represented as a power series in T.

Since $\dfrac{dE}{dT} \to 0$ as $T \to 0$, then it follows that the expansion must 'start with a term at best quadratic in T. Accordingly we can write

$$E = aT^n. \qquad (5.61)$$

The calculation of level density went through periodic refinements with the possibility of including effects of pairing energy, shell structure, rotational energy leading to expressions in order of increasing complexity. By assuming equal spacings of levels, Ericson[13] obtained the following expression:

$$\rho(E) = \frac{\sqrt{\pi}}{12a^{1/4}E_e^{5/4}} \exp\left\{2(a\,E_e)^{1/2}\right\}, \qquad (5.62)$$

where $E_e = E - \Delta$ is the effective excitation energy, $a = \dfrac{1}{6}g\pi^2$ is the level density parameter related to spacings 'g' of the single particle levels near the Fermi energy. Δ is the pairing energy $\left(= \Delta_N + \Delta_Z\right)$. $\rho\,(E)$ increases exponentially with $\sqrt{E_e}$ and also depends on A (since $a = 0.115\,A$). The level density (5.62) only specifies energy and total nucleon number. Another essential quantum number is the angular momentum J and the level density is proportional to $(2J + 1)$ for low spins and for the high spin the expression for the level density is more complicated.

In contrast with the definition of temperature employing the concept of heat bath used for large systems, the nuclear temperature can fluctuate. For small system, such as nucleus, there is no heat bath. The fluctuation in temperature puts limits on the usefulness of temperature. These limits may be determined from the density of states and the partition sum (*see* Exercise 5.11).

Existence of stochastic fluctuations in CN reaction cross sections was first predicted by Ericson[13]. The advent of Tandem accelerators with their high energy resolution led to the discovery of these fluctuations. It has been observed in a wide variety of γ-ray and particle induced reactions. In Fig. (5.10) we show nuclear reaction cross section versus energy. CN resonances shown here have a mean spacing D which is several times smaller than their average Γ. With increasing excitation energy, the average level spacings decreases

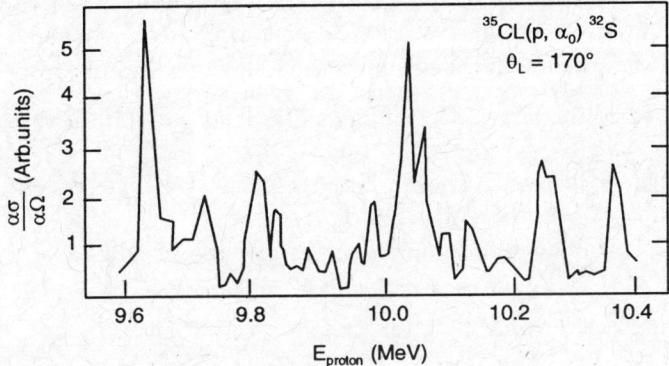

Fig. 5.10: Fluctuation in cross sections

exponentially and the average decay width increases exponentially. At few MeV above the reaction threshold, the resonances overlap strongly. Therefore, the peaks (and minima) result from constructive (or destructive) super-position of many overlapping resonances. Understanding of the Ericson's fluctuation in CN reactions was led to various theoretical developments. They are based on the fact that the amplitudes of resonances which superpose at any energy are essentially random variables, because the CN reaction is chaotic[14].

Ericson fluctuations are attributed to statistical fluctuations of complex scattering amplitudes and they are associated with fluctuations in the distribution of nuclear levels. They are observed in a wide variety of γ-ray and particle induced reactions. If we leave aside spin considerations, we may write the nuclear (scattering) amplitude at some angle θ with respect to a specified direction as

$$f(\theta) = \alpha(\theta) + i\beta(\theta)$$

and the differential cross section as

$$\sigma(\theta) = |f(\theta)|^2 = \alpha^2(\theta) + \beta^2(\theta), \qquad (5.63)$$

since α and β are real. The reason for the fluctuations may be understood in a simple way. Due to the statistical nature of the process one expects a distribution of amplitudes within the ensemble, i.e. $f = \sum_{n}^{N} f_n$, where N is the number of intermediate states. Therefore, the average cross section at a particular angle

$$\langle \sigma \rangle = \langle |f|^2 \rangle = \sum_{n,m} f_n^* f_m$$

or

$$\langle \sigma \rangle = \sum_{n} |f_n|^2 + \sum_{n \neq m} f_n^* f_m. \qquad (5.64)$$

In the above expression, the first part contains N terms and the second part contains $N(N-1)$ terms. If now it is assumed that f_n are random both in magnitude and sign, then the average cross-section comes entirely from the terms $\sum_n |f_n|^2$, since the term $f_n^* f_m$ are equally likely to be positive or negative and so average to zero. However, for large N the $N(N-1)$ terms of the second type compared with N of the first type, generate fluctuations in the cross section. These fluctuations do not tend to zero as N becomes very large and so the cross section continues to fluctuate. This may be understood by simply substituting $f_n = \pm (\sigma / N)^{1/2}$, with random signs. Then

$$\sum_n |f_n|^2 = N(\sigma / N) = \sigma, \text{ while}$$

$$\sum_{n,m} f_n^* f_m = 0 \pm \sqrt{N(N-1)} \ (\sigma / N)$$

$$= 0 \pm N \left(1 - \frac{1}{N}\right)^{1/2} (\sigma / N) = 0 \pm \sigma.$$

Thus, the cross section can fluctuate within a range from zero to $\pm \sigma$. Thus, the amplitudes fluctuate about their mean value, which may be zero, in a random way but the cross section average is not zero, i.e.

$$\langle f \rangle = \langle \alpha \rangle + i \langle \beta \rangle = 0 \quad \text{and} \quad \langle \alpha \rangle = \langle \beta \rangle = 0,$$

but $$\langle \sigma(\theta) \rangle = \langle |f(\theta)|^2 \rangle = \langle \alpha^2 \rangle + \langle \beta^2 \rangle \neq 0.$$

In view of the above it is plausible to assume that the variable α and β have independent distribution. Therefore, the distribution $P(\alpha, \beta)$ which characterizes the probability of finding α and β per unit interval within the ensemble, may be expressed as $P(\alpha, \beta) = P(\alpha) P(\beta)$ and since there is no preferment of α over β, it follows that they have a normal distribution,

$$P(\alpha, \beta) = \frac{1}{(2\pi\sigma^2)^{1/2}} e^{-(\alpha^2 + \beta^2)/2\sigma^2}. \tag{5.65}$$

From statistical theory, we may straight way conclude that, as $\sigma(\theta)$ is given by the sum of squares of two uncorrelated variables having a normal distribution with variance σ^2, its probability distribution must be a χ^2 distribution.

In Fig. 5.11 we show the normalized probability distribution of the forward ($\theta = 0$) differential cross section for $^{12}C(^{16}O, \alpha)^{24}Mg$ proceeding by way of $^{28}Si^*$ compound nucleus. The exponential curve corresponding to χ^2

distribution corresponding to 2 d.f. fits data reasonably well. To summarize, the statistical model assumes that the relative phase of these contributing amplitudes is "random". This means the average value of f is zero, i.e. $\langle f \rangle = 0$, but $\langle | f |^2 \rangle \neq 0$.

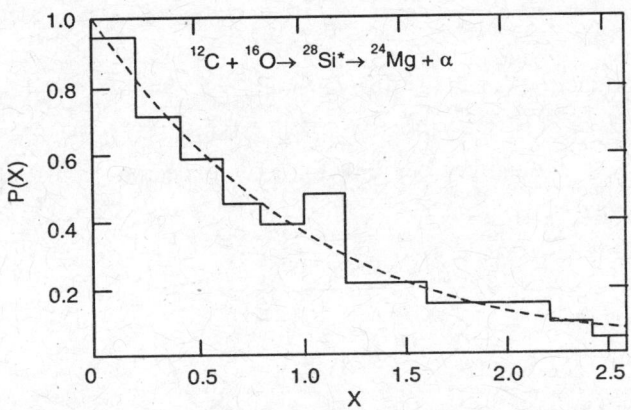

Fig. 5.11: Normalized probability distribution of forward differential cross section

B. Evaporation Probability and Cross Sections

We shall now quantitatively investigate emission of nucleons from the compound nucleus in the statistical model, first proposed by Weisskopf[4]. This author predicted that the energy spectrum of a reaction product would have the Maxwellian form:

$$\frac{d\sigma}{dE} \sim \sigma_c(E_C)\, e^{-E/T},$$

where $\sigma_C(E_C)$ is the CN formation cross section, E is the energy of the reaction product and T the temperature of the residual nucleus. The angular distribution would be isotropic. We assume in this model that several states of compound nucleus C are excited. C decays to $B+b$ and mode of decay is independent of mode of formation and there are several states of residual nucleus B in which C can decay. Let E_C and E_B be the excitation energies of the compound and residual nuclei respectively. Let ϵ_b be the kinetic energy of the emergent particle b, and let ϵ_{Cb} be the binding energy of the compound nucleus: Conservation of energy yields

$$E_C - \epsilon_{Cb} = E_B + \epsilon_b.$$

The transition probability per unit time W for C to decay into $B+b$, can be expressed in terms of the cross section for the formation of C from $B+b$ by using detailed balance: $| H_{fi} |^2 = | H_{if} |^2$.

$$W_{fi} = \frac{2\pi}{\hbar} \rho_f \mid H_{fi} \mid^2 = \sigma_{fi} \, v_i \quad \text{also} \quad W_{fi} = \frac{\rho_f}{\rho_i} \, \sigma_{if} \, v_f \, .$$

Hence, $\dfrac{W_{fi}}{\rho_f} = \dfrac{W_{if}}{\rho_i} = \dfrac{\sigma_{if} \, v_f}{\rho_i}$ or $W_{fi} = \dfrac{\rho_f}{\rho_i} \, \sigma_{if} \, v_f$. In the present case the

incident channel $i \equiv C$ and the final channel $f = Bb$ and $\rho_f = \rho_B(E_B)$. ρ_b, the density of states of b (free particle with spin s_b) is $\rho_b = (2s_b + 1) 4\pi p_b^2 / (2\pi\hbar)^3$. The transition probability per unit time for emission of particle b in the energy range $d \in_b$

$$dW_{Bb \leftarrow C} = \frac{\rho_B(E_B) \, (2s_b + 1)}{\rho_C(E_C)} \, \frac{4\pi p_b^2 \, dp_b}{(2\pi\hbar)^3} \, \sigma_{C \leftarrow Bb} \, \frac{p_b}{m_b}$$

$$= \frac{\rho_B(E_B) \, (2s_b + 1)}{\rho_C(E_C)} \, \frac{(2s_b + 1)m_b \in_b}{\pi^2 \hbar^3} \, \sigma_{C \leftarrow Bb} \, d \in_b , \qquad (5.66)$$

since, $\in_b = p_b^2 / 2m_b$ and $p_b \, dp_b = m_b \, d_b \in_b$ where m_b = mass of the particle b. We define $\sigma_c(a)$ as the cross section for the formation of the compound nucleus C through the incident channel a. If W_0 denotes the sum of probabilities of all possible modes of decay of CN, then the cross section $\sigma_{Bb \leftarrow Aa}$ for the reaction $A + a \to C \to B + b$, in accordance to independent hypothesis is given by

$$\sigma_{Bb \leftarrow Aa} = \sigma_c(a) \, W_{Bb \leftarrow C} \, / \, W_0 \text{ where } W_0 = \sum_{B'b'} W_{B'b' \leftarrow C} . \qquad (5.67)$$

Combining (5.66) and (5.67) we obtain

$$\sigma_{Bb \leftarrow Aa} = \sigma_c(a) \, \frac{\displaystyle\int_0^{\in_{bm}} d \in_b (2s_b + 1)m_b \in_b \sigma_{C \leftarrow Eb} \, \rho_B(E_B)}{\displaystyle\sum_{B'b'} \int_0^{\in'_{bm}} d \in_{b'} (2s_{b'} + 1)m_{b'} \in_{b'} \sigma_{C \leftarrow B'b'} \, \rho_{b'}(E_{B'})} , \qquad (5.68)$$

where $\in_{bm}$ and $\in'_{bm}$ are the maximum energies available to particle b and b'. We may make (5.66) more explicit by using the relation between nuclear entropy S and density and that between entropy and nuclear temperature:

$$d\dot{W}_{Bb \leftarrow C}(\in) = \frac{(2s_b + 1)m_b \in_b}{\pi^2 \hbar^3} \, \sigma_{C \leftarrow Bb}$$

$$\times \exp \left[S_B(E_C - \epsilon_{Cb} - \epsilon_b) - S_c(E_c) \right] d\epsilon_b . \qquad (5.69)$$

Since, the *CN* excitation energy is usually higher than the energy of the emitted particle, we can assume $E_C - \epsilon_{Cb} \gg \epsilon_b$. We can also assume that the density of levels in the compound nucleus and that in the residual nucleus are the same at given energy of excitation so that we can set $S_C(E) \cong S_B(E)$ and get

$$S_B(E_C - \epsilon_{Cb} - \epsilon_b) \approx S_B(E_C - \epsilon_{Cb}) - \epsilon_b \left(\frac{\partial S_B}{\partial E} \right)_{\epsilon_b = E_C = \epsilon_{Cb}}$$

$$= S_B(E_C - \epsilon_{Cb}) - \epsilon_b / T_B . \qquad (5.70)$$

In (5.70), T_B is the nuclear temperature of the residual nucleus at its maximum excitation energy and also

$$S_C(E_C - \epsilon_{Cb}) = S_C(E_C) - \epsilon_{Cb} \left(\frac{\partial S_C}{\partial E} \right)_{E_C = \epsilon_{Cb}} = S_C(E_C) - \epsilon_{Cb} / T_C ,$$

where T_C is the temperature of the compound nucleus at the excitation energy E_C. By substituting (5.70) in (5.69), and using $S_B(E_C - \epsilon_{CB}) = S_C(E_C - \epsilon_{Cb})$ we have

$$d\bar{W}_{Bb \leftarrow C}(\epsilon_b) = \frac{(2s_b + 1)m_b \epsilon_b}{\pi^2 \hbar^3} \sigma_{C \leftarrow Bb} \times \exp \left[\frac{-\epsilon_{Cb}}{T_C} - \frac{-\epsilon_b}{T_B} \right] d\epsilon_b .$$

$$(5.71)$$

If the cross section $\sigma_{C \leftarrow Bb}$ does not depend upon ϵ_b then the form of the spectrum $\sigma_{Bb \leftarrow C}(\epsilon_b) = $ constant $\epsilon_b \, e^{-\epsilon_b / T_B}$, which is Maxwellian. The statistical theory described above is applicable if the time during which the compound nucleus exists is long. The rapid energy fluctuations, narrow width of resonances and angular distribution support this assumption. Statistical model assumes that amplitude for a reaction is a sum of amplitudes denoted by specific quantum numbers (for example angular momentum and parity).

As mentioned earlier, statistical theory of CN reaction works when the compound system involves many states and the average level width is larger than the average level spacing. There are many ways in which the experimental results can be compared to predictions of statistical theory of CN. In Fig. (5.12) we compare prediction of statistical theory with the experimental cross section of Hansen and Albert[15] for $^{65}Cu(p,n)^{65}Zn$ reaction for proton energies between 5 and 11 MeV. The agreement between experimental results and prediction of theory is excellent. The predictions of theory, however, depends upon the expression of level density used in the calculation.

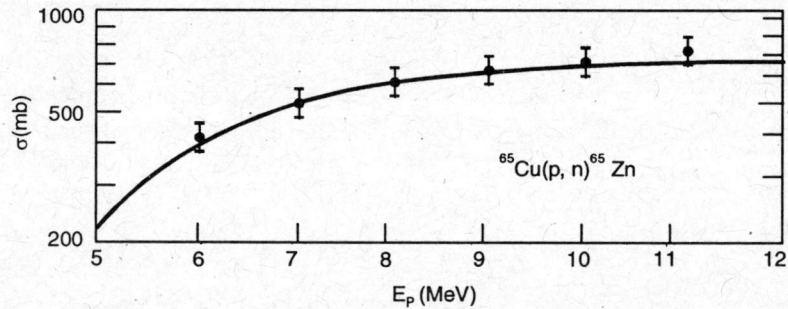

Fig. 5.12: Predictions of the statistical theory (solid line) compared with experiment

Thus, the statistical character of nuclear levels makes it possible to calculate the energy and angular distributions of particles in the continuum, where the groups of particles corresponding to particular states in the residual nucleus can not be resolved. A typical comparison between measured and calculated energy distribution of particles for the reaction $^{65}Cu(p,n)^{65}Zn$ is shown in Fig. 5.12.

Exercise 5.10. Show that $\rho(E) \sim \exp\left[a^{1/n} \dfrac{n}{n-1} E^{(1-1/n)}\right]$.

Now $dS = \dfrac{dE}{T} = a\, n\, T^{n-2} dT$. Integrating both sides we get

$$S = \frac{n}{n-1} a\, T^{n-1} = a \cdot \frac{n}{n-1} \left(\frac{E}{a}\right)^{\frac{1}{n-1}} = a^{1/n} \frac{n}{n-1} E^{\left(1-\frac{1}{n}\right)}.$$

But $S(E) \sim \log \rho(E)$ or $\rho(E) \sim e^{S(E)}$.

Hence, $\rho(E) \sim \exp\left[a^{1/n} \dfrac{n}{n-1} E^{\left(1-\frac{1}{n}\right)}\right]$.

Exercise 5.11. Find a limit on the significance of the nuclear temperature and show that it becomes more meaningful as the mass number of the nucleus and its excitation energy increase.

The limit may be determined from the thermodynamic relation

$$(\Delta E)^2 = T^2 \frac{\partial U}{\partial T} = T^2 C_V$$

where U is the internal energy and C_V is the heat capacity. Expressing T in energy units,

we get
$$\frac{\Delta T}{T} = \frac{1}{\sqrt{\partial U / \partial T}} = \frac{1}{\sqrt{C_V}}.$$

For a Fermi gas
$$U = \frac{1}{\alpha} AT^2$$

where A is the number of nucleons and α is a constant $(= \pi^2 / 4E_f)$.

Hence,
$$\frac{\Delta T}{T} = \sqrt{\frac{\alpha}{2AT}} = \left(\frac{\alpha}{4AU}\right)^{1/4}.$$

5.4 DIRECT NUCLEAR REACTIONS

From the brief description of the mechanism of direct nuclear reactions (DR) given earlier, we learnt that DR has the following characteristic features:

1. It is usually localized on the nuclear surface.
2. It is a simple one step process and involves only few degrees of freedom.
3. A small amount of angular momentum is transferred by the projectile and angular distribution of reaction product tends to be asymmetric about $\theta = \pi/2$ and be peaked forward. The peaking becomes more marked as the energy is raised and the amount of angular momentum available is increased.
4. Direct reactions are much more selective in the final states they populate, contrary to compound nuclear reactions in which formation of compound nucleus and its decay are independent. A direct reaction will feed a particular channel in a way that depends sensitively on its character. Inelastic scattering excites collective states strongly; one nucleon transfer reaction, for example, (d, p) reaction, strongly populates 'single particle states'; two nucleon transfer reaction, for example, (t, p) reaction deposits two neutrons in a nucleus such that the final states preferentially populated are those with a configuration consisting of a target plus a closely correlated neutron pair. It is thus possible to use direct reactions for testing in detail the models postulated to account nuclear structure properties.

We employ the distorted wave (DW) method for calculating cross section for the reaction of the type $A(a, b) B$ in a DR process. The total wave function is expressed as a sum of the incident wave in the incident channel α_i and the outgoing channels β_f; in this notation α, β describe how the nucleons are distributed in the entrance channel i and exit channels f respectively. $\alpha = \beta$ corresponds to inelastic scattering and $\alpha \neq \beta$ transfer reaction. The physical idea behind DW method is that elastic scattering is dominant and has to be treated fully, while non elastic events can be treated by the first order perturbation theory. We now characterize this mathematically[16-17].

A. Distorted Wave Theory

To develop a formal theory of nuclear scattering and reaction it is convenient to re-express the Schrödinger equation as an integral equation in momentum space. We start considering S. Eq. for scattering from a potential $V(r)$ which falls off more rapidly than $\dfrac{1}{r}$ as r becomes large and write

$$(H_0 + V)\psi_k = E\,\psi_k, \tag{5.72a}$$

where H_0 is the kinetic energy operator, $E = k^2$ and ψ_k is the scattering wave function with as yet unspecified boundary conditions. Let us break ψ_k into two parts

$$\psi_k = \phi_k + \chi_k, \tag{5.72b}$$

where ϕ_k is the solution of free S. Eq. with energy E:

$$H_0\phi_k = E\,\phi_k. \tag{5.73}$$

We adopt normalization condition for the scattering states

$$\langle \psi_k | \psi_k \rangle = \langle \phi_k | \phi_k \rangle = 1 \tag{5.74}$$

Inserting (5.72b) in (5.72a) and using (5.74), we may rewrite S. Eq. as

$$(E - H_0)\chi_k = V\psi_k.$$

The above equation may be solved to yield the Lippmann Schwinger (LS) integral equation for the wave function

$$\psi_k^+ = \phi_k + \lim_{\epsilon \to 0} \frac{1}{E - H_0 + i\epsilon} V\psi_k^+, \tag{5.75}$$

where the addition of $+i\epsilon$, like the separation (5.72b) is carried out to ensure that the operator $(E - H_0)$ is never zero for E real and positive and that the inverse $(E - H_0)^{-1}$ exists. This particular prescription known as "$i\epsilon$ trick", used for avoiding zero in $(E-H)$, is also reflected in the boundary condition satisfied by the eigen function (*see* Appendix B2). For an ingoing scattered wave function $+i\epsilon$ is replaced by $-i\epsilon$.

In a scattering process the transition amplitude T_{fi}, which is related to the probability that a system initially in a state $i(Aa)$ is finally found in the state $f(Bb)$ is

$$T_{fi} = \int \phi_f^*(r)\,V(r)\,\psi_i^+(r)\,dr, \tag{5.76}$$

where ϕ_f is the free particle wave function after the scattering has taken place and ψ_i^+ is the complete scattering wave function in the entrance channel. ψ_i^+ satisfies LS equation and can be written as

$$\psi_i^+(r) = \phi_i(r) + (E - H_0 + i\epsilon)^{-1} V \psi_i^+(r). \qquad (5.77)$$

Similarly for an ingoing scattered wave the boundary condition is

$$\psi_f^- = \phi_f + (E - H_0 - i\epsilon)^{-1} V^* \psi_f^-(r). \qquad (5.78)$$

The interaction potential in (5.76) can be written as $V = V_1 + V_2$ and T-matrix element can also be expressed as a sum of two parts:

$$T_{fi} = \langle \phi_f \mid V \mid \psi_i^+ \rangle = \langle \phi_f \mid V_1 + V_2 \mid \psi_i^+ \rangle. \qquad (5.79)$$

We define the solution χ_f of V_1 alone by

$$\chi_f^- = \phi_f + (E - H_0 - i\epsilon)^{-1} V_1^* \chi_f^-.$$

Then substituting ϕ_f from the above equation in (5.79), we get

$$T_{fi} = \langle \chi_f^- - (E - H_0 - i\epsilon)^{-1} V_1^* \chi_f^- \mid (V_1 + V_2) \mid \psi_i^+ \rangle$$

$$= \langle \chi_f^- \mid V_1 + V_2 \mid \psi_i^+ \rangle - \langle \chi_f^- \mid V_1 (E - H_0 - i\epsilon)^{-1} (V_1 + V_2) \mid \psi_i \rangle, \qquad (5.80)$$

since $\qquad T_{fi} = \langle \chi_f^- - (E - H_0 - i\epsilon)^{-1} V_1^* \chi_f^- \mid (V_1 + V_2) \mid \psi_i^+ \rangle.$

From Eq. 5.77, we identify that

$$(E - H_0 - i\epsilon)^{-1} \left| V_1 + V_2 \right| \psi_i^+ = \psi_i^+ - \phi_i$$

Substituting this in (5.80), We finally obtain

$$T_{fi} = \langle \chi_f^- \mid V_1 \mid \psi_i^+ \rangle + \langle \chi_f^- \mid V_2 \mid \psi_i^+ \rangle - \langle \chi_f^- \mid V_1 \mid \psi_i^+ \rangle + \langle \chi_f^- \mid V_1 \mid \phi_i \rangle$$

$$= \langle \chi_f^- \mid V_1 \mid \phi_i \rangle + \langle \chi_f^- \mid V_2 \mid \psi_i^+ \rangle, \text{ which is exact.}$$

We are considering the reaction in which the particles are incident in the channel i and outgoing in the channel f as well as the incident channel 'i'. In other words the outgoing channels 'f' contain elastic scattering as well. Now if amongst the potential V_1 and V_2, the potential V_1 has only diagonal elements non-zero, i.e. V_1 causes elastic scattering only then

$$T_{fi} = \langle \chi_f^- \mid V_1 \mid \phi_i^+ \rangle \, \delta_{fi} + \langle \chi_f^- \mid V_2 \mid \psi_i^+ \rangle.$$

For the non-diagonal elements only, $T_{fi} = \langle \chi_f^- \mid V_2 \mid \psi_i^+ \rangle.$ \qquad (5.81)

If V_2 is small, (5.81) can be given to first order in V_2 by replacing the full solution ψ_i^+ of V by χ_i^+, which is a solution of V_1 alone. This gives the T-matrix element as,

$$T_{fi} = \langle \chi_f^- | V_2 | \chi_i^+ \rangle. \tag{5.82}$$

This is called the Distorted Wave Born Approximation (DWBA) to the transition matrix T_{fi}. As we see, it is the matrix element of the potential V_2 between elastically scattered wave function in the initial and final states. Physically it means that the transition $i \rightarrow f$ dominantly proceeds through elastic channel and the potential V_2 is not so strong.

Cross-sections for Rearrangement Collisions: We will consider collisions in which outgoing systems are not identical with incoming systems. We consider the projectile $a \equiv x + b$ where x, which could be a single nucleon or a cluster, is being transferred to the target to form the residual nucleus $B (= A + x)$. If x is a cluster of nucleon then it is still treated as a single well defined entity such that interaction involved depends only on the position of the centre of mass of cluster and not on its internal coordinates.

In the rearrangement collision, the unperturbed Hamiltonian for the final state, differs from that of the initial state, when the fragments are well separated. In the rearrangement collision $a (= x + b) + A \rightarrow B (= A + x) + b$, the total Hamiltonian $H_i = H_f$. To write the Hamiltonians explicitly we choose the coordinates as shown in Fig. 5.13.

$$\vec{r}_{xA} = \vec{r}_i + \frac{M_b}{M_a} \vec{r}_{xb},$$

$$\vec{r}_f = \frac{M_A}{M_B} \vec{r}_{xA} - \vec{r}_{xb} = \frac{M_A}{M_B} \vec{r}_i - \left[\frac{M_x (M_a + M_A)}{M_a M_B} \right] \vec{r}_{xb}. \tag{5.83}$$

Let us treat x as a definite entity having spin j_x even if it contains more than one nucleon. Suppose x within 'a' has an orbital angular momentum ℓ relative to 'b' and total angular momentum $\vec{j}$ where $\vec{j} = \vec{\ell} + \vec{j}_x$ and $\vec{I}_a = \vec{I}_b + \vec{j}$. If x within the final nucleus B has angular momentum L relative to A and total angular momentum $\vec{J} (= \vec{L} + \vec{j})$ then $\vec{I}_B = \vec{I}_A + \vec{J}$. Then the transfer of angular momentum

$$\vec{J}_{BA} = \vec{I}_B - \vec{I}_A = \vec{J}, \quad \vec{j}_{ba} = \vec{I}_a - \vec{I}_b = \vec{j}$$

and

$$(L - \ell) \leq L \leq L + \ell.$$

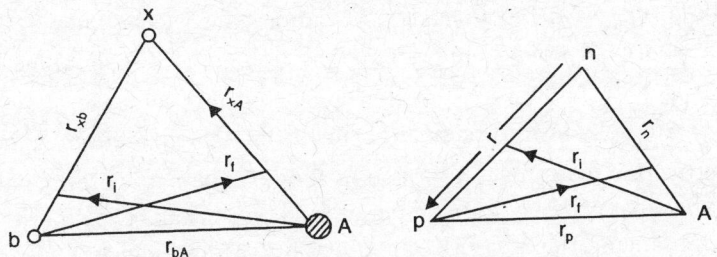

Fig. 5.13: Coordinates used for transfer reaction

Provided the particle x has even intrinsic parity, the change in internal constituent is governed by $\pi_a \pi_b = (-)^\ell$ and $\pi_A \pi_B = (-)^L$. So the overall change $\Delta \pi = \pi_A \pi_B \pi_a \pi_b = (-)^{L+\ell}$. The total Hamiltonian of the system

$$H = \epsilon_a + \epsilon_{xb} + V_{xb} + V_{xA} + V_{bA} = \epsilon_b + \epsilon_x + V_{xb} + V_{xA} + V_{bA},$$

where the ϵ's are the eigen values of a subsystem. The initial and final state Hamiltonians are

$$H_i = \epsilon_a + \epsilon_{xb} + V_{xb}(\vec{r}_{xb}) + U_{aA}(\vec{r}_i)$$

and
$$H_f = \epsilon_b + \epsilon_x + V_{xA}(\vec{r}_{xA}) + U_{b(A+x)}(\vec{r}_f), \tag{5.84}$$

where U's are optical potentials. The Hamiltonian H_{fi} can be written in either prior or post form:

$$H_{fi}(\text{Prior}) = V_{xA} + V_{bA} - U_{aA}(\vec{r}_i) \tag{5.85a}$$

and
$$H_{fi}(\text{Post}) = V_{xb} + V_{bA} - U_{b(A+x)}(\vec{r}_f) \tag{5.85b}$$

Since, interaction between b and A, i.e. V_{bA} is almost equal to the real part of the optical potential between emergent particle b and the residual nucleus $(A + x)$ then we can set $V_{bA} - U_{b(A+x)} \approx 0$, so that

$$T_{fi}(\text{Post}) = \int \psi_f^* V_{xb} \psi_i \, d\tau. \tag{5.86}$$

We employ DWBA theory, which starts with a first order interaction between a probe and a target nucleus giving a factorized amplitude

$$[T(Aa \rightarrow Bb)]_{DWBA} = \int d^3 r_i \, d^3 r_f \, \psi_b^{(-)}(\vec{k}_f, \vec{r}_f) \, \phi_b^*$$

$$\times \langle \psi_B | V_{bx} | \psi_A \rangle \, \phi_a \, \psi_a^{(+)}(\vec{k}_i, \vec{r}_i), \tag{5.87}$$

where $\phi_a(\phi_b)$ is the eigen function of the internal motion of a (b) and $\psi_a(\psi_b)$ is the eigen function for the motion of c.m of a (b) with respect to the c.m of $A(B)$. $+(-)$ denotes outgoing (incoming) scattering waves.

If the probe wave functions $\psi_a^{(+)}\phi_a$ are assumed to be known, then the measurement of T provides information through Eq. (5.87) about the relation between ψ_A, the target wave function and the residual nucleus wave function ψ_B. For example, in studies of single nucleon transfer reaction by measurement of overlap we can get information about single particle orbits in A and B. Theories that go beyond first order depend on (Aa) and (Bb) for more than the overlap integral $\langle \psi_b \ \psi_B \ |V| \ \psi_a \ \psi_A \rangle \neq 0$. This introduces a new feature; the initial and final channels in a rearrangement collision are not orthogonal other than asymptotically $\langle \psi_B \ \psi_b \ | \ \psi_A \ \psi_a \rangle \neq 0$. For example, stripping by deformed nuclei has both transfer steps and rotational excitation steps. Hence, the excitation of many members of rotational band of a deformed nucleus is to be treated in a coupled channel (CC) frame work. Finally, the internal degrees of freedom of the projectile play an important role when the projectile is fragile, say in the case of deuteron or lithium.

The single step nature of the process $A(a,b)B$ is indicated by the linear dependence of the matrix element (5.87) on the potential. This is a perturbation theory where the perturbing interaction is given by $\langle B|V_{bx}|A\rangle$. The amplitude (7.86) is an approximation to the more precise amplitude to be obtained by solving coupled equations. The amplitude (5.87) leaves out possibly important physical processes. For example, the target nucleus and/or the residual nucleus may be excited, permitting the reaction to the final state to proceed by several interfering routes as indicated in Fig. (5.14).

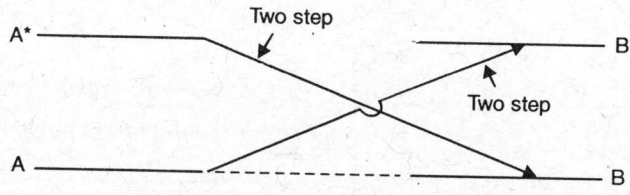

Fig. 5.14: One and two step processes

In addition to the one step processes, there are several two step amplitudes. In one, the target nucleus is excited and then makes the transition to the final state of the residual nucleus; in another, the transition is from initial target nucleus to the excited state of the residual nucleus which in the second step of the process is deexcited.

Deuteron stripping reaction

The deuteron stripping reaction is a fancy way of putting a neutron in the shell model orbit of a nucleus. Therefore, it confirms the shell model description of nuclei and provides a direct measure of the single particle character of nuclear states. We shall employ Eq. (5.22b) to evaluate cross section for (d, p) reaction. However, the method can be applied to any single nucleon transfer reaction with slight modification. The cross section for inverse reaction, namely the pick up reaction can also be evaluated in a similar manner. The cross section for (d, p) reaction (5.22b) is

$$\frac{d\sigma(d,p)}{d\Omega} = \frac{M_d M_p}{(2\pi\hbar^2)^2} \frac{k_p}{k_d} \frac{1}{(2s_d + 1)(2I_A + 1)} \sum_{\mu_B \mu_p} |T_{d-p}|^2.$$

Following Eq. (5.87) we can write explicitly

$$(T_{d-p})^{DWBA} = \sqrt{n+1} \int \psi_p^{(-)*}(\vec{k}_p, \vec{r}_p) \chi_{1/2}^{\mu_p^*} \chi_{I_B}^{\mu_B^*}(\xi, \vec{r}_n, \vec{\sigma}_n)$$

$$\times V_{np}(\vec{r}) \psi_d^{(+)}(\vec{k}_d, \vec{r}_d) \phi_d(\vec{r}) \chi_1^{\mu_d} \psi_{I_A}^{\mu_A}(\xi) d\xi d\vec{r}_n d\vec{r}_p. \quad (5.88)$$

Here $\vec{r} = |\vec{r}_n - \vec{r}_p|$ and $\psi_p^{(-)}$, $\psi_d^{(+)}$ are respectively outgoing proton wave and incident deuteron wave, which are distorted by optical potential which may not be unique. r_d is the coordinate of deuteron c.m from the target, r_p is the proton distance from the residual nucleus and χ's are the spin functions. $\psi_{I_A}^{M_A}$ and $\psi_{I_B}^{M_B}$ are respectively target and residual nucleus internal wave functions. The ξ's collectively denote all the coordinates of the target, which remain unchanged during reaction process. $\phi_d(\vec{r})$ is the radial functions of the deuteron which is assumed to be in a triplet s-state. The first factor $\sqrt{n+1}$ comes from antisymmetrization where n represents the number of neutrons in the target nucleus, which are taken into account in the calculation. We expand the wave function of the residual nucleus (target + neutron) in terms of the wave function of the target nucleus and a set of bound neutron wave functions:

$$\psi_{I_B}^{\mu_B}(\xi, \vec{r}_n, \vec{\sigma}_n) = \frac{1}{\sqrt{n+1}} \sum_{j\ell} \beta_{j\ell} \left[\psi_{I_A}^{\mu_A}(\xi) \phi_{n\ell j}^{m_j}(\vec{r}_n, \vec{\sigma}_n) \right]_{I_B}^{\mu_B}, \quad (5.89)$$

where $$\left[\psi_{I_A}^{\mu_A}(\xi) \phi_{n\ell j}^{m_j} \right]_{I_B}^{\mu_B} = \sum_{\mu_A m_j} \langle I_A j \mu_A m_j | I_B \mu_B \rangle \psi_{I_A}^{\mu_A} \phi_{n\ell j}^{m_j}$$

and $\beta_{j\ell}$ contains all the nuclear structure informations and the $\langle I_A j \mu_A m_j | I_B \mu_B \rangle$ are the C.G. coefficients. $\beta_{j\ell}$ is related to the so called spectroscopic factor $S_j = \beta_{j\ell}^2$.

In the expression (5.89) $\phi_{n\ell j}^{m_j}(\vec{r}_n, \vec{\sigma}_n)$ denotes a spin orbit function of a shell model state. The spectroscopic factor S_j is a measure of the strength with which the ground state of A i.e. $\psi_{I_A}^{\mu_A}$ is present in $\psi_{I_B}^{\mu_B}$. Or S_j is a measure of degree to which $\psi_{I_B}^{\mu_B}$ is a single particle state. For (d, p) reaction, the wave function for some state i in even-even target

$$\psi_A(i) = \left[S_j(i)\right]^{1/2} \psi_{A-1}(gs)\, \psi(j).$$

In $^{48}Ca(d,p)\,^{49}Ca$ reaction if $\ell = 1$ and $j = 3/2$, the configuration is $^{48}Ca(gs) + n(p_{3/2})$. Nucleon transfer goes preferentially to states that exhibit strong pairing correlations and so on. As the ground state of ^{48}Ca is O+, the state of ^{49}Ca formed must be $\dfrac{3}{2}+$. Similarly for (p,d) reaction the wave function for some state i in even-even target is given by

$$\psi_A(i) = \frac{1}{2j+1} \left[S_j(i)\right]^{1/2} \psi_{A-1}(gs)\, \psi(j)^h, \text{ where } h \text{ denotes hole.}$$

Substituting Eqs. (5.89) in Eq. (5.88) and summing over spin and integrating over target coordinates, we get

$$T_{d-p}(I_A\mu_A\mu_d \to I_B\mu_B\mu_p) = \sum_{\ell j} \langle I_A\, j\, \mu_A\, m_j \mid I_B\mu_B \rangle \langle \ell\, \tfrac{1}{2}\, \mu_n \mid j\, m_j \rangle$$

$$\times \langle \tfrac{1}{2}\, \tfrac{1}{2}\, \mu_n\mu_p \mid s_d\mu_d \rangle\, i^\ell\, (2\ell+1)^{1/2}\, B_\ell^m\, S_j^{1/2}, \qquad (5.90a)$$

where the overlap integral

$$B_\ell^m(k_p, k_d) = i^{(-\ell)} (2\ell+1)^{-1/2} \int \psi_p^{(-)}(\vec{r}_p, \vec{r}_p)\, \phi_{n\ell}^m(\vec{r}_n)\, V_{np}(r)$$

$$\times\, \psi_d^{(+)}(\vec{k}_d, \vec{r}_d)\, \phi_d(r)\, d\vec{r}_n\, d\vec{r}_p. \qquad (5.90b)$$

In the above we have used $\int \psi_{I_A}^{\mu_A *}(\xi)\, \psi_{I_A}^{\mu_A}(\xi)\, d\zeta = 1$. Usually one does not measure the spin directions μ_p and μ_B, and also the target and projectile are unoriented. Therefore, the cross section that is usually measured is a sum over all final orientations produced, divided by an average initial orientation:

$$\left(\frac{d\sigma}{d\Omega}\right)^{DWBA} = \frac{M_d M_p}{(2\pi\hbar^2)^2} \frac{k_p}{k_d} \frac{(2I_B+1)}{(2I_A+1)} \sum_{j\ell m} S_j |B_\ell^m|^2. \qquad (5.91)$$

Therefore, the cross section factorizes into two parts: S_j the spectroscopic factor, which is determined entirely by the properties of the nucleus and a factor $|B_\ell^m|^2$, which depends upon the detailed reaction mechanism and contains all kinematic dependence through the wave functions of the relative motion and their overlap with each other. In addition, it contains radial wave function of the transferred neutron. As a condition that the C-coefficients in Eq. (5.90a) should not vanish, the following selection rules on j and ℓ must hold: $I_B = I_A + j = I_A + \ell + \frac{1}{2}$. In addition, parity conditions must hold. The evaluation of $B_\ell^m(k_p, k_d)$ involves a six dimensional integral. To reduce it to three dimension we use the zero range approximation:

$$V_{np}(r)\,\phi_d(r) = D_0\,\delta(r), \tag{5.92}$$

where D_0 is determined for specific form of $V_{np}(r)$ and $\phi_d(r)$ with references to the coordinate Fig. (5.14). This means $r_d = r_n$ and $r_p = \frac{A}{A+1}\,r_n$, i.e. we assume that the neutron is being captured at the same place where the proton is emitted.

Usually the spectroscopic factor is experimentally determined from the relation

$$\left(\frac{d\sigma}{d\Omega}\right)_{\text{expt.}} = \frac{(2I_B+1)}{(2I_A+1)}\,S_{ij}\left(\frac{d\sigma}{d\Omega}\right)_{DWBA},$$

where the spectroscopic factor S_{ij} measures the weight of the configuration j used in the DWBA calculation in the final state i measured experimentally. $\sum_i S_{ij} = n_j$, where n_j is the total number of nucleons in a given configuraiton j and the sum embraces all states i of the product nucleus B.

To get (d, p) cross section in a more quantitative way let us consider Plane Wave Born Approximation (PWBA). The use of plane waves to describe low energy reactions may not be accurate enough but the gross feature like forward peaking is reproduced. Corrections to PWBA theory introduce quantitative changes rather than qualitative changes. If we neglect the interaction of incident deuteron with target, the interaction of outgoing proton with residual nucleus and ignore spin degrees of freedom, then we can write

$$\psi_f = \psi_p^{(-)}(k_p, r_p)\,\phi_{n\ell}^m(r_n),\quad \psi_i = \psi_d^{(+)}(k_d, r_d)\,\phi_d(r),$$

where $\psi_d^{(+)}(k_d, r_d) = e^{i\vec{k}_d \cdot \vec{r}_d} = e^{ik_d(\vec{r}_n + \frac{\vec{r}}{2})}$, $\psi_p^{(-)}(k_p, r_p) = e^{i\vec{k}_p \cdot \vec{r}_p} = e^{ik_p(\frac{A}{A+1}\vec{r}_n + \vec{r})}$,

and $\phi_{n\ell}^m(r_n) = u_{n\ell}(r_n)\,Y_\ell^m(\hat{r}_n)$. Hence,

$$T_{d-p} = \int e^{i(\vec{k}_d - \frac{A}{A+1} k_p)\cdot\vec{r}_n} e^{i(-\frac{\vec{k}_d}{2} + \vec{k}_p)\cdot\vec{r}} \times V_{np}(r)\phi_d(r)u_{n\ell}(r_n)Y_\ell^m(\vec{r}_n)d^3r_n\, d^3r\, .$$

(5.93)

We express the above the six dimensional integral as a product of two integrals (say I_1 and I_2) so that

$$T_{d-p} = \int e^{i\vec{q}\cdot\vec{r}_n} u_{n\ell}(r_n)\, Y_\ell^{m*}(\hat{r}_n)\, d^3r_n \int e^{i\vec{k}\cdot\vec{r}} V_{np}(r)\, \phi_d(r)\, d^3r, \quad (5.94)$$

where $\qquad \vec{q} = |\vec{k}_d - \dfrac{A}{A+1}\vec{k}_p|, \qquad q^2 \equiv k_d^2 + k_p^2 - 2k_d k_p\,\cos\theta$

and $\qquad \vec{k} \cong (\vec{k}_p - \vec{k}_d/2).$

We use the partial wave expansion to evaluate I_1:

$$I_1 = \sum_{\ell'=0}^{\infty} [4\pi(2\ell'+1)]^{1/2}\, i^{\ell'} \int_0^{\infty} u_{n\ell}(r_n)\, j_{\ell'}(q\, r_n)\, r_n^2\, dr_n$$

and choose the recoil direction $\vec{q}$ as the quantization axis so that

$$I_1 = \sum_{\ell'=0}^{\infty} [4\pi(2\ell'+1)]^{1/2} i^{\ell'} \int_0^{\infty} u_{n\ell}(r_n)\, j_{\ell'}(q\, r_n)\, r_n^2\, dr_n \int Y_\ell^{m*}(\hat{r}_n)\, Y_{\ell'}^0(\hat{r}_n)\, d\Omega_n.$$

$$= [4\pi(2\ell+1)]^{1/2} i^{\ell} \int_0^{\infty} u_{n\ell}(r_n)\, j_\ell(q\, r_n)\, r_n^2\, dr_n.$$

(5.95)

In order to evaluate I_1 we employ Butler cut off method[16], in which one neglects the contribution to the integral from nuclear interior such that

$$\int_0^{\infty} u_{n\ell}(r_n)\, j_\ell(qr_n)\, r_n^2\, dr_n \equiv \int_{R_0}^{\infty} u_{n\ell}(r_n)\, j_\ell(qr_n)\, r_n^2\, dr_n.$$

Dropping the index n and integrating by parts we get

$$I_1 = \int_{R_0}^{\infty} u_{n\ell}(r)\, j_\ell(qr)r^2\, dr$$

$$= \frac{R_0^2}{q^2 + \epsilon^2}\left[u_{n\ell}(R_0)\left\{\frac{d}{dr} j_\ell(qr)\right\}_{r=R_0} - j_\ell(qR_0)\left\{\frac{d}{dr} u_{n\ell}(r)\right\}_{r=R_0}\right],$$

$$= \frac{R_0^2}{q^2 + \epsilon^2}\, u_{n\ell}(R_0)\left[\left\{\frac{d}{dr} j_\ell(qr)\right\}_{r=R_0} - \frac{j_\ell(qR_0)}{u_{n\ell}(R_0)}\left\{\frac{d}{dr} u_{n\ell}(r)\right\}_{r=R_0}\right],\ (5.96)$$

provided $u_{n\ell}(r)$ vanishes at infinity, which is true for any bound state wave function. The second integral (I_2) may be simplified by using s-state radial S. Eq. for bound deuteron namely,

$$\left[\frac{-\hbar^2}{2m_d}\frac{d^2}{dr^2}+\frac{-\hbar^2}{2m_d}V_{np}(r)+|B_d|\right]u_d(r)=0, \qquad (5.97)$$

where $-|B_d|$, m_d are the deuteron binding energy and reduced mass respectively and $u_d(r)$ is the deuteron $^3S_1$ wave function such that $\int_{R_0}^{\infty} u_d^2\, dr = 1$. For the first integral, the value of the part $\int_0^{R_0} u_{n\ell}(r_n)\, j_\ell(qr_n)\, r_n^2\, dr_n = 0$. We again drop the index 'n' on neutron coordinate. The function $u_{n\ell}(r)$ and $j_\ell(qr)$ satisfy the following S.Eqs.:

$$\left[-\frac{d^2}{dr^2}+\frac{\ell(\ell+1)}{r^2}-q^2\right]r\, j_\ell(qr)=0 \qquad (5.98\text{a})$$

$$\left[-\frac{d^2}{dr^2}+\frac{\ell(\ell+1)}{r^2}\, \epsilon^2\right]r\, u_{n\ell}(r)=0, \qquad (5.98\text{b})$$

where $\epsilon^2 = \dfrac{2m_n|B_n|}{\hbar^2}$, $|B_n|=$ the last neutron binding energy. Multiplying Eq. (5.98a) by $r\, u_{n\ell}(r)$ and Eq. (5.98b) by $r\, j_\ell(qr)$ from left and then subtracting we get

$$-(q^2+\epsilon^2)\, r^2\, u_{n\ell}(r)\, j_\ell(qr)$$

$$=\frac{d}{dr}\left[r\, u_{n\ell}(r)\frac{d}{dr}r\, (j_\ell(qr)-r)\, j_\ell(qr)\frac{d}{dr}r\, u_{n\ell}(r)\right].$$

Then

$$I_2 = \int e^{i\vec{k}\cdot\vec{r}}\, V_{np}(r)\, \phi_d(r)\, d^3r$$

$$=\int\left\{\frac{\hbar^2}{2m_d}\frac{d^2}{dr^2}-|B_d|\right\}e^{i\vec{k}\cdot\vec{r}}u_d(r)\, dr$$

$$=\left[-\frac{\hbar^2 k^2}{2m_d}-|B_d|\right]\int e^{-i\vec{k}\cdot\vec{r}}\, u_d(r)\, dr$$

$$= -\frac{\hbar^2}{2m_d}(k^2 + \alpha^2) \int e^{-i\vec{k}\cdot\vec{r}} u_d(r)\, dr = \frac{\hbar^2}{2m_d}(k^2 + \alpha^2)\, P(k), \qquad (5.99)$$

where $\alpha^2 = \dfrac{2m_d \, |B_d|}{\hbar^2}$ and $P(k) = \int e^{-i\vec{k}\cdot\vec{r}} u_d(r)\, dr$.

$P(k)$ represents the probability that the momentum k is to be found in deuteron and it depends on the specific form of the deuteron wave function. The factor $P(k)$ damps the cross section at large angles because deuteron does not contain large momentum components.

If we use the fact that a normalized Hankel function of the first kind $h_\ell(i \in r)$ ($\in$ being the last neutron binding energy), satisfies the boundary condition for bound state wave function that it vanishes at infinity, then we replace $u_{n\ell}(R_0)$ by $h_\ell(i \in R_0)$ so that

$$\int_{R_0}^{\infty} u_{n\ell}(r)\, j_\ell(qr)\, r^2\, dr = \frac{R_0\, u_{n\ell}(R_0)}{q^2 + \in^2} W_\ell(qR_0),$$

$$W_\ell(qR_0) = R_0 \left[\{ j_\ell'(qr) \}_{r=R_0} - \left\{ \frac{j_\ell(qR_0)}{h_\ell(i \in R_0)}\, h_\ell'(i \in r) \right\}_{r=R_0} \right] \text{ and}$$

$$\left(\frac{d\sigma(d,p)}{d\Omega} \right)_{PWBA} = \frac{4\pi M_d M_p}{(2\pi\hbar^2)^2}\, \frac{k_p}{k_d}\, \frac{\hbar^4}{4m_d^2}\, (k^2 + \alpha^2)^2\, P^2(k)\, (2\ell + 1)$$

$$\frac{R_0^2\, u_{n\ell}^2(R_0)}{(q^2 + \in^2)^2}\, W_\ell^2$$

$$= \frac{3}{2\pi}\, \frac{(A+1)^3}{2A(A+2)^2}\, \frac{k_p}{k_d}\, P^2(k)\, \frac{1}{R_0}\, \theta_0^2(\ell)\, (2\ell + 1)\, W_\ell^2, \qquad (5.100)$$

where $\theta_0(\ell) = \dfrac{1}{3} R_0^3\, u_{n\ell}^2(R_0)$. The wave function $u_{n\ell}(R_0)$, which describes the formation of target neutron system is estimated by the knowledge of single particle potential. The Butler cut off radius R_0 is used as an adjustable parameter. Generally, R_0 extends beyond nuclear radius showing the inadequacy of PWBA theory. The expression W_ℓ^2 contains Bessel and logarithmic derivative of Hankel function and gives the angular distributions of outgoing protons, it has the properties of damped oscillations passing through several maxima and minima of diminishing amplitude corresponding to different values of ℓ. This can be seen from the properties of Bessel and Hankel functions. The first peak of $(j_\ell(qR_0)]^2$ matched the most forward peak of the measured angular distribution. In Fig. 5.15 we compare DWBA and PWBA predicted differential cross sections.

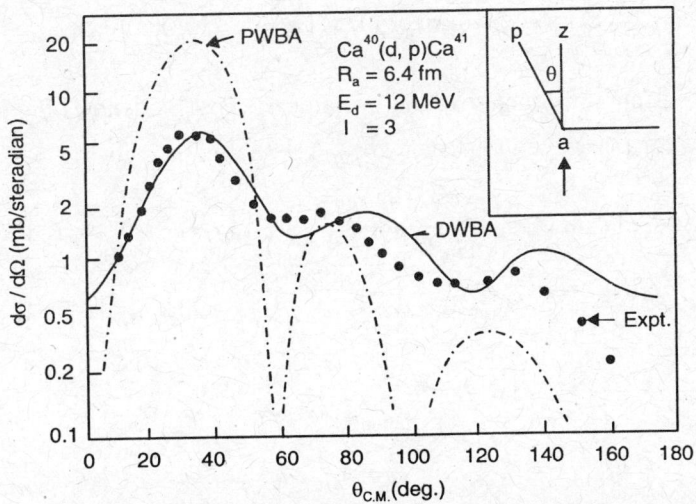

Fig. 5.15: Comparison of PWBA and DWBA differential cross sections for *Ca⁴⁰(d, p) Ca⁴¹* reaction. Inset shows the direction of the emitted proton (From Glandenning Ref. 29)

B. Evaluation of Spectroscopic Factor

The spectroscopic factor S_j is a direct measure of the single particle purity of the final (or initial) state, if $S_j = 1$ then the state is pure single particle while a value less than unity indicates that the single particle strength has been spread over two or more states. The measurement of a stripping or pickup cross section together with a calculation of B_ℓ^m, enables the spectroscopic factor to be determined. As we might expect, the purest single particle states are found in nuclei consisting of closed shells plus or minus one nucleon. As soon as the nucleus contains more than one particle outside closed shells, these particles interact together, and the resulting residual interactions perturb and split the single particle states so that their strength spread over several levels.

We can rewrite Eq. (5.91) for $I_A = 0$ and $I_B = j$ as

$$\left(\frac{d\sigma(d,p)}{d\Omega}\right)_{DWBA} = (2j+1)\,S_j\,P, \qquad (5.101)$$

where P is the nuclear reaction mechanism factor which contains the kinematics.

The spectroscopic factor[18] written in terms of wave functions, can also be written as a reduced matrix element of a particle creation operator:

$$\beta_{j\ell} = \langle I_B M_B | A | \left[I_A M_A \rangle | j m_j \rangle \right]_{I_B}^{M_B}$$

$$= \sum_{M_A m_j} \langle I_A j M_A m_j I_A M_B \rangle | a_{jm_j}^\dagger | I_A M_A \rangle,$$

where A is an antisymmetrization operator. Here $a_{jm_j}^\dagger$ denotes the creation operator for a particle in a state $| j m_j \rangle$. The matrix element

$$\langle I_B M_B | a_{jm_j}^\dagger | I_A M_A \rangle = \frac{(-)^{j+I_A-M_B}}{[I_A]^{1/2}} \langle I_B j - M_B m_j | I_A - M_A \rangle \langle I_B \| a_j^\dagger \| I_A \rangle.$$

Hence, $\beta_{j\ell} = \sum_{M_A m_j} \langle I_A j M_A m_j | I_B M_B \rangle \langle I_B j - M_B m_j \rangle | I_A - M_A \rangle$

$$\times \frac{(-)^{j+I_A-M_B}}{[I_A]^{1/2}} \langle I_B \| a_j^\dagger \| I_A \rangle$$

$$= \sum_{M_A m_j} \langle I_A j M_A m_j | I_B M_B \rangle \langle I_A j M_A m_j \rangle | I_B - M_B \rangle \frac{[I_A]^{1/2}}{[I_B]^{1/2}}$$

$$\times \frac{(-)^{j+m_j+j+I_A-M_B}}{[I_A]^{1/2}} \langle I_B \| a_j^\dagger \| I_A \rangle = \frac{(-)^{2j}}{[I_B]^{1/2}} \langle I_B \| a_j^\dagger \| I_A \rangle.$$

The partial filling of states near the Fermi level is one of the most important consequences of the pairing force. In the independent quasi-particle model the form of the spectroscopic factor is simply

$$S_j \begin{cases} = U_j^2 & \text{for } (d,p) \text{ reaction} \\ = V_j^2 & \text{for } (d,t) \text{ reaction} \end{cases}$$

That is, occupancy of a level is studied by pick up (d, t) reaction and its emptiness by (d, p) stripping reaction. Here we describe an experiment which directly measure U_j^2 and V_j^2. The occupation probability in the following example is obtained from $^{116}Sn(d,p)$ ^{117}Sn and $^{116}Sn(d,t)$ ^{115}Sn to the $d_{3/2}$ and $d_{5/2}$ states of ^{117}Sn and ^{115}Sn respectively. According to the relation between cross section $\dfrac{d\sigma(d,p)}{d\Omega}$ and spectroscopic factor S_j for transition from $j^n(v=0)$ configuration to $j^{n+1}(v=1)$, the occupation number $n_j = (2j+1)V_j^2$. Hence, $S_j = 1 - V_j^2 = U_j^2$. Accordingly $\dfrac{d\sigma(d,p)}{d\Omega} = (2j+1)$

PU_j^2 for even target. For (d, t) reaction on the same nucleus, we consider transition from j^n configuration to j^{n-1} configuration (n = even) so that

$$\frac{d\sigma(d,t)}{d\Omega} = (2j+1)TV_j^2.$$

Cohen and Price[30] have performed (d, p) and (d, t) reactions on a number of isotopes of Sn. The proton member is magic ($Z = 50$) and neutrons are smeared over Fermi levels namely $2d_{5/2}$, $1g_{7/2}$, $3s_{1/2}$ and $2d_{3/2}$ etc. For reactions under consideration the ratio of (d, p) and (d, t) cross sections to $d_{3/2}$ and $d_{5/2}$ states are:

$$\frac{[\sigma(d,p)]_{3/2}}{[\sigma(d,p)]_{5/2}} = \frac{4U_{3/2}^2}{6U_{5/2}^2} = a \text{ (say)} \quad \text{and} \quad \frac{[\sigma(d,t)]_{3/2}}{[\sigma(d,t)]_{5/2}} = \frac{4V_{3/2}^2}{6V_{5/2}^2} = b \text{ (say)}.$$

Employing $U_j^2 + V_j^2 = 1$, we get

$$V_{3/2}^2 = \left(\frac{3}{2}a - 1\right)(a - b) \quad \text{and} \quad V_{5/2}^2 = \left(a - \frac{2}{3}\right)(a - b).$$

Table (5.2) shows the values of occupation probability V_j^2 for ^{116}Sn.

Table 5.2: Values of V_j^2 for ^{116}Sn

State	Expt.	Theo.
$2d_{5/2}$	0.79	0.93
$2d_{3/2}$	0.25	0.25

These formulae assume reaction factor $P = T$, which is not quite correct because P and T depend on the Q-value which differs.

Exercise 5.13. Show that in the (d, p) reaction the captured neutron wave function must decay asymptotically with decay constant determined by the neutron binding energy.

The Hamiltonian for final state B in stripping reaction is

$$H(N) = \sum_{i=1}^{N} T_i + \sum_{i<j}^{N} V_{ij} = H(N-1) + T_N + \sum_{i=1}^{N-1} V_{iN},$$

where T_i is the kinetic energy of the ith nucleon V_{ij} are the two-body potentials acting between nucleons. The form factor is of the form $\langle \psi_A \mid \psi_B \rangle$, where integration is implied over N-1 nucleon coordinates of ψ_A and

$$\langle \psi_A \mid H(N) \mid \psi_B \rangle = E_B \langle \psi_A \mid \psi_B \rangle.$$

Using the above expression for $H(N)$

$$\langle \psi_A \mid H(N) \mid \psi_B \rangle = (E_B + T_N) \langle \psi_A \mid \psi_B \rangle + \langle \psi_A \mid \sum_{i=1}^{N-1} V_{iN-1} \mid \psi_B \rangle$$

Asymptotically, the second term in the above equation approaches zero, and we can write $(E_B + T_N) \langle \psi_A \mid \psi_B \rangle \sim T_N \langle \psi_A \mid \psi_B \rangle$. Now, $E_B - E_A$ is the separation energy B_n and T_N is the kinetic energy of the stripped particle so that

$$-\frac{\hbar^2}{2m_n} \nabla_n^2 \langle \psi_A \mid \psi_B \rangle \sim - B_n \langle \epsilon_A \mid \psi_B \rangle$$

Therefore, $\langle \psi_A \mid \psi_B \rangle \sim \exp \left(-\sqrt{2m_n B_n / \hbar^2} \; r_n \right).$

Exercise 5.14 Calculate the value of zero range parameter D_0 for a square well potential.

According to (5.92) we have

$$-D_0 \int_0^\infty \delta(r) d^3 \vec{r} = \int_0^b V_0 \frac{1}{\sqrt{4\pi}} \sqrt{2\gamma} \frac{\sin \gamma r}{r} r^2 \, dr \, d\Omega$$

or $$D_0 = V_0 \sqrt{8\pi\gamma} \int_0^b r \sin \gamma r \, dr = \frac{V_0 \sqrt{8\pi\gamma}}{\gamma^2}.$$

But $\gamma^2 = \frac{M}{\hbar^2}(V_0 - \epsilon_d) \approx \frac{MV_0}{\hbar^2}$, where M is the nucleon mass and V_0 is the depth of the square well potential in MeV. Hence,

$$D_0 = \frac{\hbar^2}{M} (8\pi\gamma)^{1/2} = 1.1 \times 10^2 \text{ MeV fm}^{3/2} \quad \text{for } V_0 = 32 \text{ MeV}.$$

Since deuteron is a loosely bound system and stays most of the time outside the range of the potential b, we use $\phi_d(r)$ from (2.38). Substituting the of $\alpha = 0.2317$ fm^{-1} we get $D_0 = 10^2$ MeV fm$^{3/2}$.

Exercise 5.15 For spin zero target $(I_A = 0)$, show that the spectroscopic factor $S_j = 1 - n_j / (2j+1)$, where n_j is the number neutrons in j shell.

We use the anticommutation relations of a_{jm_j} and $a^\dagger_{jm_j}$ namely, $a_{jm_j} a^\dagger_{jm_j} = 1 - a^\dagger_{jm_j} a_{jm_j}$. Now $\langle I_B \| a^\dagger_j \| I_A \rangle = (-)^{2j} [I_B]^{1/2} \beta_{j\ell}$ so that

$$\langle I_B M_B | a^\dagger_{jm_j} \| I_A M_A \rangle$$

$$= \frac{(-)^{j+I_A-M_B}}{[I_A]^{1/2}} \langle I_B\, j - M_B m_j | I_A - M_A \rangle \; [I_B]^{1/2} \, (-)^{2j} \, \beta_{j\ell}.$$

We multiply the above equation by its adjoint and get

$$\frac{[I_B]}{[I_A]} \beta_{j\ell} \langle I_B\, j - M_B m_j | I_A - M_A \rangle^2 = \langle I_A M_A | a_j m_j | \rangle \langle I_B M_B | a^\dagger_{jm_j} \| I_A M_A \rangle.$$

In stripping reaction I_A is the target ground state and I_B is any state in the final nucleus that can be reached by the reaction. We now obtain a rule for sum of spectroscopic factors of these final states. We first sum over M_B and m_j and then sum over I_B

$$\sum_{I_B} \beta^2_{j\ell} \frac{[I_B]}{[I_A]} = \langle I_A M_A | \sum_{m_j} a_{jm_j} a^\dagger_{jm_j} | I_A M_A \rangle.$$

We have used $\displaystyle\sum_{M_B\, m_j} \langle I_B j - M_B m_j | I_A M_A \rangle^2 = 1$ and the closure property

$\displaystyle\sum_{I_B M_B} | I_B M_B \rangle \langle I_B M_B \rangle = 1$ to get the above relation.

Hence, $\displaystyle\sum_{I_B} \frac{[I_B]}{[I_A]} S_j = \langle I_A M_A | \sum_{m_j} (1 - a^\dagger_{jm_j} a_{jm_j}) | I_A M_A \rangle.$

For spin zero target $I_A = 0$; $I_B = j$ so that $(2j+1)S_j = (2j+1) - n_j$

or $$S_j = 1 - \frac{n_j}{(2j+1)},$$

where $n_j = \langle I_A M_A | \displaystyle\sum_{m_j} a^\dagger_{jm_j} a_{jm_j} | I_A M_A \rangle$ is the number of neutrons occupying the shell j in the target ground state.

5.5 COUPLED CHANNEL THEORY

The distorted wave theory as outlined so far is a first order theory, that is, the interaction enters the scattering amplitude only once and the possibility of two step processes are ruled out. This approximation is made twice. First, the potential U_{pB} which enters the scattering amplitude T_{d-p} and determines the distorted wave is assumed to be a central optical potential, which implies that the elastic scattering is the dominant process and the second that the inelastic scattering, which for the deformed nuclei is stronger than the elastic scattering at back angles is included in an average way in the optical model. Core excitation would be still possible, however, if we had not made a second assumption.

A. Inelastic Scattering and Nuclear Collectivity

When the low lying states of the target nucleus have a strong collective nature, it is necessary to include coupling between the excited states and the ground state channels. That is to say coupling between inelastic and elastic channels. We consider the interaction of a neutral spinless particle with a nucleus that can be raised into a series of spinless excited states by the projectile target interaction. We also assume that only elastic and inelastic scattering take place, that is, nucleon transfer or any other reaction channels are neglected. The total Hamiltonian of the system is

$$H = T(r) + V(r, \xi) + H(\xi), \tag{5.102}$$

where $T(r)$ is the kinetic energy operator of the incident particle, $H(x)$ the nuclear Hamiltonian in which x represents all the nuclear coordinates, and $V(\vec{r}, \xi)$ the interaction between the incident particle and the target nucleus. The S. Eq. for the whole system is then

$$\{T(r) + V(\vec{r}, \xi) + H(\xi)\} \psi(\vec{r}, \xi) = E\psi(\vec{r}, \xi). \tag{5.103}$$

The nuclear states are defined by

$$H(\xi) \phi_\alpha(\xi) = \epsilon_\alpha \phi_\alpha(\xi). \tag{5.104}$$

$\phi_\alpha(\xi)$ form a complete orthonormal set, we ignore the breakup states of the target nucleus so that $\psi(\vec{r}, \xi)$ is expanded as

$$\psi(\vec{r}, \xi) = \sum_\alpha \phi_\alpha(\xi) \chi_\alpha(\vec{r}). \tag{5.105}$$

Putting (5.105) into (5.103), using (5.104), multiplying from the left by $\phi_\alpha(\xi)$ and integrating over the nuclear coordinates ξ, we get

$$\int \phi_\alpha^*(\xi) \{T(r) + V(r, \xi) + H(\xi)\} \sum_{\alpha'} \phi_{\alpha'}(\xi) \chi_{\alpha'}(\vec{r}) \, d\xi$$

$$= E \int \phi_\alpha^*(\xi) \sum_{\alpha'} \phi_\alpha(\xi) \chi_{\alpha'}(\vec{r}) d\xi,$$

so that

$$[T(r)+\epsilon_\alpha - E]\chi_\alpha(\vec{r}) = -\sum_{\alpha'}\int\phi_\alpha^*(\xi)\,V(r,\xi)\phi_{\alpha'}(\xi)\,d\xi\chi_{\alpha'}(r)$$

or $\qquad [E-\epsilon_\alpha - T(r)]\chi_\alpha(r) = \sum_{\alpha'=1}^{N} V_{\alpha\alpha'}(r)\,\chi_{\alpha'}(\vec{r})$ \hfill (5.106)

where $V_{\alpha\alpha'}(r) = \int\phi_\alpha^*(\xi)\,V(r,\xi)\,\phi_{\alpha'}(\xi)\,d\xi$ is the coupling potential. To integrate over the angular coordinates in (5.106) we employ the partial wave expansion

$$\chi_\alpha(\vec{r}) = \sum_{LM}\frac{u_\alpha(r)}{r}\,Y_L^M(\theta,\phi).$$ \hfill (5.107)

Substituting the above expansion in Eq. (5.106) and multiplying from left by $Y_L^{M*}(\theta,\phi)$ and integrating over angles, we get

$$\left\{\frac{d^2}{dr^2} - \frac{L(L+1)}{r^2} + K_\alpha^2 - W_{\alpha\alpha}(r)\right\}u_\alpha(r) = \sum_{\alpha\neq\alpha'}W_{\alpha\alpha'}(r)u_{\alpha'}(r),$$ \hfill (5.108)

where $\quad W_{\alpha\alpha'}(r) = \frac{2m}{\hbar^2}\sum_{LM}\int Y_{L'}^{M'}(\theta,\phi)V_{\alpha\alpha'}(r)Y_L^M(\theta,\phi)\,d\Omega.$ \hfill (5.109)

Thus, we get a set of coupled equations for the wave functions in the elastic and all the inelastic channels in the reaction being considered. If the potential $V(r,\xi)$ and $H(\xi)$ are known, then these coupled equations can be solved in principle for all the wave functions $u_\alpha(r)$, and hence by application of appropriate boundary conditions the elastic and inelastic cross sections can be calculated for any particular reaction. However, suitable approximations are made to truncate the set of equations by considering only few important channels and the effects due to remaining channels are taken into account by letting interaction potential to be complex[19].

B. Coupling of Reaction Channels

Direct reactions are affected by the structure of not only the target and residual nuclei, as discussed in the previous section but also by the properties of incident or outgoing particles. For example, it is well known that d-state component of deuteron affects (d, p) stripping angular distribution. There is yet another type of structure effects in DR namely, the virtual breakup of weakly bound projectile such as deuteron, ^{6}Li and ^{7}Li etc. The binding energy of deuteron is only 2.2 MeV. The separation energy of deuteron and α in ^{6}Li is only 1.47 MeV or a triton and α in ^{7}Li is 2.48 MeV. Such nuclei can easily be broken up into component fragments really or virtually. In the case of virtual deuteron breakup in which neutron and proton recombine to form deuteron again, it is necessary to do a coupled channel calculation coupling the virtual breakup channel with the incident deuteron or lithium channel.

There are two important reaction channels which are excited with large probability, namely, (1) the deuteron breakup and (2) deuteron stripping. It is especially important to consider breakup effects in determining optical potential for composite projectile scattered by a nucleus because breakup is a dominant source of absorption in the surface region. The deuteron-nucleus scattering system in the present form of 3-body model is assumed to consist of a proton, a neutron and an inert target nucleus. The position vectors of neutron and proton relative to the target nucleus are shown in Fig. 5.13. We ignore spins and assume that the target nucleus is infinitely massive and at rest at the origin.

In this frame of reference, the Hamiltonian H of the deuteron nucleus system takes the general form:

$$H = T(\vec{r}_n) + T(\vec{r}_p) + V_{np}(|\vec{r}_p - U_{p-A}\vec{r}_n|) + U_{p-A}(\vec{r}_p) + U_{n-A}(\vec{r}_n), \quad (5.110)$$

where T's are kinetic energies of nucleons relative to A, $U_{p-A}(\vec{r}_p)$ and $U_{n-A}(\vec{r}_p)$ are respectively proton-nucleus and neutron-nucleus optical potentials, both evaluated at half the incident deuteron energy E_d. We use the relative and c.m. coordinates [Fig. 5.13] with $\vec{r}_p$ and $\vec{r}_n$ replaced by $\vec{r}_p = \vec{R} - \frac{1}{2}\vec{r}$ and $\vec{r}_n = \vec{R} + \frac{1}{2}\vec{r}$. The Hamiltonian is then expressed as

$$H = T(\vec{R}) + H_{np}(\vec{r}) + U_p\left(\vec{R} - \frac{1}{2}\vec{r}\right) + U_n\left(|\vec{R} + \frac{1}{2}\vec{r}|\right) \quad (5.111)$$

$$= T(\vec{R}) + U_N(\vec{R},\vec{r}) + H_{np}(\vec{r}),$$

where $U_N(\vec{R}, \vec{r})$, is the sum of phenomenological neutron-nucleus and proton nucleus energy independent optical potentials given by

$$U_N(\vec{R}, \vec{r}) = U_{n-A}(\vec{r}_n) + U_{p-A}(\vec{r}_p) \quad (5.112)$$

and $T(\vec{R}) = -\dfrac{\hbar^2}{4m}\nabla_R^2$ with $\vec{R}$ denoting the distance between c.m of the deuteron with respect to the target nucleus. These optical potentials are the only input in the calculation. U_{n-A} and U_{p-A} are usually taken from (5.25). The sub-Hamiltonian

$$H_{np}(\vec{r}) = T(\vec{r}) + V_{np}(\vec{r}), \text{ where } T(\vec{r}) = -\frac{\hbar^2}{2\mu}\vec{\nabla}_r^2,$$

with $\mu = m/2$; m being the nucleon mass. V_{np} is usually chosen in the Gaussian form:

$$V_{np} = -v_0 \exp[-r/r_0^2], \text{ with } v_0 = 72.15 \text{ MeV and } r_0 = 1.484 \text{ fm}.$$

We denote the relative motion of deuteron c.m. with respect to the target by $\psi(\vec{R}, \vec{r})$, which satisfies the S. Eq.

$$[E - T(\vec{R}) - H_{np}[(\vec{r}) - U_N(\vec{R}, \vec{r})]\psi(\vec{R}, \vec{r}) = 0. \tag{5.113}$$

In the conventional folding method, one employs the so called Watanabe[20] potential

$$U_N(\vec{R}) = \int_0^\infty U_N(\vec{R}, \vec{r})|\phi_d(\vec{r})|^2 d\vec{r}, \tag{5.114}$$

where $\phi_d(\vec{r})$ is the bound deuteron wave function. Some calculations using the expression (5.114) have been made using Hulthen wave function for the deuteron internal motion and setting the depths of U_p and U_n equal. It is found that the depth of real deuteron potential is nearly the sum of depths of U_p and U_n. However, the range of the deuteron potential was also nearly the same as of the U_p and U_n. The effects of target excitation, deuteron breakup, and in antisymmetrization between nucleons in deuteron and those in the target, are included through an imaginary component and probable alteration of the real component. These modifications are, however, seem to be insufficient to improve the fit to deuteron-nucleus elastic and (d, p) stripping data. The breakup formulation pioneered by Rawitscher[21-22] which is described here, ignores rearrangement channels, the effects due to antisymmetrization between nucleons in deuteron and those in the target so that the scattering wave function of the deuteron-nucleus system can be reduced to only elastic component $\psi_{E\ell}(\vec{r}, \vec{R})$ and the breakup component $\psi_{Br}(\vec{r}, \vec{R})$:

$$\psi_{Br}(\vec{r}, \vec{R}) = \psi_{E\ell}(\vec{r}, \vec{R}) + \psi_{Br}(\vec{r}, \vec{R}). \tag{5.115}$$

The function $\psi_{Br}(\vec{r}, \vec{R})$ describes a broken up deuteron in the exit channel whereas the function $\psi_{E\ell}(\vec{r}, \vec{R})$ describes a bound deuteron in both the entrance and the exit channels. The basic idea in breakup formulation is to expand the three-body wave function $\psi(\vec{r}, \vec{R})$ in terms of the eigen states of n-p Hamiltonian H_{np} as follows:

$$\psi(\vec{r}, \vec{R}) = \phi_d(\vec{r}) f_d(\vec{R}) + \int d\vec{k}\ \phi_{\vec{k}}(\vec{r}) f_{\vec{k}}(\vec{R}), \tag{5.116}$$

where $\qquad H_{np}\,\phi_d(\vec{r})=\epsilon_d\,\phi_d(\vec{r})$ and $H_{np}\,\phi_{\bar{k}}(\vec{r})=\epsilon_k\,\phi_{\bar{k}}(\vec{r}).$

The partial wave function of the relative motion,

$$\phi_d(\vec{r}) \equiv \phi_\ell(k_i,r)\,Y_\ell^m(\hat{r})$$

is truncated to $\ell \le \ell_{max}$ with k_i chosen at n equally spaced points in the interval $0-k_{max}$. Here the index $\vec{k}$ denotes the n-p relative momentum. The eigen functions $\phi_d(\vec{r})$ and $\phi_{\bar{k}}(\vec{r})$ represent respectively a bound deuteron wave function and broken up deuteron wave functions or scattering states of n-p pair. The functions $\phi_{\bar{k}}(\vec{r})$ describe the relative motion between a neutron and proton, and hence having incoming as well as outgoing waves. They are orthogonal to each other and form a complete set:

$$\langle \phi_k(\vec{r}) \mid \phi_{k'}(r) \rangle = \delta(\vec{k}-\vec{k}')\ ,\ \langle \phi_d(\vec{r}) \mid \phi_k(\vec{r}) \rangle = 0$$

and $\qquad \mid \phi_d(r) \rangle \langle \phi_d(r') \mid + \int dk' \mid \phi_{\bar{k}}(r) \rangle \langle \phi_{\bar{k}}(r') \mid = \delta(\vec{r}-\vec{r}').$

The function $f_d(\vec{R})$ is the elastic deuteron c.m wave function defined as the projection of $\psi(\vec{r},\vec{R})$ on $\phi_d(\vec{r})$:

$$f_d(\vec{R}) = \int d\vec{r}\ \phi_d(\vec{r})\phi(\vec{r},\vec{R}), \qquad (5.117)$$

which contains both incoming and outgoing deuteron waves in the asymptotic region. The other projection $f_{\bar{k}}(\vec{R})$ are broken up deuteron c.m wave functions which asymptotically are required to behave as outgoing waves namely,

$$f_{\bar{k}}(\vec{R}) = \int d\vec{r}\ \psi(\vec{r},\vec{R})\,\phi_{\bar{k}}(\vec{r}),$$

Upon substituting Eq. (5.116) into Eq. (5.113), multiplying it respectively with functions $\langle \phi_d(\vec{r}) \mid$ and $\langle \phi_{\bar{k}}(\vec{r}) \mid$ integrating over r, we get a set of coupled integro-differential equations for $f_d(\vec{R})$ and $f_{\bar{k}}(\vec{R})$:

$$[T(\vec{R})-E+\epsilon_d]\,f_d(\vec{R})+V_{dd}(\vec{R})\,f_d(R)+\int V_{dk}(\vec{R})\,f_{\bar{k}}(\vec{R})\ d\vec{k} = 0$$

$$[T(\vec{R})-E+\epsilon_k]\,f_{\bar{k}}(R)+V_{kd}(\vec{R})\,f_d(\vec{R})+\int V_{\bar{k}\bar{k}'}(\vec{R})\,f_{\bar{k}'}(\vec{R})d\vec{k}' = 0.$$

The transition potentials $V_{dd}(\vec{R})$, for bound to bound transition, $V_{kd}(\vec{R})$ for bound to continuum transition and $V_{kk'}(\vec{R})$ for continuum to continuum transition are:

$$V_{dd}(\vec{R}) = \int \phi_d(\vec{r}) U_N(\vec{r},\vec{R}) \phi_d(\vec{r}) d\vec{r}, \qquad V_{\bar{k}d}(\vec{R}) = \int \phi_{\bar{k}}(\vec{r}) U_N(\vec{r},\vec{R}) \phi_d(\vec{r}) d\vec{r}$$

$$V_{\bar{k}\bar{k}'}(\vec{R}) = \phi_{\bar{k}'}(\vec{r}) U_N(\vec{r},\vec{R}) \phi_{\bar{k}}(\vec{r}) d\vec{r}, V_{\bar{k}d}(\vec{R}) = V_{d\bar{k}}(\vec{R}) \text{ and } V_{\bar{k}\bar{k}'}(\vec{R}) = \phi_{\bar{k}'k}(\vec{R}).$$

In the above equations the effective potential $U_N(\vec{r},\vec{R})$ is being folded between relative n-p breakup wave functions which, in the partial wave decomposition contain many relative angular momentum values $\ell = 0, 1, 2, \cdots$. Properties of these transition potentials have been examined by Rawitscher[21-22] and he found that the bound to continuum (V_{kd}) transition matrix elements for relative n-p angular momentum $\ell = 2$ are comparable to those for $\ell = 0$ for values of breakup energy ϵ_k larger than 3 MeV and they make non-negligible contribution to both deuteron elastic scattering and (d, p) cross sections. Including breakup energy and $\ell > 0$ values, he provided generalization of adiabatic calculations of Johnson and Soper[23] and Austern[24]

Effect of breakup on (d, p) stripping

Johnson and Soper[23] were the first to consider the effects of deuteron breakup in the evaluation of (d, p) transition matrix element:

$$T_{d-p} = \int \psi_p^{(-)*}(\vec{r}_p) \phi_n^*(\vec{r}_n) V_{np}(r) \Psi(\vec{r},\vec{R}) d\vec{r} \, d\vec{R} \qquad (5.118)$$

Because $V_{np}(r)$ is short ranged, the relevant part of three body scattering wave function $\psi(\vec{r},\vec{R})$ was approximated by $\psi_0(R)$, a zero range wave function, which essentially means both the replacement of H_{np} by ϵ_d (deuteron binding energy) and $U_N(\vec{r},\vec{R})$ by $\bar{U}(\vec{R})$ in Eq. (5.113). The equation for the zero range function satisfies

$$\psi_0(\vec{R}) = \int V_{np}(\vec{r}) \psi(\vec{r},\vec{R}) d^3r \qquad (5.119)$$

and

$$\left[T(\vec{R}) + \bar{U}(\vec{R}) - E_D \right] \psi_0(\vec{R}) = 0, \qquad (5.120)$$

where $\bar{U}(R)\psi_0(R) = \int_0^\infty V_{np}(r) U_N(\vec{r},\vec{R}) \psi(\vec{r},\vec{R}) d^3r$ and $E_D = E - \epsilon$.

$\psi_0(R)$ is then substituted into the expression (5.118). Though the integration over r space was restricted to small values of r yet results showed marked improvement over DWBA results due to breakup correlations embodied into the wave function $\psi_0(R)$. The "adiabatic approximation", in fact emerge from the above approximation which can be stated by the assumption that deuteron c.m motion prevails over *n-p* relative motion. Figure 5.16 compares the results of C.Ch, JS, and OM analysis[21].

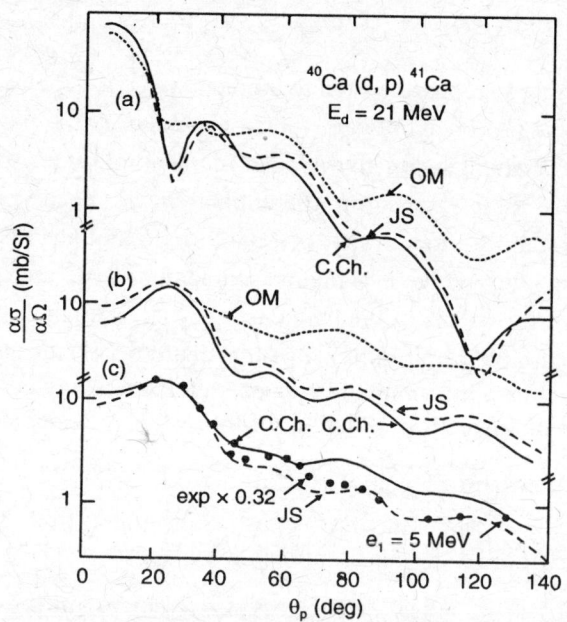

Fig. 5.16: Comparison of predictions of optical model (OM), Johnson-Soper (JS) model (Ref. 23) and Coupled channel (CCh) theory for the angular distribution of $^{40}Ca\ (d,\ p)\ ^{41}Ca$ reaction at $E_d = 21$ MeV (Rawitscher Ref. 21)

Pre-equlibrium reactions: In the recent years, it has become increasingly clear that the combination of direct interaction and compound nucleus theories is not sufficient to account for all available data, and that a more detailed account of nuclear reactions must be sought that includes the possibility of emission after direct stage but long before the attainment of statistical "equilibrium processes". This occurs in a number of steps, with emission taking place from each of them. The physical picture underlying multistep theory is based on interaction between incident nucleon and target nucleus in a number of stages of increasing complexities before the attainment of statistical equilibrium.

We have already discussed that the information on the reaction mechanisms is provided by the excitation functions, the differential cross-section at a particular angle as a function of incident energy. Analysis of some sets of experimental data[25], for example, fluctuation of cross-section of the $^{27}Al(He^3, p)^{29}Si$ reaction for several stages of ^{29}Si for incident energies from 9 to 14 MeV shows that the characteristic widths fell into two classes around 50 and 200 KeV respectively. The presence of widths intermediate between those characteristics of direct reaction and those characteristic of compound nuclear reactions show that some intermediate process takes place, subsequently called as precompound process. Non-Maxwellian shape of the high energy part of particle spectra observed further prove this conjecture.

In the initial stages of precompound process, the composite nucleus retains the memory of the projectile direction as a result of which the emission spectrum from these initial stages is preferentially forward peaked as in the case of direct reactions. However, in the later stages as the number of degree of freedom increases the memory of the projectile direction gets more and more diffused and the ejectile spectrum tends towards isotropy, indicating compound process.

Exiton Model: The pioneering work of Griffin[25], known as semiclassical exiton model, proposed much earlier of these observations, provided a quantitative theory, which could well account for experimental data. The term exciton refer to excited degree of freedom, that is, to the number of excited particles and holes. We define an exciton number n as the sum of p particles and h holes, i.e.

$$n = p + h.$$

We consider the target initially in the ground state so that all levels below the Fermi energy $\in_f$ are filled and all levels above $\in_f$ are vacant. A nucleon projectile with a given energy enters the target nucleus and forms one particle - zero hole $(1p - 0h)$ state with exciton number $n = 1$. At this stage the projectile has entered the target nucleus but is not absorbed by it. It is still in the entrance channel and can leave the nuclear force field without interacting with any individual target nucleon. This corresponds to shape elastic scattering. In order to be removed from the entrance channel and be absorbed, the projectile must interact with an individual target nucleon. Since all levels below $\in_f$ are filled, the first interaction between projectile and target nucleon will raise the latter above $\in_f$ and leave a hole below $\in_f$. Thus a $(2p-1h)$ state is formed. In other words, the absorption of the projectile nucleon by the target leads to the formation of a $n = 3$ exciton state.

After the formation of the $n = 3$ state either of the excited particles may be emitted if it has sufficient energy to escape. If, however, particle emission does not occurr then there will be further two-body interaction either between one of the two excited particles and a particle below $\in_f$, or between the two excited particles themselves. The first instance results in the formation of $3p-2h$ or $n = 5$ exciton state and the second one will lead either to a new $(2p-1h)$, with different energy configuration of particles and holes or back to the original $n = 1$ exciton state. Thus, a two-body interaction will lead to transitions in which change in exciton number is $\Delta n = \pm 2, 0$. The various

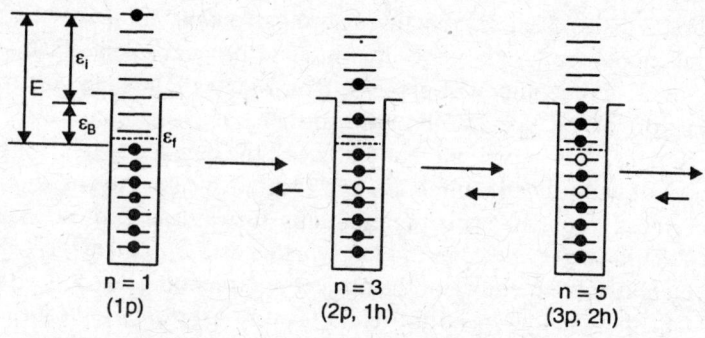

Fig. 5.17: Particle-hole excitation in Exiton Model (*see* text)

stages of exiton model are shown in Fig. (5.17). In this figure, E, n, $\in_f$, $\in_i$, B respectively stand for excitation energy, number of excitations, fermi energy, energy of incoming particle and binding energy. Particles are represented by solid circles and holes are represented by open circles. The length of the horizontal arrows illustrates the preferred direction of transition. During the step by step transition from $n = 3$ to $n = \bar{n}$ state particle emission is possible from every exiton state where a particle is in continuum, i.e. with energy greater than its separation energy. The energy-differential pre-equilibrium cross section $\sigma_{PEQ}(\in)$ is the sum of the cross-sections from each exiton state:

$$\sigma_{PEQ}(\in) = \sigma_{abs} \sum_{\substack{n=n_i \\ \Delta n=2}}^{\bar{n}} D_n P_n(\in), \qquad (5.121)$$

where σ_{abs} is the absorption cross section of the projectile by the target i.e. compound nuclear cross section. D_n is the probability of reaching n exiton state without prior emission. D_n is an overlap integral between the wave functions of the nucleons before and after the interaction leading to emission. D_n is usually called "Depletion" factor. $P_n(\in)$ is the emission probability of the ejectile with energy $\in$ from the n exiton state. The summation starts from initial exiton number n_i which is 3 if the projectile is a nucleon but is different for composite system. Usually, n_i = number of nucleons in the projectile plus 2 (1 excited particle + 1 hole). σ_{abs} is obtained from a global optical potential and P_n is expressed as a product of transmission coefficient and level density.

Feshbach, Kerman and Koonin (FKK)[26] first gave the quantum mechanical theory of pre-equilibrium emission and they divided pre-equilibrium emission spectra into multi-step direct (MSD) in which one particle always remains in the continuum, so that the memory of the incident direction is retained and the cross section is forward peaked and multi-step compound (MSC) components in which all nucleons remain bound. The MSD and MSC components are calculated separately and their sum is the total pre-equilibrium cross-section:

$$\sigma_{PEQ}(\epsilon) = \sigma_{MSD}(\epsilon) + \sigma_{MSC}(\epsilon) \tag{5.122}$$

In the FKK theory $\sigma_{MSD}(\epsilon)$ is calculated quantum mechanically and $\sigma_{PEQ}(\epsilon)$ is obtained by the exciton model (5.121). Then $\sigma_{MSC}(\epsilon)$ is obtained by subtracting $\sigma_{MSD}(\epsilon)$ and $\sigma_{PEQ}(\epsilon)$.

The FKK theoretical formulation which unified the DR and CN reactions recognized that the interaction time was of fundamental importance. A large range in the values of the interaction time separated the direct and compound region. As the interaction time grows the complexity of the nuclear states involved grows. These excitations are generated in many steps and therefore require longer interaction time.

Exercise 5.16. Show that the non-locality of the potential is equivalent to momentum dependence.

Let us consider a non-local potential of the form: $U(r, r') = U_0\, h(|\vec{r} - r'|)$.

The momentum dependent matrix element of $U(r, r')$ is

$$U(k^2) = \langle k \,|\, U(r,r') \,|\, k \rangle = U_0 \int e^{-i\vec{k}\cdot\vec{r}'}\, U(r,r')\, e^{i\vec{k}\cdot\vec{r}}\, dr\, dr',$$

where the plane wave is normalized to unit volume. Substituting $|\vec{r} - \vec{r}'| = s$, we get

$$U_0 \int e^{ik\cdot(|\vec{r}-\vec{r}'|)}\, U(r,r')\, dr\, dr' = U_0 \int e^{i\vec{k}\cdot\vec{s}}\, h(s)\, ds,$$

which is U_0 times the Fourier transform of the non-locality function $h(s)$. If $h(s)$ is chosen to be Gaussian with non-locality range b:

$$h(s) = (\pi\beta^2)^{-3/2}\, e^{-s^2/\beta^2},$$

which is normalized such that $\int\limits_0^\infty h(s)\, s^2 ds = 1$. Hence, $U(k^2) = U_0\, e^{-\frac{1}{4}\beta^2 k^2}$.

REFERENCES

1. Ilya Prigogine and Isabelle Stengers in 'Order out of Chaos', Flemingo 1985, page 68.
2. J.D. Cockcroft and Walton, Proc. Roy. Soc. (London) A**137**, 229 (1932).
3. E.O. Lawrence and M.S. Levingston, Phys. Rev. **40**, 19 (1932).

4. N. Bohr and F. Kalckar, Kgl. Denske Videnskab Mat. Fys. Medd. **14**, 10 (1937).
   V.F. Weisskopf, Phys. Rev. **52**, 295 (1937); Ibid **57**, 935 (1940).
   W. Hauser and H. Feshbach, Phys. Rev. **87**, 366 (1952).
5. J.R. Oppenheimer and M. Philips, Phys. Rev. **48**, 500 (1935).
6. S.N. Ghosal, Phys. Rev. **80**, 939 (1950).
7. H. Feshbach, C.E. Porter and V.F. Weisskopf, Phys. Rev. **96**, 448 (1954).
8. G. Breit and E.P. Wigner, Phys. Rev. **49**, 519, 642 (1936).
   P.L. Kapur and R. Peierls, Proc. Roy. Soc. (London), **A166**, 277 (1938).
9. E.P. Wigner and L. Eisenbud, Phys. Rev. **72**, 29 (1947).
10. Rama Das and S.N. Mukherjee, Pramana **24**, 715 (1985).
11. J. Halperin et al., Phys. Rev. **C21**, 545 (1980).
12. J.M. Blatt and V.F. Weisskopf in 'Theoretical Nuclear Physics', Springer-Verlag Inc., New York, 1979.
13. T. Ericson, Phys. Rev. Lett. **5**, 430 (1960).
14. H.A. Weidenmuller, in Many Body Aspects of Nuclear Physics, Ed. W.M. Alberico et al.. World Scientific 1987, page 197.
15. L.F. Hansen and R.D. Albert, Phys. Rev. **128**, 291 (1962).
16. S.T. Butler, Phys. Rev. 80, 1095 (1950), Ibid. Nature **166**, 709 (1950).
17. W. Tobockman, Phys. Rev. **94**, 1665 (1954).
18. M.H. Mac-Farlane and J.B. French, Rev. of Mod. Phys. **32**, 567 (1960).
    B.L. Cohen and R.E. Price, Phys. Rev. 121, 1441 (1961).
19. T. Tamura, Rev. Mod. Phys. **37**, 679 (1965); Rev. of Nucl. Sci. **19**, 99 (1969).
    R.J. Asquitto and N.K. Glenndenning, Phys. Rev. **181**, 1396 (1969).
20. S. Watanabe, Nucl. Phys. **8**, 484 (1958).
21. G.H. Rawitscher, Phys. Rev. **C11**, 1152 (1975).
22. G.H. Rawitscher and S.N. Mukherjee, Nucl. Phys. **A342**, 90 (1980).
23. R.C. Johnson and P.J.R. Soper, Phys. Rev. **C1**, 976 (1970).
24. N. Austern et al., Physics Reports, **154**, No. 3 (1987).
    G.H. Rawitscher and S.N. Mukherjee, Ann. of Phys. (N.Y.) **68**, 57 (1971).
25. J.J. Griffin, Phys. Rev. Lett. **17**, 478 (1966).
    M. Blann, Ann. Rev. Nucl. Sc. **25**, 123 (1975).
26. H. Feshbach, A. Kerman and S. Koonin, Ann. Phys. N.Y. **125**, 429 (1980).
27. Haro von Butlar in 'Nuclear Physics - An Introduction' Academic Press, New York 1968.

SUGGESTED BOOKS FOR FURTHER READING

28. Direct Nuclear Reactions, N.K. Glendenning, Academic Press, 1980.
29. Nuclear Reactions and Nuclear Structure, P.E. Hodgson, Clarendon Press-Oxford 1971.
30. Nuclear Reaction Theory, H. Feshbach, Wiley, 1996.
31. Direct Nuclear Reaction, G.R. Satchler, Oxford 1983.
32. Nuclear Reaction, Daphne Jackson, Methuen and Co. Ltd. London 1970.

6

Heavy Ion Collisions

"Heavy Ion Science: Gateway to the Unknown".

– D.A. Bromley[1]

6.1 NUCLEI UNDER DIFFERENT STRESS DIMENSION

With the realization that fundamental domain of nuclear physics is virtually inaccessible so long as one is limited to the low energy accelerators, there has been an increase in the maximum energy of the accelerators by a factor of ten every six years. A network of large and impressive heavy-ion and particle accelerator facilities is coming up with important science goal – to provide a new basis for the understanding of new form of nuclear matter and ultimate particle structure.

Nuclei can be stressed in many ways. This is shown in Fig. 6.1, in which three axes denote different "stress dimension". The nuclear ground state corresponds to the origin. The hemisphere centered at the origin denotes the modest stress induced at the low excitation energy and angular momentum for nuclei that are stable (or near stability), which can be studied with interactions of stable accelerated beams and targets. This arena has provided the locus for much of nuclear physics to date, complemented by forays along each axis. Heavy ion projectile fragmentation experiments allow us to study nucleus as a function of temperature. The nuclear excitations at high spin are focused around nuclear structure studies with many discoveries like back bending phenomena and super deformation. The former showing interplay of single particle and collective motions in nuclei and the latter opening up new regions of deformation. The nuclear matter under high temperature and high particle density may lead to the formation of quark gluon plasma (Fig. 6.1 left).

An average estimate of neutron-proton composition of β-stable nuclei can be obtained from (1.40) as

351

$$(N - Z)_{\beta-stable} = A^{5/3} \frac{a_3}{4a_4} \left[1 - A^{-2/3} (m_n c^2 - m_p c^2)/a_3 \right]$$

Substituting a_3 and a_4 from (1.39b) we get

$$(N - Z)_{\beta-stable} = 76 \times 10^{-4} A^{5/3} \left[1 - 1.8 A^{-2/3} \right] \qquad (6.1)$$

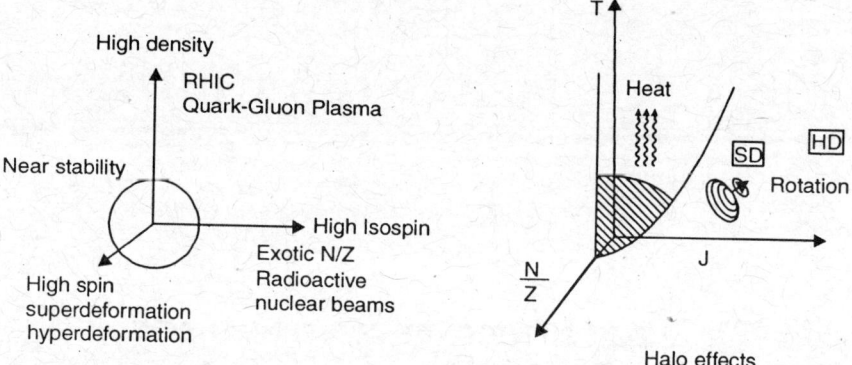

Fig. 6.1: Nuclei under different stress (left) and nuclei under high excitation (T), high spin (J) and high N/Z ratio (right)

For small A, $N \approx Z$, and when A exceeds 40, significant neutron excess is preferred. Eq. (1.41) used here gives an average estimate because shell effects and odd-even effects are ignored.

The valley of β-stability (6.1) represents nuclei of especially high nuclear binding energy. The opposite extreme can also be investigated by use of the mass formula. The nucleon separation energy given by (1.44) can be written as

$$S_n(N \cdot Z) \approx \left| \frac{\partial BE(N,Z)}{\partial N} \right|_Z \quad \text{and} \quad S_p(N \cdot Z) \approx \left| \frac{\partial BE(N,Z)}{\partial Z} \right|_N .$$

By adding more and more nucleons, the separation energy decreases (*see* 1.38) and eventually we reach a point where an extra nucleon is not bound any more. It does not stick to the nucleus; it falls off or "drips" off by itself. This point is characterized by separation energy. The resulting curves in a neutron proton diagram are called "drip lines". Analytic expressions can easily be derived by using (1.38). They are shown in Fig. 6.2. Within the driplines are the stable nuclei with the valley of β-stability in the middle. This is the basis of the familiar nuclear wall chart. The stable nuclei have similar N, Z numbers, and thus extend diagonally across the chart. Territory of nuclear physics is indicated in Fig. 6.2 by the limits of nuclear stability. The proton and neutron drip lines are shown by wavy curves.

Many fascinating new results have come up as we are now able to trace systematically over large changes in neutron number, 20 to 30 neutrons. The properties of both ground, excited collective and single particle states of a given element show interesting behaviour. A new region of deformation with $50 < N < 82$ and $28 < Z < 50$ shows unexpectedly large deformation. Prolate and oblate shape coexistence in Mercury isotopes $^{184}Hg - ^{190}Hg$ has been observed.

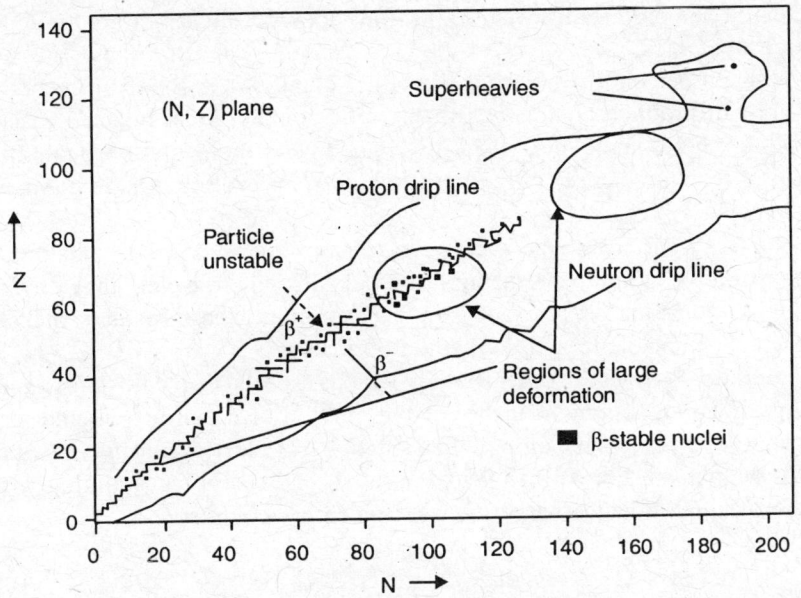

Fig. 6.2: Nuclear Wall Chart (*see* text)

Another important area of research is the extension of our knowledge of the periodic table in the region of actinides beyond the new elements including the theoretically predicted superheavy elements (Fig. 6.2). Such studies are only possible with heavy ions.

Heavy-ion fusion reactions have given us one route out of the valley of stability and go beyond BE/A versus A curve shown in the beginning (Fig. 1.3). Technological break through now allows us to produce "exotic" β-unstable nuclei with abnormal N/Z ratio. A novel structural feature called the neutron halo has been found, which shows that recently measured radii in light neutron rich nuclei are anomalously large compared to their stable counterpart. Near the dripline, bindings are very weak and radii greatly expand. Recently, new-generation facilities of radioactive ion beams in GANIL (France), GSI (Germany), MSU (USA) and RIKEN (Japan) have provided us an opportunity to study the nuclear structure properties of light nuclei far from the stability line and to examine nuclear reactions relevant to nucleosynthesis at various sites of the universe. Historically, every time in past when new nuclei have become accessible, new phenomena and new understanding have emerged.

A. PHYSICS OF HEAVY IONS

Heavy Ion (HI) accelerators have been providing us in the last decade with beams of stable nuclei in the energy range MeV/nucleon to GeV/nucleon. Research has produced a wealth of information on nuclear structure and reaction dynamics.

With proper choice of projectile-target combination, heavy ion induced fusion can produce nuclei far from stability unlike alpha induced reaction which produces nuclei near stability. High angular momentum brought in HI induced reaction can strongly populate high spin levels and large number of channels at a given energy.

The large ion mass, size and electric charge involved in HI collisions exhibit strong Coulomb interaction, short de Broglie wavelength with large orbital angular momentum and strong absorption. These features make HI collisions amenable to simple method of analysis.

Coulomb Barrier: In order to study various aspects of HI collisions, enough energy must be supplied for the relative motion of the projectile and target nuclei to overcome the Coulomb barrier (V_{Coul}) of electrostatic energy, which holds them apart. For a projectile of mass M_P and charge Z_P, colliding with a target of mass M_T and charge Z_T, we have for touching spheres:

$$V_{Coul} = \frac{Z_P Z_T e^2}{r_P + r_T} = \frac{Z_P Z_T}{A_P^{1/3} + A_T^{1/3}} \frac{1.44}{a_0} \text{ MeV,} \qquad (6.2a)$$

where $a_0 \sim (1.3\text{-}1.4)$. Usually $(r_P + r_T)$ is denoted by r_C.

The centre of mass energy of the projectile

$$E_{c.m} = \frac{1}{2} \frac{M_P M_T}{M_P + M_T} (\vec{v}_P - \vec{v}_T)^2 = E_{LAB} \frac{M_T}{M_P + M_T},$$

for the stationary target $(\vec{v}_T = 0)$. Here $\vec{v}_P$ and $\vec{v}_T$ are the projectile and target velocities in laboratory respectively, and the reduced mass $\mu = \dfrac{M_P M_T}{M_P + M_T}$.

The large ion mass makes its wave packet easier to localize. The energy spread ΔE is obtained from the uncertainty relation $\Delta p \cdot \Delta R \cong \hbar/2$ as

$$\Delta E = (\Delta p)^2 / 2M_{ion} = \frac{\hbar^2}{8(Am)\,(\Delta R)^2} \approx \frac{5}{A \cdot (\Delta R)^2}, \qquad (6.2b)$$

where ΔE and ΔR are expressed respectively in MeV and fm.

If the energy spread per nucleon ≈ 0.5 MeV, then the ion can be localized within $\Delta R = (\sqrt{10}/A)$ fm $= 0.2$ fm in the case of oxygen ion ($A = 16$), and this distance is very small compared to the nuclear radius. Hence, the classical concept of impact parameter and trajectory may remain valid for the analysis of HI collisions. Thus, the localization gives the concept of a well defined trajectory. The orbit concept also requires that the uncertainty in the angle of scattering $\Delta\theta$ should be small compared to θ.

The total angular momentum in projectile nucleus c.m. can reach large values:

$$\ell_{max} \approx Ak \times 2r \approx 12A^{4/3}\left(\frac{k}{1 \text{ GeV/c}}\right)\hbar \gg \hbar, \qquad (6.2c)$$

where k is the c.m. momentum per nucleon, A is number of nucleons in the projectile and $r(= 1.2\,A^{1/3})$, the nuclear radius.

The elastic scattering angular distributions in HI collisions show a rich variety of structure, depending on incident energy, masses and charges of interacting nuclei. Our aim is to understand these structures. We consider collisions between two heavy ions at energies well above the Coulomb barrier. At the interaction barrier r_C, the local wave length of the relative motion can be obtained as follows:

$$k = \left[\frac{2\mu}{\hbar^2}\,(E_{LAB} - V_{Coul})\right]^{\frac{1}{2}} = \left[\frac{A_P A_T}{A_P + A_T}\,\frac{2Mc^2}{(\hbar c)^2}\,(E_{LAB} - V_{Coul})\right]^{\frac{1}{2}},$$

so that $\quad \lambda = k^{-1} = \left[\frac{A_P + A_T}{A_P A_T}\,\frac{20 \text{ MeV}}{E_{LAB} - V_{Coul}}\right]^{\frac{1}{2}}$ fm, $\qquad (6.2d)$

where E and V are in MeV.

Another parameter characterizing the Coulomb interaction, is the Sommerfeld parameter n defined by

$$n = (Z_P Z_T e^2 / \hbar v) = (Z_P Z_T e^2 \mu)/\hbar^2 k. \qquad (6.2e)$$

For the collision, $Ar + Th$, at 50 MeV above the interaction barrier, we find $\lambda = 0.1$ fm, which is small compared to the characteristic nuclear length ($r \approx 12$ fm) of colliding nuclei. This again justifies the classical treatment of HI collisions. In Table 6.1 we give characteristic values of the parameters, V_{Coul}, λ, n and r/λ for some heavy ion systems.

Table 6.1: Characteristic parameters of HI system

System	E_{LAB} (MeV)	V_{Coul} (MeV)	λ (fm)	n	r/λ
$_{8}^{16}O \rightarrow _{28}^{58}Ni$	60	36	0.25	24	31
$_{18}^{40}Ar \rightarrow _{20}^{40}Ca$	208	54	0.08	20	111
$_{35}^{79}Br \rightarrow _{82}^{208}Pb$	398	289	0.056	320	235

According to the classical scattering diagram Fig. 6.3, we distinguish qualitatively three different types of collisions. They are determined by the impact parameter b or equivalently the relative angular momentum $\ell = kb$, where k denotes the relative wave number. We define an interaction radius $r(= r_P + r_T)$, which is given by the closest distance of approach between the centres of two nuclei. The grazing collision corresponds to $b = b_g$ and it leads to the nuclear interaction between projectile and target due to which the direct reaction processes dominate.

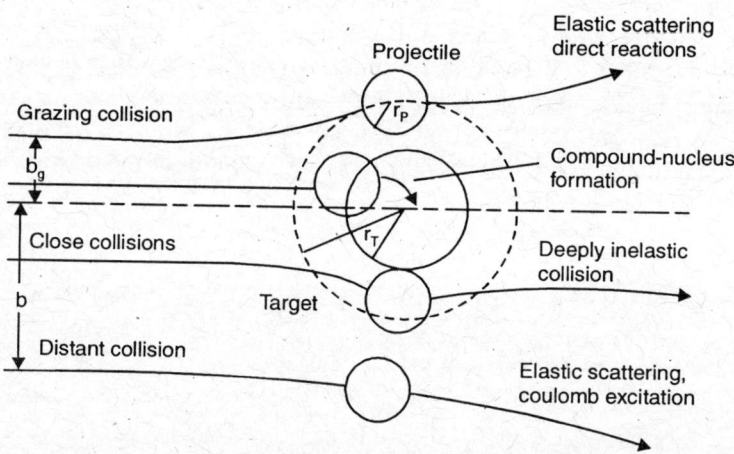

Fig. 6.3: The classical picture of HI collisions. (Adopted from Hodgson Ref. 35)

When $b = b_g$, the nuclear interaction is negligible and we have the pure Coulomb or Rutherford scattering. For $b << b_g$, we have close collision leading to the formation of compound nucleus. In particular, $b = 0$ corresponds to the head on collision. The dependence of the character of reactions induced by HI on impact parameter b is further classified. There are three major categories of the reactions:

(*i*) Fusion, which is the dominant inelastic process for light nuclei and small impact parameter.

(*ii*) Deep inelastic collision, which is characterized by large energy loss, while some of the features of the two-body interaction are preserved. For the small impact parameter deep inelastic collision merges into fusion.

(*iii*) Quasi elastic scattering dominates at large impact parameter where the Coulomb field at the edge of the nuclear field is responsible for the excitation. Excitation involves few MeV of energy loss, populating states near the ground state. The major features of such reactions are determined by the dynamics of two-body interaction via Coulomb plus nuclear potential.

A schematic picture of the angular momentum dependence of the various types of reactions with the corresponding cross section is shown in Fig. 6.4. The decomposition of the reaction cross section with respect to the impact parameter is usually displayed by considering differential cross section with respect to b or with respect to the angular momentum $\ell = kb$, i.e.

$$\frac{d\sigma}{d\ell} = k^{-1}\frac{d\sigma}{db} = k^{-1}(2\pi b), \text{ since } d\sigma = 2\pi b\, db,$$

or
$$\frac{d\sigma}{d\ell} = 2\pi k^{-2}\ell = 2\pi \lambdabar^{2}\ell. \qquad (6.3)$$

Plotting $d\sigma/d\ell$ as function of ℓ, we obtain a straight line as shown in Fig. 6.4. Qualitatively, the reaction cross section under the curve can be divided into three parts: (*i*) compound nuclear reactions, which includes compound

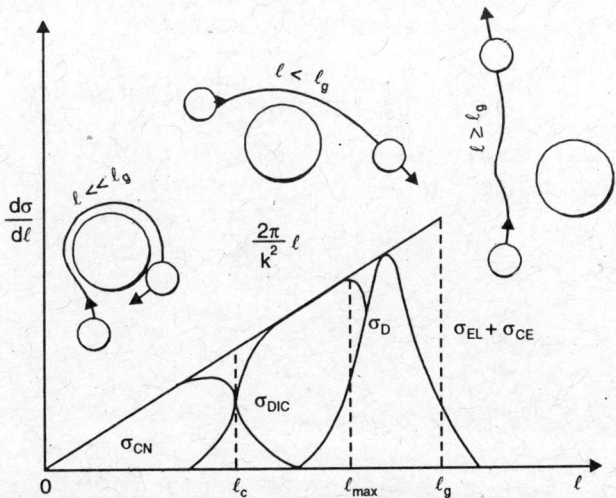

Fig. 6.4: Angular momentum decomposition of reaction cross section ($\ell_g = kb_g$) (Adopted from W. Norenberg, Ref. 2)

evaporation, compound fission and fusion for $\ell = \ell_c$, (*ii*) direct reactions, which includes quasi-elastic scattering for $\ell_c < \ell < \ell_{max}$ and (*iii*) elastic scattering and Coulomb excitation for $\ell > \ell_{max} \leq \ell_g$, where hardly any reaction occurs. In the above classification ℓ_c, ℓ_{max} and ℓ_g respectively denote critical angular momentum, maximum angular momentum and grazing angular momentum.

The effective potential between two ions $V_e(\ell, r)$ is the sum of Coulomb, centrifugal and nuclear potentials (Eq. 5.17):

$$V_e(\ell, r) = V_{Coul}(r) + V_\ell(r) + V_N(r) = V_c(r) + V_\ell(r), \qquad (6.4)$$

where $\qquad V_c(r) = V_{Coul}(r) + V_N(r)$.

The Coulomb term $V_{Coul}(r)$ is repulsive, the nuclear term $V_N(r)$ is attractive with short range and the centrifugal term $V_\ell(r)$ accounts for the increasing difficulty for ions with higher orbital angular momentum ℓ to approach each other. This dependence of the potential on ℓ, gives rise to a series of effective potentials for a typical colliding system (Fig. 5.5). When the potential given by (6.4) is real, the simplest version JWKB method gives a good description of scattering phase shift. However, we present here a simple semi-classical description of the scattering processes.

Employing the conservation of energy and angular momentum, we can write the equation of orbit as

$$\frac{1}{2}\mu\dot{r}^2 + \frac{1}{2}\mu r^2\dot{\phi}^2 + V_c(r) = E_{c.m.}$$

or $\qquad$ $$\dot{r}^2 = \frac{2}{\mu}\left[E_{c.m} - V_c(r) - \frac{\ell^2}{2\mu r^2}\right] \text{ and } \dot{\phi} = \ell/\mu r^2.$$

Dividing we get $\dot{\phi}/\dot{r} = \dfrac{\ell}{r^2} \dfrac{1}{\left[2\mu E_{c.m} - 2\mu V_c(r) - \dfrac{\ell^2}{r^2}\right]^{1/2}}$

or $\qquad$ $$\phi = \int \frac{\ell}{r^2} dr \left[2\mu\{E_{cm} - V_e(\ell, r)\}\right]^{-1/2},$$

where $V_e(l, r)$ is given by (5.17).

$$\theta(\ell) = \pi - 2\int_D^\infty \frac{\ell}{r^2} dr \left[2\mu\{E_{cm} - V_e(\ell, r)\}\right]^{-1/2}, \qquad (6.5)$$

where D is the distance of closest approach.

To get the Coulomb deflection angle, $V_e(\ell, r)$ is replaced by $V_{Coul}(r)$ and the asymptotic Coulomb scattering angle follows from (6.5):

$$\theta_{Coul}(\ell) = 2 \, arc \tan[n/\ell].$$

The differential cross section is obtained from the deflection function as

$$\frac{d\sigma}{d\theta} = 2\pi b \, \frac{db}{d\theta} = \frac{2\pi}{k^2} \frac{\ell \, d\ell}{d\theta} \quad or \quad \frac{d\sigma}{d\Omega} = \frac{b}{\sin\theta} \left| \frac{db}{d\theta} \right|, \qquad (6.6)$$

since $\dfrac{d\sigma}{d\Omega}$ is a +ve quantity, we take the absolute value.

The deflection function diagram (θ versus b) or equivalently (θ versus ℓ) is shown in Fig. 6.5. The deflection function follows $\theta_{Coul}(\ell)$ for large ℓ. At $\ell \cong \ell_{CR}$, the balance between the Coulomb and nuclear forces produces a local maximum in the deflection angle, which is called the Coulomb rainbow angle, denoted by θ_r^C. For small ℓ values, the deflection function develops a second rainbow for $\ell = \ell_{NR}$ and the corresponding angle, called the nuclear rainbow angle, is denoted by θ_r^N. The attractive nuclear potential pulls the trajectory to a maximum negative angle as shown in Fig. 6.5 (left). The Coulomb and the nuclear rainbows are shown in Fig. 6.5 (right) for $^{209}Bi + {}^{136}Xe$ collision. One contrast between light and heavy ion scattering is the prominence of the nuclear rainbow in the former and the Coulomb rainbow in the latter. We emphasize that (6.5) also holds for the dissipative heavy ion collision, in which case a friction force enters in the classical equation of motion.

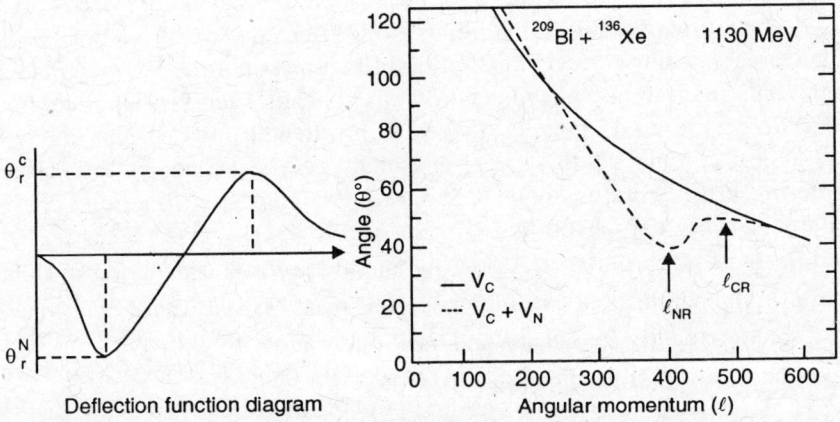

Fig. 6.5: Right: Coulomb rainbow (CR) and nuclear rainbow (NR) Left: Deflection function w.r.t. impact parameter b (Adopted from Schröder and Huizenga Ref. 3)

Elastic scattering determines the potential depth at strong absorption radius. This is defined as the radius of the classical turning point for a trajectory in which half of flux is absorbed. A rough formula for the strong absorption radius determined from the elastic scattering is $r_{sA} = 1.4 \, (A_T^{1/3} + A_P^{1/3})$.

The deflection function diagram discussed above is also useful to categorise types of reactions involved in HI collisions. For large b, θ is +ve because of Coulomb scattering. Any region where it is flat means a large amount of flux will go into small angular range. The strongly damped collision data suggest that it is nearly flat for $b \leq r_{sA} = 1.4 (A_T^{1/3} + A_P^{1/3})$.

B. Diffraction Pattern and Deep Inelastic Collisions

Localization of HI collisions enabled us to describe elastic scattering in terms of classical deflection function, though they are significantly modified due to the quantal and the absorption effects. Because of their mutual interaction both the projectile and target can be excited along their path. Therefore, the elastic cross section is reduced, i.e. we have absorption with respect to the elastic channel. The quantal aspects associated with the absorptive properties of HI interaction due to the presence of many non-elastic channels, manifest themselves in quantum diffraction effects. It is demonstrated here, how effects such as the Fresnel and Fraunhoffer diffractions arise due to the interference of diffracted waves from the target in different domain of HI elastic scattering. The optics of HI scattering is determined by two parameters: n (Sommerfeld parameter) and ℓ_c (critical angular momentum).

For the strongly absorbed particles many features of HI collisions are given by simple diffraction theories. The condition of diffraction in configuration space requires that the wave length of beam should be small compared to the size of the diffracting object. Interference of diffracted waves from both sides of the target nucleus gives a Fraunhoffer diffraction pattern. For large grazing angles this interference phenomenon is very weak. Therefore, Fraunhoffer diffraction is observed only in the scattering of light ions. In the limit of negligible Coulomb field ($n \ll 1$) and high energies ($\ell_c \gg 1$), the elastic scattering cross-sections show features of Fraunhoffer diffraction. These features can be reproduced by a simple partial wave analysis with certain modifications of partial wave scattering amplitude for the scattering of light ions at high energies. At very high energies, many partial waves contribute, δ_ℓ is predominantly imaginary and varies continuously with ℓ. Hence, we make the following modifications of $f(\theta)$:

(1) Replace $\ell + \dfrac{1}{2}$ by ℓ where ℓ is now a continuous variable.

(2) Assume continuous variation of phase shift with ℓ, i.e. replace δ_ℓ by phase shift function $\delta\,(\ell)$.

(3) Replace $P_\ell\,(\cos\theta)$ by an asymptotic form for large ℓ and small θ, i.e. $P_\ell\,(\cos\theta) \cong J_0\{(2\ell+1)\sin\theta/2\} \approx J_0(\ell\theta)$, where J_0 is a Bessel function.

(4) Finally replace summation over ℓ by integration.

Hence, the scattering amplitude $f\,(\theta)$ takes the form:

$$f(\theta) = \frac{1}{ik}\int \ell\,d\ell\,J_0(\ell\theta)\,(e^{2i\delta(\ell)}-1). \tag{6.7}$$

In the sharp cut off approximation, $e^{2i\delta(\ell)}=0$ for $\ell<\ell_c$, there is complete scattering. We evaluate the integral (6.7) to get the diffractive cross section:

$$f(\theta) = -\frac{1}{ik}\int k^2 b\,db\,J_0(kb\theta),\quad \text{where } \ell = kb \text{ and } b \text{ is the impact parameter,}$$

which varies from 0 to $r_0\,(=r_P + r_T)$. Hence,

$$f(\theta) = -\frac{1}{ik\theta^2}\int_0^{r_0}(kb\theta)\,d(kb\theta)\,J_0(kb\theta) = -\frac{1}{ik\theta^2}\int_0^{kr_0\theta} x\,J_0(x)\,dx,$$

where $x = (kb\theta)$. Since $\dfrac{d}{dx}(x\,J_1(x)) = x\,J_0(x)$, we get

$$f(\theta) = -\frac{1}{ik\theta^2}\,kr_0\theta\,J_1(kr_0\theta)$$

or $\qquad \dfrac{d\sigma(\theta)}{d\Omega} = |f(\theta)|^2 = (kr_0^2)^2\left[\dfrac{J_1(kr_0\theta)}{kr_0\theta}\right]^2. \tag{6.8}$

Thus, the diffraction cross section has an oscillatory behaviour with spacings $\Delta\theta = \hbar/kr_0$ since $\Delta\ell = kr_0$. An example of such cross sections is shown in Fig. 6.6. The simple formula given above reproduces peaks in the angular distributions at the correct angles but the oscillations in the theoretical cross sections are much more pronounced than seen in the experimental data. However, an exact fit is obtained by the optical model analysis.

When the Coulomb effects are strong ($n \gg 1$) and in the limit of high energies ($\ell_c \gg 1$), the angular distribution of elastic scattering is Fresnel type as shown in Fig. 6.7 (inset), which has the following characteristic features: an illuminated region for $\theta < \theta_c$ where $\sigma_{el}(\theta) \approx \sigma_{Ruth}(\theta)$ and a shadow region for $\theta > \theta_c$ where $\sigma_{el}(\theta) \approx 0$; indicating a strong absorption from elastic channel. Usually, the rather flat region of $\sigma_{el}(\theta) \approx \sigma_{Ruth}(\theta)$ at small scattering angles is modified by diffraction of the matter waves.

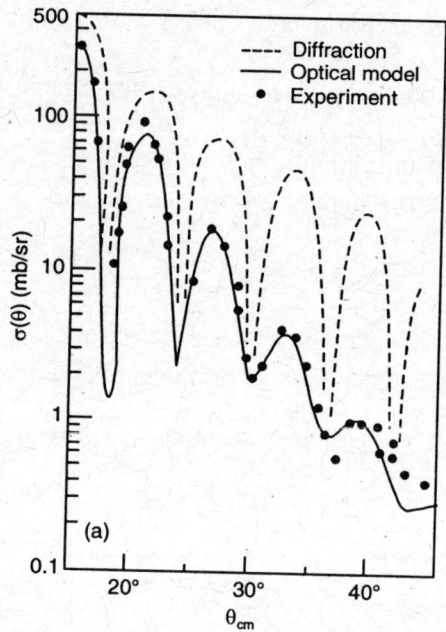

Fig. 6.6: Fraunhoffer diffraction pattern seen in HI elastic scattering Data (Phys. Rev. B134, 1964)

The evaluation of the elastic scattering amplitude is again based on an approximation of partial wave series by an integral over the continuous variable ℓ, in which the elastic scattering S-matrix elements $S_\ell \doteq \eta_\ell\, e^{i2\delta_\ell}$ are interpolated by the continuously differentiable functions $S_\ell = \eta(\ell)e^{i2\delta(\ell)}$, where $\eta(\ell)$ are reflection functions and $\delta(\ell)$ are phase shift functions. The asymptotic value of the Legendre polynomials will change because of the Fresnel scattering condition as

$$P_\ell(\cos\theta) \cong \sqrt{\frac{2}{\pi\ell\sin\theta}}\ \cos(\ell\theta - \pi/4)$$

$$\cong \sqrt{\frac{1}{2\pi\ell\sin\theta}}\ \left[e^{i(\ell\theta-\pi/4)} + e^{-i(\ell\theta-\pi/4)}\right]. \tag{6.9}$$

We adopt a convenient parametrization of the elastic scattering reflection function over the region in ℓ-space contributing to nuclear elastic scattering:

$$|\eta_\ell| = \left[1 + \exp\left(\frac{\ell - \ell_{max} - 1/2}{\Delta}\right)\right]^{-1},$$

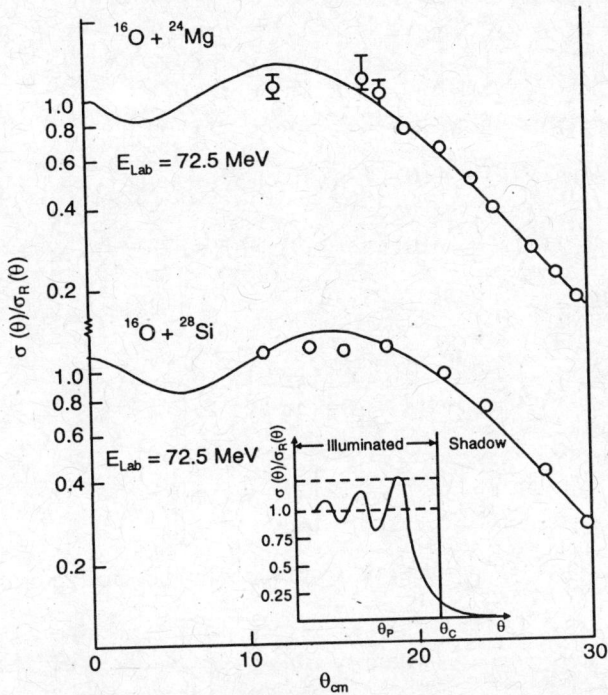

Fig. 6.7: ^{16}O elastic scattering on ^{24}Mg and ^{28}Si targets showing Fresnel pattern (Mukherjee and Pandey Ref. 4) Inset - Schematic diagram of Fresnel diffraction

where the parameter Δ describes the width of the ℓ window contributing to elastic scattering not specified by the model. With (6.9) and the 'sharp cut off' of angular momentum space, one gets a simple expression for the scattering amplitude,

$$f(\theta) = \frac{1}{2} f_c(\theta) \, erfc \, (e^{i\pi/4} u)$$

where $f_c(\theta)$ is the Coulomb amplitude and '$erfc$', the error function given by

$$erfc \, (z) = \frac{2}{\sqrt{\pi}} \int_z^\infty e^{-t^2} dt$$

and $erfc(e^{i\pi/4} u) = [1 - \{C(u) + S(u)\} - i\{C(u) - S(u)\}],$
where $C(u)$ and $S(u)$ are respectively Fresnel Cosine and Sine integrals with

$$u = \left(\frac{\ell_c}{2\sin\theta_c} \right)^{1/2} 2\sin\left(\frac{\theta - \theta_c}{2} \right) = \sqrt{n} \, \text{cosec} \, \frac{\theta_c}{2} \sin\left(\frac{\theta - \theta_c}{2} \right),$$

since $\ell_c = n \cot\theta_c / 2.$

We define
$$y = \left(\frac{2}{\pi}\right)^{1/2} u = \left(\frac{2n}{\pi}\right)^{1/2} \operatorname{cosec} \frac{\theta_c}{2} \sin\left(\frac{\theta - \theta_c}{2}\right) \qquad (6.10)$$

and express $f(\theta)$ in terms of $C(y)$ and $S(y)$ with argument y as

$$f(\theta) = f_c(\theta)\{1 - [C(y) + S(y)] - i\,[C(y) - S(y)]\}.$$

Let us write $f(\theta) = \dfrac{1}{2} f_c(\theta)(1 - a - ib)$, where $a = C(y) + S(y)$ and $b = C(y) - S(y)$ so that

$$\frac{d\sigma_{e\ell}(\theta)}{d\sigma_{Ruth}(\theta)} = \frac{|f(\theta)|^2}{|f_c(\theta)|^2} = \frac{1}{4}[1 - 2a + a^2 + b^2]$$

$$= \frac{1}{2}\left\{\left[\frac{1}{2} - C(y)\right]^2 + \left[\frac{1}{2} - S(y)\right]^2\right\}. \qquad (6.11)$$

At $\theta = \theta_c$, the Fresnel integrals $C(y)$ and $S(y)$ become zero and we get

$$\frac{d\sigma_{e\ell}(\theta)}{d\sigma_{Ruth}(\theta)} = \frac{1}{4}, \qquad \text{where} \quad \left(\frac{d\sigma}{d\Omega}\right)_{Ruth} = \left(\frac{Z_P Z_T e^2}{4E}\right)^2 \frac{1}{\sin^4 \theta/2}.$$

Blair[5], therefore, denoted θ_c by $\theta_{1/4}$ and called it a quarter point. The value of $\theta_{1/4}$ calculated numerically by Blair[5] and/or obtained from the Fresnel formula agrees well with the value obtained from experiment.

Eq. (6.11) is the limiting expression for the simplest case of a completely absorbing sharp edged nucleus with cut off angular momentum ℓ_C and critical angle θ_c. Although, in actual scattering situation, the ideal pattern (6.11) is somewhat modified by the diffuseness of nuclear surface and by the real nuclear phase shift. The ratio

$$\frac{d\sigma_{e\ell}(\theta)}{d\sigma_{Ruth}(\theta)} \quad \text{has a large peak for } y = -\sqrt{\frac{3}{2}}.$$

If the angle at which this peak occurs is θ_p, then

$$-\sqrt{\frac{3}{2}} = \left(\frac{2n}{\pi}\right)^{1/2} \operatorname{cosec} \frac{\theta_c}{2} \sin\left(\frac{\theta_p - \theta_c}{2}\right).$$

In other words, $\sin\left(\dfrac{\theta_c - \theta_p}{2}\right) \operatorname{cosec} \dfrac{\theta_c}{2} = \left(\dfrac{3\pi}{4n}\right)^{1/2}$. If we define

$$\Delta = 2\left(\frac{n}{3\pi}\right)^{1/2} \sin(\theta_c - \theta_p)\Big/\sin\frac{\theta_c}{2},$$

then for the Fresnel cross section, $\Delta = 1$. Inserting the experimental values of θ_c and θ_p, we get a measure of the extent to which the cross section is Fresnel type. The angle $\delta\theta$ between the maximum of the diffraction pattern and the critical angle θ_c can be estimated as $\delta\theta = \theta_c (3\pi/4n)^{1/2}$, since $(\theta_c - \theta_p)$ is generally quite small and $\Delta = 1$.

The assumption of a sharp boundary between elastic scattering and total absorption has been relaxed by Frahn[6,7] in his generalized Fresnel model, which allows reflection of partial waves above the barrier, absorption of partial wave below the barrier, deviation from the Rutherford orbit caused by the real part of the nuclear potential and deformed nuclear shapes[4]. In Fig. 6.7 the results of generalized Frenel model[4] is compared with experiment.

Deep Inelastic Collisions: The deep inelastic collisions correspond to a band of impact parameters larger than b_c where compound nucleus formation begins to take place, and smaller than grazing impact parameter b_g where direct reactions occur (*see* Fig. 6.3). The experimental results on deep inelastic collisions show following characteristic features:

(1) The identity of projectile and target respectively is essentially preserved, although considerably amount of mass can be transferred.

(2) They involve a large amount of energy loss from the relative motion into the intrinsic degrees of freedom.

(3) Angular distribution shows sideways peaking (strongly non-isotropic) characteristic of a direct reaction with interaction time

$$\tau_{\text{int}} \approx (10^{-22} - 10^{-24}) \text{ sec.}$$

(4) They involve lot of nucleon exchanges among the interacting nuclei.

(5) Because of the large level densities of the highly excited fragments, only mean quantities which are averaged over many outgoing channels are observed.

(6) The cross-sections for deep inelastic collisions increase with the increasing bombarding energy and increasing mass of the colliding nucleus or increase essentially to an increasing $\ell \leq \ell_g$ and a decreasing ℓ_c.

An example of the energy distribution of HI collisions is shown in Fig. 6.8 for $^{84}_{36}Kr + ^{209}_{83}Bi$ at 712 MeV projectile Lab. energy. The Coulomb energy of the two touching sphere $V_{Coul} = \dfrac{36 \times 83 \times 1.44}{14.4} \approx 299$ MeV. In the experiment the initial $E_{c.m} = \dfrac{209}{209 + 84} \times 712 = 499$ MeV. The energy loss, if the fragments

separate with no kinetic energy would be = 200 MeV. This is indicated by an arrow in Fig. 6.8.

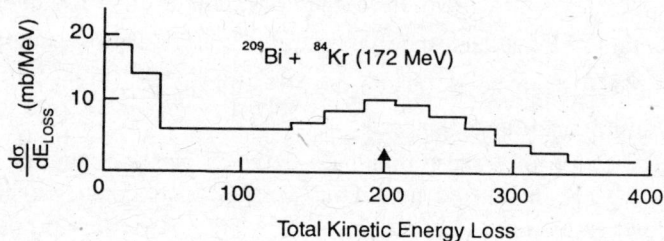

Fig. 6.8: Energy distribution of heavy ion collisions for ^{84}Kr on ^{209}Bi at 712 MeV projectile energy. The energy loss is indicated by an arrow around 200 MeV (Gross Ref. 8)

To explain the phenomena of damped nuclear collisions several phenomenological models have been developed. It is necessary that any such model should consider the geometry of the trajectories properly, and should provide mechanisms for energy loss and fluctuations of the dynamical variables which reproduce the data. The available models are generally classified into two categories – (*i*) the classical trajectory models, and (*ii*) the diffusion models.

A realistic classical trajectory calculation must include the nuclear dissipation that links radial and tangential frictions, which are responsible for loss of energy and angular momentum. Here four coordinates are used to describe a collision: the distance r between mass centres of projectile and target, their relative orientations θ_P and θ_T respectively, and orientation θ of the total system. Equation of motion of the above system is obtained from the well known Lagrange-Rayleigh equation, expressed in terms of a set, $\{q_i\}$ of n collective coordinates as[9]

$$\left[\frac{d}{dt} \left(\frac{\partial}{\partial \dot{q}_i} \right) - \frac{\partial}{\partial q_i} \right] L = -\frac{\partial F}{\partial \dot{q}_i}, \tag{6.13}$$

where $i = 1...4$, $L = T - V$ and T (V) is kinetic energy (potential energy). F is the Rayleigh dissipation function related to the rate of energy dissipation $(\partial E/\partial t)$, where $E = T + V$.

F is proportional to velocity and is given by

$$F(r) = f(r) \left[\dot{r}^2 + \frac{1}{2} u_\theta^2 \right], \tag{6.14}$$

where $\dot{r}$ and $u_\theta = \dot{r}\theta$ are the radial and the tangential velocities respectively. The dissipation function $f(r)$ are approximated in terms of friction coefficients

$C_r(r)$ and $C_\theta(r)$ corresponding respectively to radial and tangential (sliding) motions. The latter describes the friction occurring at the surface of two nuclei in the grazing collision.

One can solve equations of motion for different values of the relative angular momentum ℓ or the impact parameter b and obtain the corresponding classical trajectories followed by the system. For a given ℓ value, when the trajectory enters the interaction region, the frictional force acts, some relative kinetic energy and orbital angular momentum are lost in a continuous manner. The trajectory thus gets modified compared to those obtained without the presence of the frictional force. Some of these trajectories can escape from interaction region, leading to deep inelastic collisions. The kinetic energy in this case is then lower than the original energy in the entrance channel. This classical model can also allow one to calculate the differential cross section from the deflection function $\theta (\ell)$.

Interaction time of the damped reaction process may be estimated from measurable quantity like deflection function/angle. Typically, one can write for the interaction time

$$\tau(\theta) = \frac{\theta_g - \theta}{\overline{\omega}} = \frac{\mathscr{I}(\theta_g - \theta)}{\hbar \overline{\ell}}, \tag{6.15}$$

where $\mathscr{I}\overline{\omega} = \hbar\overline{\ell}$. The average angular momentum, $\overline{\ell}$ is calculated using the relation

$$\overline{\ell} = \frac{\displaystyle\int_{\ell_f}^{\ell_g} [\ell(\ell+1)]\, d\ell}{\displaystyle\int_{\ell_f}^{\ell_g} \ell\, d\ell} \cong \frac{2}{3} \frac{\ell_g^3 - \ell_f^3}{\ell_g^2\, \ell_f^2}. \tag{6.16}$$

Exercise 6.1 A 60 MeV beam of ^{16}O collides with ^{64}Cu target. Estimate the angle of the first diffraction minimum.

The first minimum will occur at an $\angle\theta$ such that the scattered waves from the centre of the target and from its edge are out of phase by $\lambda/2$. Therefore, the first minimum occurs at $r\theta = \dfrac{\lambda}{2}$ where $r = 1.2 \times 4$ fm $= 4.8$ fm and $\lambda = \dfrac{2\pi\hbar}{p}$ with $p = \sqrt{2ME} = 1342$ MeV/c. Therefore,

$$\lambda = \frac{2\pi(197 \text{ MeV fm})}{1354 \text{ MeV}} = 0.9222 \text{ fm.}$$

Hence, $4.8 \text{ fm} \times \theta$ radian $= 0.4611$ fm or $\theta \cong 5.5°$.

Exercise 6.2 Show that the equation of orbit for a charged particle when scattered by a Coulomb field is $D = a(1 + \text{cosec } \theta/2)$, where $D(D_0) = $ distance of closest approach for impact parameter $b(0)$. $a = n/k$ and $n = kD_0/2$.

Let the incident particle whose initial velocity is v attains a velocity v_0 at $r = D$. Conservation of energy leads to

$$\frac{1}{2}\mu v^2 = \frac{1}{2}\mu v_0^2 + \frac{Z_T Z_P e^2}{D}.$$

Dividing the above equation by $\frac{1}{2}\mu v^2$, we get $1 = \left(\frac{v_0}{v}\right)^2 + \frac{D_0}{D}$.

Conservation of angular momentum leads to $\mu v b = \mu v_0 D$. Hence,

$$1 = \frac{b^2}{D^2} + \frac{D_0}{D} \quad \text{or} \quad b^2 = D^2 - D_0 D.$$

Now, $\quad \ell = kb = \frac{k D_0}{2}\cot \theta/2 = n\cot \theta/2$ and $\ell^2 + n^2 = (kD - n)^2$.

Hence, $kD = \sqrt{\ell^2 + n^2} + n = n(1 + \text{cosec } \theta/2)$

or $\quad D = a(1 + \text{cosec } \theta/2)$.

Exercise 6.3. Consider semiclassical picture of the reaction $A(a, b) B$. Let Z_a, Z_A, Z_b, Z_B be the atomic numbers of the nuclear fragments in the incoming and outgoing channels. The relative motion trajectories are determined by the c.m energy ϵ, wave number k, Sommerfeld parameter n and orbital angular momentum ℓ. The incoming channel parameters $(\epsilon_i, k_i, n_i, \ell_i)$ are usually not the same as outgoing channel parameters $(\epsilon_f, k_f, n_f, \ell_f)$. Incoming trajectories join as smoothly as possible on the outgoing trajectories. In quantitative term, this could mean that the distance of closest approach for incoming and outgoing channel parameters must be the same. Find $\dfrac{\epsilon_f}{\epsilon_i}$ and hence show that for the transfer reaction $^9Be(^{16}O, \, ^{17}O) \, ^8Be$ in a head on collision, the Q-value of the reaction is zero.

$$D = \frac{n_i}{k_i}\left[1 + \left\{1 + \frac{\ell_i(\ell_i+1)}{n_i^2}\right\}^{1/2}\right] = \frac{n_f}{k_f}\left[1 + \left\{1 + \frac{\ell_f(\ell_f+1)}{n_f^2}\right\}^{1/2}\right]$$

$$\frac{\epsilon_f}{\epsilon_i} = \frac{\hbar^2 k_f^2}{2M_f} \times \frac{2M_i}{\hbar^2 k_i^2} = \frac{M_i}{M_f}\frac{k_f^2}{k_i^2} = \frac{M_i}{M_f} \times \frac{k_f}{k_i} \times \frac{n_f}{n_i}\rho,$$

where $\quad \rho = \left[1 + \left\{1 + \frac{\ell_f(\ell_f+1)}{n_f^2}\right\}^{1/2}\right] \Big/ \left[1 + \left\{1 + \frac{\ell_i(\ell_i+1)}{n_i^2}\right\}^{1/2}\right].$

Now $\dfrac{n_f}{n_i} = \dfrac{Z_b Z_B e^2}{\hbar v_f} \times \dfrac{\hbar v_i}{Z_a Z_A e^2} = \dfrac{Z_b Z_B}{Z_a Z_A}\dfrac{v_i}{v_f}.$ Hence,

$$\frac{\epsilon_f}{\epsilon_i} = \frac{M_i}{M_f}\frac{k_f}{k_i}\frac{Z_b Z_B}{Z_a Z_A}\frac{v_i}{v_f}\rho$$

$$= \frac{M_i}{M_f}\frac{v_i}{v_f}\frac{k_f}{k_i}\frac{Z_b Z_B}{Z_a Z_A}\rho = \frac{k_i}{k_f}\frac{k_f}{k_i}\frac{Z_b Z_B}{Z_a Z_A}\rho = \frac{Z_b Z_B}{Z_a Z_A}\rho.$$

For head on collision $\rho = 1$ and for the transfer reaction $^9Be(^{16}O,\ ^{17}O)\ ^8Be$,

we have $\dfrac{Z_b Z_B}{Z_a Z_A} = 1$, so that $\epsilon_f / \epsilon_i = 1$ or $\epsilon_f - \epsilon_i = Q = 0.$

6.2 HEAVY ION FUSION

Nuclear fusion is the process in which two nuclei combine to form a single compound nucleus. This compound nucleus will be in general in a highly excited state and will decay by gamma or particle emission. At small excitation energies one has discrete levels forming rotational bands of different character. The analysis of the very high energetic gamma ray has allowed the study of the nuclear shapes as a function of excitation energy and angular momentum.

Some typical features of fusion cross sections shown in Fig. 6.9 with respect to energy are as follows[10]:

(1) At low energies but above the barrier (region I) fusion cross sections σ_F follow the total reaction cross section σ_R, whereas for higher energies (region II) they fall below it. This fall is very pronounced for lighter systems. For very high energies (region III), the fission process dominates, while fusion cross section tends towards zero.

(2) Fusion cross sections can be very large, they reach maximum values of more than 1 barn in reaction of relatively light nuclei.

(3) The ratio of fusion to reaction cross section drops strongly with the product of charges of two interacting nuclei. This is shown in Fig. 6.10.

(4) Fusion cross section in general is a smooth function of energy, but it shows pronounced oscillations for light nuclei.

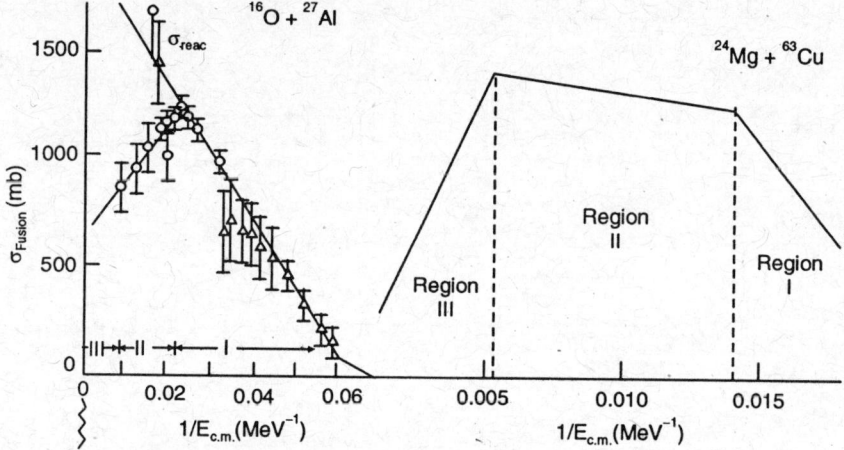

Fig. 6.9: Characteristics of sfusion versus $1/E_{c.m}$ (Adopted from Mosel Ref. 10)

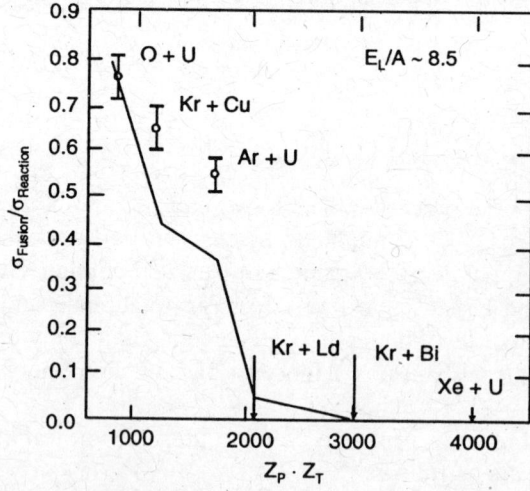

Fig. 6.10: Ratio σ fusion σ_R with respect $Z_p \times Z_T$ (Adopted from Mosel Ref. 10)

Production of High Spin States and Superheavy Nuclei: Many new elements have been created by fusion-evaporation reactions. This is a two-step process. In the first step projectile and target nucleus have to amalgamate to form a mono-nuclear system, in the second step the nuclear system has to dissipate the excitation energy gained in the fusion process and come to the ground state. The excitation energy of the heavy system is dissipated by γ-emission and neutron evaporation. For each step of neutron evaporation there is a strong fission competition which strongly increases towards the heaviest system which are highly fissile. The survival probability of the compound nucleus is of the order of 10^{-2} to 10^{-3} per evaporation step. The production cross section can be described as:

$$\sigma = \sigma_{fusion}\,(E - V_B)\,P_x(E^*), \qquad\qquad (6.17)$$

where E is the c.m projectile energy, V_B the fusion barrier, P_x the survival probability depending on the number 'x' of evaporated neutrons and E^*, the excitation energy of the compound system. Two types of fusion reactions are used successfully. In the first one, the compound nucleus is set into rapid rotation to learn about the properties of nuclear high spin states using γ-spectroscopy.

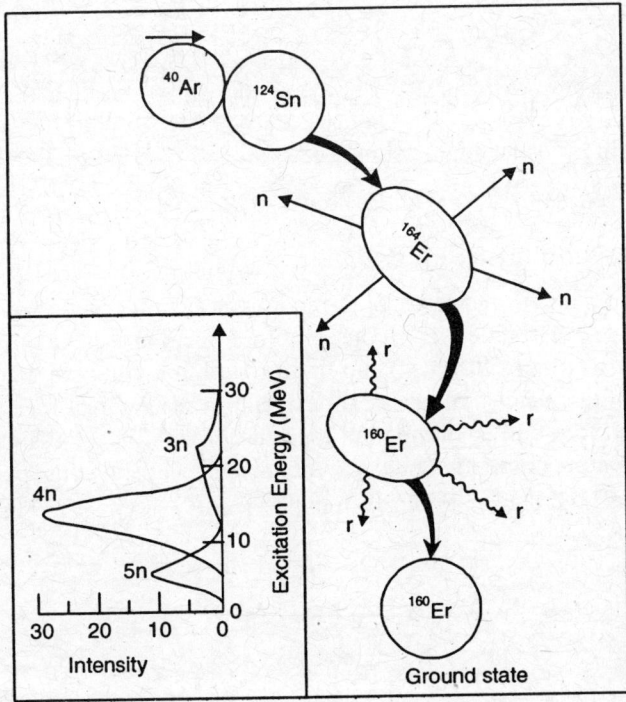

Fig. 6.11: Gamma emission and neutron evaporation in heavy ion fusion

We illustrate this in Fig. 6.11 for the case of ^{40}Ar accelerated into ^{124}Sn scattering a compound nucleus ^{160}Er at high excitation energy. This hot system evaporates four neutrons before ending up as a bound nucleus ^{160}Er, which then cools further through the emission of γ-rays. States upto $40\hbar$ in normal nuclei and $60\hbar$ in super deformed nuclei have been identified.

In the second one, the limits of nuclear stability in mass and charge are explored to learn about the limitation of number of elements and the status of superheavy elements, which are far in the transuranium region, located near the next double shell closure ($Z = 114$, $N = 184$) above lead. At present, elements of $Z = 107$ upto $Z = 112$ have been synthesized and their shear existence shows that the nuclear stabilizing shell effects are strong enough to counterbalance the liquid drop fission channel. The heaviest elements unambiguously identified are the elements with $Z = 112$ at GSI and $Z = 113$ at RIKEN.[39] The other group found an island of more neutron rich species. These nuclei have been produced at JINR Dubna using actinide targets from uranium to californium, using beams of ^{48}Ca (The island of super heavy element is shown in Fig.6.2). Evaporation residue yields for the fusion reactions leading to the synthesis of new elements is given by

$$\sigma_{ER}(E) = \sum_{\ell=0}^{\infty} (2\ell+1) \, T(\ell,E) \, [1 - P_f(\ell,E^*)] \, ,$$

where $T(\ell, E)$ is the capture transmission coefficient and $[1 - P_f(\ell,E^*)]$ is the probability of surviving fission. The synthesis of $Z = 113$ element is done via $^{209}Bi(^{70}Zn,n) \, ^{278}X_{113}$ with $E(^{70}Zn) = 349$ MeV.

A. Barrier Penetration Model

The development of a macroscopic model for fusion reaction involves the calculation of potential energy (conservative forces) and calculation of dissipation energy (frictional forces) and finally, setting up and solving equations of motion in classical/semi-classical or quantum mechanical form. If the energy loss is large due to dissipation then the probability of fusion becomes high. To form the compound nucleus, the incident nucleus must penetrate the effective potential barrier V_e of two ions: $V_e = V_{Coul} + V_N + V_\ell$. The Coulomb potential has the form:

$$V_{Coul}(r) = \frac{Z_P Z_T e^2}{r} \left(3 - \frac{r}{r_c^2}\right) \text{ for } r \leq r_c \tag{6.18a}$$

$$= \frac{Z_P Z_T e^2}{r} \text{ for } r > r_c \tag{6.18b}$$

The nuclear potential has form :

$$V_N(r) = -V_0 / [1 + \exp(r - r_N)/a] \qquad\qquad (6.18c)$$

and
$$V_\ell(r) = \frac{\hbar^2(\ell+1)\ell}{2\mu r^2}. \qquad\qquad (6.18d)$$

Classically, the nuclei fuse if the energy of the incident ion equals at least to the effective potential barrier (Eq. (6.4)) and we get in the sharp cut off model $\sigma_F = \sigma_{CN}$.

The nuclear part $V_N(r)$ would be zero at large distances, but attractive when the two nuclear densities start to overlap each other. By analyzing a large amount of experimental data for fusion above the barrier, Bass (*see* Ref. 14) showed that the one dimensional nuclear potential can be described by an attractive surface energy contribution of the following form:

$$V_N(s) = -4\pi\gamma \left\{ r_P r_T / (r_P + r_T) \right\} f(s), \qquad\qquad (6.18e)$$

where $s = r - r_P - r_T$ with r being the distance between the centres of two spherical nuclei, $f(s)$ is a universal function given by $F(s) \propto e^{-s/d}$, with d as the range parameter and γ the surface tension constant (Eq. 1.30). In Fig. 6.12

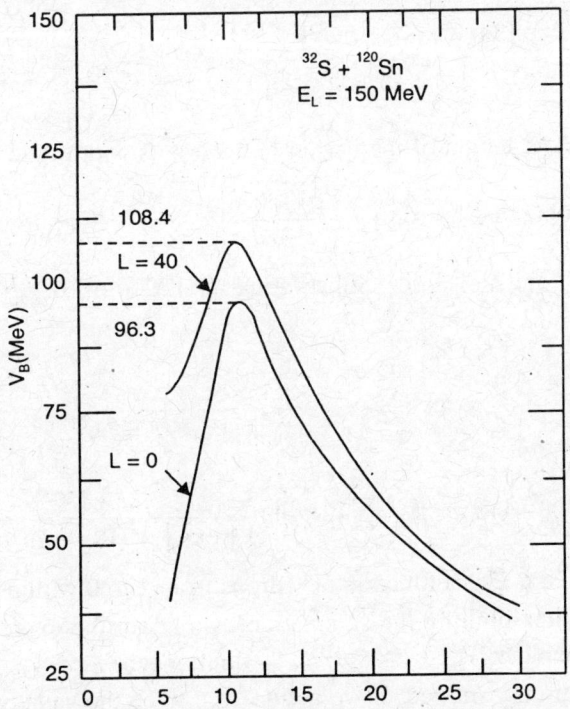

Fig. 6.12: Bass potential for the system ^{32}S + ^{120}Sn at E = 150 MeV

we show the Bass potential for the system $^{32}S + ^{120}Sn$ at 150 MeV projectile energy. The barrier height at $\ell = 0$ is $V_{B\ell} = 96.3$ MeV corresponding to the barrier radius $r_B = 10.5$ fm. The barrier height at $\ell = 40$ is $V_{B\ell} = 108.4$ MeV.

The above description of fusion in angular momentum space which is same for the formation of compound nucleus, is known as critical distance or critical angular momentum model for fusion. Analysis of low energy fusion data reveals that $r_c = r_{0c}(A_P^{1/3} + A_T^{1/3})$, where $r_{0c} = (1.0 \pm 0.07)$ fm. However, a close inspection of fusion cross-section data at low energies, i.e. at large $1/E$ shows that the measured cross sections are larger than the predicted values. This enhancement is due to the quantum mechanical penetrability at energies below the potential barrier. The quantum mechanical expression of σ_F, which allows penetration through potential barrier, is obtained as follows:

$$\sigma_F = \pi\lambda^2 \sum_{\ell=0}^{\infty}(2\ell+1)\, T_\ell P_\ell, \tag{6.19a}$$

where T_ℓ is the transmission probability through the barrier and P_ℓ is the fusion probability once the barrier is penetrated. For the parabolic barrier

$$T_\ell(E) = \frac{1}{1 + \exp\left[2\pi(V_{B\ell} - E)/\hbar\omega_\ell\right]}, \text{ where } V_{B\ell} = \left[V(r) + \frac{\hbar^2\ell(\ell+1)}{2\mathscr{I}}\right] \tag{6.19b}$$

and $r = r_{B\ell}$ is the height of the barrier and $\hbar\omega_\ell$ its width. The radius of the barrier $r_{B\ell}$ is obtained by setting $\left[\dfrac{\partial V_{B\ell}(r)}{\partial r}\right]_{r=r_{B\ell}} = 0$. For simplicity, we assume $r_{B\ell} = r_B$ and $\hbar\omega_\ell = \hbar\omega$. Substituting $T_\ell(E)$ from (6.19b) and $P_\ell = 1$ (sharp cut off) we get:

$$\sigma_F = \pi\lambda^2 \sum_{\ell=0}^{\ell_c} \frac{2\ell+1}{1 + \exp\left[2\pi(V_{B\ell} - E)/\hbar\omega\right]}. \tag{6.20a}$$

$$\sigma_R = \pi\lambda^2 \sum_{\ell=0}^{\infty}(2\ell+1)\,T_\ell = \pi\lambda^2 \sum_{\ell=0}^{\infty}(2\ell+1)\frac{2\ell+1}{1 + \exp\left[2\pi(V_{B\ell} - E)/\hbar\omega\right]}. \tag{6.20b}$$

$T_\ell(E)$ for fixed E as a function of ℓ drops from 1 to 0 within the range of few ℓ values, thus limiting the summation over 'ℓ' in the expression of σ_R. $T_\ell(E) = 1$, if $\exp\left[2\pi(V_{B\ell} - E)/\hbar\omega\right] \ll 1$, which means $E \gg V_{B\ell}$ and that all partial waves are transmitted. If we define ℓ_{max} to be the value when $T_\ell(E)$ drops to 0.5, then $\exp\left[2\pi(V_{B\ell} - E)/\hbar\omega\right] = 1$. Employing (6.19b) we get

$$\ell_{max}(\ell_{max}+1) = \frac{2\mathcal{I}_B}{\hbar^2}(E-V_{B\ell}), \qquad (6.21a)$$

or
$$\ell_{max}\,\hbar = r_B\sqrt{2\mu(E_{cm}-V_B)}, \qquad (6.21b)$$

where $= V_B$ nuclear plus Coulomb potential, $\mathcal{I}_B = \mu r_B^2$ and r_B is the distance of $\ell = 0$ barrier having height $V_B(= V(r_B))$. The above angular momentum ℓ_{max} used to get the reaction cross section σ_R has to be compared to ℓ_c of fusion satisfying

$$\ell_c(\ell_c+1) = \frac{2\mathcal{I}_c}{\hbar^2}(E-V_c), \qquad (6.22)$$

where $V_c = V(r_c)$. Experimental data show that at high energies the reaction cross section σ_R is much higher than the fusion cross section (σ_F). Hence, the problem is to understand the observed limitation of fusion cross section at high energies, i.e. sudden drop of σ_F/σ_R ratio.

A plot of $\ell_c(E)$ and $\ell_{max}(E)$ with respect to energy is shown in Fig. 6.13. At the low energy, all incident partial waves are captured so that $\ell_{max} \approx \ell_c$ and $\sigma_F \approx \sigma_R$. As energy increases, there comes a point where $\ell_{max}(E)$ is too large to be accepted by the nucleus to remain fused and the limit of fusion is set by ℓ_c that can be accepted before flying apart and not by ℓ_{max}. All partial waves below $\ell_1 = \ell(E_1)$ feel an attractive force ($V_C \geq V_B$) even at energies above barrier, pulling them into critical distance r_c and thus to fusion. In order

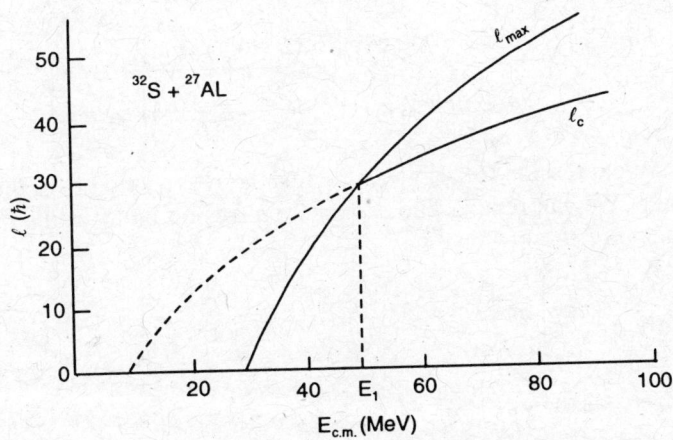

Fig. 6.13: Plot of critical angular momentum (ℓ_c) and maximum angular momentum (ℓ_{max}) with respect to the incident energy

to make quantitative comparison of σ_F with observed values we replace summation over l by integration and use (6.20a) and get:

$$\sigma_F = \int_0^{\ell_c} \pi\lambda^2 \frac{(2\ell+1)\,d\ell}{1+\exp\left[2\pi\left\{V_B - \dfrac{\ell(\ell+1)\hbar^2}{2\mathscr{I}_B} - E\right\}\Big/\hbar\omega\right]}. \qquad (6.23)$$

In (6.23), the numerator can be made an exact differential of the denominator.

Let $y = 2\pi(E - V_B)/\hbar\omega$, $x = \dfrac{\pi\hbar\ell(\ell+1)}{\mathscr{I}_B\omega} - y$ so that $dx = \dfrac{\pi\hbar(2\ell+1)\,d\ell}{\mathscr{I}_B\omega}$.

Then σ_F takes the form:

$$\sigma_F = \pi\lambda^2 \int_0^{\ell_c} \frac{(2\ell+1)d\ell \, \exp\left[y - \pi\hbar\ell(\ell+1)/\mathscr{I}_B\omega\right]}{1+\exp\left[y - \pi\hbar\ell(\ell+1)/\mathscr{I}_B\omega\right]}.$$

Hence, $$\sigma_F = \lambda^2 \int_{x_1}^{x_2} \frac{\mathscr{I}_B\omega}{\hbar}\,\frac{e^{-x}}{1+e^{-x}}\,dx = \frac{\hbar\omega\mathscr{I}_B}{2\mu E}\int_{x_2}^{x_1}\frac{-e^{-x}}{1+e^{-x}}\,dx,$$

where $x_1 = -y = 2\pi(V_B - E)/\hbar\omega$. Following (6.22) we get

$$x_2 = \pi\hbar\ell_c(\ell_c+1)/\mathscr{I}_B\omega + 2\pi(V_B - E)/\hbar\omega = \frac{2\pi}{\hbar\omega}\left[\frac{\mathscr{I}_C}{\mathscr{I}_B}(E - V_C) + (V_B - E)\right]$$

$$\sigma_F = \frac{\hbar\omega\mathscr{I}_B}{2\mu E}\log\frac{1 + \exp\{2\pi(E - V_B)/\hbar\omega\}}{1+\exp\left[2\pi\left\{E - V_B - \dfrac{\mathscr{I}_C}{\mathscr{I}_B}(E - V_c)\right\}\Big/\hbar\omega\right]}. \qquad (6.24)$$

The above expression (6.24) gives the fusion cross section over the whole energy range. We can reduce it to separate expression both for the high and low energies. For the low energies

$$\exp\left[2\pi\left\{E - V_B - \frac{\mathscr{I}_C}{\mathscr{I}_B}(E - V_c)\right\}\Big/\hbar\omega\right] \ll 1$$

or $$E(1 - \mathscr{I}_C/\mathscr{I}_B) - \left(V_B - \frac{\mathscr{I}_C}{\mathscr{I}_B}V_C\right) \ll 0.$$

We can neglect the exponential in the denominator of (6.24) and get

$$\sigma_F = \frac{\hbar\omega \mathscr{I}_B}{2\mu E} \log\left[1 + \exp\{2\pi(E - V_B)/\hbar\omega\}\right]. \qquad (6.25)$$

If $E - V_B \gg \hbar\omega$ then

$$\sigma_F = \frac{\hbar\omega \mathscr{I}_B}{2\mu E} \cdot \frac{2\pi(E - V_B)}{\hbar\omega} \approx \pi r_B^2\left(1 - \frac{V_B}{E}\right). \qquad (6.26)$$

For the high energies

$$\exp\left(2\pi\left\{E - V_B - \frac{\mathscr{I}_C}{\mathscr{I}_B}(E - V_C)\right\} \Big/ \hbar\omega\right) \gg 1$$

or

$$E\left(1 - \frac{\mathscr{I}_C}{\mathscr{I}_B}\right) - \left(V_B - \frac{\mathscr{I}_C}{\mathscr{I}_B} V_C\right) \gg 0.$$

We can neglect 1 both in the denominator and the numerator of the expression (6.24) and get

$$\sigma_F = \frac{\hbar\omega \mathscr{I}_B}{2\mu E} \log \frac{\exp[2\pi(E - V_B)/\hbar\omega]}{\exp\left[2\pi\left\{E - V_B - \frac{\mathscr{I}_C}{\mathscr{I}_B}(E - V_C)\right\} \Big/ \hbar\omega\right]}$$

$$= \frac{\hbar\omega \mathscr{I}_B}{2\mu E} \frac{(E - V_C)}{\hbar\omega} = \pi r_c^2\left(1 - \frac{V_C}{E}\right). \qquad (6.27)$$

Thus, we can see from Eq. (6.26) that at the low energies the barrier parameters r_B and V_B with $r_B = r_{0B}(A_P^{1/3} + A_T^{1/3})$ can fit the fusion data well by taking $r_{0B} = 1.45$ fm. For high energies, σ_F is determined by V_C and r_c. The plot of σ_F versus $1/E$ shows picture of two straight lines with different slope intercept with bend in between (Fig. 6.13). This bend occurs at $\ell_c = \ell_{\max}$.

In the Table 6.2 we compare the BPM predicted fusion cross sections obtained both at the low energy and high energy, using respectively the relations (6.26) and (6.27), with the experimental values for $^{32}S + ^{24}Mg$ system.

Table 6.2: BPM predicted fusion cross sections compared with experiment. The parameters are taken from the work of Glass and Mosel[11] and Galin et al.[11] or Gutbrod et al.[12]

$\frac{1}{E}$ MeV^{-1}	Theory σ_F (mb)	Experiment σ_F (mb)	Parameters
0.03	359	351	$V_B = 28.3$ MeV, $r_B = 8.7$ fm
0.01	1076	1135	$V_C = 8.0$ MeV, $r_C = 6.1$ fm

B. Effect of Nuclear Friction

The classical barrier penetration model (BPM) discussed above assumes that the incident nucleus looses energy due to nuclear friction before it reaches the barrier. Let us assume that an ion with the initial angular momentum ℓ_0 and the centre of mass energy E_0 has lost the amounts $\Delta\ell$ and ΔE on angular momentum and energy respectively by the time it reaches the barrier. Thus, the classical condition (Eq. 6.21) for just reaching the barrier position r_B reads:

$$(\ell_0 - \Delta\ell)\ (\ell_0 - \Delta\ell) = \frac{2\mu r_B^2}{\hbar^2}[E_0 - \Delta E - V_B]$$

or

$$\ell_0^2 \cong \frac{2\mu r_B^2}{\hbar^2}\frac{\hbar^2}{2\mu\lambda^2}\left[1 - \frac{V_B + \Delta E}{E_0}\right] + 2\ell_0\Delta\ell.$$

Hence,

$$\sigma_F = \pi\lambda^2\ell_0^2 = \pi r_B^2\left[1 - \frac{V_B + \Delta E}{E_0}\right] + \frac{\pi\hbar^2}{\mu E_0}\ell_0\Delta\ell. \qquad (6.28)$$

One sees that the energy loss decreases the fusion cross section because the ion may not have enough energy left to surpass the barrier and be trapped. On the other hand, angular momentum dissipation increases the fusion cross section because it lowers the relative angular momentum and hence the effect of centrifugal barrier. Both effects counteract each other. The difference between fusion cross sections with and without friction is

$$\Delta\sigma_F = -\pi r_B^2\frac{\Delta E}{E_0} + \frac{\pi\hbar^2}{\mu E_0}\ell_0\Delta\ell. \ \Delta\sigma_F, \text{which will be zero if}$$

$$\frac{\Delta E}{\Delta\ell} = \frac{\hbar^2\ell_0}{\mu r_B^2}\frac{2(E_0 - V_B)}{\ell_0}. \qquad (6.29)$$

Hence, if condition (6.29) holds, we get back expression for friction free case, and the dissipation marks itself as far as fusion cross section is concerned. Dissipation can be incorporated into BPM by shifting critical ℓ value to still higher values. The maximum possible angular momentum loss is given by so called sticking limit. If two nuclei stick together they rotate with a single frequency. The initial angular momentum $L_i(= \ell_i) = \mu r^2\dot{\theta}_i$ and the final angular momentum,

$$L_f = \ell_f + L_P + L_T = \left(\mu r^2 + \frac{2}{5}M_P r_P^2 + \frac{2}{5}M_T r_T^2\right)\dot{\theta}_f.$$

The conservation of angular momentum gives $\ell_i = \ell_f + L_P + L_T$, so that the difference in angular momentum

$$\Delta\ell = L_P + L_T = \left(\frac{2}{5} M_P r_P^2 + \frac{2}{5} M_T r_T^2\right)\dot{\theta}_f.$$

If we define sticking moment of inertia by $\mathscr{I}_s = \mu r^2 + \frac{2}{5} M_P r_P^2 + \frac{2}{5} M_T r_T^2$

then $\Delta\ell = (\mathscr{I}_s - \mathscr{I}_{ns})\dot{\theta}_f$, where $\mathscr{I}_{ns} = \mu r^2$ and $\mathscr{I}_s\dot{\theta}_f = \mathscr{I}_{ns}\dot{\theta}_i$. Finally,

$\Delta\ell = \ell_i \dfrac{\mathscr{I}_s - \mathscr{I}_{ns}}{\mathscr{I}_s}$. For a touching system $(r = r_P + r_T)$ and for symmetric

system

$$(M_P = M_T,\ r_P = r_T)$$

we get

$$\Delta\ell = \frac{2}{7}\,\ell_i.$$

Thus, if potential for ℓ_i does not exhibit a pocket anymore but for $\Delta\ell = (2/7)\ell_i$ still does, then fusion is still possible due to angular momentum dissipation effects.

Another quantity of interest, calculated within the framework of one dimensional barrier penetration model, is the time T for complete fusion above the Coulomb barrier. It is customary to regard T as T_{CN} spent by the *HI* system within the fusion distance r_F while moving on the classical Coulomb-Nuclear (CN) trajectories at energy E above the barrier height V_B:

$$T_{CN} = 2\int_{r_m}^{r_F} \frac{dr}{\dot{r}}, \tag{6.30a}$$

where r_m is the distance of closest approach where kinetic energy vanishes. Now

$$\frac{1}{2}\mu\,\dot{r}^2 = E - V_N(r) - V_c(r) - V_\ell(r)$$

so that

$$T_{CN} = \sqrt{2\mu}\int_{r_m}^{r_F} \frac{dr}{\left[E - V_N(r) - V_c(r) - V_\ell(r)\right]^{1/2}} \tag{6.30b}$$

where μ is the reduced mass of the *HI* system and $V_N(r), V_c(r), V_\ell(r)$ are defined in Eq. (6.18). The integral (6.30b) has been evaluated analytically by

Menon et al.[13] employing Woods – Saxon potential within the framework of the one dimensional BPM for low partial waves including the radial frictional force given by Bass[14] in the form

$$F = -Q f(r) \dot{r} \tag{6.31}$$

where $f(r) = \delta (r - r_N)$ and Q is a +ve coefficient. The Newton's equation of motion is then

$$\mu \frac{d^2 r}{dt^2} = -\frac{d}{dr} V_e(r) - Q \frac{d}{dt} \theta (r - r_N), \tag{6.32}$$

where θ is a step function. It was found by Menon et al.[13] that the inclusion dissipation reduces the fusion time T_{CN} substantially by a factor of 1/4 to 1/8.

Exercise 6.4. Estimate ℓ_{max} and ℓ_c for a heavy ion reaction $^{16}_{8}O + ^{27}_{13}Al$ at $E_{LAB} = 100$ MeV using Fig. 6.10 (left) for the cross sections.

We have $V_{Coul} = 20.87$ MeV and according to (6.2a) $k = 6.2$ fm^{-1}. Hence, $\ell_{max} = k(a_0 A_P^{1/3} + a_0 A_T^{1/3}) \cong 44$, where $a_0 = 1.3$ fm. Employing (6.6) we can also estimate ℓ_{max} from the measured reaction cross section (use Fig. 6.10 left). Accordingly, we have

$$\sigma_R = 1625 \text{ mb} = \frac{\pi}{k^2} \ell_{max}^2 . \text{ Hence, } \ell_{max} = 44.$$

Similarly, the experimental fusion cross section yields

$$\sigma_F = 875 \text{ mb} = \frac{\pi}{k^2} \ell_c^2 \text{ which gives } \ell_c = 37.$$

Exercise 6.5. Calculate the maximum value of angular momentum which can contribute to the fusion process $^{32}S + ^{120}Sn$ at $E_{LAB} = 150$ MeV for the Bass potential (Fig. 6.12) for relative angular momentum $\ell = 0$ and $\ell = 40$.

We employ Eq. 6.21(b) and substitute for $\ell = 0$ and $\ell = 40$ respectively $V_B = 96.3$ MeV and 108.4 MeV at $r_B = 10.5$ fm. Hence, for $\ell = 0$ and $E_{c.m.} = 118.4$ MeV:

$$\ell_{max} \cdot 197 = 10.5 \left[\frac{2 \times 32 \times 120 \times 938.9 \times 22.1}{152} \right]^{1/2} .$$

Hence, $\ell_{max} = 54.$

For $\ell = 40, E_{c.m} = 118.4$ and $V_B = 108.4$ MeV, $\ell_{max} = 37.$

6.3 NUCLEI FAR FROM STABILITY

"Nuclei far from stability have never been so close to us as they are now —
——— *thanks to radioactive ion beam facility".[15]*

With beams of particles impinging on a heavy nucleus, one can literally "blow" a nucleus to pieces and thus form a large range of nuclei in which the neutron-to-proton ratio (N/Z) is located far away from the region of beta stability. By then separating the so formed isotopes with the electric and magnetic fields of an isotope separator system, and bringing them rapidly to a measuring station one can study their properties within the short life time ($T_{1/2} < 1$ day) for which such nuclei survive.

The development of intense radioactive ion beams produced through high energy projectile fragmentation in heavy ion reactions has led to a dramatic progress in the study of light neutron rich (*N/Z* ~ 2.5) unstable nuclei such as $^{11}Li, ^{11}Be, ^{14}Be$ and ^{17}B etc. Large production cross-section and very small velocity broadening of the fragments helped experimentalists to make A/Z separation of fragments.

These experimental studies included: (1) measurement of interaction cross section, fragmentation and electromagnetic disassociation cross sections, and (2) measurement of spectra of particles from fragmentation. The measurement of first type gives global properties of exotic nuclei, e.g., matter radii, an extended neutron distribution. The measurements of second type e.g., the differential distribution of fragments by coincidence measurement provide information on two neutron correlation. These experimental studies have revealed the following unexpected features[16-18]:

1. Two neutron (2n) separation energy is quite small.
2. These nuclei have an unusually large size, much larger than predicted by $r_0 A^{1/3}$.
3. Fragmentation studies show that two neutron removal cross sections are strikingly large. For example, the experimental two neutron removal cross section for ^{11}Li projectile on ^{208}Pb target for beam energy 800 MeV/A is around 800 mb.

Further, in such studies transverse momentum distribution of neutrons and longitudinal momentum distribution of ^{9}Li fragments have been found strikingly narrow and a narrow momentum distribution corresponds to large spatial distribution in accordance to the uncertainty principle. The structural properties of halo nuclei have been extensively studied within the framework of various three-body models.[19-20]

The nucleon-matter radii of β-unstable nuclei are determined from an interaction cross section measurements in transmission experiment with high energy radioactive nuclear beams. The interaction cross section σ_I is defined as the total probability of one or more nucleon removal from the projectile.

The reaction cross section σ_R, on the other hand, is defined as the difference of the total cross section σ_T and the elastic scattering cross section σ_E, i.e. $\sigma_R = \sigma_T - \sigma_E$.

The interaction cross section $\sigma_I < \sigma_R$ because σ_I does not include target excitation. However, at high energy collision $\sigma_R \approx \sigma_I$. The interaction radius is defined on the basis of a simple geometrical model as

$$\sigma_I = \pi \left[r_I(P) + r_I(T) \right]^2, \tag{6.33}$$

where P and T indicate projectile (P) and the target (T). The interaction radius defined above neglects the difference ($\sigma_R - \sigma_E$) and it is essentially radius of a completely absorbing sphere (black sphere). The most used model for the calculation of interaction cross section σ_I and the reaction cross section σ_R is the Glauber model,[20] which is discussed in the next section in the context of the high energy heavy ion collisions. The reaction cross section in the Glauber model is given by

$$\sigma_I \approx \sigma_R = 2\pi \int_0^\infty \left[1 - T(b) \right] b\,db, \tag{6.34}$$

where $T(b)$ is the transmission for an impact parameter b and is calculated from the nucleon density distribution $\rho(r)$ and the total nucleon-nucleon (NN) scattering cross section within the frame work the Glauber model[21]:

$$T(b) = \exp\left[-\sum_{i,j} \sigma_{ij} \int ds\, \rho_{ti}(s)\, \rho_{pj}(|\vec{b} - \vec{s}|) \right],$$

where summation indices i, j run over protons and neutrons. σ_{ij} are the experimental N-N cross section at given energy[23,24], $\rho_t(s)$ and $\rho_p|\vec{b} - \vec{s}|$ are respectively, z-integrated target and projectile densities

$$\rho(\omega) = \int_{-\infty}^\infty \rho(\omega^2 + z^2)^{1/2}\, dz$$

with $\omega = x^2 + y^2$. Here we are considering the scattering of a projectile nucleon with a target nucleon located at r with impact parameter b, where $\rho_p|\vec{b} - \vec{s}|$ where $\vec{r} = \vec{s} + \hat{n}z$.

σ_R is parametrized by taking into account the volume and surface contribution as

$$\sigma_R = \pi \, (r_{vol} + r_{surf})^2 \left[1 - \frac{V_{coul}}{E_{c.m}} \right],$$

where $(r_{vol} + r_{surf})$ are the radii of the interaction region separated into volume and surface contributions and V_c is the Coulomb barrier given by (6.1).

Here
$$r_{vol} = \gamma_0 (A_P^{1/3} + A_T^{1/3}) \text{ and } r_{surf} = \gamma_0 \left(\frac{A_P^{1/3} A_T^{1/3}}{A_P^{1/3} + A_T^{1/3}} - c \right), \tag{6.35}$$

where the parameters γ_0, a and c are determined by fitting many data of σ_R.

Let us look at equations (6.34) in more detail. We first separate the density of the nucleus into the core and the halo parts,

$$\rho(r) = \rho_c(r) + \rho_h(r), \tag{6.36}$$

After substituting into (6.34), the interaction cross section (σ_I) is decomposed into a contribution from the core (σ_c) and that from the halo (σ_h),

$$\sigma_I = 2\pi \int b \, db \, [1 - T_c(b)] + 2\pi \int b \, db \, T_c(b) \, [1 - T_h(b)] = \sigma_c + \sigma_h, \tag{6.37}$$

where the relation $T(b) = T_c(b) T_h(b)$ is used.

σ_h is the reaction cross section of the halo neutrons by the target but without core. σ_h is modified by σ_c, i.e. by $T_c(b)$.

By extending the Glauber model[20] to calculate the two-nucleon removal cross section (σ_{-2n}) and the interaction cross section (σ_I), one finds that the halo nucleus is not modified when the two neutrons are removed from it. If we represent the ground state of the halo nucleus and the core by Ψ_0 and Φ_0 respectively, and by ϕ_0, the wave functions of the two halo neutrons, then we can write for a decoupled system $\Psi_0 = \Phi_0 \phi_0$. One can then obtain the relation

$$\sigma_{-2n}(\Psi_0) = \sigma_I(\Psi_0) - \sigma_I(\Phi_0), \tag{6.38}$$

where it is assumed no bound state exists between nuclei ψ_0 and Φ_0. Measured cross sections suggest that the above relation is satisfied. For example, for the fragmentation reaction $^{11}Li + ^{12}C \rightarrow ^{9}Li + 2n + ^{12}C$ at $790 \frac{MeV}{A}$ energy of ^{11}Li beam, we get

$$\sigma_{-2n} = (220 \pm 10) \text{ mb, } \sigma_I(^{11}Li) - \sigma_I(^{9}Li) = (260 \pm 20) \text{ mb.}$$

To get more insight into the structure of halo nuclei, it is quite important to understand for example, mechanism of two neutron removal cross section. Let us consider the role played by the Coulomb field in the direct fragmentation process:

$$a(b+c)+A \rightarrow b+c+A, \tag{6.39}$$

where the projectile '*a*' is formed by the fragments *b* and *c*. For example, in the case of ^{11}Li projectile, $^{11}Li \equiv {}^9Li + 2n$ *i.e.*, $b \equiv {}^9Li$ and $c = 2n$. The transition matrix for the above reaction can be calculated within the frame work post form DWBA, discussed in Chapter 5. The transition matrix in the present case is given by

$$T_{fi}^{(+)} = \langle \chi^{(-)}(q_b, r_b) \, \chi^{(-)}(q_c, r_c) \, | \, V_{bc}(r_{bc}) \, | \, \phi_a(r_{bc}) \, \chi^{(+)}(q_a, r_a) \rangle \tag{6.40}$$

where χ's are the distorted waves in the respective channels, ϕ_a is the intrinsic wave function of the projectile, q's are the relative momenta of $a + A$ and $c + A$ system, r's are the respective relative distances, and V_{bc} is the interaction between *b* and *c*.

The two neutron removal cross section (break up cross section) for ^{11}Li projectile is written approximately as a sum of Coulomb breakup and nuclear breakup cross sections:

$$\sigma_{-2n} = \sigma_{-2n}^C + \sigma_{-2n}^N, \tag{6.41}$$

neglecting the interference. The theoretical DWBA analysis of Shyam et al.[21] reveals that the Coulomb breakup can approximately be 80% of the two neutron removal cross section in the case of ^{11}Li on ^{197}Au collisions at 29 MeV/ A incident projectile energy.

Another plausible interpretation of the observed enhanced two-neutron removal cross section is to assume a soft E1 giant resonance resulting from the oscillation of weakly bound halo neutrons relative to the core. In either case the interpretation is based on the decoupling of halo neutrons from the core. The decoupled neutrons are sometimes treated as a 'di-neutron' cluster to represent correlative aspects of two neutrons.

We summarise in Table 6.3 properties of some of these halo nuclei.

Table 6.3: Properties of Halo Nuclei

Nucleus	N/Z	Configuration	S_{2n} (MeV)	R_{rms}^m (fm)	σ_{-2n} (mb)
$^{11}_{3}Li$	2.67	$n + n + {}^{11}Li$	0.294 ± 0.03	3.10-3.15	220 ± 10
$^{6}_{2}He$	2.0	$n + n + {}^4He$	0.97	2.46-2.52	189 ± 14
$^{19}_{5}B$	2.8	$n + n + {}^{17}B$	0.87	–	–
$^{17}_{5}B$	2.4	$n + n + {}^{15}B$	2.45	3.0	–
$^{14}_{4}Be$	2.5	$n + n + {}^{12}Be$	1.28	3.01-3.11	210 ± 10

In Table 6.3 the neutron removal cross sections (σ_{-2n}) are with a carbon target and radioactive beam energy $(E/A) = 790$ MeV. We see from Table 6.3 that the two neutron separation energy is quite small. In the case of ^{11}Li one neutron separation energy which is 1.05 MeV, is also quite small compared to the common (6-8) MeV value in stable nuclei. The matter radius is also nearly 1.5 times larger than the value obtained from the $A^{1/3}$ relation. These facts show that the two outer neutrons are loosely bound to the core and hence forming a halo. The halo nucleon has been squeezed out from the core by the Pauli principle. The neutron density distribution in such loosely bound nuclei shows an extremely long tail. Although the density of a halo is very low, it strongly affects the reaction cross section and leads to new properties in such nuclei. These radioactive halo nuclei have short half life. For example, $t_{1/2} = 10^{-3}$ sec for ^{11}Li, which decays as $^{11}Li(3/2, g.s) \rightarrow^{11} Be(1/2, 0.32\ MeV)$ $+ \beta^- + \overline{\nu}$. The near equality of magnetic moment (μ) and quadrupole moment (Q) for ^9Li and ^{11}Li as shown in Table 6.4 indicates that no great difference in shape exists between ^9Li and ^{11}Li and no drastic change in the spin structure. In other words, the charge (or proton) distribution does not visibly change from ^9Li and ^{11}Li. All the measurements carried out so far are consistent with the persistence of the proton distribution before and after the formation of a neutron halo.

Charge-changing cross sections (σ_{cc}) and the reaction cross sections (σ_R) of Li isotopes at 80A MeV on a carbon target indicate that the reaction cross section increases with the neutron number N, but the charge changing cross section (σ_{cc}) remains constant. This again shows that the charge distribution does not change greatly from ^9Li to ^{11}Li.

Table 6.4: Comparison of electromagnetic moments of *Li* isotopes

Nucleus	$(I^\pi)_{gs}$	$\mu\ (nm)$	$Q\ (mb)$
^9Li	$\dfrac{3}{2}-$	3.4931 ± 0.0006	-27.4 ± 1.0
^{11}Li	$\dfrac{3}{2}-$	3.6678 ± 0.0025	-31.2 ± 4.5

When the matter rms radii (R_{rms}^m) of halo nuclei determined from the interaction cross section (σ_I) are plotted with respect to the total isospin T_3, we can see an isospin dependence of R_{rms}^m. The isospin dependence of nuclear matter radii (Fig. 6.14) provides a new systematics that could not be studied without radioactive nuclear beams. Nuclei with larger isospins show larger radii.

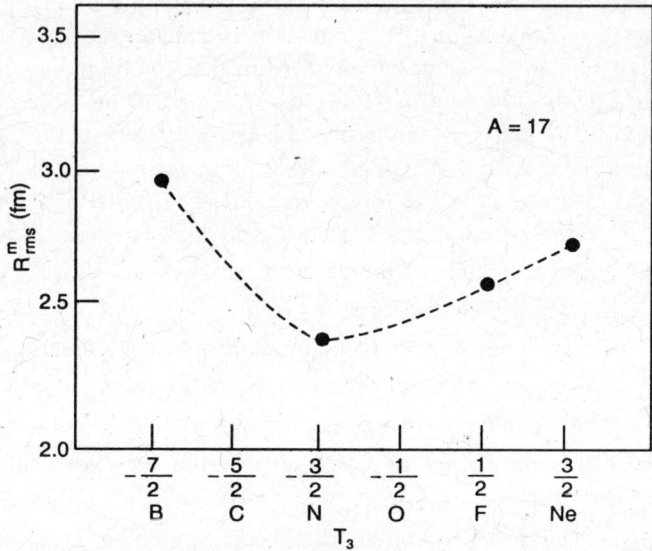

Fig. 6.14: Isospin dependence of rms radius

A. Physics of Halo Nuclei

The halo nuclei are expected to have loose three-body structure and is called "Borromean" - a word coined from the Greek mythology. It is supposed to be a system of three interlocking rings, in which cutting any one ring will free the other two as shown in Fig. 6.15.

In other words, it means that none of the two body sub-systems of the three-body system is bound. For example, ^{11}Li is bound but ^{10}Li is unbound by about 0.5 eV. The size of the two-body system, represented by the magnitude of s-wave scattering length a_S, should be much larger than the range a_S of the N-N force and that the absolute energy of the three-body system should be much smaller than $|E| << \hbar^2/\mu r_0^2$ where μ is the reduced mass and r_0 the radius of a halo nucleus. These conditions are both met for halo nuclei.

The apparent reason for the formation of a neutron halo is simple and can be understood in terms of a potential model[23]. Let us assume that a nucleus has a neutron loosely bound to an inert core, for example,

^{11}Be ($S_n = 0.503$ MeV) or ^{17}C ($S_n = 0.739$ MeV).

We assume for simplicity that the interaction potential between the neutron and the core is a square well potential having a range equal to the radius of

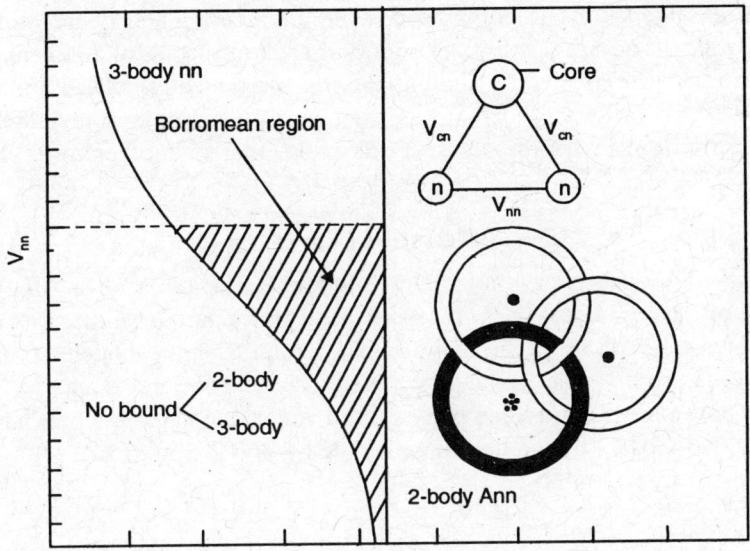

Fig. 6.15: Schematic diagram of 3-body ($A+n+n$) system with respect to V_{cn} and V_{nn} (*see* text). Adopted from P.G. Hansen and A.S. Jensen (Ref. 18)

the core nucleus. The unnormalized s-state wave function of the neutron outside the potential range is

$$\psi(r) \sim (e^{-kr}/r). \tag{6.42}$$

Using this wave function, the density distribution of the neutron is written as $\rho(r) = |\psi(r)|^2$. The parameter k, which determines the slope of the density tail, is related to the separation energy of the neutron (S_n) by $(\hbar k)^2 = \mu S_n$, where μ is the reduced mass of the system. As can be seen from these equations, when S_n decreases k becomes smaller and thus, the tail of the distribution become longer.

Although the surface diffuseness is known to be equal for all stable nuclei, that constancy is now understood simply as a reflection of the nearly-constant nucleon separation energy (6-8 MeV) for stable nuclei. In fact, the asymptotic slope of the density distribution of stable nuclei is consistent with the slope calculated by (6.42). In general, the surface diffuseness is expected to depend on the nucleon separation energy. The neutron halo is the most pronounced case for a small separation energy (< 1 MeV). The momentum distribution [$f(p)$] of the neutron is expressed by the Fourier transform of the wave function (6.42),

$$f(p) = \frac{C}{(p^2 + k^2)}, \tag{6.43}$$

where p is the Cartesian component of the momentum and C is a constant. The width of the momentum distribution is again related to parameter k. In contrast with the density distribution, the smaller ϵ mean smaller k and larger r. This is obviously a reflection of Heisenberg's uncertainty principle: when the distribution in coordinate space is wide, that in momentum space is narrow.

B. A Simple Three-Body Model of ^{11}Li

In this section, we present a simple variational calculation for a three-body model of a halo nucleus - ^{11}Li for example. This method is similar to the one already described (*Sec.* 2.3b) in the context of three identical nucleons except that in the present case we have two identical neutrons and a core.[22]

We consider ^{11}Li as a three body system consisting of a ^{11}Li core (labelled 3) and two loosely bound neutrons, labelled 1 and 2 (Fig. 6.16).

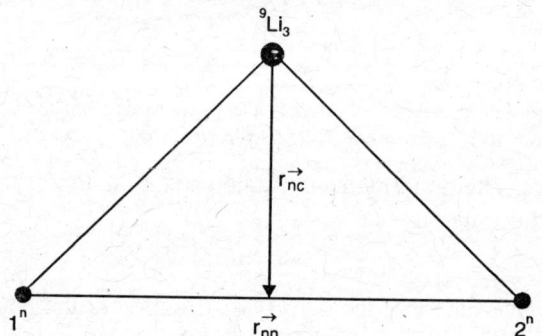

Fig. 6.16: Jacobi coordinates defined with respect to $\vec{r}_{nc}$ and $\vec{r}_{nn}$

The Hamiltonian of the system is

$$H = -\frac{\hbar^2}{2m_1}\nabla_{r_1}^2 - \frac{\hbar^2}{2m_2}\nabla_{r_2}^2 - \frac{\hbar^2}{2m_3}\nabla_{r_3}^2 + V_{nn}(r_{12}) + V_{cn}(r_{23}) + V_{cn}(r_{31}),$$

$$(6.44)$$

where m_i and r_i are the mass and position vector of the ith particle ($i = 1,2,3$). The n-n and the core-n interaction potentials are denoted by V_{nn} and V_{cn}, respectively. In order to separate the c.m motion we define Jacobi coordinates $\vec{\xi}_1$ and $\vec{\xi}_2$ and c.m. coordinate $\vec{R}$ as follows:

$$\vec{\xi}_1 = \left[\frac{m_1 m_2}{m(m_1 + m_2)}\right]^{1/2} (\vec{r}_2 - \vec{r}_1) = \sqrt{\frac{m_1}{2m}}(\vec{r}_2 - \vec{r}_1),$$

$$\vec{\xi}_2 = \left[\frac{m_3(m_1 + m_2)}{mM} \right]^{1/2} \left(\vec{r}_3 - \frac{m_1\vec{r}_1 + m_2\vec{r}_2}{m_1 + m_2} \right) = \sqrt{\frac{2m_3m_1}{mM}} \left[\vec{r}_3 - \frac{1}{2}(\vec{r}_1 + \vec{r}_2) \right],$$

$$\vec{R} = \frac{m_1\vec{r}_1 + m_2\vec{r}_2 + m_3\vec{r}_3}{M} \tag{6.45}$$

where
$$m_1 = m_2, \ M = \sum_{i=1}^{3} m_i = 2m_1 + m_3$$

and
$$m = \left\{ \sum_{i,j>1} m_i m_j \right\} M^{-1} = \frac{m_1(m_1 + 2m_3)}{2m_1 + m_3}.$$

Note that the coordinate ξ_1 is antisymmetric and ξ_2 is symmetric under the exchange of r_1 and r_2. Also $\frac{1}{3} \sum_{i,j>1} r_{ij}^2 = \xi_1^2 + \xi_2^2$. With the above choice of the Jacobi coordinates, the total Hamiltonian of the three-particle system can then be resolved into the c.m motion and relative motion:

$$H = -\frac{\hbar^2}{2M} \nabla_R^2 + \left[-\frac{\hbar^2}{2m}(\nabla_{\xi_1}^2 + \nabla_{\xi_2}^2) + V(\xi_1, \xi_2) \right],$$

where $V(\xi_1, \xi_2)$ is the sum of three pairwise interaction expressed in terms of the relative vectors ξ_1 and ξ_2.

In the above, we are not interested in the centre of mass motion, which gives the motion of the nucleus as free particle. Hence, we consider only the relative motion:

$$\left[-\frac{\hbar^2}{2m}(\nabla_{\xi_1}^2 + \nabla_{\xi_2}^2)\psi + V(\xi_1, \xi_2)\psi \right] = E\psi.$$

We choose three particle space wave function for the ground state of ^{11}Li (in s-state) as

$$\psi = Ne^{-\alpha(\xi_1^2 + \xi_2^2)}, \tag{6.46a}$$

where N is the normalization constant and α the variation parameter.

Now,
$$(4\pi)^2 N^2 \int_0^\infty e^{-2\alpha\xi_1^2} \xi_1^2 \, d\xi_1 \int_0^\infty e^{-2\alpha\xi_2^2} \xi_2^2 \, d\xi_2 = 1.$$

We consider the standard integral of the form:

$$\int_0^\infty x^{2n} e^{-ax^2} dx = \frac{1.3 \cdots (2n-1)}{2^{n+1} a^n} \sqrt{\frac{\pi}{a}}. \tag{6.46b}$$

In the present case $n = 1$, $a = 2\alpha$, hence $N = \left(\dfrac{2\alpha}{\pi}\right)^{3/2}$. Expectation value of the kinetic energy

$$\left\langle \psi_{\text{space}}(\xi_1, \xi_2) \middle| -\frac{\hbar^2}{2m} (\nabla_{\xi_1}^2 + \nabla_{\xi_2}^2) \middle| \psi_{\text{space}}(\xi_1, \xi_2) \right\rangle$$

is calculated as follows (for the s-state):

$$\nabla_{\xi_1}^2 = \frac{d^2}{d\xi_1^2} + \frac{2}{\xi_1} \frac{d}{d\xi_1}$$

$$(\nabla_{\xi_1}^2 + \nabla_{\xi_2}^2) e^{-\alpha(\xi_1^2 + \xi_2^2)} \equiv \left[-12\alpha + 4\alpha^2 (\xi_1^2 + \xi_2^2) \right] e^{-\alpha(\xi_1^2 + \xi_2^2)}.$$

Hence, $\left\langle \psi_{\text{space}}(\xi_1, \xi_2) \middle| -\dfrac{\hbar^2}{2m} (\nabla_{\xi_1}^2 + \nabla_{\xi_2}^2) \middle| \psi_{\text{space}}(\xi_1, \xi_2) \right\rangle$

$$= -\frac{\hbar^2}{2m} 16\pi^2 \frac{8\alpha^3}{\pi^3} \left[-\frac{3\pi}{64\alpha^2} \right] = \frac{3\hbar^2 \alpha}{m}. \tag{6.47}$$

Next, we calculate the expectation value of the potential energy as follows:

The *n-n* interaction V_{nn} is chosen as a standard singlet *n-n* interaction given by

$$V_{nn}(r_{12}) = V_{10} e^{-\mu_1 c^2 \xi_1^2} \text{ where } c = \sqrt{\frac{2m}{m_1}}. \tag{6.48}$$

Here $V_{10} = -31$ MeV and $\mu_1 = 0.3086$ fm^{-2} which yield respectively, the effective range and scattering length $r_{0s} = 2.76$ fm and $a_s = 23.75$ fm, obtained from N–N scattering data. The core-nucleon potential (V_{cn}) is not accurately known – there are no direct experimental informations about this interaction. We assume a sum of two Gaussians:

$$V_{cn}(r_{23}) = \sum_i V_{i0} e^{-\mu_i r_{23}^2}; \ i = a \text{ (attractive)}, r \text{ (repulsive)}$$

with $r_{23} = \left(d\xi_2 - \dfrac{c}{2} \xi_1 \right)$ where $d = \sqrt{\dfrac{mM}{2m_3 m_1}}$ and $\mu_a = 0.153786$ fm^{-2}.

V_{a0} is varied to produce two neutron separation energy. A short range $\mu_r = 0.6$ fm^{-2} and a strongly repulsive $V_{r0} = 65$ MeV are included in V_{cn} to simulate the Pauli exclusion principle between the valence neutrons and core nucleons. In choosing the parameters μ_2 and V_{r0}, we ensure that the two body $^9Li - n$ system is unbound by 0.5 MeV. Calculation of the expectation value of the potential energy is straight forward. Noting

$$V_{cn}(r_{23}) + V_{cn}(r_{13}) = 2V_{cn}(r_{23}),$$

we get

$$E(\alpha) = \langle \psi \mid H \mid \psi \rangle = \frac{3\hbar^2 \alpha}{m} + V_{10}\left[1 + \frac{\mu_1}{\alpha}\frac{m_1 + 2m_3}{2m_1 + m_3}\right]^{-1/2}$$

$$+ \sum_i 16V_{i0}\left[4 + \frac{2\mu_i}{\alpha}\frac{(m_1 + 2m_3)(m_1 + m_3)}{m_3(2m_1 + m_3)}\right]^{-1/2},$$

$$i = a, r. \tag{6.49}$$

Employing the Rayleigh-Ritz variational principle, the variational parameter α is determined by the condition

$$\frac{\partial E}{\partial \alpha} = 0. \tag{6.50}$$

The differentiation with respect to α is done analytically and the resulting equation is solved numerically to obtain α. Our calculation yields $\alpha = 0.055$ fm^{-2} and $V_{a0} = -23.75$ MeV corresponding to minimum for $^9Li + n + n$ system. The above 3-body model predicted BE of ^{11}Li, which compares well with the experimental value as shown below:

$$BE(^{11}Li)_{Theory} = 297 \text{ keV} \qquad BE(^{11}Li)_{Expt.} = (295 \pm 35) \text{ keV}.$$

Next we calculate the mean square halo radius defined as

$$(R_{halo})^2_{gs} = \langle \mid \frac{1}{2}(r_{13}^2 + r_{23}^2)\mid \rangle_{gs} = \langle \mid d^2\xi_2^2 + \frac{c^2}{4}\xi_1^2 \mid \rangle_{gs} = \frac{3}{4\alpha}\left(\frac{c^2}{4} + d^2\right)$$

$$= \frac{3}{4\alpha}\frac{(m_1 + 2m_3)(m_1 + m_3)}{(2m_1 + m_3)m_3}.$$

The numerical value of $(R_{halo})_{gs} = 5.11$ fm, which compares well with $(R_{halo})_{Expt.} = 5.1$ fm. The mean square matter radius of a halo nucleus (core + 2n) is

$$(R_{rms}^m)^2 = \frac{A-2}{A}(R_{rms}^c)^2 + \frac{2}{A}(R_{halo})_{gs}^2 .$$

For ^{11}Li $\qquad (R_{rms}^m)^2 = \frac{9}{11}(R_{rms}(^9Li))^2 + \frac{2}{11}(R_{halo})_{gs}^2.$

Taking $(R_{rms} \, ^9Li) = 2.5$ fm, we get $(R_{rms}^m,(^{11}Li))_{Theory} = 3.14$ fm, which compares well with the measured value

$$(R_{rms}^m,(^{11}Li))_{Expt.} = (3.10 \pm 0.17) \text{ fm}.$$

Exercise 6.6. ^{11}Li nucleus decays into ^9Li core and two neutrons. Express the excitation energy of ^{11}Li in terms of the relative v_{2n-9} and v_{nn}.

The excitation energy, $E = E_d + S_{2n}$, where $S_{2n} \, (= 0.34 \pm 0.05$ MeV) is the two neutron separation energy and E_d is the decay energy given by

$$E_d = \frac{1}{2}\mu_{cn} \, v_{2n-9}^2 + \frac{1}{2} \, \mu_{nn} \, v_{nn}^2,$$

where $\qquad \mu_{cn} = \dfrac{m_9(2m_n)}{m_9 + (2m_n)} \quad$ and $\quad \mu_{nn} = \dfrac{m_n}{2}.$

Exercise 6.7. Employing the asymptotic wave function (6.42) show that the rms radius diverges as the separation energy S_n decreases to zero.

$$\langle r^2 \rangle = \frac{\displaystyle\int_0^\infty r^2 \, e^{-2kr} \, dr}{\displaystyle\int_0^\infty e^{-2kr} \, dr} = \frac{1}{2k^2} = \frac{\hbar^2}{4\mu S_n}, \text{ so also } S_n \to 0 \text{ as } \langle r^2 \rangle \to \infty.$$

Exercise 6.8. Calculate $\langle r_{nn}^2 \rangle^{1/2}$ and $\langle r_{nc}^2 \rangle^{1/2}$ of ^{11}Li ground state employing the wave function (6.46) with $\alpha = 0.055$ fm^{-2}.

$$\langle r_{nn}^2 \rangle_{gs} = \langle c^2 \xi_1^2 \rangle = N^2 c^2 (4\pi)^2 \int e^{-2\alpha(\xi_1^2 + \xi_2^2)} \times \xi_1^4 \, \xi_2^2 \, d\xi_1 \, d\xi_2$$

$$= \frac{8\alpha^3}{\pi^3} \cdot \left(\frac{2m}{m_1}\right) 16\pi^2 \times \frac{3}{32\alpha^2}\sqrt{\frac{\pi}{2\alpha}} \frac{1}{8\alpha}\sqrt{\frac{\pi}{2\alpha}} = \frac{3m}{2m_1\alpha}.$$

Substituting the value of α, we get $\langle r_{nn}^2 \rangle_{gs}^{1/2} = 6.86$ fm. Following the same method we get

$$\langle r_{nc}^2 \rangle_{gs} = \frac{3}{4} \left(\frac{mM}{2m_3 m_1} \right) \frac{1}{\alpha} \cdot \text{Hence}, \langle r_{nc}^2 \rangle_{gs}^{1/2} = 3.79 \text{ fm}.$$

6.4 HIGH ENERGY HEAVY ION COLLISIONS

The characteristic features of the interactions between nuclei at high energies from a geometrical concept, depend sensitively on the impact parameter of the collision. In particular, the limiting values of the impact parameter $b(= r_T + r_P)$ and the incident projectile energy can be used to provide us a guide map for the various reaction processes. Figure 6.17 shows the reaction cross section (essentially relative impact parameter) as a function of energy. Before we begin the detailed interpretation of data, it is useful to illustrate what nuclear collisions look like.

(1) Collisions occurring at large impact parameter in which projectile and target nuclei barely touch each other, the energy momentum transfer between these two nuclei are relatively small so that projectile nucleus breaks up into few fragments with velocities close to the beam velocity. These are called "peripheral collisions".

(2) Collisions in which large number of particles are observed at large angles suggest two types of mechanism: one in which projectile nucleus interacts weakly with target producing particles in the forward direction and the

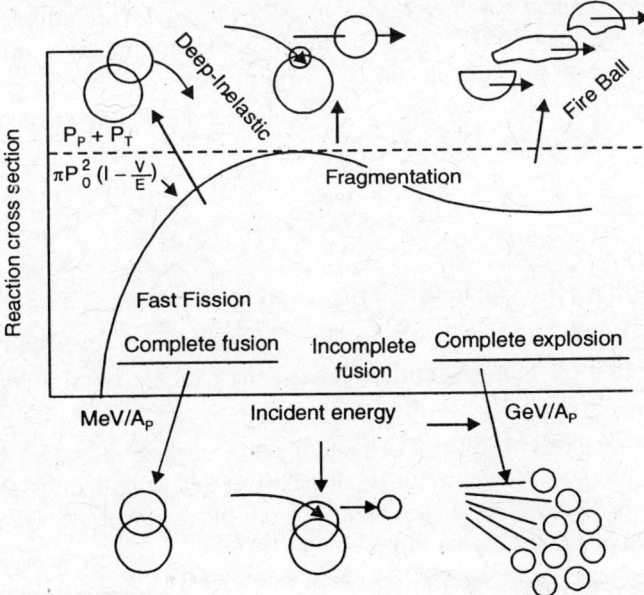

Fig. 6.17: Heavy ion reaction cross section shown schematically with respect to the incident energy

other in which the remaining part of the projectile interacts strongly with the target creating particles at the large angles.

(3) Central collisions (violent collisions) which originate at a small impact parameter and produce particles over wide range of angles. In such a process, projectile interacts strongly with the target nucleons; the available energy is shared amongst all participating nucleons and pions are produced in the collision.

An interesting question is whether the classical concepts can be applied to calculate inclusive or exclusive differential cross sections. In general, there are distinct interference effects between different partial amplitudes in a differential cross section. However, with nuclei, such interference effects can be expected to be less important because of the enormous number of final channels accessible.

The outer boundary in Fig. 6.17 is defined by $\sigma_R = \pi r^2 (1 - V/E)$ [Eq. 5.19], where $r = r_T + r_P$ and $V =$ the barrier height. For large impact parameter, the cross section is dominated by the grazing processes, such as simple transfer and inelastic scattering. For closer collisions at low energies, the deeply inelastic process takes over. For close collisions, when the impact parameter is of the order of 1.00 $(A_T^{1/3} + A_P^{1/3})$ fm, the deeply inelastic reactions merge into the complete fusion of a compound nucleus.

For a given impact parameter, we can ask next, how quantum phenomena affect the dynamical path? While the total angular momentum is large, the relative angular momentum in each nucleon-nucleon (N-N) collision is not. Taking the force range $r = \hbar / m_\pi c = 1.4$ fm, the relative angular momentum is only

$$\ell = \left(\frac{k}{m_\pi c} \right) \hbar.$$

For $k = 500$ MeV/c, we have $\ell = 3\hbar$, which can be regarded as large even initially. After only one N-N collision, the subsequent N-N collision will have $\ell < \hbar$. This implies that quantum mechanical treatment of individual N-N collision is necessary.

Projectile Fragmentation: The phenomenon of the production of many complex fragments in the intermediate energy nuclear collisions may take place either through (*a*) multistep binary decay like sequential fission of the excited nuclei or (*b*) through a more violent 'single step' true multifragmentation. Experiments with heavy ions at a few GeV/A projectile energy on various targets showed that the fragments observed at forward angles have a Gaussian form in the rest frame of the projectile.

To interpret such a momentum distribution of fragments, a simple statistical model has been introduced by Goldhaber[23]. In this model, F nucleons are picked up from the projectile to make the fragment. The momentum width of a fragment is divided as follows:

The dispersion of the longitudinal (Z-direction) momentum distribution in the projectile rest frame is given by the expectation value of the squared momentum of F nucleons, i.e.

$$\sigma^2 = \left\langle \left\{ \sum_i^F p_z(i) \right\}^2 \right\rangle = F\langle p_z^2(1)\rangle + F(F-1)\langle p_z(1)\,p_z(2)\rangle. \quad (6.51a)$$

The first term in the above expression (6.51a) can be estimated in the Fermi gas model (Eq. 1.36b):

$$\langle p_z^2 \rangle = \frac{1}{3}\left(\frac{3}{5}\,p_F^2\right) = \frac{1}{5}\,p_F^2, \quad (6.51b)$$

where p_f is the Fermi momentum in the projectile. To calculate the second term in Eq. (6.51a) we use the fact that the total momentum of the projectile is zero so that

$$\left\langle \left\{ \sum_{i=1}^A p_z(i) \right\}^2 \right\rangle = A\langle p_z^2(1)\rangle + A(A-1)\langle p_z(1)\,p_z(2)\rangle = 0$$

or
$$\langle p_z(1)\,p_z(2)\rangle = \frac{1}{A-1}\,\langle p_z^2(1)\rangle. \quad (6.51c)$$

Combining (6.51a) and (6.51c) we find $\sigma^2 = \dfrac{F(A-F)}{A-1}\,\dfrac{p_f^2}{5}$. With σ^2 defined above we can write

$$\frac{d\sigma}{dp} = e^{-p^2/2\sigma^2}, \text{ where } p \text{ is the horizontal momentum.} \quad (6.52)$$

Thus we see that the momentum distribution of fragments has a Gaussian form.

Glauber Model: The semiclassical method for quantum scattering problem assumes a rather simple form in the domain of high energies where it may safely be assumed that the projectile follows a straight line trajectory within the interaction region. This straight line approach has been extensively studied and developed for high energy nucleus-nucleus collisions by Glauber and others and it is known in the literature by the Glauber model.[20]

To describe the dynamics of nucleus-nucleus collision process, it is useful to consider collision at nucleon level. After an incident projectile nucleon suffers a collision, the resultant energetic nucleon like object can be treated loosely as projectile object, which continue to make further collisions along the direction of the projectile. The Glauber model for multiple collision process provides a quantitative consideration of the geometrical configuration of the nuclei when they collide. The model is based on the following assumptions:

(1) The concept of mean free path is used with the assumption of nucleon cross section.

(2) We assume that the basic nucleon-nucleon (NN) cross section remains the same throughout the passage of the nucleon through the other nucleus.

With the straight line assumption, the average number of collisions that an incident nucleon suffers in a target at impact parameter b is

$$<n(b)> = \int dz \rho(z, b) \, \sigma_T^{NN},$$

where $\rho(z, b)$ is the target nucleon density distribution and σ_T^{NN} is the total nucleon-nucleon (NN) cross section[24].

Let us consider nucleus-nucleus collision with an impact parameter b as shown in Fig. 6.18. Let the impact parameters for a nucleon in the nucleus 1 and that for a nucleon in the nucleus 2 are respectively, b_1 and b_2. We neglect the transverse motion of nuclei when they pass each other. Then the reaction cross section in the Glauber theory framework is

$$\sigma_R = \int_0^\infty b \, db \left[1 - \exp\left\{ -\sigma_T^{NN} \, \chi(b) \right\} \right], \qquad (6.53)$$

where $\chi(b) \sigma_T^{NN}$ is the total probability of the occurrence of *NN* collisions. We are considering a *NN* collision within the transverse area element db when one nucleon is situated at an impact parameter b relative to another nucleon. For two nuclei in the form of two spheres, the Glauber thickness function $\chi(b)$ can be written as

$$\chi(b) = \int_{-\infty}^{\infty} d^2 b_1 \, \rho_z^1(|\vec{b}_1|) \, \rho_z^2(|\vec{b}_1 - \vec{b}|), \qquad (6.54)$$

where $\rho_z(r)$ is the Z direction integrated nucleon density distribution at the location r,

$$\rho_z(r) = \int dz \, \rho(r^2 + z^2).$$

In terms of the transparency function $T(b)$, i.e. the probability that at the impact parameter b, the projectile will pass through the target without interaction, σ_R can be expressed according to (6.34).

We shall now calculate $\chi(b)$ analytically making some assumptions. For spherical and unpolarized nuclei $\chi(b)$ depends on magnitude of b so that we can write $\chi(\vec{b}) = \chi(b)$. We assume for simplicity that the density distribution ρ_z has the Gaussian form:

$$\rho_z^1(\vec{b}_1) = \sqrt{\pi}\ r_1 \rho_1(0)\ e^{-b_1^2/r_1^2}, \tag{6.55a}$$

$$\rho_z^2(|\vec{b}_1 - \vec{b}\,|) = \sqrt{\pi}\ r_2 \rho_2(0)\ e^{-(\vec{b}_1 - \vec{b})^2/r_2^2}, \tag{6.55b}$$

with the normalization $\displaystyle\int_{-\infty}^{\infty} \rho_z^1(\vec{b}_1)\,db_1 = \int_{-\infty}^{\infty} \rho_z^2(|\vec{b}_1 - \vec{b}\,|)\,db = 1.$ (6.56)

For light nuclei r_i is related to root mean square radius (r_{rms}) by

$$r_i = \sqrt{\frac{1}{1.5}}\ r_{rms} \quad \text{and}\quad \rho_i(0) = A_i/(r_i\sqrt{\pi})^3;\ i = 1, 2.$$

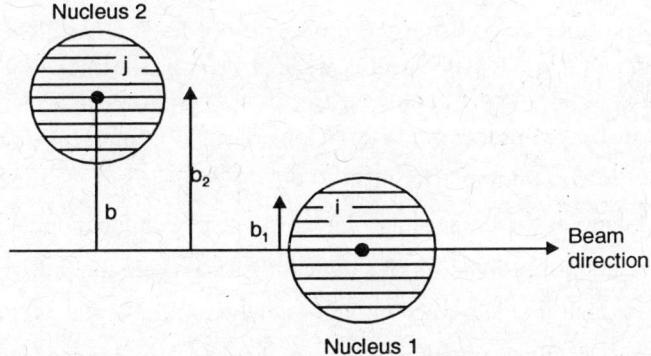

Fig. 6.18: Schematic diagram of a nuclear collision with an impact parameter b. The impact parameter for a nucleon in nucleus 1 and that for a nucleon in nucleus 2 are, respectively, b_1 and b_2

We now evaluate $\chi(b)$ explicitly to get σ_R as follows:

$$\chi(b) = C \int_{-\infty}^{\infty} \exp\left(-\frac{b_1^2}{r_1^2} - \frac{(\vec{b}_1 - \vec{b})^2}{r_2^2}\right) d^2 b_1 \quad \text{where}\ C = \pi r_1 r_2\ \rho_1(0)\rho_2(0)\ \ (6.57)$$

Now, $$\chi(b) = C \exp(-b^2/r_2^2)\ I(b_1),$$

where $$I(b_1) = \int_{-\infty}^{\infty} \exp(-b_1^2 p^2 + \vec{b}_1 \cdot \vec{b} q^2)\,d^2 b_1$$

with $p^2 = \dfrac{1}{r_1^2} + \dfrac{1}{r_2^2}$ and $q^2 = \dfrac{2}{r_2^2}$. Integrating we get

$$I(b_1) = \frac{\pi r_1^2 r_2^2}{(r_1^2 + r_2^2)} \exp\left\{\frac{r_1^2 b^2}{r_2^2(r_1^2 + r_2^2)}\right\}, \text{ using the standard integral,}$$

$$\int_{-\infty}^{+\infty} \exp(-p'^2 z^2 + q'z)\, d^2z = \frac{\pi}{p'^2} \exp(q'^2/4p'^2).$$

Hence, $\chi(b) = \dfrac{\pi^2 r_1^3 r_2^3 \rho_1(0)\, \rho_2(0)}{(r_1^2 + r_2^2)} \exp\left\{\dfrac{-b^2}{r_1^2 + r_2^2}\right\}.$

Thus, transparency function $T(b)$ gets reduced to a simple form:

$$T(b) = \exp\left[-\sigma_T^{NN}\pi^2 \frac{\rho_1(0)\, \rho_2(0)\, r_1^3 r_2^3}{(r_1^2 + r_2^2)} \exp\left\{\frac{-b^2}{r_1^2 + r_2^2}\right\}\right]. \tag{6.58}$$

For nucleon-nucleus collision the probability of having n collisions at an impact parameter b is calculated as follows. Let $\chi(b)$ be the thickness function of the nucleus at the impact parameter b and let N be number of nucleons in the nucleus and n the number of NN collisions. The $^N C_n$ represents the number of combination for finding n collisions out of N possible NN encounters. $[\chi(b)\, \sigma_{in}]^n$ denotes the probability of occurrence of n independent collisions (being the product of probability of single collision) where the indices 'in' are used to distinguish incident nucleon and nucleons in the target. Accordingly $[1 - \chi(b)\, \sigma_{in}]^{N-n}$ gives the probability of having N-n misses. Hence the required probability

$$P(n, b) = N_{C_n}[\chi(b)\, \sigma_{in}]^n [1 - \chi(b)\, \sigma_{in}]^{N-n}$$

$$\approx \frac{N^n}{n!}[\chi(b)\, \sigma_{in}]^n \left[e^{-\chi(b)}\, \sigma_{in}^{(N-n)}\right], \text{ since } \frac{N!}{n!\,(N-n)!} = \frac{N^n}{n!}.$$

$$\approx \frac{(\chi(b)\, \sigma_{in}^N)^n}{n!}\, e^{-\chi(b)}\sigma_{in}^N.$$

Now, $<n(b)> \approx \bar{n}(b) = \chi(b)\, \sigma_{in}^N$. Hence, $P(n,b) = \dfrac{[\bar{n}(b)]^n}{n!}\, e^{-\bar{n}(b)},$ $\quad$ (6.59) which is a Poisson distribution and $\bar{n}(b)$ is the average number of collisions at the impact parameter b. Employing the above distributions we can express

$$\sigma_{AB} = \int d^2b \ B(b) \ \int d^2b_A \ d^2b_B \ \delta(b - b_A - b_B) \ \sigma_T^{NN}, \qquad (6.60)$$

where m, n are respectively nucleons in projectile A and target B. $B(b)$ is the impact parameter distribution.

Hydrodynamical Model: The dynamics of the collision becomes much simpler if all the microscopic interactions only serve to establish equilibrium. In this case, they do not have to be followed in detail, and one use a description in terms of more macroscopic variables such as density or temperature. A derivation of hydrodynamics from the Boltzman equation shows that the size of the mean free path between collisions plays a decisive role together with the relaxation time, i.e. the time required to distribute microscopic disturbances amongst many particles such as to establish local equilibrium.

The application of hydrodynamical model ideas to high energy heavy ion collisions is centred on the question of compression of nuclear matter in such collisions so that the projectile velocity v_p becomes greater than the velocity of sound in the nuclear matter v_s, resulting the formation of nuclear shock waves. The velocity of sound in nuclear matter is given by $v_s = \sqrt{\dfrac{K}{9M}}$ or $K = 9 M v_s^2$, where M is the nucleon mass and K is the nuclear compressibility defined by

$$K = \left[9\rho^2 \frac{\partial^2 W(\rho)}{\partial \rho^2} \right]_{\rho_0}. \qquad (6.61)$$

Here, $W = E/A$ is the nuclear matter energy per nucleon and ρ is the nuclear density. $W(\rho)$ has a minimum at $\rho = \rho_0 = 0.14 \ \text{fm}^{-3}$ with $W_0 = -16 \ \text{MeV}$. The value of K is about $(210 \pm 20) \ \text{MeV}$. K is not a directly measurable quantity, but derivable from experimental observables like giant monopole resonance and model construction. Substituting the above value of K we get $v_s = 0.17c$. If v_s is greater than the projectile velocity v_p, then the initial compression of the front layer under impact will distribute itself through the rest of the ions in an adiabatic fashion. On the other hand if $v_p > v_s$, then before the compression of the front layer can be dissipated into succeeding layers, the latter would have rammed into the front layers and come to stop. This is the situation of shock front propagation.

In high energy collision, since the incoming projectile velocity is larger than the sound velocity (supersonic), the matter piles up in the overlap or interaction zone, thus, building up higher density. We can expand $W(\rho)$ near the minimum as

$$W(\rho) = W_0 + (\rho - \rho_0)\left(\frac{\partial W}{\partial \rho}\right)_{\rho_0} + \frac{1}{2}(\rho - \rho_0)^2 \left(\frac{\partial^2 W}{\partial \rho^2}\right)_{\rho_0}. \qquad (6.62)$$

The first derivative of energy per particle at equilibrium density ρ_0 vanishes so that

$$W(\rho) = W_0 + \frac{1}{2}(\rho - \rho_0)^2 \left(\frac{\partial^2 W}{\partial \rho^2}\right)_{\rho_0}.$$

Therefore, according to (6.61) we get,

$$W(\rho) - W_0 = \frac{1}{2}\left(\frac{\rho - \rho_0}{\rho_0}\right)^2 \cdot \frac{K}{9}.$$

If the available extra energy per nucleon in the ion-ion collision is about 250MeV/A, while $K \approx 200$ MeV, then $\dfrac{\rho - \rho_0}{\rho_0} \cong 5$ or $\rho = 6\,\rho_0$.

This is assuming that all the incident energy goes upto compression. This building up higher density is resisted by the nuclear compression modulus. At the end, the tremendous compressional energy is released through decompression of piled up matter. For peripheral collision, this deflects the projectile and target matter sideways. Equations governing the heavy ion collisions will be simply classical equation of hydrodynamics which can be formulated as conservation equations for mass momentum and energy.

To complete the set of equations of motion in the hydrodynamic model, an equation of state has to be specified. The nuclear matter energy per nucleon at zero entropy as a function of density is set equal to the binding energy $W_0 = W(\rho)$. For finite entropy per nucleon s, the corresponding excitation energy of a Fermi gas is added

$$W(\rho, s) = W_0(\rho) + W_F(\rho, s). \qquad (6.63a)$$

Accordingly, the temperature and pressure are given by

$$T = \left(\frac{\partial W}{\partial s}\right)_\rho \quad \text{and} \quad P = \rho^2 \left(\frac{\partial W}{\partial \rho}\right)_s. \qquad (6.63b)$$

(6.61) and (6.63b) lead to another definition of $K = 9\left(\dfrac{\partial P}{\partial \rho}\right)_{\rho_0}$.

Pressure (P), temperature (T) and volume (V) are called state variables in thermodynamics. They are not fully independent; a relationship of the type $f(P, V, T) = 0$, restricting their independence is what we could call an equation of state (EOS). For the nuclear system, the information that we have in the entire W-ρ-T space is at $T = 0$, $W \sim -16$ MeV for $\rho = \rho_0 = 0.14$ fm^{-3} (equilibrium density at $T = 0$). To have more information we have to perturb the nucleus by subjecting it to compression and to a higher temperature and measuring the responses of the nucleus towards these perturbations.

Participant Spectator Model: In this model[28] we estimate an average number of participant and spectator nucleons in HI collision on the basis geometrical model (Sec. 5.2) : $a + b \to c + x$, where A_P and A_T are, respectively, number of nucleons in the projectile and target. If a proton inside the projectile hits the target, it is classified as a participant, otherwise it remains as spectator.

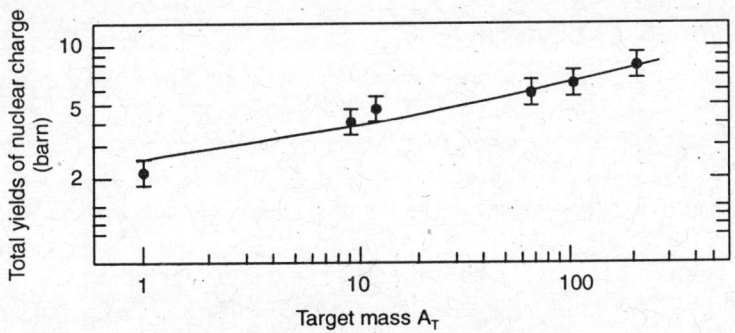

Fig. 6.19: Total yields versus target mass

The average number of participant protons from the projectile is Z_P multiplied by the ratio of the target cross section to σ_G:

$$\langle Z_{proj}^{parti} \rangle = Z_P \frac{\pi a_0^2 A_T^{2/3}}{\sigma_G} = \frac{Z_P A_T^{2/3}}{(A_P^{1/3} + A_T^{1/3})^2},$$

where Z_P is the number of protons in the projectile. Similarly, we have

$$Z_{target}^{parti} = \frac{Z_T A_P^{2/3}}{(A_P^{1/3} + A_T^{1/3})^2},$$

where Z_T is the number of protons in the target. The total number of proton participants :

$$(1) \quad Z_{tot}^{parti} = \langle Z_{proj}^{parti} \rangle + \langle Z_{target}^{parti} \rangle = \frac{Z_P A_T^{2/3} + Z_T A_P^{2/3}}{(A_P^{1/3} + A_T^{1/3})^2},$$

Similarly, the total number of protons assigned to the projectile spectator and target spectator are respectively, given by

$$(2) \quad Z_{tot}^{proj\ spect} = Z_P - \langle_{proj}^{parti} \rangle = \frac{Z_P (A_P^{2/3} + 2 A_P^{1/3} A_T^{1/3})}{(A_P^{1/3} + A_T^{1/3})^2}.$$

$$(3) \quad Z_{tot}^{targ\ spect} = \frac{Z_T (A_T^{2/3} + 2 A_P^{1/3} A_T^{1/3})}{(A_P^{1/3} + A_T^{1/3})^2}.$$

The total integrated inclusive cross section of nuclear charge for each of these three regions ($i = 1, 2, 3$) is therefore, expressed as

$$\sigma_{tot}^{Ch}(i) = Z_{tot}^{(i)} \times \sigma_G \qquad (6.64)$$

In Fig. 6.19 the above formula (6.64) is compared with data for ^{16}O projectile of beam energy 2.1 GeV/A. The observed projectile and target mass dependence are well reproduced by (6.64) for $a_0 = 0.95$ fm.

A. Thermodynamic Model

The models developed on the assumption of rapid thermalization through multiple collisions are called the thermodynamic models for relativistic heavy ion collisions. The advantage of the thermodynamic approach is that, dynamics of collision does not have to be solved. Many complex dynamical processes lead to thermodynamics. The thermodynamic models originally developed by the Berkley group[26] are the *fireball* and *firestreak* models. Following assumptions are made: (1) All hadrons are in thermal equilibrium in the nuclear fireball, (2) all strong interactions are turned off when the hadron number density reaches some critical value and (3) non-interacting gas formula can be applied at the critical density.

Nuclear Fireball Model : The basic premise of nuclear fireball model is the following: For a given impact parameter b there will be an overlap between

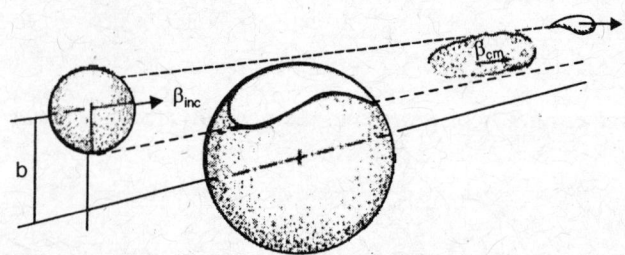

Fig. 6.20: Schematic representation of the nuclear fireball model. (Adopted from Ref. 26)

target and projectile. After collision, the overlapping regions of the target and projectile fuse together and come to rest in the c.m. of the fireball (Fig. 6.20). The available kinetic energy before the collision goes into creating thermal motion after the collision. Particles are then emitted isotropically in the rest frame of the fireball. Because the c.m. of the fireball is moving in the laboratory, the inclusive spectra in the laboratory will in general be anisotropic. Thus, the fireball model has three essential ingredients: geometry to calculate the number of nucleons in the fireball; kinematics to calculate the velocity of the fireball and its excitation energy, and thermodynamics to describe its decay.

Geometry: In the fireball model the target and projectile are assumed to make clean cylindrical cuts through each other having target and projectile spectator residues. Assuming the projectile and target to be uniform density spheres (radius = $1.2A^{1/3}$), one can calculate as a function of impact parameter b, the participating volume $V_i(b)$ of each nucleus ($i = T$ or P) from the extrapolated straight line trajectories.

The participating target nucleons = $V_T(b)\,\rho$, where ρ is the nucleon density. Participating protons from target = $V_T(b)\,\rho\,\dfrac{Z_T}{A_T}$.

The total participating protons

$$N(b) = \rho\left[V_T(b)\frac{Z_T}{A_T} + V_P(b)\frac{Z_P}{A_P} \right] \qquad (6.66)$$

or

$$N_{protons} = \int_{0}^{r_T + r_P} 2\pi b N(b)\, db. \qquad (6.67)$$

Kinematics: Employing relativistic kinematics one can calculate the velocity of the centre of mass of the participant nucleons in the lab as $\beta_{c.m} = P_{lab}/E_{lab}$ where $\dot{P}_{lab}$ is the momentum of the system in the lab and E_{lab} is the total energy of the system in the lab. If t_i is the projectile incident kinetic energy per nucleon and m' is the rest mass of the bound nucleon (931 MeV) then

$$P_{lab} = N_P\,[t_i(t_i + 2m')]^{1/2}, \ E_{lab} = (N_P + N_T)m' + N_P t_i,$$

where N_P and N_T respectively denote participating nucleons from the projectile and target. The total energy in the centre of mass of the fireball

$$E_{c.m} = (E_{lab}^2 - P_{lab}^2)^{1/2}$$

$$= \left[(N_P + N_T)^2 m'^2 + N_P^2 t_i^2 + 2N_P t_i (N_P + N_T)m' - N_P^2 \{t_i^2 + 2m't_i\} \right]^{1/2}$$

$$= \left[(N_P + N_T)^2 m'^2 + 2N_P N_T m't_i \right]^{1/2} = m' \left[N_P^2 + N_T^2 + 2\gamma N_P N_T \right]^{1/2},$$

$$(6.68)$$

where $\left(\dfrac{t_i}{m'} + 1 \right) = \gamma$. We assume that enough interaction occurs during the initial formation and subsequent expansion of the fibreball that there is a mechanism ($\pi + N$ and $N + N$ collisions) to randomize the available kinetic energy. The available kinetic energy in the centre of mass system is

$$\epsilon = \frac{E_{c.m}}{(N_P + N_T)} - m = \frac{m'}{1+f}(1 + f^2 + 2f\gamma)^{1/2} - m,$$

$$(6.69)$$

where m is the mass of the free nucleon taken to be 939 MeV, $f = \dfrac{N_T}{N_P}$ and $(\epsilon + m)$ is the total energy of a nucleon in centre of mass. For equal mass nuclei, $f = 1$ and $\epsilon = m' [(1+\gamma)/2]^{1/2} - m$. Randomization of available kinetic energy leads thermal equilibrium amongst all the hadrons. A prerequisite for the establishment of thermal hadronic system is that the mean free path L is less than the size r of the system. The mean free path between collisions based on nuclear density and N-N cross section is $1/\rho\sigma$. Assuming initial nuclear density ($r = 0.14$ fm^{-3}) and N-N cross section σ at this energy to be 25 mb, one gets $L = 2.9$ fm. The condition $r > L$ then leads to $A^{1/3} > 2.4$ where A is the number of nucleons in the fireball. Therefore, the model will break down for peripheral collision of light nuclei.

   Thermodynamics: We use some concepts of statistical thermodynamics to describe the decay of the fireball and also the formation of nuclei. The first step is to assume thermal equilibrium amongst nucleons in the fireball and define a temperature τ, which for a relativistic ideal gas of nucleons is

$$\frac{E_{c.m}}{(N_P + N_T)}\frac{1}{\tau} = 3 + \frac{m}{\tau}\frac{K_1(m/\tau)}{K_2(m/\tau)},$$

$$(6.70)$$

where K's are the Macdonald functions, whose explicit expressions are given below:

$$K_\nu(z) = \sqrt{\frac{\pi}{2z}}\, e^{-z} \left\{ 1 + \frac{\mu - 1}{8z} + \frac{(\mu - 1)(\mu - 9)}{2!\,(8z)^2} \right.$$

$$\left. + \frac{(\mu - 1)(\mu - 9)(\mu - 25)}{3!\,(8z)^3} + \cdots \right\}$$

$$(6.71a)$$

$$K_\nu(z) = (z/2)^\nu \frac{\Gamma(1/2)}{\Gamma(\nu + 1/2)} \int_0^\infty e^{-z\cosh t} \sinh^{2\nu} t \, dt$$

with $\quad 4\nu^2 = \mu$ and $z = m/\tau.$ $\hfill$ (6.71b)

$$K_2(z) \cong \sqrt{\frac{\pi}{2z}} \, e^{-z} \left\{ 1 + \frac{15}{8z} + \frac{105}{128z^2} \right\}, \quad K_1(z) \cong \sqrt{\frac{\pi}{2z}} \, e^{-z} \left\{ 1 + \frac{3}{8z} - \frac{15}{128z^2} \right\},$$

and $\quad K_0(z) \cong \sqrt{\dfrac{\pi}{2z}} \, e^{-z} \left\{ 1 - \dfrac{1}{8z} + \dfrac{9}{128z^2} \right\}.$

Employing (6.69) and (6.70) we get

$$\frac{\in + m}{\tau} = 3 + \frac{m}{\tau} \frac{K_1(z)}{K_2(z)} \quad \text{or} \quad \frac{\in}{\tau} = 3 + \left[\frac{K_1(z)}{K_2(z)} - 1 \right] z. \qquad (6.72)$$

Neither the temperature nor the composition of the initial fireball is observable. Because any conceivable expression must look at the product of the collision after the fireball has disassembled. It is the expansion of the fireball that carries, may be the most of the particles to the counting apparatus. At some point during expansion when density falls below certain critical value, the thermal contact between particles is broken and particles no longer interact. This is called freeze out point. Presumably the freeze out density would not be less than one particle per pion wavelength,

$$\rho_\pi = \left[\frac{4\pi}{3} (1.4)^3 \right]^{-1} \approx 0.085 \ \text{fm}^{-3}, \text{which is less than the normal nuclear}$$

density, $\rho_N = \left[\dfrac{4\pi}{3} (1.2)^3 \right]^{-1} = 0.14 \ \text{fm}^{-3}.$

The critical density ρ_c can assume any value between $(1/2 - 1/3)\rho_N$ where ρ_N is the normal nuclear density. ρ_c is the only free parameter of the model[27]. The expansion time scale is of the order of 5×10^{-23} sec. The next step is to use the non-interacting gas formulas to describe each hadron type, which we expect to describe each statistically significant component of the fireball, when it decays. The distribution function

$$f(\in) = 1/e^{(\in - \mu)/\tau} \pm 1,$$

where $+(-)$ corresponds to FD(BE) distribution. The distribution of particles in momentum space is $(\hbar = c = k = 1)$

$$dN_i = \frac{d_s V}{2\pi^2} \frac{p^2 dp}{\{e^{(\epsilon_i - \mu_i)/\tau} \pm 1\}}. \tag{6.73a}$$

Here d_s is the spin degeneracy factor $= 2S_i + 1$, S_i is the spin, and μ_i is the chemical potential for particle of ith type. V is the volume of the fireball at the instant of decay and $\pm$ refers to fermion/boson. ϵ_i is the total energy (including rest mass energy) per particle in the c.m of the fireball. We confine ourselves to proton distribution and hence write (dropping index i)

$$dN = \frac{(2S+1)V}{2\pi^2} \frac{p^2 dp}{\{e^{(\epsilon - \mu)/\tau} + 1\}}. \tag{6.73b}$$

Employing the relativistic value for $\epsilon = \sqrt{p^2 + m^2}$ we get the total number of particles

$$N = \frac{(2S+1)V}{2\pi^2} \int\limits_0^\infty \frac{p^2 dp}{1 + e^{-\beta\mu} \, e^{\beta\sqrt{p^2 + m^2}}}, \text{ where } \beta = \frac{1}{\tau}. \tag{6.73c}$$

When $e^{(\epsilon - \mu)/\tau} \gg 1$; then neglecting 1 in the denominator we get the relativistic Maxwell-Boltzman (MB) distribution:

$$N = \frac{(2S+1)Ve^{\beta\mu}}{2\pi^2} \int\limits_0^\infty e^{-\beta m\sqrt{1 + p^2/m^2}} \, p^2 dp. \tag{6.74a}$$

Let $\cosh^2 t = 1 + p^2/m^2$ so that $\sinh^2 t = p^2/m^2$. Therefore,

$$N = \frac{(2S+1)Ve^{\beta\mu}}{2\pi^2} m^3 \int\limits_0^\infty e^{-\beta m \cos ht} \, \sinh^2 t \cosh t \, dt. \tag{6.74b}$$

We rewrite (6.73b) as

$$\frac{d^2 N}{p^2 dp \, d\Omega} = \frac{V}{4\pi} \frac{(2S+1)}{2\pi^2} \frac{1}{1 + e^{(\epsilon - \mu)/\tau}}. \tag{6.75}$$

For Maxwell-Boltzman distribution $e^{[(\epsilon - \mu)/\tau]} \gg 1$, we get

$$\frac{d^2 N}{p^2 dp \, d\Omega} = \frac{V}{4\pi} \frac{(2S+1)}{2\pi^2} e^{\beta\mu} e^{-\beta\epsilon}.$$

Substitution of N eliminates μ and we get (*see* Ex 6.13)

$$\frac{d^2N}{p^2 dp\, d\Omega} = \frac{N}{4\pi}\frac{e^{-\beta\epsilon}}{m^2\tau\, K_2(m/\tau)}$$

$$= \frac{N}{4\pi m^3}\frac{e^{-\epsilon/\tau}}{\left[\dfrac{\tau}{m}K_0(m/\tau) + 2(\tau/m)^2\, K_1(m/\tau)\right]} \tag{6.76a}$$

A good approximation to the full fireball model calculations involving the summation of energy spectra overall impact parameter $(0 \le b \le r_p + r_T)$ can be converted to a cross section as:

$$\frac{E_L}{p_L^2}\frac{d^2\sigma}{dp_L\, d\Omega_L} = \int_0^{r_p+r_T} \epsilon\frac{d^2N}{p^2 dp\, d\Omega}(2\pi b)\, db. \tag{6.76b}$$

A comparison of fireball predicted proton inclusive spectra (6.76b) for Ar on Pb at 800 MeV/A is shown in Fig. 6.21. The dotted lines give the calculated values for critical density $\rho_c = 0.092$ fm^{-3} $(V_c = (3/2)\, V_{FB})$, the solid line for $\rho_c = 0.068$ $(V_c = 2V_{FB})$. Here V_C is the critical volume and V_{FB} is the volume of the fireball. It is remarkable that the fireball model with so many drastic assumptions can describe the gross features of proton inclusive spectra.

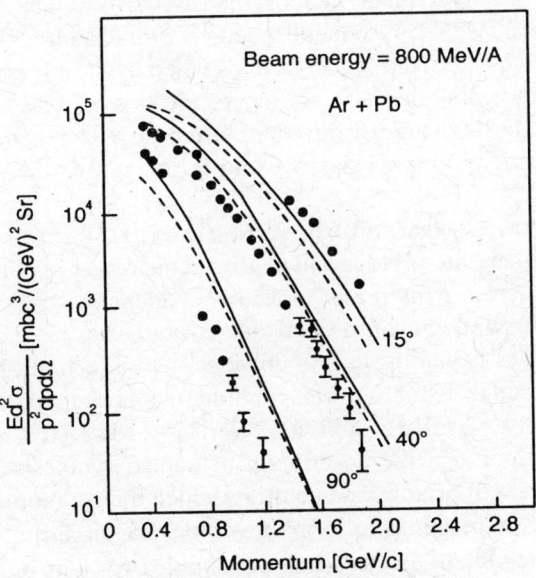

Fig. 6.21: Comparison of fireball model predicted proton inclusive spectra for Ar on Pb at 800 MeV/A incident energy (Pandey and Mukherjee Ref. 27)

B. The 'Cosmic' Connection

Our universe was created nearly 13 ± 2 billion years ago with a big explosion called the "Big Bang" that created space and time. Just after the Big Bang, the universe was radiation dominated containing a hot plasma of photons, neutrinos, electrons, positrons, quarks and gluons. It was expanding in accordance to $H_0 = \dot{r}(t_0)/r(t_0)$ where H_0 is called the Hubble constant and the index 0 represents the current value at time $t = t_0$. $r(t)$ is interpreted as the horizon size at time t. The density of particles, n was very high so that the rate of the reaction ($\propto n\sigma$) was higher than the expansion rate $\dot{r}/r$, resulting thermodynamic equilibrium amongst particles. Creation of a proton or neutron from photon requires a temperature of 1.1×10^{13} K at which the photon wavelength $\approx$ nucleon size.

The adiabatic expansion of the universe let to cooling according to $r(t) \propto (1/T)$, where T is the temperature of the universe. In about 10^{-5} seconds after the Big Bang, the universe underwent a first order phase transition to the matter dominated universe. Around 1GeV energy transition from free quarks to quarks confined in mesons baryons took place (*see* section 6.5).

Primordial and Stellar nucleosynthesis led to a universe whose baryonic content is dominated by hydrogen and helium. It allowed stars like our Sun to burn hydrogen($4^1H_1 \leftrightarrow {}^4He_2$+energy) steadily for billions of years for the evolution of life on earth. The synthesis of light elements was practically completed by the Big Bang Nucleosynthesis (BBN) when the temperature dropped below 30 KeV. Subsequently, density of the helium nuclei merge into carbon, oxygen etc. and the sequence ends when the iron is produced. We can use nuclear physics as the basis to understand this process[40]. The emphasis here is on synthesis of nuclei through thermo-nuclear reactions in the temperature and pressure conditions existing in the stars rather than how stars evolve.

Identified mass in the universe is mostly in the form of atomic nuclei. About 75% of these are protons and nearly all the rest are helium with a small amount of deuteron, carbon and other heavy elements. We use the fireball model to understand the primordial nucleosynthesis, i.e. how the nuclear physics processes began, when the universe was 1 second old, having temperature greater than 10^{10} °K, corresponding to an average kinetic energy per particle of nearly 1 MeV ($\epsilon = kT = 0.86 \times 10^{-10}$ MeV °K$^{-1} \times 10^{10}$ °K).

The starting point of the description of nucleo synthesis is a dense and heated initial state of protons and neutrons which then combine through complex set of nuclear reactions to form the various nuclei during the space time evaluation of the fireball. Let us consider a simple two-body reaction, radiative capture, for example: $n + p \rightleftharpoons d + \gamma$.

The mechanism of deuteron production is too slow in heavy ion case from the point of view of expansion time scale of the order of 5×10^{-23} sec. The inverse reaction can take place through giant resonances whose width ΔE is several MeV (~ 5 MeV) and whose maximum cross section is 1 mb at an energy of 10 MeV. At the fireball temperature ~ 50 MeV the cross section will be still smaller. The reaction rate

$$\Delta t = \frac{\hbar}{\Delta E} = \frac{\hbar c}{5 \text{ MeV} \times 3 \times 10^{23} \text{ fm sec}^{-1}} \approx 10^{-20} \text{ sec.}$$

which is quite slow to produce any deuteron. Therefore, the first step must then be a three body reaction of the type: $p + n + N \rightleftharpoons d + N$, the nucleon N takes away the excess energy. Another possible reaction is $p + p \rightarrow d + e^+ + \nu$.

We employ chemical kinetics and statistical thermodynamics to understand the formation of nuclei. A chemical reaction in a mixture of reacting substances ultimately leads to the establishment of an equilibrium state, in which the quantity of each of the species that takes part in the reaction, no longer changes. This case of thermodynamic equilibrium is called chemical equilibrium. We consider a homogeneous system of energy E and volume V. Let N_i be the number of nuclei of type i. The entropy S of the system is a function of following variables.

$$S = S(E, V, N, \cdots, N_m). \tag{6.77}$$

These variables of course can change in a general process:

$$dS = \left(\frac{\partial S}{\partial E}\right)_{V,N} dE + \left(\frac{\partial S}{\partial V}\right)_{E,N} dV + \sum_{i=1}^{m} \left(\frac{\partial S}{\partial N_i}\right)_{E,V,N} dN_i. \tag{6.78}$$

Here the subscript N denotes the fact that all numbers $N_1, \cdots, N_m$ are kept constant in taking the partial derivative and in the case of $(\partial S/\partial N_i)$, all numbers $N_1, \cdots, N_{i-1}, \cdots N_m$ except N_i are kept constant. We examine the thermodynamical relations when all numbers are kept constant, i.e. $dN_i = 0$. We know

$$dS = \frac{dQ}{T} = \frac{dE + PdV}{T} = \frac{dE}{T} + \frac{P}{T} dV. \tag{6.79}$$

By comparison of (6.78) and (6.79) we get

$$\left(\frac{\partial S}{\partial E}\right)_{V,N} = \frac{1}{T} \text{ and } \left(\frac{\partial S}{\partial V}\right)_{E,N} = \frac{P}{T}.$$

Let us use the abbreviation $\mu_j = -T\left(\dfrac{\partial S}{\partial N_j}\right)_{E,V,N}$.

The quantity μ_j is called the chemical potential per particle of the jth species and it has been so defined that it has dimension of energy. Finally Eq. (6.78) takes the form

$$dS = \frac{1}{T}\, dE + \frac{P}{T}\, dV + \sum_{i=1}^{m}\left(-\frac{\mu_i}{T}\right)_{E,V,N} dN_i$$

or
$$dE = TdS - PdV + \sum_{i=1}^{m}\mu_i\, dN_i. \tag{6.80}$$

The changes in number of nuclei must be proportional to N_i, the number of nuclei in the system. N_i can change as a result of nuclear reaction between nuclei. But they cannot change in arbitrary way because the conservation of nucleons requires that $\sum b_i N_i = 0$ be satisfied, where b_i is some small integer number of N_i nuclei involved in the nuclear transformation. At equilibrium, entropy S is maximum and $dS = 0$. Under the assumed condition of constant E and V, we get

$$\sum_{i=1}^{m}\mu_i\, dN_i = 0 \quad \text{or} \quad \sum_{i=1}^{m}b_i\,\mu_i = 0,$$

which is the general condition for chemical equilibrium. For the nuclear reaction under consideration, d-n-$p = 0$, we can write also

$$b_1\,\mu_p + b_2\,\mu_n + b_3\,\mu_d = 0.$$

On comparing we get $\mu_d = \mu_n + \mu_d$, since $b_1 = b_2 = -1$ and $b_3 = 1$.

Next, we employ ensemble theory to compute the abundances of the various species. The number of particles of a given specy $n_i = e^{\beta\mu_i} Z_i$, where Z_i is the partition function of the ith specy. Now $\log n_i = \beta\mu_i - \log Z_i$ or $\beta\mu_i = \log n_i - \log Z_i$, $\beta\mu_d = \log n_d - \log Z_d$, $\beta\mu_p = \log n_p - \log Z_p$, $\beta\mu_n = \log n_n - \log Z_n$. We can combine the above relations as

$$\log\frac{n_d}{Z_d} = \log\frac{n_p}{Z_p} + \log\frac{n_n}{Z_n} = \log\frac{n_p n_n}{Z_p Z_n} \quad \text{or} \quad n_d = \frac{n_p n_n}{Z_p Z_n} Z_d. \tag{6.81}$$

The above equation together with some conserved quantities such as baryon number, charge, etc., allow us to compute the abundances of various species. We fix the number particles as follows:

$$n_p + n_d = N_p = \text{total number of protons}$$
$$n_n + n_d = N_n = \text{total number of neutrons.}$$

Now we briefly outline steller evolution in astrophysical context.

With the estimated temperature of the steller bodies, it is not possible to synthesize enough quantities of He and deuteron nuclei from the hydrogen nuclei to account for the present abundances, and therefore, these nuclei must have been produced in the very early stage of the universe, when its temperature was extremely high during the first seconds of the beginning of the universe. The standard model of the expanding universe suggests that the size of the universe scales with time and due to the adiabatic expansion, the universe cools. In the early periods when the universe was dominated by radiation over matter, the size of the universe scaled with the square root of time, i.e. $r(t) \propto t^{1/2}$ and the temperature T scaled as inverse of the size parameter $r(t)$, i.e. $T(t) \propto 1/r(t)$. After the era of premodial nucleosynthesis, the universe continued to expand eventually for nearly 10^6 years as a mixture of nuclei, photons, electrons and neutrinos and it eventually cooled to temperature 30 keV. Gradually, many nuclei and electrons combined together to form atoms and molecules. As many of the atoms and molecules clustered together due to mutual attraction of their gravity, they eventually gathered enough mass to become Stars[30].

Let us consider some remarkable coincidences that have made formation of a star. May be the best known example is the presence of narrow state in the excitation spectrum of ^{12}C at 7.65 MeV. The location and width of this state control the fusion rate of three ^4He nuclei into ^{12}C nucleus and thus are determining factors of the abundances of carbon and heavier elements in the universe. Microscopic calculations have shown that the abundant production of ^{12}C in steller fusion requires the fine tuning of the strength of the N-N force to the precision of $\pm 4\%$. Another salient example of apparent fine-tuning is the neutron-proton mass difference $(\Delta m/m) = 1.8 \times 10^{-3}$, which is a subtle balance of the difference between the masses u and d-quarks and electromagnetic self energies of the proton and neutron. To understand steller evolution, we first define gravitational coupling constant for proton

$$\alpha_p = \frac{Gm_p^2}{\hbar c},$$

where m_P is the mass of proton and G is Newton's gravitational constant Substituting proton mass (= 1.67×10^{-27} kg), we get

$$\alpha_p = 4.17 \times 10^2 \, \frac{\text{MeV} \cdot \text{meter}}{\text{kg}^2} \times \frac{(1.67 \times 10^{-27} \text{ kg})^2}{197 \text{ MeV} \times 10^{15} \text{ meter}} = 5.9 \times 10^{-39}$$

The existence of stable, very long-lived stars with significant energy output, like our Sun, relies on the near coincidence of α_p with α, the fine structure constant:

$$\alpha_p = \alpha^{12}(m_e/m_p)^4,$$

where m_e and m_p denote respectively electron and proton mass.

We mentioned earlier that Stars are mostly from the gas cloud containing protons and the number of protons this cloud will contain $N = M/m_p$, where M is the mass of the cloud. The average separation between protons, r, is given by

$$\frac{4\pi}{3} r^3 \frac{M}{m_p} = \frac{4\pi}{3} R^3 \text{ or } r = \left(\frac{m_p}{M}\right)^{1/3} R = RN^{-1/3}.$$

Therefore,
$$V_G = \frac{GM^2}{R} = \frac{Gm_p^2 N^{5/3}}{r} = \alpha_p \frac{\hbar c}{r} N^{5/3}.$$

The temperature of the Star follows from

$$NkT = V_G = \alpha_p N^{5/3} \frac{\hbar c}{r} \text{ or } kT = \alpha_p N^{2/3} \frac{\hbar c}{r}.$$

As the Star contracts, i.e. r decreases, the temperature of the Star increases. But the gravitational collapse is prevented by reaching equilibrium between thermal pressure due to nuclear fuel and gravitational pressure. At the end of the nuclear fuel a total collapse can be avoided by electronic pressure. In the cloud, electrons are also localized. An electron confined within the distance r has minimum momentum given by the uncertainty relation as $p_e = \dfrac{\hbar}{r}$. Hence, minimum average energy of electron

$$E_e = \frac{p_e^2}{2m_e} = \frac{\hbar^2}{2m_e r^2}.$$

Another electron attempting to enter this volume must go into excited state in accordance to the Pauli exclusion principle and the resulting electron degeneracy pressure will balance gravitational pressure when

$$N \frac{\hbar^2}{m_e r^2} = \alpha_p N^{5/3} \frac{\hbar c}{r} \text{ i.e., } r = \left(\frac{\hbar}{m_e c}\right) \frac{1}{\alpha_p N^{2/3}}.$$

So the maximum temperature that can be achieved

$$kT = (\alpha_p \, N^{2/3} \, \hbar c)\left(\frac{\alpha_p \, N^{2/3} \, m_e \, c}{\hbar}\right) = \alpha_p^2 \, N^{4/3} \, m_e \, c^2. \qquad (6.82)$$

We can say that two opposite forces balance each other to keep the Star in equilibrium. First, there is gravity which tends to contract the Star and second there is an internal pressure generated by thermal energy of gas particles which resists the contraction. If gravity is greater than the internal pressure, star will explode.

Thus, the energy liberated in these reactions yields a pressure in the plasma, which opposes compression due to gravitation and an equilibrium is reached. The Sun is a star in its initial phase of evolution. The hydrogen to helium fusion reaction:

$$4^1 H_1 \rightleftharpoons {}_2^4 He_2 + 2e^+ + 2\nu_e + \Delta E$$

taking place in the sun, releases nulear energy $\Delta E \approx 25$ MeV per reactions. The cross section involved in the above reaction near the coulomb barrier (~1 MeV) is of the order of a millibarns. One can obtain a characteristic luminosity L_c based on the above reaction as follows:

$$L_c = \in N\Delta E/\tau_c,$$

where τ_c is the characteristic time scale of the reaction, $\in$ the fraction of total number of solar nuclei N. τ_c is given by

$$\tau_c = \frac{1}{n\sigma v},$$

where n is the density of nuclei and v is the relative velocity of reactant.

Exercise 6.10. Calculate the characteristic luminosity of a sun like star from the following data:

$$N = 10^{57}, \in = 10^{-2}, \sigma = 1 \text{ mb}, \Delta E = 25 \text{ MeV}, n = 10^{26} \text{ cm}^{-3}$$

and $v = 10^9$ cm s^{-1}.

We have
$$\tau_c = \frac{1}{10^{26} \text{ cm}^{-3} \times 10^{-27} \text{cm}^2 \times 10^4 \text{cm sec}^{-1}}$$

$$= 10^{-8} \text{ sec.}$$

$$L_c \sim \frac{\in N\Delta E}{\tau_c} = \frac{10^{-2} \times 10^{57} \times 25 \text{ MeV}}{10^{-8} \text{ sec}}$$

$$= 25 \times 10^{63} \text{ MeV sec}^{-1}.$$

Exercise 6.11. Show that in the non-relativistic limit $\in = \dfrac{3}{2}\tau$.

$$\frac{K_1}{K_2} \cong \left(1 + \frac{3}{8z}\right)\left(1 + \frac{15}{8z}\right)^{-1} \cong \left(1 - \frac{3}{2z}\right).$$

Hence, $\left(\dfrac{K_1}{K_2} - 1\right)z \cong -\dfrac{3}{2}$. Therefore, $\dfrac{\in}{\tau} = 3 - \dfrac{3}{2} = \dfrac{3}{2}$ that is $\in = \dfrac{3}{2}\tau$.

6.5 NEW STATE OF NUCLEAR MATTER

A new state of nuclear matter of high energy density created in a Relativistic Heavy Ion Collider (RHIC) will enable us to study phase transition from hadronic matter to Quark Gluon Plasma (QGP). Let us consider a gas of nucleons at low temperature and density. The nuclear ground state density $\rho_0 = 0.14$ fm^{-3}.

Let us imagine that we can change the density in some way and change ρ_0 to a state where nucleons begin to touch one another, i.e. to a density equal to that of a nucleon.

$$\rho_N = \frac{M}{\dfrac{4\pi}{3}(0.8)^3 A \, fm^3} = 0.44 \text{ GeV fm}^{-3} = 3\rho_0$$

If we increase ρ beyond ρ_N, the quark wave functions in neighbouring nucleons begin to overlap and a quark may find itself in a neighbouring nucleon, that is, it has broken out of its confinement. A condition like $\rho \gg \rho_N \approx 3\rho_0$ certainly must be fulfilled to create a quark gluon plasma. This deconfined plasma was the original state of all nuclear matter in the universe (Fig 6.1 left). Recent QCD calculations with nucleon number zero suggest that $\rho \approx 10 \, \rho_0$ is needed. Such densities may be reached in high energy heavy ion collisions, which may also lead to upper limit of hadronic matter temperatures of 150-200 MeV. The search for QGP is one of the most notable examples of international collaboration for large scale experiments done in U.S.A. at Brook Heaven National Laboratory with Alternative Gradient Synchrotron (BNL AGS) and BNL Relativistic Heavy Ion Collider (BNL RHIC) and also in Europe at CERN with super proton synchrotron (SPS) and large hadron collider (LHC). International collaborative experiments called: PHINIX, STAR, ALLICE etc. have been designed to see the prospect of QGP formation in the coming years. The scenario of the relativistic heavy ion collision is shown in Fig. 6.22.

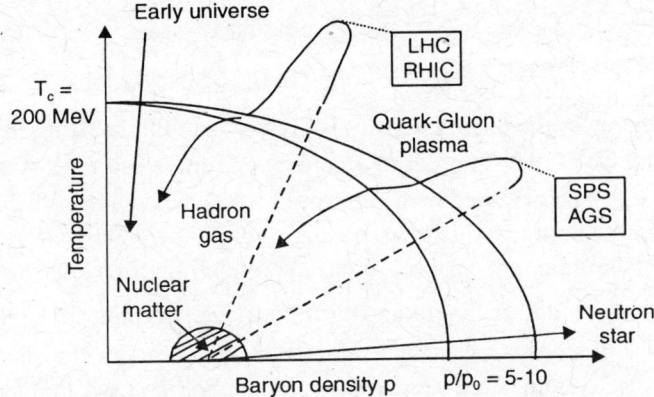

Fig. 6.22: Scenario of the relativistic heavy ion collisions

The ultra-relativistic heavy ion collisions offer an opportunity of creating an extended matter with very high particle and high energy density in the laboratory, having density $\rho = (10 \text{ to } 15) \, \rho_N$. The available energy is equipartioned amongst accessible degrees of freedom, thus generating a hot dense matter usually called a "Fireball". The physical variables characterizing a fireball are energy density ϵ, baryon number density n_b and volume V. In such nucleus-nucleus collisions it may be possible to create regions of deconfined matter so energetic that forces that confine quarks and gluons into individual hadrons are overcome. However, deconfinement does not mean that quarks in the QGP are set free as the quarks are separated, they are only given a large cage to roam about. During phase transition hadronic bag fuse together and hadrons loose their identity resulting in·a high density of color sources over macroscopic distances. The quarks and gluons are not locally confined but are free to roam over the entire system.

As a concrete example, we may consider high energy heavy-ion collisions at the Relativistic Heavy Ion Collider (RHIC), at Brookhaven National Laboratory. In this collider the nuclei are accelerated to an energy of 100 GeV per nucleon. For collision of gold on gold in such a collider, the energy carried by each nucleus is about 100 × 197 GeV or 19.7 TeV and the c.m. energy is about 2 × 19.7 = 39.4 TeV. At CERN the large Hadron Collider (LHC) can accelerate a nucleus to a c.m energy of 3 TeV per nucleon, which will lead to c.m. energy of about 1262 TeV for collision of lead on lead. As far as the energy deposition in the vicinity of c.m is concerned, the effect of the many inelastic nucleon-nucleon (*N-N*) collisions in the nucleus-nucleus reaction is roughly additive in nature. The more the number of inelastic N-N collisions, the greater the amount of energy deposited in the vicinity of c.m in a cooperative manner. The energy density existing in a proton,

$$\epsilon_p = m_p \bigg/ \frac{4\pi}{3} r_p^3 = 0.5 \text{ GeV/fm}^3.$$

QCD predicts the existence of a phase transition from Hadron Phase (HP) to QGP whenever the energy density is greater than that existing in a proton, i.e. $\epsilon_p = 0.5$ GeV/fm^3. In order to obtain a very crude estimate of the energy density that can be obtained in the ultra-relativistic heavy ion collisions, let us consider collision of two nuclei with kinetic energy per nucleon E/A in the lab system. In the c.m system, both nuclei are Lorentz contracted by a factor $\gamma = 1 \big/ \sqrt{1 - v^2/c^2}$ and they will appear initially as two thin disks. As a result these N-N collisions occur at the same time and in about the same special region of c.m system. If two equal mass nuclei collide (A-A collisions) with each other and come to a stop when they overlap completely, then the energy density in the c.m before expansion starts is $\epsilon = 2 \epsilon_N \gamma^2$ where $\epsilon_N = 0.14$ GeV/fm^3 is the normal nuclear matter density in its ground state. For $\gamma = 2$, i.e. $v/c = \sqrt{3}/2$, we get $\epsilon/\epsilon_N = 8$ or $\epsilon = 8 \epsilon_N$. For $^{28}Si + {}^{197}Au$ collision at 14.6 GeV/A, $\gamma = 1.8$ and we have $\epsilon = 1$ GeV/fm^3. This energy density, which is of an order of magnitude higher than ϵ_N, is deposited in a small region of space in a short duration of time and therefore, may favour the formation of a deconfined state of quarks and gluons. A large pressure of quark matter can arise in two ways: (1) when the temperature of the matter is high and/or (2) when baryon number density is large.

A. Thermodynamics of Quark Gluon Plasma

We consider the case of a quark-gluon system in thermal equilibrium at high temperature T, within a volume V. For simplicity, we examine the case where quarks and gluons are idealized to be non-interacting and massless. However, we examine the situation where the baryon density is high. We adopt a phenomenological approach in calculating properties and signals of QGP. However, it is not a substitute for more formal QCD based theory. The partial pressures arising from the quarks and antiquarks can be calculated by employing statistical thermodynamics[30].

If the sizes of colliding nuclei are large enough, then the fireball formed will consist of a large number of particles. The process of particle production and annihilation continue till the fireball disassembles. In order to describe the properties of such a system, we can employ statistical mechanics, which is applicable to a system involving large number of particles. To investigate the thermodynamical properties of QGP within the frame work of statistical mechanics, our first task is to derive equation of state of QGP. This task is considerably simplified by the observation that the gluons are massless and light u and d quarks $(m_u \approx 5 \text{ MeV}, m_d' \approx 10 \text{ MeV})$ are essentially massless particles, at least on the scale of energies available in the hot plasma, i.e. 200 MeV.

Let us first count the number of degrees of freedom associated with the constituents in QGP. Gluons carry color and spins and so do quarks which in addition come in two flavors u and d in a nucleon. We therefore have the following multiplicity:

Gluons: $\qquad d_g = 2 \text{ (spin)} \times 8 \text{ (colors)} = 16$

Quarks: $\qquad d_q = 2 \text{ (spin)} \times 3 \text{ (colors)} \times 2 \text{ (flavours)} = 12$

All we have to do now is to calculate the energy density (energy per unit volume) residing with each degree of freedom. We begin with the massless gluons which upon neglect of all interactions, i.e. by setting gluon chemical potential $\mu_g = 0$ form an ideal relativistic Bose gas of temperature $(= 1/\beta)$. We use the natural units $k = \hbar = c = 1$, and express the gluon energy density per degree of freedom as

$$\epsilon_g = \int_0^\infty \frac{p}{(e^{\beta p} - 1)} \frac{d^3 p}{(2\pi)^3}. \qquad (6.83a)$$

Substituting $\beta p = x$ and $pd^3 p = p(4\pi p^2 dp) = 4\pi T^4 x^3 dx$, we get

$$\epsilon_g = \frac{4\pi T^4}{(2\pi)^3} \int_0^\infty \frac{x^3 dx}{e^x - 1}. \qquad (6.83b)$$

Now $\qquad \displaystyle\int_0^\infty \frac{x^{n-1}}{e^x + 1} dx = (n-1)! \, \xi(n)$

and $\qquad \displaystyle\int_0^\infty \frac{x^{n-1}}{e^x + 1} dx = (1 - 2^{1-n})(n-1)! \, \xi(n), \qquad (6.83c)$

where $\xi(n)$ is Riemann's zeta function. Some useful values are

$$\xi(2) = \frac{\pi^2}{6}, \; \xi(3) = 1.202, \; \xi(4) = \frac{\pi^4}{90}.$$

Hence, inserting $\xi(4)$ in (6.83b), we find that

$$\epsilon_g = \frac{\pi^2 T^4}{30}. \qquad (6.84a)$$

The number of gluons per degree of freedom

$$n_g = \frac{1}{(2\pi)^3} \int_0^\infty \frac{d^3 p}{(e^{\beta p} - 1)} = \frac{1}{2\pi^2} \int_0^\infty \frac{p^2 dp}{(e^{\beta p} - 1)}$$

$$= \frac{T^3}{2\pi^2} \int_0^\infty \frac{x^2 dx}{e^x - 1} = \frac{T^3}{2\pi^2} \, 2! \, \xi(3) = \frac{1.202}{\pi^2} \, T^3, \text{ since } \beta p = x. \qquad (6.84b)$$

To calculate the energy density for quarks and antiquarks, we have to introduce the chemical potential μ_q because there will be in general surplus of quarks over antiquarks in QGP. The QGP produced in nuclear collisions will have a net baryon number excess. At zero temperature, the meaning of μ_q is the energy required to add another quark to the plasma. Since no antiquarks are present at $T = 0$, the energy necessary to add an antiquark is zero. This does not imply $\mu_q = 0$, however, the additional antiquark may annihilate one of the quark from the top of the Fermi sea and release the energy μ_q. The chemical potential for antiquarks must therefore be chosen as $\mu_{\bar{q}} = -\mu_q$.

To calculate the energy density per degree of freedom carried by quarks, we use formula for the relativistic Fermi gas and treat quarks as massless.

$$\epsilon_q = \int_0^\infty \frac{p}{(e^{\beta(p - \mu_q)} + 1)} \frac{d^3 p}{(2\pi)^3}.$$

We substitute $\beta(p - \mu_q) = x$ and get

$$\epsilon_q = \frac{T^4}{2\pi^2} \int_{-\beta\mu_q}^\infty (x + \beta\mu_q)^3 \{e^x + 1\}^{-1} \, dx. \qquad (6.85a)$$

The only change occurring in the same expression for antiquarks is the replacement of the chemical potential $\mu_{\bar{q}} \rightarrow -\mu_q$, which is compensated by the variable change $x = \beta(p + \mu_q)$. Hence,

$$\epsilon_{\bar{q}} = \frac{T^4}{2\pi^2} \int_{\beta\mu_q}^\infty (x - \beta\mu_q)^3 (e^x + 1)^{-1} \, dx. \qquad (6.85b)$$

Splitting the integral as $\displaystyle\int_{\beta\mu_q}^\infty = \int_0^\infty - \int_0^{\beta\mu_q}$ and making the substitution $\mu_q \rightarrow -\mu_q$, we get:

$$\epsilon_q + \epsilon_{\bar{q}} = \frac{T^4}{2\pi^2} \left[\int_0^\infty dx \{ (x+\beta\mu_q)^3 + (x-\beta\mu_q)^3 \} (e^x+1)^{-1} + \int_{-\beta\mu_q}^0 (x+\beta\mu_q)^3 \, dx \right].$$

Expanding and evaluating the integral term by term we get

$$\epsilon_{q\bar{q}} = \epsilon_q + \epsilon_{\bar{q}} = \frac{7\pi^2 T^4}{120} + \frac{\mu_q^2 T^2}{4} + \frac{\mu_q^4}{8\pi^2}. \tag{6.86}$$

Assuming the same chemical potential for up and down quarks and multiplying ϵ_g and $\epsilon_{q\bar{q}}$ by respective degrees of freedom, we get

$$\epsilon_{QGP} = d_g \, \epsilon_g + d_q(\epsilon_{q\bar{q}}) = 16 \cdot \frac{\pi^2 T^4}{30} + 12 \left[\frac{7\pi^2 T^4}{120} + \frac{\mu_q^2 T^2}{4} + \frac{\mu_q^4}{8\pi^2} \right]$$

$$= \frac{37\pi^2 T^4}{30} + 3\mu_q^2 T^2 + \frac{3}{2\pi^2}\mu_q^4, \tag{6.87}$$

since gluon and quark multiplicities are respectively 16 and 12.

In order to determine the value of the quark chemical potential μ_q in a given physical situation, one has to know its relation to the baryonic density n_b, which is one third of the difference between the density of quarks and antiquarks. The density of quarks and antiquarks are:

$$n_q = \int_0^\infty \left[e^{\beta(p-\mu_q)} + 1 \right]^{-1} \frac{d^3 p}{(2\pi)^3} \text{ and } n_{\bar{q}} = \int_0^\infty \left[e^{\beta(p+\mu_q)} + 1 \right]^{-1} \frac{d^3 p}{(2\pi)^3}.$$

The number density of free quarks and antiquarks per degree of freedom are:

$$n_q = n_{\bar{q}} = \frac{1}{2\pi^2} \int_0^\infty \frac{p^2 dp}{1+e^{\beta p}} = \frac{1}{2\pi^2} T^3 \frac{3}{2} (1.202). \tag{6.88a}$$

Utilizing the same method as employed in the evaluation of the energy density, we can obtain a relation between n_b and μ_q as follows:

$$n_q - n_{\bar{q}} = \frac{T^3}{2\pi^2} \left[\int_0^\infty \{ (x+\beta\mu_q)^2 - (x-\beta\mu_q)^2 \} (e^x+1)^{-1} \, dx + \int_{-\beta\mu_q}^\infty dx (x+\beta\mu_q)^2 \right]$$

$$= \frac{\mu_q T^2}{6} + \frac{\mu_q^3}{6\pi^2}.$$

Multiplying by the number of degrees of freedom $d_q = 12$ and dividing by 3, we get

$$n_b = 4(n_q - n_{\bar{q}}) = \frac{2}{3} \mu_q T^2 + \frac{2}{3\pi^2} \mu_q^3. \qquad (6.88b)$$

At zero temperature Eq. 6.88b gives the well known expression for the degenerate Fermi-gas, but for high temperature, the first term causes the chemical potential to drop like T^{-2} if the baryon density is kept constant. Differentiating $\in_{q\bar{q}}$ given by (6.86) with respect to μ_q and using the result of (6.88b), we get

$$\frac{\partial \in_{q\bar{q}}}{\partial \mu_q} = \frac{\mu_q}{2} T^2 + \frac{1}{2} \frac{\mu_q^3}{\pi^2} = \frac{3}{4} \left[\frac{2}{3} \mu_q T^2 + \frac{2}{3\pi^2} \mu_q^3 \right] = \frac{3}{4} n_b.$$

or

$$n_b = \frac{4}{3} \left(\frac{\partial \in_{q\bar{q}}}{\partial \mu_q} \right). \qquad (6.89)$$

The above relation has a fundamental thermodynamic origin and remains valid if interaction amongst the quarks are taken. For equal numbers of quarks and antiquarks (i.e., $\mu_q = 0$), we get $n_b = 0$. Similarly, the entropy density s and the pressure P in the ideal gas of quarks and gluons are given by:

$$s = \frac{1}{3} \frac{\partial \in_{QGP}}{\partial T} = \frac{74}{45} \pi^2 T^3 + 2\mu_q^2 T. \qquad (6.90)$$

For massless fermions and bosons it can be shown that the pressure P is related to the energy density $\in$ by

$$P = \frac{1}{3} \in_{QGP} = \frac{37}{90} \pi^2 T^4 + \mu_q^2 T^2 + \frac{1}{2\pi^2} \mu_q^4. \qquad (6.91)$$

In the deconfined state quarks are not necessarily free, instead, they get a bigger cage to move around freely. We employ the bag model (Sec 2.5 A, page 129-131) to get the critical baryon density in the QGP state $(n_{bc})_{QGP}$ from the bag pressure B.

$(n_{bc})_{QGP}$ at $T = 0$ can be estimated from (6.91) by setting $P = B$ so that

$$B = \frac{1}{2\pi^2} \mu_{qc}^4, \qquad (6.92)$$

where μ_{qc} is the critical quark chemical potential. Therefore, $(\mu_{qc}) = (2\pi^2 B)^{1/4}$ and

$$(n_{bc})_{QGP} = \frac{2}{3\pi^2} \mu_{qc}^3 = \frac{2}{3\pi^2} (2\pi^2 B)^{3/4} = \frac{4}{3\pi^2} \left(\frac{1}{2\pi^2}\right)^{1/4} B^{3/4}. \qquad (6.93)$$

B. Hadron to Quark Phase Transition

To get some numerical values of the critical baryon number density at which the compressed hadron matter become a QGP with high baryon content at $T = 0$, we use $B^{1/4} = 206$ MeV and get $(n_b)_{QGP} = 0.72/\text{fm}^3$. Now the conservation of baryon number for the whole system indicates that $(N_q/3) + N_b =$ constant, where N_q is the excess of quarks over antiquarks, and N_b is the excess of nucleons over antinucleons. We thus get $dN_q = -3dN_b$. We use the fundamental thermodynamic relation

$$ds = \frac{dQ}{T} = \frac{dE + PdV}{T} \quad \text{or} \quad dE = Tds - PdV. \qquad (6.94a)$$

This can be written in a more generalized form in terms of chemical potential $dE = Tds - PdV + \mu dN$, when only one type of species is present. We write the Gibbs free energy

$$G = E - Ts + PV \quad \text{or} \quad dG = dE - Tds - sdT + PdV + VdP. \qquad (6.94b)$$

or $dG = Tds - PdV + \mu dN - Tds - sdT + PdV + VdP = -sdT + VdP + \mu dN.$

Hence, $\left(\dfrac{\partial G}{\partial N_q}\right)_{T,P} = \mu_q$, $\left(\dfrac{\partial G}{\partial N_b}\right)_{T,P} = \mu_b$ and $\mu_q = \dfrac{\mu_b}{3}$.

It should be noted that QGP is a continuum concept. Even though QGP is expected to be deconfined at high temperature, the deconfinement extends only over the region within the boundary of the hot quark matter. Because of this boundary, the pressure as given by Eq. (6.91) in the plasma will be subject to the negative bag pressure B arising from the boundary

$$P_{(QGP)_{Boundary}} = \frac{37}{90} \pi^2 T^4 + \mu_q^2 T^2 + \frac{1}{2\pi^2} \mu_q^4 - B \qquad (6.95)$$

and the energy density is correspondingly altered,

$$\in_{(QGP)_{Boundary}} = \frac{37}{90} \pi^2 T^4 + 3\mu_q^2 T^2 + \frac{3}{2\pi^2} \mu_q^4 + B. \qquad (6.96)$$

However, the expressions for entropy density and baryon density are not changed. We can combine (6.95) and (6.96) into a single equation and write

$$\in_{QGP} = 3P_{QGP} + 4B.$$

The above equation implies that we have added the latent heat density $4B$ for the deconfinement phase transition. Here the purpose of the latent heat density is to liberate the quarks. The phase diagram for equilibrium phase transition is constructed by means of the Gibbs criteria: the pressure, temperature and chemical potentials must remain constant across phase boundary and these give conditions for the mechanical, thermal and chemical equilibrium:

$$P_{QGP} = P_{HP}, T_{QGP} = T_{HP} \quad \text{and} \quad \mu_{QGP} = \mu_{HP},$$

where *HP* refers to hadron phase. Here we have used the third condition on $\mu_B = 3\mu_q$. Equations (6.95) and (6.96) are often used in the literature[28] and are called the phenomenological bag model equation of state (EOS) for QGP. The Fermi momentum $\mu_{u,d} = 434$ MeV. The values of $(n_{bc})_{QGP}$ and $(\mu_{u,d})_{QGP}$ should be compared to the nucleon number density $n_b = 0.14/\text{fm}^3$ and nucleon Fermi momentum of 251 MeV for normal nuclear matter at equilibrium. Thus, critical baryon density is about 5 times the normal nuclear matter density. When the density of baryons exceeds this density, the baryon bag pressure is not strong enough to withstand the pressure due to degeneracy of quarks, and the confinement of quarks within individual baryon bags will not be possible leading to the formation of a state of deconfined quarks.

In Fig. 6.23 we get the thermodynamic view of this phase transition, where the energy density changes discontinuously at the critical temperature T_C and jumps by 4*B*, which gives the latent heat per unit volume required to go from the HP to the QGP phase. In fact, this latent heat is the energy released by the relativistic mass of particles. Using the relation (6.87) we find that the energy density of massless quarks and gluons is $\dfrac{\epsilon}{T^4} = \dfrac{37\pi^2}{30}$ in the central region

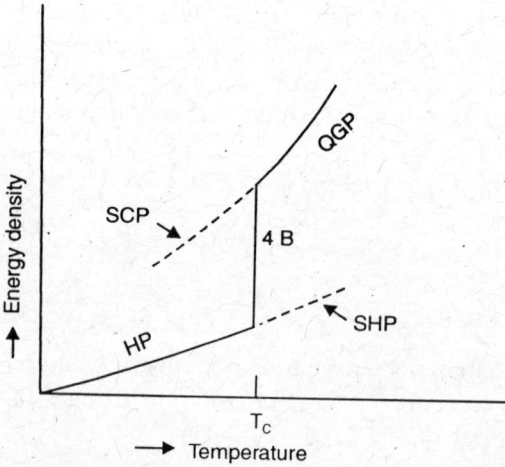

Fig. 6.23: Thermodynamic view of hadron to quark phase transition. Energy density changes discontinuously and jumps by 4B

where $\mu_g = 0$. An ideal gas of massless pions gives $\dfrac{\in}{T^4} = \dfrac{\pi^2}{10}$ *(see* Ex. 6.12).

The figure also demonstrates that unless the transition between the two phases occurs very slowly, it is a very complicated non-equilibrium process. Imagine that the system is in QGP phase and is cooled rapidly down to below the critical temperature. If it were to remain in equilibrium, it would change to HP. However, it needs time for quarks to hadronize (the formation time of 1 fm in the quark rest frame). So, rather than passing to HP quickly, the system will continue for a while in the QGP phase and it become supercooled (SCP). Likewise, when the system is heated rapidly from HP, it becomes superheated (SHP).

Efforts are going on to devise unique and unambiguous signals for the formation of QGP and to study the space time history of QGP formed in ultrarelativistic heavy ion collision experiments. Photon multicipty detector data coming from CERN SPS, STAR experiment at RHIC and ALICE experiment at LHC (CERN) show a great promise in studying various aspects of the reaction mechanism of phase transition from hadronic matter to QGP. The field of QGP is emerging as good region for exploration by experimentalists and theorists in both the areas of particle physics, nuclear physics and astrophysics. There are many intriguing ideas concerning some realistic models of QGP formation with distinctive and testable predictions, yet to be confirmed unambiguously by experiments[31-33].

Exercise 6.12. Employing the Stirling formula derive the Stefan Boltzman relation from (6.83b).

Since, $\dfrac{x^3}{e^x - 1} = \dfrac{e^{-x} x^3}{1 - e^x} \cong x^3 e^{-x} \left[1 + e^{-x} + e^{-2x} + \right] \approx \sum\limits_{n=1}^{\infty} x^3 e^{-nx}$, then

$$\in_g = \frac{T^4}{2\pi^2} \sum_{n=1}^{\infty} \int_0^{\infty} e^{-nx} x^3 dx = \frac{T^4}{2\pi^{2i}} \sum_{n=1}^{\infty} \frac{1}{n^4} \int_0^{\infty} e^{-nx} (nx)^3 d(nx)$$

$$= \frac{T^4}{2\pi^2} \sum_{n=1}^{\infty} \frac{1}{n^4} \int_0^{\infty} e^{-y} y^3 dy, \text{ where } y = nx.$$

We use the Stirling formula, $\int\limits_0^{\infty} e^{-y} y^3 dy = 3!$ and $\sum\limits_{n=1}^{\infty} \dfrac{1}{n^4} = \dfrac{\pi^4}{90}$, and finally get

$$\in_g = \frac{T^4}{2\pi^2} \cdot 3! \cdot \frac{\pi^4}{90} = \frac{\pi^2 T^4}{30} \quad \text{(Stefan Boltzman relation)}$$

Exercise 6.13. Express the relativistic MB distribution given by (6.74b) in terms of the Macdonald function (6.71).

We integrate by parts the integral in (6.74b):

$$\int_0^\infty (\sinh t \ \cosh t) \, e^{-\beta m \cosh t} \sinh t \, dt$$

$$= \left[\sinh t \ \cosh t) \int e^{-\beta m \cosh t} \sinh t \ dt \right]_0^\infty$$

$$- \int_0^\infty \frac{d}{dt}(\sinh t \ \cosh t) \int_0^\infty e^{-\beta m \cosh t} \sinh t \ dt$$

$$= \frac{1}{\beta m} \left[\int_0^\infty e^{-\beta m \cosh t} \, dt + 2 \int_0^\infty \sinh^2 t \, e^{-\beta m \cosh t} dt \right]$$

$$= \frac{1}{\beta m} \left[K_0(\beta m) + K_1(\beta m) \right] \quad \text{according to (6.71b).}$$

Hence,
$$N = \frac{(2S+1)Ve^{\mu/\tau}}{2\pi^2} m^2 \tau \left[K_0(m/\tau) + \frac{2\tau}{m} K_1(m/\tau) \right].$$

Let us look at the energy:

$$E = \frac{(2S+1)Ve^{\beta\mu}}{2\pi^2} \int_0^\infty \epsilon \, e^{-\beta\epsilon} \, p^2 dp$$

$$= \frac{(2S+1)Ve^{\beta\mu}}{2\pi^2} \int_0^\infty \cosh^2 t \, \sinh^2 t \, e^{-\beta m \cosh t} \, dt$$

$$= \frac{(2S+1)Ve^{\beta\mu}}{2\pi^2} m^3 \tau \left[K_1(m/\tau) + \frac{3\tau}{m} K_2(m/\tau) \right].$$

Exercise 6.14. Find the non-relativistic expressions for (6.76a):

$$\left(\frac{d^2N}{p^2 dp \, d\Omega} \right) = N (2\pi m\tau)^{-3/2} e^{-p^2/2m\tau}$$

where $\dfrac{d^2N}{p^2 dp\, d\Omega}$ is the centre of mass momentum distribution and

$\in = \gamma_{c.m}(E_L - \beta_{c.m}\, P_L \cos\theta_L)$. Here P_L, E_L and θ_L are the momentum, total energy and angle in the laboratory respectively and $\gamma_{c.m.} = 1/(1-\beta_{c.m})^{1/2}$.

Exercise 6.15. Express the chemical potential μ_i in terms of Gibbs free energy:

$$G = E - TS + PV$$

or

$$G = TdS - PdV + \sum_{i=1}^{m}\mu_i\, dN_i - TdS - SdT + PdV + VdP.$$

Hence, $\mu_i = \left(\dfrac{\partial G}{\partial N_i}\right)_{T,PN}$. Therefore the chemical potential per particle is just Gibbs free energy per particle.

Exercise 6.16. Calculate the energy density, pressure, gluon and quark number densities in a QGP treated as a free gas of massless quarks of 3 flavours with a temperature of 200 MeV, which is of the order of QGP transition temperature.

We employ (6.84) and (6.86) and set $\mu_q = 0$ to get

$$\in_{QGP} = d_g \in_g + d_q \in_{q\bar{q}} = 16\cdot\frac{\pi^2 T^4}{30} + 18\cdot\frac{7\pi^2 T^4}{120} = \frac{127}{120}\pi^2 T^4$$

$$= \frac{127}{120}(3.141)^2\cdot\frac{16\times 10^8\, MeV^4}{(197)^3\ MeV^3 fm^3} = 2.2\,\frac{GeV}{fm^3},$$

after using the conversion factor for $\hbar = c = 1$.

Hence, $\qquad P_{QGP} = \dfrac{1}{3}\in_{QGP} = 0.71\ \text{GeV/fm}^3$.

Employing (6.84b) we get for the *QGP* in question; the gluon number density

$$= d_g n_g = \frac{16}{\pi^2}1.202\, T^3 = \frac{16}{(3.141)^2}\frac{1.202\times 8\times 10^6 MeV^3}{(197)^3\ MeV^3 fm^3} = \frac{2.0}{fm^3}.$$

Employing (8.74b) we get for the QGP in question; the quark or antiquark number density

$$= d_q \, n_q = \frac{18}{\pi^2} \, 1.202 \, T^3 = \frac{18}{2(3.141)^2} \, \frac{8 \times 10^6 MeV^3 \times 1.5 \times 1.202}{(197)^3 \; MeV^3 fm^3} = 1.7/fm^3.$$

Exercise 6.17. Show that in a relativistic pion gas the ratio of the entropy of the system to the number of pions in the system is a constant, independent of temperature and pressure.

$dE_\pi = P_\pi \, dV + T ds_\pi$ and the entropy density, $\dfrac{ds_\pi}{dV} = \dfrac{\epsilon_\pi + P_\pi}{T}$. Entropy per

pion $\dfrac{ds_\pi}{dN_\pi} = \dfrac{ds_\pi}{n_\pi dV} = \dfrac{\epsilon_\pi + P_\pi}{n_\pi T}$. Here N_π is the pion density. Now pion comes

in three flavours (π^+, π^-, π^0) i.e. $d_\pi = 3$. Pion gas is a massless boson gas and therefore we can use (8.70) and write:

$$\epsilon_\pi = d_\pi \left(\frac{\pi^2 T^4}{30} \right) = \frac{\pi^2 T^4}{10}, \; P_\pi = \frac{1}{3} \epsilon_\pi = \frac{\pi^2 T^4}{30} \; \text{and} \; n_\pi = \alpha_\pi \cdot \frac{1.202}{\pi^2} \, T^3.$$

Therefore, $\dfrac{ds_\pi}{dN_\pi} = \dfrac{4\pi^2}{30} T^4 \left(\dfrac{\pi^2}{3 \times 1.202 T^4} \right) = 3.6$, a constant independent of

pressure and temperature.

Exercise 6.18. Estimate the critical temperature at which the quark-gluon pressure is equal to the bag pressure B for a special case where there is no net baryon number.

The number quarks and the number of antiquarks in the system are

equal ($\mu_q = 0$; $n_b = 0$) gives $\dfrac{37}{90} \pi^2 T_c^4 = B$ or $T_c = \left(\dfrac{90}{37\pi^2} \right)^{\frac{1}{4}} B^{1/4}$.

Thus, for $B^{1/4} = 206$ MeV (*see* Chapter 2), we have $(T_c)_{QGP} \cong 144$ MeV.

REFERENCES

1. D.A. Bromley, Nucl. Phys. **A400**, 3C (1983).
2. W. Norenberg, J. De. Phys. **C537**, 141 (1976).
3. W.U. Schroder and J.R. Huizenga, in Treatise on Heavy Ion Sciences Ed. D.A. Bromley (1984) Vol. 2 page 115.
4. S.N. Mukherjee and L.N. Pandey, J. De. Phys. **C6**, 445, (1984), Phys. Rev. C29, 1326 (1984).
5. J.S. Blair, Phys. Rev. **95**, 1218 (1954); J.S. Blair, in Lectures in Theoretical Physics Vol. Ill c Ed. P.D. Kunz et al., University of Colorado Press, Colorado, (1966), page 343.

6. J. Randrup, Ann. of Phys. (N.Y.) 112, 356 (1978), Nucl. Phys. **A307**, 319 (1978). W.E. Frahn, Nucl. Phys. **A302**, 281 (1978). W.E. Frahn and K.E. Rehm, Phys. Reports **37**, 1 (1978).

7. W.E. Frahn, in Treatise on Heavy Ion Sciences, Ed. D. Allan Bromley Premn Press 1984, Vol. 1, page 135.

8. D.H.E. Gross, Nuclear Physics **A240**, 472 (1975). J. Randrup and W.J. Swiaticki, Ann. Phys. (N.Y.) **124**, 193 (1980).

9. J. Blocki et al., Annals of Phys. **113**, 330 (1978).

10. U. Mosel, in Treatise on Heavy Ion Sciences. Ed. D.A. Bromley, Vol. 2, 1984.

11. D. Glas and U. Mosel, Nucl. Phys. **A237**, 429 (1975). J. Galin et al., Phys. Rev. (**C9**, 1018 (1974).

12. H.H. Gutbrod et al., Nucl. Phys. **A213**, 267 (1973).

13. V.J. Menon, S.N. Mukherjee, C.S. Shastry and B. Sahu, Phys. Rev. **C41**, 1031 (1990).

14. R. Bass, Nucl. Phys. **A231**, 45 (1974).

15. J. Dobaczewski et al., Phys. Rev. Lett. **72**, 981 (1994).

16. I. Tanihata et al., Phys. Lett. **B160**, 380 (1985); Ibid **B206**, 592 (1988).

17. I. Tanihata, Phys. Nucl. Part, Phys. **22**, 157 (1996); Ibid. Prog. Part. Nucl. Phys. **35**, 505 (1995).

18. P.G. Hansen and A.S. Jensen, Ann. Rev. Nucl. Part. Sc. 591 (1995).

19. I. Mazumdar, V. Arora and V.S. Bhasin, Phys. Rev. **C61**, 051303(R) (2000); Phys. Rev. **69**, 061301(R) 2004. B.V. Damlin et al., Sov. J. Nucl. Phys. **53**, 71 (1991).

20. J. Glauber, in Lectures in Theoretical Physics, Ed. W.E. Brittin and L.G. Dunham (Interscience, N.Y. 1959) Vol. **1**, p. 315.

21. R. Shyam, P. Banerjee and G. Baur, Nucl. Phys. **A540**, 341 (1992).

22. B. Chakraborty, T.K. Das and S.N. Mukherjee, Fizika B (Zagreb) **9**, 1 (2000). S. Mahapatra, J. Nag, D.P. Sural and S.N. Mukherjee, Pramana, J. of Phys. **57**, 717 (2001). M.K. Pal, S.N. Mukherjee, T.K. Das and J. Nag, J. Phys. G: Nucl. Part. Phys. **24**, 1513 (1998).

23. A.S. Goldhaber, Phys. Lett. **53B**, 306 (1974).

24. G. Giacomellic, in Total Cross-section Measurements: Progress in nuclear physics Vol. **12**, Pergamon Press, 1970.

25. P.J. Simens and J.O. Rasmussen, Phys. Rev. Lett. **42**, 880 (1979).

26. J. Goset et al., Phys. Rev. **C16**, 629 (1977). S. Nagamiya, Phys. Lett. **81B**, 147 (1979). S. Das Gupta and A.Z. Mekjian, Phys. Reports **72**, 131 (1981);

27. S.N. Mukherjee and L.N. Pandey, Prog. in Theoret. Phys. (Japan) **68**, 684 (1982).

28. S. Nagamiya and M. Gyulassy, Advances in Nuclear Physics, Plenum Press, Vol. **13** (1984).

29. P.D.B. Collins, A.D. Martin and E.J. Squires, in Particle Physics and Cosmology. John Wiley and Sons, 1989 page 44c.

30. C.P. Singh, Intl. J. Mod. Phys. **A7**, 7185 (1992), Physics Reports, 232 (1993).

31. Luciano Maianai, CERN Bulletin, 712000, February 14, 2000.
32. M.C. Abreu et. al. NA50 Collaboration, Phys. Lett. B477, 28 (2000).
33. Itzhak Tserruya in Proc. Intl. Conf. on Physics and Astrophysics of Quark Gluon Plasma, Editors : B. Sinha, D. Srivastava and Y.P. Viyogi. Pramana **60**, 577 (2003).

SUGGESTED BOOKS FOR FURTHER READING

34. Introduction to the Theory of Heavy Ion Collisions
    W. Nörenberg and H.A. Widenmüler, Springer Verlag Berlin (1976) Vol. 51.
35. Nuclear Heavy Ion Reactions, P.E. Hodgson, Clarendon Press, Oxford (1978).
36. Treatise on Heavy Ion Science, Ed. D.A. Bromley Plenum Press (1984) Vol. 1, and Vol. 2.
37. Introduction to High Energy Heavy Ion Collision, C.Y. Yong, World Scientific Singapore, 1994.
38. Superheavy Elements – Status and Prospects Gottfried Münzenberg, Proc. DAE-BRNS 50th Symposium on Nuclear Physics. Editors S. Kailas, Suresh Kumar and L.M. Pant. Bhabha Atomic Research Centre, Mumbai India (2005).
    S. Hofmann and G. Münzenberg, Rev. Mod. Phys. **72**, 733 (2000).
39. R. Vandenbosch, Ann. Rev. Nucl. and Part Sci. **42**, 447 (1992).
40. Nuclear Physics in a nutshell C.A. Bertulani Princeton University Press, Princeton and Oxford, 2007.

Weak Interaction and Neutrino Physics

Symmetries dictate dynamics

-Marciano[39]

7.1 BETA DECAY AND TESTS OF FUNDAMENTAL PROCESSES

In this chapter we shall concentrate mainly on aspects that pertain in large part about intrinsic neutrino properties (β-decay, double β-decay, neutrino oscillations etc.) and symmetry tests for the invariance of the basic interactions. Wolfgang Pauli in 1930 suggested that neutrino (invisible then and remained so for another 25 years) is emitted along with electron in nuclear beta decay where the elementary processes are: $p \to n + e^+ + \nu_e$, $n \to p + e^- + \bar{\nu}_e$. Pauli's neutrino (or antineutrino) is a neutral, massless, spin half particle and its presence rescues apparent violation of both energy and angular momentum conservations in the beta decay process.

Subsequently it was found by particle physicists that the neutrino comes in three flavours: ν_e, ν_μ and ν_τ. Schewartz, Steinberger and Lederman shared the Nobel prize in 1988 for the discovery of the muon neutrino by the neutrino beam method. Experimental confirmation of neutrino required many years of effort but the discovery of antineutrino ($\bar{\nu}_e$) in reverse beta decay: $\bar{\nu}_e + p \to n + e^+$ by Reines and Cowan[1] led to the Nobel prize for them in 1995. Martin Perl also shared the prize for his discovery of the τ lepton. The 2002 Nobel Prize was awarded to Raymond Davis Jr[2] and Koshiba[3] for their painstaking experiments of increasing size and number, which proved neutrinos have mass.

The experimental observations on beta decay are as follows:

(1) The energy spectrum of emitted β-particles is continuous. The continuous β-spectrum (Fig. 7.1a) displays a characteristic end point energy that corresponds exactly to the level energy difference (usually between ground states) in the parent and daughter. However, electron energies below the end point energy (E_0) seems to involve "missing energy", thus violating energy conservation.

(2) Linear and angular momenta are also not found conserved in β-decay.

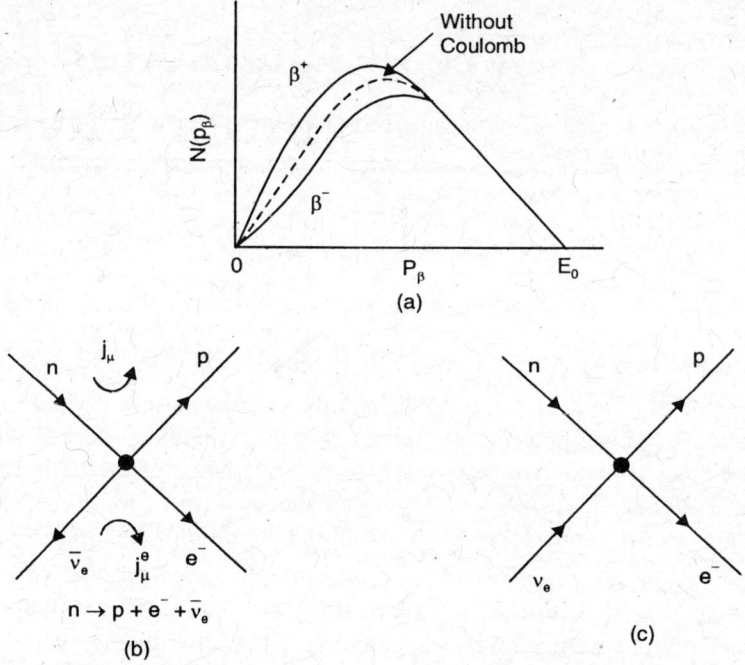

Fig. 7.1: (*a*) Continuous β-spectra, (*b*) Four Fermion interaction and (*c*) Inverse β-decay

In β-decay the direction of the emitted electrons and that of the recoiling nuclei can be observed: they are almost never exactly opposite as required by the linear momentum conservation. The non-conservation of angular momentum follows from the known spins of electron, proton and neutron. The β-decay involves conversion of a nuclear neutron into a proton: $n \rightarrow p + e^-$. Since the spin of each particle involved is 1/2, this reaction cannot take place if spin (hence angular momentum) is to be conserved.

A. Energy Spectrum and the Fermi Theory[4]

We shall give a simplified presentation of the theoretical formulation of β-decay. A formal theory of β-decay within the framework of particle physics and general theory of weak interaction is discussed in the next section. We shall only show that even a crude approach reproduces the shape of the β-spectrum. Since the interaction responsible for β-decay is weak, the standard perturbation theory can be applied. The transition rate is given by the Golden Rule:

$$dw_{fi} = \frac{2\pi}{\hbar} |\langle f|H_\beta|i\rangle|^2 \rho(E),$$ (7.1)

where $\rho(E)$ is the density of the final states available to the system and H_β is the Hamiltonian responsible for β-decay. For the simplest β-decay, $n \rightarrow p + e^- + \bar{\nu}$, we can say that the initial state (i) consists of a neutron and a full sea of negative energy neutrinos. The final state (f) consists of a proton, positive energy electron and a hole in the negative energy neutrino sea. In other words, a positive energy electron has been created and a negative energy neutrino has been destroyed.

The number of states available to an electron in the momentum interval p to $p + dp$ per unit volume, $dN_e = 4\pi p^2 dp/(2\pi\hbar)^3$ and to an antineutrino in the momentum interval q and $q + dq$, $dN_{\bar{\nu}} = 4\pi q^2 dq/(2\pi\hbar)^3$. Therefore,

$$\rho(E) = \frac{dN}{dE_0} = \frac{16\pi^2 q^2 p^2}{(2\pi\hbar)^6} \frac{dp\, dq}{dE_0},$$ (7.2a)

where the sum over spin states is included in H_β and $E_0 = E_e + E_{\bar{\nu}} + E_R$. In our approximation, we take the recoil energy $E_R = 0$ and drop the subscript on E_e:

$$E_0 = E + E_{\bar{\nu}},$$ (7.2b)

where E_0 is the total lepton energy including the rest masses. E_0 is the end point energy and also the maximum energy of the β-particle for $E_{\bar{\nu}} = 0$. Sometime the energy conservation is taken into account by including a term $\delta(E_0 - E - E_{\bar{\nu}})$ in (7.2a). For the massless neutrino, $E_{\bar{\nu}} = cq$. For fixed p, we have $dE_0 = dE_{\bar{\nu}} = cdq$ or $\frac{dq}{dE_0} = \frac{1}{c}$ so that

$$\rho(E) = \frac{16\pi^2}{(2\pi\hbar)^6 c} p^2 q^2 dp = \frac{16\pi^2}{(2\pi\hbar)^6 c^3} p^2 (E_0 - E)^2 dp.$$ (7.3)

Substituting $\rho(E)$ from (7.3) in (7.1), we get the beta spectrum:

$$dw_{fi} = \frac{1}{2\pi^3 c^3 \hbar^7} |\langle f|H_\beta|i\rangle|^2 p^2 (E_0 - E)^2 dp. \qquad (7.4)$$

$$\left(\frac{dw_{fi}}{p^2 dp}\right)^{1/2} = \text{const.} \left[|\langle f|H_\beta|i\rangle|^2\right]^{1/2} (E_0 - E) \qquad (7.5a)$$

or alternately,
$$\left[\frac{N(p)}{p^2}\right]^{1/2} = \text{const.} \left[|\langle f|H_\beta|i\rangle|^2\right]^{1/2} (E_0 - E), \qquad (7.5b)$$

where $N(p)$ is the number of β-particles coming with momentum p. If the expression on the left hand side of (7.5b) is determined experimentally and plotted against electron energy E, a straight line cutting the x-axis at $E = E_0$ results, when the matrix element is momentum independent. Such a plot is called the "Kurie Plot". Figure 7.2a shows the Kurie Plot for the neutron decay. It is indeed a straight line over most of the energy range.

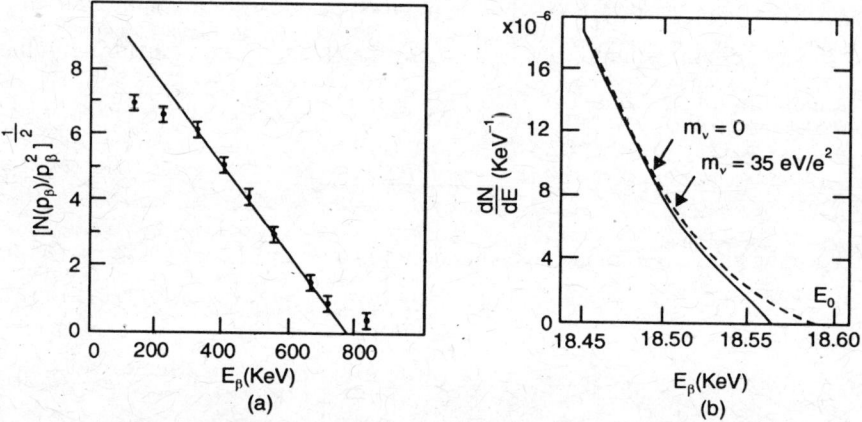

Fig. 7.2: (a) Kurie plot for the free neutron decay and (b) Tritium β spectrum

Coulomb Effects: If a nucleus with charge Ze decays by e^- emission, it will experience a Coulomb force once it has left the nucleus. This force will decelerate e^- and accelerate e^+ (in the case of positron emission). So the spectrum will be distorted (Fig. 7.1a). There will be more positrons of high energy and more electrons of low energy than predicted by (7.5). The effect of Coulomb potential on emitted electrons can be computed easily. The effect is particularly important at low electron (positron) energies and for high Z nuclei. The Coulomb correction introduces an additional factor in (7.4), which becomes

$$dw_{fi} = \frac{1}{2\pi^3 c^3 \hbar^7} \left[|\langle f|H_\beta|i\rangle|^2\right]^{1/2} F(\mp, Z, E) p^2 (E_0 - E)^2 dp,$$

where $F(\mp,Z,E)$ is called the Fermi function[5] and it has the form:

$$F(\mp,Z,E) \qquad (7.6)$$

where $x = \mp\dfrac{2\pi Zc}{137v}$ and v is the velocity of electron (positron) and the sign indicates whether it applies to electrons or positrons.

Beta-Decay Lifetime: Information about the magnitude of the matrix element can be obtained from the mean lifetime τ of a beta emitter. Since the matrix element is considered to be momentum independent, the total transition rate

$$w_{fi} = \frac{1}{\tau} = \frac{1}{2\pi^3 c^3 \hbar^7} \Big[|\langle f|H_\beta|i\rangle|^2\Big] \int\limits_0^{p_0} dp\, p^2 (E_0 - E)^2\, F(\mp,Z,E),$$

where p_0 is the maximum momentum. For very large energies we can neglect electron rest energy and write $E_0 \cong cp_0$, and for small Z, we set $F \approx 1$. The integral

$$w_{fi} = \frac{1}{\tau} = \frac{1}{2\pi^3 c^3 \hbar^7} \Big[|\langle f|H_\beta|i\rangle|^2\Big] \int\limits_0^{p_0} dp\, p^2 (E_0 - E)^2\, F(\mp, Z, E),$$

Therefore, $\qquad \dfrac{1}{\tau} = \dfrac{1}{c^6 \hbar^7} |\langle f|H_\beta|i\rangle|^2 \, \dfrac{E_0^5}{60\pi^3}. \qquad (7.8)$

This decrease of lifetime as the fifth power of E_0 is called the Sargents Rule.

The above value of the integral is useful only for a rough estimate. However, accurate value of the integral is needed for making a meaningful comparison with data. The accurate value of the integral has been tabulated with the following abbreviation:

$$\int\limits_0^{p_0} F(\mp, Z, E)\, p^2 (E_0 - E)^2 \, dp = m_e^5 c^7 f(E_0), \qquad (7.9)$$

where m_e is the electron rest mass. The factor $m_e^5 c^7$ has been inserted to make $f(E_0)$ dimensionless. Usually instead of τ, we use $t_{1/2} = \tau \log_e 2$ to get

$$\left|\langle f|H_\beta|i\rangle\right|^2 = \frac{2\pi^3}{f t_{1/2}} \log_e^2 \frac{\hbar^7}{m_e^5 c^4}, \qquad (7.10a)$$

$$|\langle f|H_\beta|i\rangle|^2 \equiv G_F^2 |\mathbf{M}|^2, \qquad (7.10b)$$

where G_F is a four fermion coupling constant and M is the matrix element which involves an integral over the interaction volume of four fermion wave function. It is customary to use $ft_{1/2}$ and not ft in tabulations. $ft_{1/2}$ is called the *comparative half life*. This is so called because all β-decaying states would have the same value of $ft_{1/2}$ if all matrix elements are equal. However, there is a wide range of $ft_{1/2}$ values, ranging from 10^3 to 10^{23} sec. Normally, therefore the base ten logarithm of its value (in seconds), $\log ft_{1/2}$ value is quoted.

The β-transitions are governed by selection rules, depending on whether the β-decay is allowed or forbidden. Let us consider β-transition from initial nuclear state (I_i, π_i) to a final state (I_f, π_f), the change in the total angular momentum I is

$$\Delta I = I_f - I_i \text{ and } \Delta I = \Delta L + \Delta S \text{ where } \Delta L \text{ and } \Delta S$$

are respectively change in the orbital and spin angular momenta of the nucleus in β transition. The beta interaction used by Fermi is a four fermion interaction, evaluated at the same space point $(r = 0)$. Therefore, for the allowed transition, the orbital angular momentum carried by electron neutrino pair must be $\Delta L = 0$ and $\Delta I = I_f - I_i$. The electron and neutrino are spin 1/2 particles. Their spins can be aligned either antiparallel or parallel to each other so that the total spin angular momentum carried by them can be $S = 0$ or $S = 1$. In the first case, the spin change of nuclear states will be $\Delta I = \Delta S = 0$. This corresponds to the Fermi (F) selection rule. The second case, $\Delta I = \Delta S = 0, \pm 1 \ (0 \to 0 \text{ forbidden.})$, corresponds to the Gamow-Teller[6] (GT) selection rule. Therefore, in the Fermi (F) transition, emitted $e^-(e^+)$ and $\overline{\nu}_e(\nu_e)$ have antiparallel spins. In the Gamow Teller (GT) transitions $e^-(e^+)$ and $\overline{\nu}_e(\nu_e)$ have parallel spins. In short we can write:

1. $I_i = I_f = 0$ for Pure F Supper allowed - $\log ft_{1/2} = 3$ to 4

2. $I_i \neq I_f$ for Pure GT Allowed - $\log ft_{1/2} = 4.5$ to 5

3. $I_i = I_f \neq 0$ for F or GT Forbidden - $\log ft_{1/2} = 7$ to 9.

The quantity $ft_{1/2}$ (Eq. 7.10a) gives information about the matrix element **M**, which depends on the overlap of the initial and final nuclear wave functions. For allowed transitions, where the overlap is essentially complete, $|M|^2 = 1$.

From (7.10a) and (7.10b) we may calculate G_F for a pure Fermi transition: ${}^{16}_{8}O \rightarrow {}^{14}_{7}N^* + \beta^+ + \nu_e$ with $ft_{1/2} \approx 3100$ sec. On substitution we get,

$$G_F = \left[\frac{2\pi^3 \hbar^7}{m_e^5 c^4} \frac{\log_e^2}{f(E_0)t_{1/2}} \right]^{1/2}$$

$$= \left[\frac{2\pi^3 (197 \text{ MeV fm})^7 0.693}{(0.51 \text{ MeV})^5 \times 3 \times 10^{23} \text{ fm/sec} \times 3100 \text{ sec}} \right]^{1/2}$$

$$\cong 1.02 \frac{10^{-5}}{M_P^2} \text{GeV}^{-2}, \text{ where } M_p \text{ is the mass of proton in GeV.} \tag{7.11}$$

Table 7.1: Super allowed β-transition

Transition	ΔI	E_0 (MeV)	Δp	$ft_{1/2}$ (sec)	$\log ft_{1/2}$ (sec)	Type
${}^{6}_{2}He \xrightarrow{\beta^-} {}^{6}_{3}Li$ $(I_i = 0^+ \rightarrow I_f = 1^+)$	1	3.50	No	810	2.91	GT
${}^{14}_{8}O \xrightarrow{\beta^+} {}^{14}_{7}N^*$ $(I_i = 0^+ \rightarrow I_f = 0^+)$	0	1.81	No	3100 ± 20	3.52	F
${}_0 n^1 \xrightarrow{\beta^-} {}_1 H$ $\left(I_i = \frac{1}{2}^+ \rightarrow I_f = \frac{1}{2}^+\right)$	0 or 1	0.78	No	1080 ± 16	3.08	F + GT

In the Fermi theory of β-decay, the neutrino is assumed to be massless. Since the careful examination of experimental electron spectra constitutes an important measure of the antineutrino (neutrino) mass, we must distinguish between m_e and $m_{\bar{\nu}}$, the electron and antineutrino masses respectively and modify accordingly the β-spectra.

Assuming $m_{\bar{\nu}} \neq 0$ and distinguishing it from the electron mass m_e, we can find $\rho(E)$ as follows: For the antineutrino,

$$q^2 c^2 + m_{\bar{\nu}}^2 c^4 = E_{\bar{\nu}}^2 = (E_0 - E)^2 \tag{7.12}$$

or
$$q^2 \frac{dq}{dE_0} = \frac{(E_0 - E)}{c^3} \left[(E_0 - E)^2 - m_{\bar{v}}^2 c^4\right]^{1/2}.$$

Hence, according to (7.2a)

$$\rho(E) = \frac{16\pi^2}{(2\pi\hbar)^6 c^3} p^2 dp (E_0 - E) \left[(E_0 - E)^2 - m_{\bar{v}}^2 c^4\right]^{1/2}. \qquad (7.13)$$

Employing the above value of $\rho(E)$, we get

$$\left(\frac{dw_{fi}}{p^2 dp}\right) = \frac{m_e^2 c}{2\pi^3 \hbar^7} |\langle f|H_\beta|i\rangle|^2 (\epsilon_0 - \epsilon) \left[(\epsilon_0 - \epsilon)^2 - \mu^2\right]^{1/2} F(\epsilon),$$

where $\quad \epsilon = \dfrac{E}{m_e c^2}, \epsilon_0 = \dfrac{E_0}{m_e c^2}$ and $\mu = \dfrac{m_{\bar{v}}}{m_e}$ $\qquad\qquad (7.14)$

or
$$\left[\frac{N(p)}{p^2}\right]^{1/2} = \text{constant} \left[|\langle f|H_\beta|i\rangle|^2 F(\epsilon)\right]^{1/2}$$

$$\times (\epsilon_0 - \epsilon)^{1/2} \left[(\epsilon_0 - \epsilon)^2 - \mu^2\right]^{1/2}. \qquad (7.15)$$

The above equation reduces to (7.5b) for $F(\epsilon) = 1$ and $m = 0$. Clearly, the effect of finite neutrino mass becomes visible if $(\epsilon_0 - \epsilon) \sim \mu$, that is very near the threshold. It is found that the spectral shape near the end point for the β-decay of ^3H is distorted and the distortion is very sensitive to neutrino mass[6]. This is shown on the Kurie plot (rate of reaction plotted against the electron energy). The end point shifts (Eq. 7.15) and the rate near end point is depressed (Fig. 7.2b) (for details see Sec. 7.4B).

Fermi assumed that the electron neutrinos were created during the process of β-decay. So he patterned his theory taking ideas from the quantum theory of radiation. He made the following modification to distinguish the weak interaction from the electromagnetic interaction:

(1) The coupling constant e is replaced by the weak coupling constant g.

(2) The electromagnetic current $\vec{j}_{em}$ is replaced by the weak current $\vec{j}_w$.

Subsequently Yukawa[7] and later Klien[8] suggested that the exchanged particle is not massless and neutral but massive and electrically charged. In the case of EM interaction the exchanged particle (photon) is neutral and massless.

Although the theories of Yukawa and Klein were overshadowed by greater success of Fermi's theory, the idea of weak interaction being mediated by the exchange of vector bosons nevertheless had a fundamental influence in formulation of the modern gauge theory of electroweak interaction.

In order to evaluate the β-decay matrix element $\langle f \,|\, H_\beta \,|\, i\rangle$, we must know H_β, the basic interaction inside the nucleus which changes neutron to proton (or vice versa). Fermi assumed H_β to be r independent (zero range contact interaction). In the Fermi transition (F) only a neutron is changed to a proton (or vice-versa). This change can be brought by raising operator t_+ (or t_- when a proton is changed to neutron). Therefore, for the Fermi transition the matrix element is the simple overlap integral

$$M_F = \langle p\, e^- \nu_e |H_\beta| n\,\rangle \approx \int \psi_p^*(\vec{r})\, t_+ \,\psi_n(\vec{r})\, d\tau, \qquad (7.16)$$

where ψ includes isospin and spin functions of the nucleon as well and t_+ operates in the i-spin space.

In the Gamow-Teller transition (GT), a neutron is changed into a proton and in addition, its spin is changed. Therefore, GT matrix element is given by

$$M_{GT} = \int \psi_p^*(\vec{r})\, \vec{\sigma}\, t_+ \,\psi_n(\vec{r})\, d\tau. \qquad (7.17)$$

Both integrals are taken over to the volume of the nucleus. These expressions are obtained by assuming that the variation of neutrino and electron eigen functions over the nuclear volume can be neglected. We can write for the nuclear β-decay

$$|M_F|^2 = |\langle\, \psi_f^* \,| \sum_i t_\pm^i \,|\psi_i\rangle|^2 = |\langle\, \psi_f^* \,|\, T_\pm \,|\psi_i\rangle|^2, \qquad (7.18a)$$

where $T_\pm = T_1 \pm iT_2$, the raising and lowering operators for the total i-spin of the nucleus. When the i-spin magnitude is a good quantum number, M_F is non-zero only if the initial and final states have the same T value.

Hence, $\qquad |M_F|^2 = T(T+1) - T_{3i}\, T_{3f}.$

The above result assumes that all other characteristics of the states are identical.

For ${}_8^{14}O_6 \xrightarrow{\ \beta^+\ } {}_7^{14}N_7$ decay:

$$T = 1,\ T_{3i} = -1,\ T_{3f} = 0,\ \text{so that } |M_F|^2 = 2. \qquad (7.18b)$$

For free neutron decay:

$$|M_F|^2 = |\langle\, \psi_f^* \,| \sum_i t_\pm^i \,|\psi_i\rangle|^2 = |\langle\, \psi_f^* \,|\, T_\pm \,|\psi_i\rangle|^2,\ \text{so that } |M_F|^2 = \frac{3}{4} + \frac{1}{4} = 1. \qquad (7.18c)$$

It is not difficult to calculate values of $|M_{GT}|$ using the extreme single particle shell model. If a single nucleon is involved, transitions are allowed only by states of the same ℓ, since the operator $\vec{\sigma}$ can at most flip the spin. Let us consider the mirror decay $^{14}_{8}O_6 \xrightarrow{\beta^+} {}^{14}_{7}N_7$: In this process one neutron outside the closed shell is transformed into a proton in a level with the same j. We shall treat the case $j = \ell + 1/2$ and consider the decay of the state $m_j = j$. According to (7.17), the square of the matrix element,

$$|M_{GT}|^2 = \sum_{m'_j} \sum_{k=1}^{3} |\int \psi_j^{*m'_j} \vec{\sigma}_k \ \psi_j^j \ d\tau|^2.$$

The state $m_j = j$ can decay into the state with $m'_j = j$ owing to the term σ_3. This contribution to the square of the matrix element is 1 for $j = \ell + 1/2$. In addition, it can decay into $m'_j = j - 1$ with the two other components $\sigma_1 = \dfrac{1}{2}(\sigma_+ + \sigma_-)$ and $\sigma_2 = \frac{1}{2i}(\sigma_+ - \sigma_-)$. Also σ_+ operating on a state with maximum m_j gives zero. Thus,

$$|\int \psi_j^{*j-1} \sigma_1 \ \psi_j^j \ d\tau|^2 + |\int \psi_j^{*j-1} \sigma_2 \ \psi_j^j \ d\tau|^2$$

$$= \frac{1}{2} |\int \psi_j^{*j-1} \sigma_- \ \psi_j^j \ d\tau|^2 = \frac{1}{2} \left| \frac{1}{j} \ \sqrt{2j} \right|^2 = \frac{1}{j}. \tag{7.19}$$

The total matrix element is

$$|M_{GT}|^2 = 1 + \frac{1}{j} = \frac{j+1}{j} \text{ for } j = \ell + 1/2. \tag{7.20a}$$

Similarly we can derive,

$$|M_{GT}|^2 \begin{cases} = \dfrac{j}{j+1} & \text{for } j = \ell + 1/2. \quad \Delta j = 0. \tag{7.20b} \\[2ex] = \dfrac{2(2j+1)}{2\ell+1} & \text{for } j = \ell - 1/2. \quad \Delta j = 1. \tag{7.20c} \end{cases}$$

For the β-decay under consideration ($j = 1/2$, $\ell = 0$, $m_j = 1/2$), we get $|M_{GT}|^2 = 3$.

Let G_V and G_A be respectively the strength of the Fermi (vector) and Gamow-Teller (axial vector) transitions. Then for the free neutron decay, which is a mixed F and GT transitions, we have:

$$\frac{1}{\tau_n} = \frac{1}{c^7 \hbar^7} \left[G_V^2 \left| M_F \right|^2 + G_A^2 \left| M_{GT} \right|^2 \right] \frac{E_0^5}{60 \pi^3}. \qquad (7.21a)$$

The conservation of angular momentum prevents any interference between vector and axial vector transitions. Substituting $\left| M_F \right|^2 = 1$ (shown earlier) and $\left| M_{GT} \right|^2 = 3$, we get

$$\left(\frac{1}{\tau_n} \right)_{\text{expt.}} = \frac{1}{c^6 \hbar^7} \left[G_V^2 + 3 G_A^2 \right] \frac{E_0^5}{60 \pi^3}. \qquad (7.21b)$$

The factor 3 appears because of the three orientations of the lepton spins in GT transition. Employing the data given in Table 7.1 for ^{14}O and neutron β-decay, we obtain the value of G_A/G_V as follows:

$$^{14}O \rightarrow {}^{14}N + e^+ + \nu_e$$

which is a pure Fermi transition, whereas $n \rightarrow p + e^- + \bar{\nu}_e$ is a mixed Fermi (F) and Gamow Teller (GT) transitions. Hence, employing (7.18) and (7.21b) and neglecting the variation of neutrino and electron wave functions over the nuclear volume, we get

$$\frac{(ft_{1/2}) \,^{14}O}{(ft_{1/2})n} = \frac{G_V^2 + 3 G_A^2}{2 G_V^2} = \frac{3100 \pm 20}{1080 \pm 16}$$

or

$$\frac{1}{2} + 1.5 \frac{G_A^2}{G_V^2} = \frac{3100 \pm 20}{1080 \pm 16}.$$

Hence, $\quad \dfrac{G_A^2}{G_V^2} = 1.58 \pm 0.04 \quad$ or $\quad \left| \dfrac{G_A}{G_V} \right| = 1.26 \pm 0.02. \qquad (7.21c)$

To determine individually G_V and G_A and the sign of the ratio G_A/G_V, we need to measure a second quantity in addition to τ_n. The decay asymmetry of polarized neutrons could be one of such measurements. The number of electrons that are emitted in the direction of the neutron spin $N \uparrow \uparrow$ is smaller than the number $N \uparrow \downarrow$ emitted in the opposite direction. The asymmetry A is defined by

$$A = \left[\frac{N \uparrow \uparrow - N \uparrow \downarrow}{N \uparrow \uparrow + N \uparrow \downarrow} \right] \frac{1}{\beta}, \text{ where } \beta = \frac{v}{c}.$$

The asymmetry is connected to $g_A = \dfrac{G_A}{G_V}$ by $A = -\dfrac{g_A (g_A + 1)}{1 + 3 g_A^2}.$

The measurement of asymmetry yields $A = -0.115 \pm 0.001$. Combining this information, we have

$$G_A/G_V = -1.26 \pm 0.02 .$$ (7.22)

Another excellent example of mixed transition is k-capture process, for which the same Fermi's β-decay theory can be applied. Let us consider the following k-capture process: $^7_4Be_3 + e^- \rightarrow {}^7_3Li_4 + \nu_e$.

The total energy of the initial state is simply the mass of the atom $M(A,Z)$, the energy of the final state of emitted neutrinio and the residual atom mass $M(A, Z-1)$. Therefore, the energy of the emitted neutrino

$$E_\nu = [M(A, Z) - M(A, Z - 1)] c^2 - \epsilon_k ,$$

where ϵ_k is the binding energy of the electron in k-shell. On the assumption of zero neutrino mass, the criterion for k-electron capture is

$$[M(A, Z) - M(A, Z - 1)] c^2 \geq \epsilon_k .$$

The g.s. of $^7_4Be_3$ lies 0.86 MeV above $^7_3Li_4$, since

$$\Delta E = \frac{6}{5} \frac{Ze^2}{r} - (m_n - m_p)c^2 = \left(\frac{6}{5} \times \frac{3 \times 1.44}{1.25 \times 1.913} - 1.293 \right) = 0.86 \text{ MeV}.$$

The excitation energy of $(1/2)^-$ in ^7Li is 0.48 MeV. Two modes of electron capture are possible:

(1) $(3/2)^- \rightarrow (3/2)^-$: $\Delta j = 0$, $\Delta \pi = 0$, which is a combination of F and GT type transitions and

(2) $(3/2)^- \rightarrow (1/2)^-$: $\Delta j = 1$, $\Delta \pi = 0$, which is a pure *GT* transition.

For the F type transition $(3/2)^- \rightarrow (3/2)^-$, the initial and final wave functions are similar, so that

$$|M_F|^2 = T(T+1) - T_{3i} \, T_{3f} = \frac{3}{4} + \frac{1}{4} = 1.$$

For the GT type transition the single particle model (7.20a) gives

$$|M_{GT}|^2 = \frac{j+1}{j} = \frac{3/2+1}{3/2} = \frac{5}{3} \text{ for } \Delta j = 0.$$

For the GT type transition $(3/2)^- \rightarrow (1/2)^-$ Eq. (7.20c) gives

$$|M_{GT}|^2 = \frac{4\ell}{2\ell + 1} = \frac{4}{3} \text{ for } \Delta j = 1, \ell = 1.$$

$$k\text{-capture decay rate} \quad \frac{\lambda_k(\frac{3}{2}^- \to \frac{3}{2}^-)}{\lambda_k(\frac{3}{2}^- \to \frac{1}{2}^-)} = \frac{|M_{GT}|^2_{3/2} + \frac{G_V^2}{G_A^2}|M_F|^2}{|M_{GT}|^2_{1/2}} \frac{E_{\nu_1}^2}{E_{\nu_2}^2}$$

$$= \frac{\frac{5}{3} + \left(\frac{G_V}{G_A}\right)^2 |M_F|^2}{(4/3)} \times \left(\frac{0.86}{0.86 - 0.48}\right)^2 = 8.82$$

Hence, the branching ratios are:

$$B\left(\left(\frac{3}{2}\right)^- \to \left(\frac{3}{2}\right)^-\right) = \frac{8.82}{9.82} \approx 90\% \quad \text{and} \quad B\left(\left(\frac{3}{2}\right)^- \to \left(\frac{1}{2}\right)^-\right) \approx 10\%.$$

Inverse Beta Decay

The inverse β-decay $(\nu_e p \to n e^+)$ cross section can be calculated by employing Eq. (5.22a) as follows: In units of $\hbar = c = 1$, we have

$$\sigma = \frac{p^2}{\pi} \frac{1}{v_i v_f} |\langle f | H_\beta | i \rangle|^2 = \frac{G_F^2}{\pi} |M_F|^2 \frac{p^2}{v_i v_f},$$

where v_i, v_f are respectively the relative velocities of particles in the initial and final states $(v_i = v_f = c)$ and p is the magnitude of the momentum of the neutron in the c.m. In the present case we are dealing with a mixed transition, $|M_F|^2 = 1$ and $|M_{GT}|^2 = 3$. Therefore,

$$\sigma = \frac{G_F^2}{\pi} \left(|M_F|^2 + |M_{GT}|^2\right) p^2 = \frac{4 G_F^2 p^2}{\pi}.$$

For neutrinos in the MeV range above the threshold $(Q = 1.8 \text{ MeV})$ and for G_F given by (7.11) one gets

$$\sigma = \frac{4}{3.141} \frac{1.04 \times 10^{-10} \text{GeV}^{-2}}{M_P^4} (1 \text{MeV})^2 \approx 10^{-43} \text{ cm}^2,$$

which corresponds to a mean free path of antineutrino absorption in water,

$$L = \frac{1}{\dfrac{6 \times 10^{23}}{18 \text{ cm}^3} \times 10^{-43} \text{ cm}^2} \cong 3 \times 10^{20} \text{ cm}.$$

B. Parity Violation and Weak Interaction

We already mentioned earlier (Ch. I) that the parity conservation is associated with the invariance of a physical process to space reflection and that it was proposed first by Lee and Yang[9] for the decay of K^+ meson, which was difficult to reconcile with parity conservation. K^+ meson decays into $\pi^+\pi^0$ with branching ratio 21.1% and also K^+ decays into $\pi^+\pi^+\pi^-$ with branching ratio 5.6%. The remaining four other decay modes add up to the value of branching ratio 73.3%. Since π-meson is pseudoscalar, its intrinsic parity is -1. The spin of K-meson or π-meson is zero in either case. Therefore, for $\pi^+\pi^0$ decay the orbital angular momentum of the system is zero because of total angular momentum conservation and thus the parity of $\pi^+\pi^0$ system is $+1$. On the other hand, for $\pi^+\pi^+\pi^-$ system, the total angular momentum is given by the sum of the orbital angular momentum of two π^+ and the one between the centre of mass of two π^+ and the remaining one π^-. The sum should be zero because it should be equal to the spin of the K-meson which is zero. This can be realized when the magnitude of these two orbital angular momenta are equal. Therefore, parity of this system due to orbital angular momentum becomes $+1$. Since, the intrinsic parity of π is -1, the intrinsic parity of three pion system is $(-1)^3 = -1$ and, hence, parity of $\pi^+\pi^+\pi^-$ is -1.

Then, if parity is conserved in these decays, we must have two independent decay modes of K^+ mesons, say θ^+ and τ^+; one is of even parity and another is of odd parity. θ^+ decays as $\pi^+\pi^0$ and τ^+ decays as $\pi^+\pi^+\pi^-$. However, it seems to be accidental that these two particles θ^+ and τ^+ have the same mass, the same life-time, the same spin and so on. This problem is known as τ–θ puzzle. To solve this puzzle, Lee and Yang suggested that K^+ meson decays through weak process in which parity conservation is violated. If this is the case for any weak interaction process, the effect must appear in the β-decay of nucleus. The effect was observed in the experiment of Wu et al.[10].

They proposed to observe the correlation between the spin $\vec{j}$ of ^{60}Co and the momentum $\vec{p}_e$ of electron produced in the β-decay of ^{60}Co:

$$^{60}Co\,(J^\pi = 5^+) \rightarrow {}^{60}Ni(J^\pi = 4^+) + e^- + \bar{\nu}_e.$$

The linear momentum $p(= m\,dr/dt)$ is a polar vector and it changes sign on reflection. But the angular momentum $(\vec{r} \times \vec{p})$ is an axial vector and does not change sign under reflection. This is also true for spin angular momentum. The spin of ^{60}Co is first aligned in a Z-direction perpendicular to the X–Y plane at some point z (Fig. 7.3a). The direction of spin $\vec{j}$ of ^{60}Co in the mirror image is the same as that of the real ^{60}Co because of axial vector

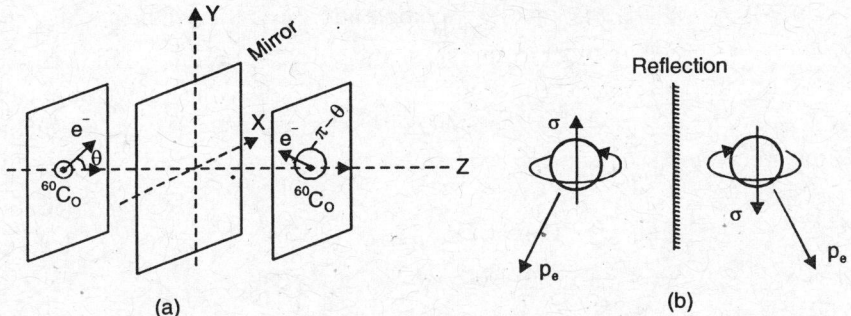

Fig. 7.3: (*a*) Schematic diagram of Wu's experiment and (*b*) Helicity

nature of $\vec{J}$. Then when an electron is produced in a direction, forming an angle θ relative to $\vec{J}$, the electron in the mirror image is produced in a relative angle $(\pi - \theta)$ because of vector nature of $\vec{p}_e$, as shown in Fig. 7.3(*a*). Therefore, if the process is parity invariant, the probability for finding electrons produced with the angle θ and $\pi - \theta$ must be the same. Wu and her collaborators[10] observed that the probability for finding electrons produced with $\theta = \pi$ was much larger than one with $\theta = 0$. This was the first demonstration of parity violation, i.e. parity non-conservation.

In the real decay of ^{60}Co electron spin is antiparallel to its momentum direction. But in the mirror image electron spin is parallel to its momentum as shown in Fig. 7.3*b*. The same is true for antineutrino as well. Though the polarization of β-particles observed in the β-decay of polarized nuclei is not a general characteristic of electrons, it is regarded as basic property of the antineutrino (also of the neutrino). In the case of pm decay muons are generally polarized. T invariance forbids a component of polarization normal to decay plane, specified by $(\vec{p}_\pi \times \vec{p}_\mu)$. Such a polarization would indicate that the average of the term $\vec{s}_\mu \times (\vec{p}_\pi \times \vec{p}_\mu)$ is non-zero, i.e. muon spin $\vec{s}_\mu$ would have a component along the direction of $(\vec{p}_\pi \times \vec{p}_\mu)$. Such a term is odd under T and should average to zero if there is no preferred direction of time.

It is customary to describe the polarization of spin $\frac{1}{2}$ particle not by $\vec{s} \cdot \vec{p}$ but by a helicity operator defined by

$$\hat{h}(p) = \frac{\vec{\sigma} \cdot \vec{p}}{|p|} = \vec{\sigma} \cdot \hat{n}, \qquad (7.24)$$

where $\hat{n}$ is the unit vector in the direction of $\vec{p}$. In other words, helicity is the projection of spin in the direction of momentum (or motion). Helicity is a pseudoscalar quantity.

$\langle \hat{h}(p) \rangle \neq 0$ is a sign of parity violation and $\langle \hat{h}(p) \rangle = 0$ indicates that the parity is conserved. In order to appreciate the physical meaning of the helicity, let us consider the Dirac equation of a free particle. The four component Dirac spinor $\psi \equiv \begin{pmatrix} \varphi \\ \chi \end{pmatrix}$ satisfies

$$(\vec{\sigma} \cdot \vec{p})\chi + m\phi = E\phi$$
$$(\vec{\sigma} \cdot \vec{p})\phi - m\chi = E\chi, \qquad (7.25a)$$

where ϕ and χ are two component spinors. Let us define $\phi_R = (\phi + \chi)/2$ and $\phi_L = (\phi - \chi)/2$. Then it is easy to see from (7.25a) that we get coupled equations:

$$(\vec{\sigma} \cdot \vec{p})\phi_R + m\phi_L = E\phi_R$$
$$(\vec{\sigma} \cdot \vec{p})\phi_L - m\phi_R = -E\phi_L. \qquad (7.25b)$$

For the special case of zero mass particle ($m = 0$), ϕ_L, ϕ_R become uncoupled spinor in (7.25a). The Dirac equation then reduces to a two component description and we can write:

$$\hat{h}\phi_R = \frac{E}{|p|}\phi_R \text{ and } \hat{h}\phi_L = -\frac{E}{|p|}\phi_L. \qquad (7.26)$$

For the positive energy solution, the spin $\vec{\sigma}$ and momentum $\vec{p}$ are aligned in φ_R and antialigned in ϕ_L as shown in Fig. 7.3b. For this reason, ϕ_R is called the right handed and ϕ_L the left handed spinor. For massless particle, $E = |p|$, ϕ_R is an eigen state of the helicity operator with eigen value +1 and ϕ_L an eigenstate with eigen value -1. If a neutrino or an antineutrino is produced by W-exchange, the neutrino helicity is negative, while the antineutrino helicity is positive. In fact all experiments are consistent with neutrinos being always left handed and antineutrinos right handed.

Non-zero mass electrons could be both left handed and right handed. It has been conclusively established by experimental results that the electrons emitted in β^--decay have left handed helicity more often than right handed helicity. Indeed, at elevated relativistic electron energies ($\beta \to 1$) the state of emitted electron asymptotically approaches pure left handed helicity $[\langle h \rangle = -v/c = -1]$ resulting in parity violation in the β-decay process.

The very low interaction cross section of neutrino with matter makes the measurement of neutrino helicity extremely difficult. Goldhaber et al.[16] developed an elegant method to measure neutrino helicity. For details see reference (11). The helicity of neutrino determined from this experiment is

$$\langle \hat{h}(p) \rangle \approx -1.0 \pm 0.3.$$

Exercise 7.1. Show that the helicity operator commutes with the free particle Dirac Hamiltonian.

The free particle Dirac Hamiltonian, $H = \hat{\alpha} \cdot \vec{p} + \vec{\beta} \cdot m$,

where $\alpha = \begin{pmatrix} 0 & \vec{\sigma} \\ \vec{\sigma} & 0 \end{pmatrix}$ and $\beta = \begin{pmatrix} I & 0 \\ 0 & -I \end{pmatrix}$.

Hence, $\qquad [H, \hat{h}(p)] \equiv [H, \vec{\sigma} \cdot \vec{p}] = \hat{\alpha} \cdot \vec{p} \ \ \vec{\sigma} \cdot \vec{p} - \vec{\sigma} \cdot \vec{p} \ \ \hat{\alpha} \cdot \vec{p}$

$$= \begin{pmatrix} 0 & \vec{\sigma} \cdot \vec{p} \\ \vec{\sigma} \cdot \vec{p} & 0 \end{pmatrix} \vec{\sigma} \cdot \vec{p} - (\vec{\sigma} \cdot \vec{p}) \begin{pmatrix} 0 & \vec{\sigma} \cdot \vec{p} \\ \vec{\sigma} \cdot \vec{p} & 0 \end{pmatrix} = 0.$$

Exercise 7.2 Find the parity of π^- from the reaction $\pi^- + d \rightarrow n + n$, employing s-wave π^-.

It is well known that spin-parity of deuteron

$$I_d^\pi = 1^+ \text{ so that } P(\pi^-) = P^2(n)(-)^\ell, \ell$$

being the orbital angular momentum of the relative motion of two neutrons. Since n-n system is fermionic system and so is exchange antisymmetric,

$\ell = 1, I = 1$, giving $P(\pi^-) = -1$.

7.2 MODIFICATION OF THE FERMI THEORY

With the establishment of non-conservation of parity, the Fermi theory required modification. The most successful extension was put forward by Feynman and Gell-Mann[12] and in somewhat different form by Sudarshan and Marshak[13]. But the modified theory is not far away from Fermi's original version. 'Two component theory of neutrino' is proposed as the modified version in which a neutrino (ν_e or ν_μ) only exists as a particle of negative helicity (left handed) and an antineutrino as a particle with positive helicity (right handed). If the neutrino were a normal Dirac fermion it would have to be represented by a four component state function with two helicity states for the particle and two for the antiparticle. The neutrino, according to this theory, is itself an object that does not confirm to reflection symmetry. A simple reflection reverses the momentum vector but not spin component and gives a neutrino with helicity (+1), a state that according to the two component neutrino theory does not exist.

A. Current-Current Interaction

In order to write β interaction in a more formal relativistic covariant form, we employ relativistic description of spin 1/2 particles. In addition we must consider various restrictions imposed on H_β by the experimental results, namely the parity violation and the existence of only left handed neutrinos with helicity -1 in nature. Interaction operator originally proposed by Fermi is a four fermion interaction, the spinors (field operators) for the four fermions $(n, p, e^-, \bar{v}_e)$ being evaluated at the same space point. The decay is conventionally analyzed in terms of the crossed process: $n + v_e \rightarrow p + e^-$ in which an outgoing antineutrino becomes an incoming neutrino.

The exchanged particles contribute a propagator term to the transition matrix element (Fig. 7.4). The propagator has the general form:

$$\frac{1}{Q^2 + m_w^2 c^2}. \tag{7.27}$$

Here, Q^2 is the square of the four momentum transferred $(= -q^2)$ in the interaction and m_W is the mass of the exchanged particle. In the case of *EM* interaction mediated by a virtual photon, this results in a factor $1/Q^2$ in the amplitude because photon is massless. However, in the case of weak interaction the large mass of the exchanged particle, called vector boson, causes the second term in the denominator of the propagator to dominate and we can write $(c = 1)$:

$$M \propto \frac{g}{\sqrt{2}} \frac{1}{Q^2 + m_W^2} \frac{g}{\sqrt{2}} \xrightarrow{Q^2 \rightarrow 0} \frac{g^2}{2 m_W^2}. \tag{7.28}$$

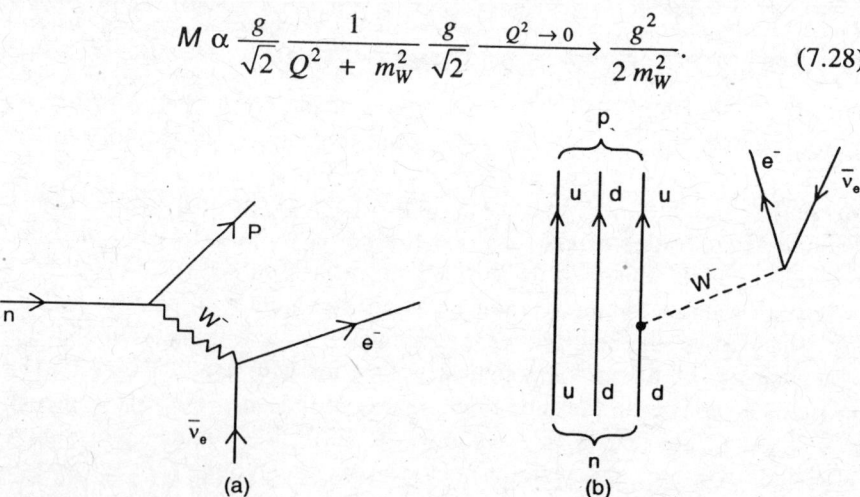

Fig. 7.4: (*a*) Exchange modification of the Fermi and (*b*) Quark view of neutron decay interaction in neutron decay

The difference to an EM interaction is seen in the finite mass of the exchanged particle. The very large mass of exchanged boson at small Q^2 causes weak interaction to be much smaller than the electromagnetic interaction. The large mass also means that its range $r_W = \hbar/(m_W c)$ is also very limited. The vector bosons, $W^\pm$, Z^0 have spin 1 and negative parity. $W^\pm$ bosons carry a single unit of electric charge so that they can mediate beta interaction. The mass of the vector bosons are estimated to be around 105 GeV. Therefore, the weak interaction mediated by vector bosons has a range $r_W \approx 2 \times 10^{-3}$ fm, which is negligible on nuclear scale. Since no nucleus has enough excitation energy to produce $W^\pm$bosons, these quanta appear only indirectly as virtual particles in intermediate states. CERN's linear electron positron (LEP) colliders produces real Z or W boson via[14]

$$e^+ + e^- \rightarrow Z^0 \qquad \text{when} \quad \sqrt{s} = 2E_e > m_z c^2$$

and $\qquad e^+ + e^- \rightarrow W^+ + W^- \qquad$ when $\quad \sqrt{s} = 2E_e > 2m_W c^2,$

where $\sqrt{s}$ is the c.m. energy. LEP's direct W-mass determination gives $m_W = (80.350 \pm 0.056)$ GeV/c^2.

After several years of preparation combining thousands of measurements of five different experiments from LEP at CERN and SLAC in USA, a precision value of the mass of Z^0 boson has been reported[15]:

$$M_Z = (91.1875 \pm 0.0021) \text{ GeV/c}^2.$$

LEP experiments give also a measure of number of neutrinos of active flavour.

The original Fermi interaction between an electron and proton, for example, can be described in the lowest order by the Feynman diagram (Fig. 7.1*b*). The matrix element M for this process is obtained by considering the interaction between electron four vector current j^e_μ with the electromagnetic field A_μ generated by the proton four vector current j^p_μ :

$$M = j^e_\mu A_\mu \approx \frac{1}{Q^2} j^e_\mu \cdot j^\mu_p, \qquad (7.29)$$

where $c = 1$ and $j_\mu(x) \equiv (i\rho(x), \vec{j}(x))$, $\mu = 0, 1, 2, 3$. The zeroeth (time) component $j_0(x)$ is the charge density $\rho(x)$ generating the space part $\vec{j}(x)$. The identification of the electromagnetic potential A_μ with source j^μ_p is made through the Maxwell's equation $\Box A^\mu = j^\mu_p$. The term $1/Q^2$, where Q^2, the square of the four momentum transferred to the photon, is essentially a photon propagator. The interaction is viewed as current-current interaction in terms of the Dirac spinors. The matrix element being contraction of two vector currents is a scalar: $j_e \cdot j^p = \sum_\mu j^\mu_e j^p_\mu.$

Since weak interaction is taking place at a point, Fermi expressed the beta interaction without the propagator term as the scalar product of two currents j_W^h and j_W^ℓ, where j_W^h and j_W^ℓ are respectively weak hadron currents and weak lepton currents. He wrote the matrix element as

$$M_F = \frac{G_F}{\sqrt{2}} (\overline{\Psi}_p \gamma^\mu \Psi_n)(\overline{\Psi}_{e^-} \gamma_\mu \Psi_\nu), \tag{7.30}$$

where $j_w^h = \overline{\Psi}_p \gamma^\mu \Psi_n$ and $j_w^\ell = \overline{\Psi}_{e^-} \gamma_\mu \Psi_\nu$ each respectively carrying $+e$ and $-e$ charge so that charge is conserved. The factor $1/\sqrt{2}$ is used by convention. In this expression, the four component spinors are interpreted as follows:

$\overline{\Psi}_p$ creates a proton, Ψ_n destroys a neutron, $\overline{\Psi}_{e^-}$ creates an electron and Ψ_ν destroys a neutrino (creates an anti-neutrino). G_F is the Fermi constant and it is related to the square of the weak charge g.

The Fermi current (7.30) is a polar vector (V) in four-space, so it is customary to denote Fermi vector coupling, G_F by G_V. In the Gamow Teller matrix element containing weak axial vector hadron current and lepton current we have:

$$M_{GT} \sim \frac{G_A}{\sqrt{2}} (\overline{\Psi}_p \gamma_\mu \gamma_5 \Psi_n)(\overline{\Psi}_{e^-} \gamma_\mu \gamma_5 \Psi_\nu), \tag{7.31}$$

where the coupling constant is denoted by G_A. The current-current interaction (7.30) and (7.31) are Lorentz scalars and, hence, cannot account for the parity violation. In order to account for the parity violation the only essential change to Fermi's original proposal is the replacement of γ^μ by $\gamma^\mu(1-\gamma^5)$, called *V-A* interaction, as we shall see next.

B. Two Component Theory of Neutrinos

The two component theory of neutrinos formulated on the basis of left handed (or right handed) component can be best understood with the introduction of another operator called "chirality" or handedness operator. The chiral transformation is defined by

$$\psi'_\nu(x) = \gamma_5 \psi_\nu(x),$$

where γ_5 is the chiral operator. We can project the left handed or the right handed part as follows:

$$\Psi_{\nu L} = \frac{1}{2}(1-\gamma_5)\psi_\nu \neq 0 \text{ and } \Psi_{\nu R} = \frac{1}{2}(1+\gamma_5)\psi_\nu \neq 0. \tag{7.32}$$

Therefore, the normalized projection operators P_L and P_R are

$$P_L = \frac{1}{2}(1 - \gamma_5) \text{ and } P_R = \frac{1}{2}(1 + \gamma_5), \text{ such that } P_L + P_R = 1.$$

Left and right decomposition of ψ_ν is $\psi_{\nu L} + \psi_{\nu R} = \psi_\nu$.

Again,
$$P_L^2 = \frac{1}{4}\left[1 - 2\gamma_5 + \gamma_5^2\right] = \frac{1}{2}(1 - \gamma_5) = P_L,$$

$$P_R^2 = \frac{1}{4}\left[1 + 2\gamma_5 + \gamma_5^2\right] = \frac{1}{2}(1 + \gamma_5) = P_R,$$

and
$$\overline{\psi}_{\nu L} = \psi_\nu^\dagger \frac{1}{2}(1 - \gamma_5)\gamma_0 = \overline{\psi}_\nu \frac{1}{2}(1 + \gamma_5), \tag{7.33a}$$

where $\overline{\psi}_\nu$ is the row spinor defined by $\overline{\psi}_\nu = \psi_\nu^\dagger \gamma_0$, $\qquad$ (7.33b)

and $\psi_\nu^\dagger$ is the adjoint spinor. ψ is the solution of the Dirac equation in the covariant form. The change of sign in front of γ_5 occurs because it anti-commutes with γ_0. Similarly $\overline{\psi}_{\nu R} = \overline{\psi}_\nu \frac{1}{2}(1 - \gamma_5)$, so that

$$\overline{\psi}_{\nu L}\gamma_\mu \psi_{\nu R} = \overline{\psi}_\nu \frac{1}{2}(1 + \gamma_5)\gamma_\mu \frac{1}{2}(1 + \gamma_5)\psi_\nu$$

$$= \overline{\psi}_\nu \left[\frac{1}{4}(\gamma_\mu - \gamma_\mu\gamma_5)(1 + \gamma_5)\right]\psi_\nu$$

$$= \overline{\psi}_\nu \left[\frac{1}{4}\gamma_\mu(1 - \gamma_5)(1 + \gamma_5)\right]\psi_\nu = 0, \text{ since } (1 + \gamma_5)(1 - \gamma_5) = 0.$$

Similarly $\overline{\psi}_{\nu R}\gamma_\mu \psi_{\nu L} = 0$. Thus, fermionic vector (also axial vector) currents couple to either left handed fermion or to right handed ones. Since only currents appear in the gauge theories, one can consider the left and right handed fermions as separate entity.

Further,
$$\overline{\psi}_\nu \psi_\nu = \overline{\psi}_\nu \left[\frac{1 - \gamma_5}{2} + \frac{1 + \gamma_5}{2}\right]\left[\frac{1 - \gamma_5}{2} + \frac{1 + \gamma_5}{2}\right]\psi_\nu$$

$$= \overline{\psi}_{\nu L}\psi_{\nu R} + \overline{\psi}_{\nu R}\psi_{\nu L}. \tag{7.33c}$$

Similarly, we find

$$\overline{\psi}_\nu \gamma_\mu \psi_\nu = \overline{\psi}_{\nu L}\gamma_\mu \psi_{\nu L} + \overline{\psi}_{\nu R}\gamma_\mu \psi_{\nu R}. \tag{7.33d}$$

So the scalar term mixes R and L fermions (7.33c) whereas the vector term does not (7.33d). In the case of electromagnetic current ($\overline{\psi}\gamma_\mu \psi$), the left-

and right handed pieces couple to the photons with equal proportions. This is to be expected since photon does not violate parity. The presence of only left handed neutrinos indicates that the weak interaction term can only contain the left handed current, which is given by

$$j_\mu^L = \overline{\psi}_{vL} \, \gamma_\mu \, \psi_{vL} = \frac{1}{4}\left[\{(1-\gamma_5)\,\psi_v\}^\dagger \gamma_0 \gamma_\mu (1-\gamma_5)\,\psi_v \right] \text{(since } \overline{\psi}_v = \psi^\dagger \gamma_0)$$

$$= \frac{1}{2}\,\overline{\psi}_{vL}\,\gamma_\mu(1-\gamma_5)\,\psi_v \equiv j_\mu^V - j_\mu^{AV}.$$

The current j_μ^{AV} does not change under parity transformation, so that the combination $j_\mu^V - j_\mu^{AV}$ is not parity conserving (*see* Appendix C). Therefore, for weak interaction, the gauge interaction must couple to left handed fermion current only. The other major difference from electromagnetism is that the interaction couples fermions of different types, for example, the electron and its neutrino. This is taken into account by defining a lepton doublet:

$$\psi_{vL} = \begin{bmatrix} v_e \\ e \end{bmatrix}.$$

The charge raising hadronic current has the form:

$$j_\mu^\dagger = \overline{\psi}_p \, \gamma_\mu \, \frac{1}{2}(1-\gamma_5)\psi_n.$$

Hence, the charge conserving current-current interaction gives the beta decay interaction Hamiltonian in the form:

$$H_\beta(n \to p\, e^-\, \overline{v}_e) = \frac{G_F}{\sqrt{2}}\,[\overline{\psi}_p\,\gamma^\mu(1-\gamma^5)\,\psi_n][\overline{\psi}_{v_e}\,\gamma_\mu\,(1-\gamma_5)\,\psi_e]$$

$$= \frac{4G_F}{\sqrt{2}}\,j^\mu\,j_\mu^\dagger;$$

the factor 4 appears because the currents are defined with the normalized projection operators.

With the help of Dirac bilinear covariants O_i (*see* Appendix C1) we can write the general form of the interaction Hamiltonian density for β-decays. The general form of the four fermion local interaction, which is parity conserving and invariant under Lorentz transformation can be written as

$$H_\beta(x) = \sum_i C_i \{\overline{\psi}_p(x)\, O_i \psi_n(x)\} \{\overline{\psi}_e(x)\, O_i\, \psi_{v_e}\} + h.c. \qquad (7.34)$$

where the summation is over $i = S, V, T, A, P$ and C_i are coupling constants which are in general complex. The O_i represent the operators (Appendix C1) which act on the nucleon wave functions and on the lepton wave functions. The sum over i refers to the five forms of interaction: scalar (S), vector (V), tensor (T), axial vector (A) and pseudoscalar (P). C_i are adjustable constants and refer respectively to interaction terms which forbid change of parity in the lepton states. In order to accommodate parity violating effects we use different coefficients C_i' for parity violating terms. The occurrence of both the C_i and C_i' terms mean that parity is not conserved. The adjoint wave function $\overline{\psi}$ is related to the Hermitian conjugate wave function $\psi^\dagger$ by $\overline{\psi} = \psi^\dagger \gamma^0$. Since β-decay does not respect parity, the general form of parity non conserving interaction Hamiltonian density

$$H_\beta(x) = \frac{1}{\sqrt{2}} \sum_i \{\overline{\psi}_p(x)\, O_i \psi_n(x)\}\, \{\overline{\psi}_e(x)\, O_i (C_i - C_i'\gamma_5) \psi_{v_e}(x)\} + h.c.$$

$$h.c. = \frac{1}{\sqrt{2}} \sum_i \{\overline{\psi}_n(x)\, O_i \psi_p(x)\}\{\overline{\psi}_{v_e}(x)\, (C_i^* + C_i'^* \gamma_5)\, O_i \psi_e(x)\},$$

since $O_i^\dagger = O_i$ and $\gamma_5^\dagger = \gamma_5$. The factor $1/\sqrt{2}$ is included for historical reasons. The number of independent C_i's can be reduced when the invariance of $H_\beta(x)$ under charge conjugation (C), parity (P) and time reversal (T) is separately taken into account. The condition $C_i = C_i'$ corresponds to maximal parity violation.

Since $(1-\gamma_5)(1+\gamma_5) = 0$, it follows that the Dirac bilinear covariants have the following matrix elements:

$$\overline{\psi}_{vL}\, O_i\, \psi_{v'L} = \frac{1}{4}\, \overline{\psi}_v (1+\gamma_5)\, O_i (1-\gamma_5) \psi_{v'}$$

$$= 0 \qquad \text{if } i = S, T, P$$
$$\neq 0 \qquad \text{if } i = V, A.$$

Here ψ_{vL} and $\psi_{v'L}$ are left handed projected spin 1/2 fermion fields. Experimental information regarding the ten complex coupling constants C_i and C_i' is obtained from a series of experiments designed to detect parity violating effects. A summary of result is

$$C_i = C_i' = 0 \qquad \text{when } i = S, T, P$$

$$C_V = C_V' \equiv G_V, \quad C_A = C_A' \equiv G_A \quad \text{and} \quad C_A = -(1.26)\, C_V.$$

We obtain therefore the following $V - A$ (vector minus axial vector) weak interaction Hamiltonian density by

$$H_\beta(x) = \frac{G_F}{\sqrt{2}} \{\bar{\Psi}_p(x)\gamma_\mu(1+g_A\gamma_5)\Psi_n\}\{\bar{\Psi}_e(x)\gamma^\mu(1-\gamma_5)\Psi_{v_e}(x)\} + h.c.$$

$$(7.35)$$

where $G_F = G_V$ and $g_A = -(G_A/G_V) = 1.26$.

We see that, if $G_V = -G_A$ i.e. the axial vector and vector couplings are equal in magnitude but opposite in sign, we obtain the famous $V - A$ interaction.

Exercise 7.3. Show that $\frac{1}{2}(1-\gamma_5)\Psi_v$ projects only left handed component.

Let $\Psi_v = \begin{pmatrix} \phi \\ \chi \end{pmatrix}$, then $\frac{1}{2}(1-\gamma_5)\Psi_v$

$$= \frac{1}{2}\left[\begin{pmatrix} I & 0 \\ 0 & I \end{pmatrix} - \begin{pmatrix} -I & 0 \\ 0 & I \end{pmatrix}\right]\begin{bmatrix} \phi \\ \chi \end{bmatrix} = \begin{pmatrix} I & 0 \\ 0 & 0 \end{pmatrix}\begin{pmatrix} \phi \\ \chi \end{pmatrix} = \begin{pmatrix} \phi \\ 0 \end{pmatrix}.$$

Exercise 7.4. Show that the charge lowering leptonic current of the form

$$j^\mu = \Psi_e \gamma^\mu \frac{1}{2}(1-\gamma^5)\Psi_{v_e}$$ in ev_e scattering involves only left handed electron.

The factor $\frac{1}{2}(1-\gamma^5)$ automatically relates a left handed neutrino. Now

$$\bar{\Psi}_e \gamma^\mu \frac{1}{2}(1-\gamma^5)\Psi_{v_e} = \bar{\Psi}_e \frac{1}{2}(1-\gamma^5)\gamma^\mu \Psi_{v_e} = \bar{\Psi}_{eL}\gamma^\mu\Psi_{v_e},$$

since according to (7.33b):

$$\bar{\Psi}_{eL} = \Psi_e^\dagger \frac{1}{2}(1+\gamma_5)\gamma_0 = \Psi_e^\dagger \gamma_0 \frac{1}{2}(1-\gamma_5) = \bar{\Psi}_e \frac{1}{2}(1-\gamma_5).$$

Exercise 7.5. Show that the Fermi interaction violates the unitarity bound in neutrino-electron scattering.

The amplitude of scattering is obtained by using the trace theorem (Appendix C1) as follows:

$$M = \frac{G_F}{\sqrt{2}}\left[\bar{\Psi}_{v_e}(k')\gamma_\mu(1-\gamma_5)\gamma_{e^-}(p)\right]\left[\bar{\Psi}_{e^-}(p')\gamma^\mu(1-\gamma_5)\gamma_{v_e}(k)\right]$$

$$\frac{d\sigma}{d\Omega} = \frac{|M|^2}{64\pi^2 s} = \frac{1}{64\pi^2 s}\frac{1}{2}G_F^2 \, 32 \, s^2 = \frac{G_F^2}{4\pi^2}s$$

$$(1)$$

where $\sqrt{s} = 2E$ and E is the c.m energy of the $\nu_e + e^-$ system. Hence, the total cross section $\sigma = \dfrac{G_F^2}{\pi} s$.

Again the partial wave decomposition of elastic scattering cross section (Appendix B2, Eq. 13) is

$$\left(\frac{d\sigma}{d\Omega}\right) = | f(\theta) |^2 = \left| \frac{1}{2ik} \sum_{\ell=0}^{\infty} (2\ell+1)(1-\eta_\ell) P_\ell(\cos\theta) \right|^2,$$

where η_ℓ is the partial wave amplitude for the ℓth partial wave. For the point like Fermi interaction, only η_0 contribute and we have

$$\frac{d\sigma}{d\Omega} = \frac{1}{4E^2} |1-\eta_0|^2 .$$

However, since unitarity requires $|\eta_0| \leq 1$ for every partial wave, we obtain the upper bound for the differential cross section as $\dfrac{d\sigma}{d\Omega} \leq \dfrac{1}{4E^2}$ and for the total cross section as $\sigma \leq \pi/E^2$. Therefore, the prediction of (Eq. 1), being in the lowest order of the Fermi interaction violates unitarity bound at some high energy

$$E \approx \frac{1}{\sqrt{2}} \sqrt{\frac{\pi}{G_F}} \approx \sqrt{\frac{3.141}{2 \times 1.166 \times 10^{-5}}} \text{ GeV} = 367 \text{ GeV}.$$

Exercise 7.6. Show that for a massless fermion, the chirality equals the helicity. For a massless fermion the Dirac equation reads

$$\gamma^\mu p_\mu \psi = 0. \tag{1}$$

Multiplying (1) by $\gamma^5 \gamma^0 = -i\, \gamma^1 \gamma^0 \gamma^2 \gamma^3$ we get $\gamma^5 \gamma^0 \gamma^\mu p_\mu \psi = 0$

Now $\quad \gamma^5 \gamma^0 \gamma^i = \gamma^5 \gamma^0 \begin{pmatrix} 0 & \sigma_i \\ -\sigma_i & 0 \end{pmatrix} = \gamma^5 \begin{pmatrix} 0 & \sigma_i \\ \sigma_i & 0 \end{pmatrix}$

$$= \begin{pmatrix} 0 & 1 \\ 1 & 0 \end{pmatrix} \begin{pmatrix} 0 & \sigma_i \\ \sigma_i & 0 \end{pmatrix} = \begin{pmatrix} \sigma_i & 0 \\ 0 & \sigma_i \end{pmatrix} = \sum^i$$

Hence, $\sum \cdot \hat{p}\, \psi = \gamma^5 p^0 \psi$. Since for a massless particle $p^0 = |p|$ then in accordance to (7.24) we get: $\sum \cdot \dfrac{\hat{p}}{|p|} \psi = \gamma^5 \psi$. Thus, for a massless particle the chirality equals the helicity.

7.3 TYPES OF WEAK DECAYS

We have so far discussed nuclear β-decay, the oldest and the best known examples of weak interaction. There are other examples of weak interaction, which can be straight forwardly divided up into three categories: pure leptonic processes, semi-leptonic processes and non-leptonic processes. In the leptonic processes only leptons are involved. For example, the muon decay $\mu^- \rightarrow e^- + \bar{\nu}_e + \nu_\mu$ with $t_{1/2} \approx 2.2$ μ-sec and maximum electron energy ≈ 53 MeV. The other examples are: $\tau^- \rightarrow \mu^- + \bar{\nu}_\mu + \nu_\mu$ and $\tau^- \rightarrow e^- + \bar{\nu}_e + \nu_\tau$.

In the semi-leptonic processes one particle in the decay products is hadron and the other particle is (are) lepton (leptons):

$$\tau^- \rightarrow \pi^- + \nu_\tau, \quad n \rightarrow p + e^- + \bar{\nu}_e, \quad \pi^+ \rightarrow \pi^0 + e^+ + \nu_e.$$

The pion decay life time $t_{1/2} \approx 1.8$ sec and the maximum positron energy ≈ 4.1 MeV.

Another example is the hypercharge changing semi leptonic decay $\sum^- \rightarrow n + e^- + \bar{\nu}_e$, with $t_{1/2} \approx 0.95 \times 10^{-7}$ sec and maximum electron energy ≈ 230 MeV.

Non-leptonic processes, also called hadronic weak decays, are those in which only hadrons are involved. Examples are:

$K_{2\pi}$ or θ decay	$K^\pm \rightarrow \pi^\pm + \pi^0$
$K_{3\pi}$ or τ decay	$K^\pm \rightarrow \pi^\pm + \pi^+ + \pi^-$
Hyperon decay	$\Lambda^0 \rightarrow p - \pi^-$ and $\Lambda^0 \rightarrow n + \pi^0$

Hadronic weak processes obey hyperchange selection rule $|\Delta Y| = 1$.

Why all the processes mentioned above are called weak regardless of whether they involve leptons, hadrons or both? Justification comes from the fact that the strengths of the interaction responsible for various processes are the same. Restricting exclusively to electrons, muons and their neutrinos, one can write four basic interactions having same coupling constant. Now we have three families of leptons:

$$\begin{pmatrix} \nu_e \\ e \end{pmatrix}, \quad \begin{pmatrix} \nu_\mu \\ \mu \end{pmatrix}, \quad \begin{pmatrix} \nu_\tau \\ \tau \end{pmatrix}.$$

A. Universality of Weak Interactions

The discovery of muonic neutrino and tau neutrino as new leptons distinct from the electron neutrino led to an extension of concept of universality of weak interactions. The universal character of weak interaction is expressed by writing the total weak current of leptons as a sum of an electron, a muon and a tau current with equal weight as,

$$j_\mu^\ell = j_\mu^e + j_\mu^\mu + j_\mu^\tau$$

$$= \overline{\psi}_{\nu_e} \gamma_\mu (1-\gamma_5)\psi_e + \overline{\psi}_{\nu_\mu} \gamma_\mu (1-\gamma_5)\psi_\mu + \overline{\psi}_{\nu_\tau} \gamma_\mu (1-\gamma_5)\psi_\tau,$$

which can be written as

$$j_\mu^\ell = (\overline{\psi}_{\nu_e} \; \overline{\psi}_{\nu_\mu} \; \overline{\psi}_{\nu_\tau}) V \begin{pmatrix} \psi_e \\ \psi_\mu \\ \psi_\tau \end{pmatrix}, \text{ where } V = \gamma_\mu (1-\gamma_5).$$

The explicit form of V shows the lepton universality. For purely leptonic decay, $\mu^- \rightarrow e^- + \overline{\nu}_e + \nu_\mu$, considered as four fermion point interaction, we get

$$H(\mu^- \rightarrow e^- \overline{\nu}_e \nu_\mu) = \frac{G_\mu}{\sqrt{2}} \left[\overline{\psi}_{\nu_\mu} \gamma^\alpha (1-\gamma_5)\psi_\mu \right] \left[\overline{\psi}_e \gamma_\alpha (1-\gamma_5)\psi_{\nu_e} \right],$$

$$(7.36)$$

which is a pure *V-A* interaction.

The value of the weak coupling constant G_μ is determined from the measured value of mean life of muon decay: $\mu^+ \rightarrow e^+ \nu_e \overline{\nu}_\mu$. We choose muon decay because muon decay involves no hadrons so that complications due to hadronic currents do not enter into the QED calculations, and also muon life time has been accurately measured. QED calculation of muon decay transition rate is given below:[49]

Let p_μ, p_e, k_{ν_e} and k_{ν_μ} are respectively four momentum of μ, e, ν_e and ν_μ. Then

$$H(\mu^- \rightarrow e^- \overline{\nu}_e \nu_\mu) = \frac{G_\mu}{\sqrt{2}} \left[\overline{\psi}_{\nu_\mu} (k_{\nu_\mu}) \gamma^\alpha (1-\gamma_5) \psi_\mu (p_\mu) \right]$$

$$\times \left[\overline{\psi}_e (p_e) \gamma_\alpha (1-\gamma_5) \psi_{\nu_e} (k_{\nu_e}) \right],$$

where ψ and $\overline{\psi}$ denote spinors of particles and antiparticles and α denotes Lorentz index. The transition rate is given by $(\hbar = c = 1)$,

$$d\omega = \frac{1}{2E_\mu} \overline{|H|^2} d\rho, \text{ where } d\rho \text{ is the Lorentz invariant phase space given by}$$

$$d\rho = \frac{d^3 p_e}{(2\pi)^3 \, 2E_e} \frac{d^3 k_{\nu_e}}{(2\pi)^3 \, 2\omega_{\nu_e}} \frac{d^3 k_{\nu_\mu}}{(2\pi)^3 \, 2\omega_{\nu_\mu}} \times (2\pi)^4 \, \delta^4 (p_\mu - p_e - k_{\nu_\mu} - k_{\nu_e})$$

$$= \frac{1}{(2\pi)^5} \frac{d^3 p_e}{2E_e} \frac{d^3 k_{\nu_e}}{2\omega_{\nu_e}} \theta \, (E_\mu - E_e - \omega_{\nu_e}) \, \delta((p_\mu - p_e - k_{\nu_e})^2), \qquad (7.37)$$

where $\delta^4 (p_\mu - p_e - k_{\nu_\mu} - k_{\nu_e})$ is a delta function conserving four momentum, θ is the step function, $\theta \, (x) = 1$ for $x > 0$ and $= 0$ for $x < 0$ and $\overline{|H|^2}$ denotes the probability averaged over the spin of initial particles and summed over spin of final particles. In the present case

$$\overline{|H|^2} = \frac{1}{2} \sum_{\text{spin}} |H|^2$$

$$= \frac{1}{2} \frac{G_\mu^2}{2} \sum_{\text{spin}} \left[\overline{\Psi}_{\nu_\mu} (k_{\nu_\mu}) \, \gamma^\alpha (1 - \gamma_5) \, \psi_\mu (p_\mu) \right.$$

$$\left. \times \overline{\Psi}_\mu (p_\mu) \, \gamma^\beta (1 - \gamma_5) \, \psi_{\nu_\mu} (k_{\nu_\mu}) \right]$$

$$\times \sum_{\text{spin}} \left[\overline{\Psi}_e (p_e) \, \gamma_\alpha (1 - \gamma_5) \, \psi_{\nu_e} (k_{\nu_e}) \, \overline{\Psi}_{\nu_e} (k_{\nu_e}) \, \gamma_\beta (1 - \gamma_5) \, \psi_e (p_e) \right]$$

$$= \frac{G_\mu^2}{4} \, \text{Trace} \left[k_{\nu_\mu} \, \gamma^\alpha (1 - \gamma_5) \, (p_\mu - m_\mu) \, \gamma^\beta (1 - \gamma_5) \right]$$

$$\times \, \text{Trace} \left[p_e \, \gamma_\alpha (1 - \gamma_5) \, k_{\nu_e} \, \gamma_\beta (1 - \gamma_5) \right],$$

where $p = \gamma_\mu \, p^\mu$. Employing the trace theorem (Appendix C) we get,

$$\overline{|H|^2} = \frac{G_\mu^2}{4} \times 256 \, (k_{\nu_\mu} \cdot p_e) \, (p_\mu \cdot k_{\nu_e}). \qquad (7.38)$$

Here we neglect the electron mass $(m_e \leq m_\mu / 200)$. In the rest frame of muon, $p = (m_\mu, 0, 0, 0)$, we have

$$(k_{\nu_\mu} \cdot p_e) \, (p_\mu \cdot k_{\nu_e}) = \frac{1}{2} \, \omega_{\nu_e} \, m_\mu (m_\mu^2 - 2m_\mu \, \omega_e).$$

Thus, we can write the decay rate as

$$d\omega = \frac{1}{2E_\mu} \, 32G_\mu^2 \, \omega_{v_e} m_\mu (m_\mu^2 - 2m_\mu \, \omega_e) \times \frac{1}{32\pi^5} \frac{d^3 p_e}{2E_e} \frac{d^3 k_{v_e}}{2\omega_{v_e}}$$

$$\times \delta \, (m_\mu^2 - 2m_\mu E_e - 2m_\mu\omega_{v_e} + 2E_e\omega_{v_e}(1-\cos \theta)),$$

where θ is the angle between the emitted electron and electron antineutrino. Using $d^3 p_e \, d^3 k_{v_e} = 4\pi E_e^2 \, dE_e \cdot 2\pi\omega_{v_e}^2 \, d\omega_{v_e} \, d(\cos \theta)$, integrating over $\cos \theta$

and using delta function, we obtain $d\omega = \dfrac{G_\mu^2}{2\pi^3} \, dE_e \, d\omega_{v_e} (m_\mu^2 - 2m_\mu \, \omega_{v_e})$.

From the condition $-1 \le \cos \theta \le 1$, the energy ω_{v_e} and E_e are constrained as $(m_\mu/2) - E_e \le \omega_{v_e} \le m_\mu/2$ and $0 \le E_e \le m_\mu/2$, respectively. Then the energy spectrum of electron is obtained by integrating ω_{v_e} over this region. The rate $d\omega_\mu(E_e)$ in muon decay for the emission of electron with energy between E_e and $E_e + dE_e$ becomes for $E_e \gg m_e$,

$$d\omega_\mu(E_e) = \frac{G_\mu^2 \, m_\mu^2 \, E_e^2}{4\pi^3} \left[1 - \frac{4}{3} \frac{E_e}{m_\mu} \right] dE_e. \tag{7.39a}$$

The total transition rate for muon decay is obtained as follows:
We integrate (7.39a) and get

$$\omega_\mu = \frac{G_\mu^2 \, m_\mu^2}{4\pi^3} \int_0^{E_{max}} dE_e \, E_e^2 \left[1 - \frac{4}{3} \frac{E_e}{m_\mu} \right]$$

$$= \frac{G_\mu^2 \, m_\mu^2}{4\pi^3} \int_0^{m_\mu/2} dE_e \, E_e^2 \left[1 - \frac{4}{3} \frac{E_e}{m_\mu} \right] = \frac{G_F^2 \, m_\mu^5}{192\pi^3}, \tag{7.39b}$$

since two neutrinos share equally the muon rest energy so that $E_{max} = m_\mu/2$. Experimentally also it is found that the positron energy spectrum is strongly peaked around the maximum allowed value $(m_\mu/2)$.

Hence, $\qquad \dfrac{1}{\omega_\mu} = \tau_\mu = \dfrac{192\pi^3}{G_\mu^2 \, m_\mu^5}.$

Using $\tau_\mu \cong 2.2 \times 10^{-6}$ sec, $m_\mu \cong 105$ MeV, 1 sec $= 0.15 \times 10^{22}$ MeV^{-1} and $(\hbar = c = 1)$, we get $G_\mu = 1.167 \times 10^{-5}$ GeV$^{-2} \approx 10^{-5}/m_p^2$, $\qquad$ (7.40) where m_p is the mass of the proton which is of the order of 1 GeV. Note that G has a dimension of mass and $g^2/\hbar c$ is dimensionless; $g/2\sqrt{2} = \sqrt{4\pi\alpha}$, where α is the fine structure constant and $e^2 = 4\pi\alpha$ for $\hbar = c = \epsilon = 1$.

Again τ^- decays in three different channels, two leptonic, namely,

$$\tau^- \to e^- \overline{\nu}_e \nu_\tau, \ \tau^- \to \mu^- \ \overline{\nu}_\mu \nu_\tau$$

and one semi-leptonic, $\tau^- \to \pi^- \nu_\tau \ (\equiv \overline{u} \ d \nu_\tau)$.

Assuming $G_\mu = G_\tau$, we can write $\Gamma_{\tau e}/\Gamma_{\mu e} = \tau_\mu/\tau_\tau = (m_\tau/m_\mu)^5$.

Now $\Gamma_{\tau e} = \Gamma_{\tau\mu}$ and $\Gamma_{\tau\overline{u}d} = 3\Gamma_{\tau e}$, considering the decay at the quark level. The factor 3 follows from $\overline{u}d$ pair appearing in three different colors $(r\overline{r}, \ b\overline{b}, \ g\overline{g})$. Hence, the τ lepton lifetime

$$\tau_\tau = \frac{1}{\Gamma_{\tau e} + \Gamma_{\tau\mu} + \Gamma_{\overline{u}dv_\tau}} = \frac{1}{5\Gamma_{\tau e}} = \frac{\tau_e}{5}$$

$$= \frac{1}{5}\left(\frac{m_\mu}{m_\tau}\right)^5 \tau_\mu \approx 3.1\times 10^{-13} \ \text{sec.} \tag{7.41}$$

Now, $(\tau_\tau)_{\text{expt.}} = (2.956 \pm 0.031)\times 10^{-13}$ which once again confirms the universality of weak interaction. The near equality of coupling constant involved in weak decay of muon, τ lepton and β-decay, i.e. $(G_\tau)_{\tau-decay} = (G_\mu)_{\mu-decay} = (G_F)_{\beta-decay}$, shows that there is a *universal* weak interaction in all weak decays. In other words, it suggests a universality of weak charge; the value of weak charge is the same for all particles which possess it. It is customary, therefore, to represent universal coupling constant G by G_F called the Fermi constant, which we do from now on.

The difference between $G_A/G_V = -1$ and $G_A/G_V = -1.26$ could then be interpreted as due to renormalization effects of strong interaction. A better understanding of this difference led to the formulation of Conserved Vector Current (CVC) and Partially Conserved Axial-vector Current (PCAC). Furthermore, the axial vector current of the nucleon, corresponding to the term $\gamma_\mu \gamma_5$ has contributions from the virtual pions around the nucleon, according to the principle of PCVC. When the wave functions have long wavelengths compared to the nucleon size, this effect can be included by multiplying the $\gamma_\mu \gamma_5$ term for nucleons with an additional factor, conventionally denoted by G_A/G_V. The hypothesis of CVC implies that no such correction is necessary for the vector current term γ_μ. The value of G_A/G_V obtained from muon decay is close but not equal to 1.

Following (7.27) and normalized projection operators we can now express muon decay matrix element in terms of W^- boson propagator for $Q^2 \ll m_W^2$ as

$$M = \left(\frac{g}{\sqrt{2}} \, \psi_{\nu_\mu} \, \gamma^\alpha \frac{1}{2} (1 - \gamma^5) \, \psi_\mu \right) \times \frac{1}{m_W^2} \left(\frac{g}{\sqrt{2}} \, \overline{\psi}_e \, \gamma_\alpha \frac{1}{2} (1 - \gamma_5) \, \psi_{\nu_e} \right).$$

(7.42a)

Comparing (7.42a) with (7.36) we get $G / \sqrt{2} = g^2 / 8 \, m_W^2$. (7.42b)

B. Cabibbo Currents and Quarks

The universal Fermi interaction in the current-current form with V-A currents is quite successful in describing observed leptonic weak decays. However, the situation is less clear in the treatment of semileptonic and hadronic weak decays. Experiments show that the strangeness non-conserving weak decays are relatively suppressed as compared to the strangeness conserving weak decays. This is shown in Table 7.2.

Table 7.2: Comparison of $ft_{1/2}$ values for weak decays without and with change of strangeness

Strangeness conserving	Strangeness non-conserving	$\dfrac{ft_{1/2}(\Delta S = 1)}{ft_{1/2}(\Delta S = 0)}$
$\pi^+ \to \pi^0 e^+ \nu_e$	$K^+ \to \pi^0 e^+ \nu_e$	50
$n \to p \, e^- \, \overline{\nu}_e$	$\Lambda^0 \to p \, e^- \, \overline{\nu}_e$	17
$\Sigma^- \to \Lambda^0 e^- \, \overline{\nu}_e$	$\Sigma^- \to n \, e^- \, \overline{\nu}_e$	18

A modification of the weak current that explains the observed enhancement of comparative half lives of $\Delta S = 1$ transitions with respect to $\Delta S = 0$ transitions, was proposed by Cabibbo[16]. Instead of introducing new couplings to accommodate strange particle decay, he tried to keep universality by modifying the hadronic current. He assumed that the total weak current of hadrons flows into the $\Delta S = 0$ and $\Delta S = 1$ branches, keeping the total weak current of hadrons

$$\vec{j}_w^h = a \, \vec{j}_w^0 + b \, \vec{j}_w^1.$$

Here $\vec{j}_w^0$ and $\vec{j}_w^1$ are currents corresponding to $\Delta S = 0$ and $\Delta S = 1$ transitions respectively. $\vec{j}_w^0$ and $\vec{j}_w^1$ are normalized so that the strength of the corresponding transitions are given by the coefficients a and b. Since the total weak hadronic current is not changed, we have $|a|^2 + |b|^2 = 1$. It is customary to write $a = \cos \theta_C$, $b = \sin \theta_C$. The normalization condition is then automatically satisfied, and the weak current of hadron becomes

$$\vec{j}_w^h = \cos\theta_c \, \vec{j}_w^0 + \sin\theta_c \, \vec{j}_w^1.$$

To find the approximate value of the Cabibbo angle θ_c, we note that the transition rate is proportional to $|\langle f | H_w | i \rangle|^2$. The rate of $\Delta S = 0$ transition is then proportional to $G_F^2 \cos^2 \theta_c$ and that for $|\Delta S| = 1$ to $G_F^2 \sin^2\theta_c$. The ratio of $ft_{1/2}$ values in Table 7.2 gives $\cot^2 \theta_c$:

$$\cot^2 \theta_c = \frac{ft_{1/2}(|\Delta S| = 1)}{ft_{1/2}(\Delta S = 0)}. \tag{7.43}$$

For purely leptonic decay, the transition rate is proportional to G_F^2 only.

We can calculate the value of θ_c for $n \to p \, e^- \, \bar{\nu}_e$ and $\Lambda^0 \to pe^-\bar{\nu}_e$ decays. Employing the branching ratio from Table 7.2 we get

$$\cot^2 \theta_c = 17, \text{ which gives } \theta_c = 0.24 \text{ radians} \cong 13^0.$$

Because of the small values of θ_c, those decays whose amplitudes are proportional to the $\cos \theta_c$ are known as Cabibbo favoured decays while those with amplitude proportional to $\sin \theta_c$ are Cabibbo suppressed.

Let us now understand Cabibbo's proposal at the lepton-quark level[12,17]. The β decay process $n \to pe^- \nu_e$ at a quark level is $d \to u \, e^- \nu_e$ (Fig. 7.5 left): one of the d quarks in the neutron (ddu) transforms into a u quark with remaining u and d quarks acting as spectators. In contrast, β-decay process $\Lambda^0 \to p \, e^- \, \bar{\nu}_e$ in which Λ^0 has quark content uds, the strange quark in Λ^0 transforms into a u quark (Fig. 7.5 right). Again this involves charge changing weak decay but in this case there is also a change of strangeness at the baryon vertex and hadronic current is therefore called strangeness changing $\Delta S = 1$ weak current. The quark current has the same V-A structure:

$$j_\mu^{q \to q'} \approx \bar{\psi}_{q'} \, \gamma^\mu (1 - \gamma^5) \, \psi_q.$$

Further, we retain the universality and use the same coupling constant for $\Delta S = 1$ and $\Delta S = 0$ decays. Therefore, all particles, quarks as well as leptons carry a weak charge g, but the quarks are mixed. In weak interactions with

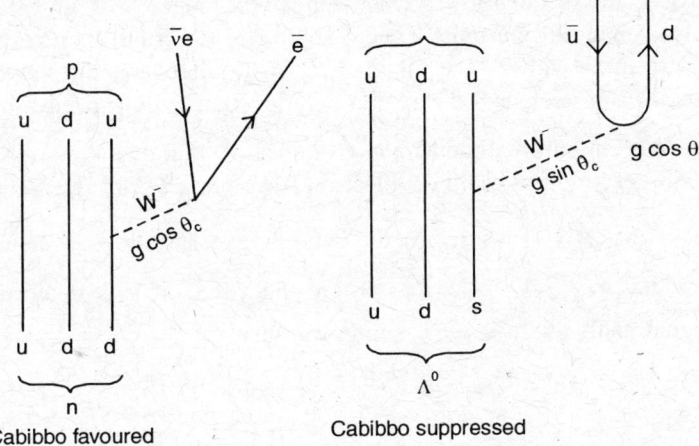

Fig. 7.5: Neutron decay at the quark level

charged currents, leptons can only be transformed into their partners in the same doublet; e.g., $e^- \leftrightarrow \nu_\mu$. Similarly we can group the quarks into families $\begin{pmatrix} u \\ d \end{pmatrix}, \begin{pmatrix} c \\ s \end{pmatrix}$. Quark transitions are observed not only within a family but, to a lesser degree, from one family to another. For charged currents, the "partner" of the flavour eigen state $|u\rangle$ is not therefore the flavour eigen state $|d\rangle$ but a linear combination of $|d\rangle$ and $|s\rangle$. We call this linear combination $|s'\rangle$. Similarly partner of the c-quark is orthogonal to $|s'\rangle$ and we call it $|d'\rangle$. The coefficients of these linear combinations can be written as

$$|d'\rangle = \cos\theta_c \, |d\rangle + \sin\theta_c \, |s\rangle$$

and $$|s'\rangle = \cos\theta_c \, |s\rangle + \sin\theta_c \, |d\rangle \qquad (7.44)$$

The quark mixing is described by a single parameter θ_c. Eq. (7.44) can be written in the matrix form[18] usually called CKM matrix:

$$\begin{pmatrix} |d'\rangle \\ |s'\rangle \end{pmatrix} = U \begin{pmatrix} |d\rangle \\ |s\rangle \end{pmatrix}, \text{ where the matrix } U = \begin{pmatrix} \cos\theta_c & \sin\theta_c \\ -\sin\theta_c & \cos\theta_c \end{pmatrix} \qquad (7.45)$$

Applying the operator T_+ to the flavour eigen state $|s\rangle$ yields a linear combination of $|u\rangle$ and $|c\rangle$. Just as was the case for the β-decay of the neutron, the matrix element contains a Cabibbo factor, here $\sin\theta_c$.

The observation that strangeness non-conserving weak interactions are relatively suppressed as compared to the strangeness conserving weak interactions, led to the further extension of the concept of universality to involve weak hadronic currents. Thus, Cabibbo theory establishes quark lepton universality and removes the slight discrepancy between the value of G_F determined from nuclear β decay compared to that from μ-decay.

Exercise 7.7. On purely dimensional ground show that the total cross section for neutrino elastic scattering off electron for a point Fermi interaction is

$$\sigma \, (\nu_e \; e^-) \sim G_F^2 \, s \,, \quad \text{where } s \text{ is the c.m energy of } \nu_e - e^- \text{ system.}$$

Let us assume $\sigma(E) \sim G_F^2 \, E^k$, where the value of k is determined by dimensional analysis. In $\hbar = c = 1$ units we have

$$|E| = M, \, [\hbar c] = [ML] = 1.$$

Now, $$\sigma = [L^2] = M^{-2} \text{ and } G_F = \frac{10^{-5}}{M_P^2},$$

so that $[G_F] = M^{-2}$. Therefore $M^{-2} = M^{-4} \cdot M^k$.

Hence, $k = 2$ and $\sigma(E) \sim G_F^2 E^2 \sim G_F^2 s$.

Exercise 7.8. Calculate the invariant amplitude for $\nu_e \; e^- \to \nu_e \; e^-$ process using *V-A* interaction and, hence, show that

$$\sigma(\nu_e \; e^-) = \frac{G_F^2 \, s}{\pi}.$$

The neutrino-electron scattering process for charged current V-A interaction gives[40]

$$H = \frac{G_F}{\sqrt{2}} \left[\overline{\psi}_{\nu_e}(k') \; \gamma^\mu (1 - \gamma_5) \; \psi_e(p) \right] \times \left[\overline{\psi}_e(p') \gamma_\mu (1 - \gamma_5) \; \psi_{\nu_e}(k) \right].$$

The calculation proceeds along the line of muon decay discussed in Sec. 7.3A. The neutrino-electron scattering cross section is calculated as follows:

$$\frac{1}{2} \sum_{spin} |H|^2 = \overline{|H|^2} = \frac{G_F^2}{4} \text{Trace} \left[\gamma^\mu (1 - \gamma_5) p \, \gamma^\nu (1 - \gamma_5) k' \right]$$

$$\times Trace \, [\gamma_\mu (1 - \gamma_5) k \; \gamma_\nu (1 - \gamma_5) p'] = \frac{G_F^2}{4} \cdot 256 (p \cdot k) \, (k' \cdot p').$$

For elastic scattering in the centre of mass system we have in the relativistic limit $m_e = 0$:

$$s = (p + k)^2 = 2p \cdot k = (k' + p')^2 = 2k' \cdot p'.$$

Hence,

$$\frac{1}{2} \sum_{\text{spin}} |H|^2 = \frac{64 G_F^2}{4} s^2 = 16 \, G_F^2 s^2$$

$$\frac{d\sigma}{d\Omega}(\nu_e e^-) = \frac{1}{64 \pi^2 s} \overline{|H|}^2 = \frac{1}{64 \pi^2 s} 16 G_F^2 s^2$$

$$\sigma(\nu_e e^-) = \frac{G_F^2 \, s}{\pi}.$$

Exercise 7.9. Estimate the mass of the intermediate vector boson m_W employing (7.40) and (7.42)

$$\frac{G}{\sqrt{2}} = \frac{g^2}{8 m_W^2} = \frac{(2\sqrt{2})^2 (4\pi\alpha)}{8 m_W^2}, \text{ where a = fine structure constant} = \frac{1}{137}.$$

Hence,

$$m_W = \left[\frac{\sqrt{2} \, (4\pi\alpha)}{G} \right]^{1/2} = 105 \text{ GeV}.$$

7.4 NEUTRINO PHYSICS – LONG TERM PERSPECTIVES

We describe here how the ethereal neutrino gradually assumed a tangible form and played a pivotal role in our quest to understand intrinsic particle properties. In the road map for the exploration of neutrino mass, new experiments are coming on line and new facilities are planned to find the nature of the neutrino mass, the possible cosmological relevance of neutrino and the origin of all matter in the Universe.

A. Sources and Detection of Neutrinos

Besides the major production of neutrinos in nuclear reactions through β-decay of fission products, which are neutron rich and transform into more stable elements producing electron neutrinos, neutrinos are amply produced by the interaction of the primary cosmic ray flux with the earths atmosphere and in the solar burning process where protons fuse into heavier elements like deuterium and helium (Fig. 7.6).

Long before, Hans Bethe examined all reactions that might lead to hydrogen to helium. He found a cyclic phenomenon starting from $^{12}C - ^{14}N - ^{16}O - ^{12}C$ plus ^4He and showed that CNO cycle and the *p-p* chain supply about equal amount of energy at a temperature $T = 16 \times 10^6$ K. CNO reactions are more temperature sensitive than *p-p* chain due to high Coulomb barrier for CNO reactions.

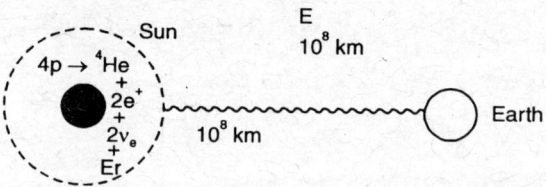

Fig. 7.6: Neutrinos reaching earth from the Sun

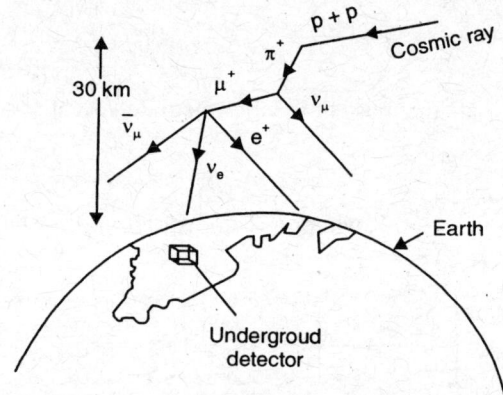

Fig. 7.7: Neutrinos reaching earth from upper atmosphere

Conversion of 1 kg of hydrogen to helium yields an energy production of 6×10^{14} Joules.

The total amount of energy released by the Sun per sec, called solar luminosity L_O, is 3.845×10^{26} Watt. Therefore, the total produced mass of ^4He through hydrogen burning during solar age ($= 4.5 \times 10^9$ years) is

$$M(^4He) = \frac{3.845 \times 10^{26} \times 4.5 \times 10^9 \times 3 \times 10^7}{6 \times 10^{14}} \text{ kg} = 8.6 \times 10^{28} \text{ kg}.$$

The total mass of the Sun $= M_o = 2 \times 10^{30}$ kg.

Atmospheric Neutrinos

The primary cosmic ray (mainly high energy protons) flowing towards the earth and interacting with the earths atmosphere produces pions and kaons:

$$p + p \rightarrow n + p + \pi^+, \quad p + p \rightarrow \Lambda + k^+ + p.$$

Pions decay as

$$\pi^\pm \rightarrow \mu^\pm + \nu_\mu(\overline{\nu}_\mu) \text{ and } \mu^\pm \rightarrow e^\pm + \nu_e(\overline{\nu}_e) + \overline{\nu}_\mu(\nu_\mu)$$

Thus, high energy *p-p* collisions via strong interaction produces pions, which decay via weak interaction producing neutrinos with flux ratio

$$\frac{flux \; \nu_\mu}{flux \; \nu_e} = 1 \text{ and } \frac{flux \; \nu_\mu + flux \; \overline{\nu}_\mu}{flux \; \nu_e} = 2 \text{ or } \frac{flux \; \nu_\mu + flux \; \overline{\nu}_\mu}{flux \; \nu_e + flux \; \overline{\nu}_e} = 2.$$

The decay rate of $\pi^+ \to e^+ + \nu_e$ is strongly suppressed compared to that of $\pi^+ \to \mu^+ + \nu_\mu$ by a factor of $(m_e/m_\mu)^2$. Measurements of atmospheric neutrino flux are done separately for the low energy neutrinos (≤ 1 GeV) and for neutrinos of higher energies. Inputs required for the neutrino flux calculation using Monte Carlo (MC) simulations are: (*i*) primary fluxes of cosmic ray nucleons, (*ii*) hadronic interaction with air nuclei, (*iii*) rate of energy loss of particles in the atmosphere and (*iv*) decay kinematics of pions, kaons and muons. Though the absolute values of ν_e and ν_μ fluxes have some uncertainties in their calculations (around 20%) resulting from the uncertainties in the inputs, yet in the ratio of these fluxes $(\nu_\mu + \overline{\nu}_\mu)/(\nu_e + \overline{\nu}_e)$, such uncertainty is considerably reduced. The above decay chain suggests that the ratio is roughly 2. Therefore, the usual practice is to present the atmospheric neutrinos as the ratio of ratios

$$R = \frac{[(\nu_\mu + \overline{\nu}_\mu)/(\nu_e + \overline{\nu}_e)]_{observed}}{[(\nu_\mu + \overline{\nu}_\mu)/(\nu_e + \overline{\nu}_e)]_{MC \; predicted}}.$$

For the solar neutrinos this ratio is usually defined in terms of the measured and calculated electron neutrino flux. Another ratio is that of the total, flavour independent, neutrino flux and the calculated electron neutrino flux.

Also a quantity of interest is the up down asymmetry:

$$A = (N_{up} - N_{down} / N_{up} + N_{down})_{\mu-like},$$

where the 'up' neutrinos are those neutrinos travelling from the bottom of the earth and reaching the detector placed on the surface of the earth, covering a distance equal to the diameter of the earth ($= 4.8 \times 10^4$ km), and the 'down' neutrinos are those neutrinos travelling from the upper atmosphere and reaching the detector placed on the surface of the earth covering a distance equal to (15-30) km.

Solar neutrinos produced via nuclear reactions are shown in Table 7.3a. The reaction type along with the maximum neutrino energy E_ν and flux 'f' for each reactions are given. We see from Table 7.3a that *pp* neutrino dominates the solar neutrino flux, though its energy is relatively low and its detection not easy, while ^8Be and ^8B have higher neutrino energies and are detectable under controlled laboratory conditions. Neutrino energy spectrum is shown in Table 7.3b.

Table 7.3: (a) Solar Neutrino Flux

Reaction	$f_n \times 10^{10}$ cm^{-2} sec^{-1}
$p + p$	6.10
$p + e^- + p$	1.5×10^{-2}
$e^- + {}^7Be$	4×10^{-1}
${}^8B \rightarrow$	5×10^{-2}

Table 7.3: (b) Neutrino Spectrum

Source	Energy E_n (eV)
Cosmic ray Microwave Background (CMB)	10^{-5}
Sun	10^6
Supernova	10^7
Atmospheric	10^9
Altra Galatic Neutrino	10^{15}
Low Energy Accelerator	10^7
High Energy Accelerator	10^9
Reactor	10^{10}

Solar neutrino flux coming out from the reaction chain (Fig. 7.6) can be calculated as follows. We get one neutrino for every 13.4×10^6 eV energy, whereas the energy received by the earth's surface from the Sun is

$L_\circ = 0.135$ Joules/cm^2/sec $= 0.135 \times 6 \times 10^{18}$ eV/cm^2/sec.

Therefore, the solar neutrino flux received by the earth in day and night is

$$f_\nu = \frac{0.81 \times 10^{18} \text{ eV/cm}^2/\text{sec}}{13.4 \times 10^6 \text{ eV}} = 6 \times 10^{10} /\text{cm}^2/\text{sec}.$$

In the standard hot Big Bang Nucleosynthesis (BBN), the neutron number gets frozen at a temperature T when the inverse reactions $e^- + p \rightarrow n + \nu_e$ and $\bar{\nu}_e + p \rightarrow e^+ + n$ get suppressed and the neutron/proton ratio (n/p) is fixed at

$$\frac{n}{p} = e^{-(m_n - m_p)/T} . \tag{7.46}$$

The temperature T is governed by the expansion rate yielding $T = 0.66$ MeV. The fractional number of ${}^4He$ nuclei is

$$x = \frac{N({}^4He)}{p - n} = \frac{n/2}{p - n} = \frac{1}{2} \frac{1}{\left(\dfrac{p}{n} - 1 \right)}.$$

Substituting the value of $\dot{T} = 0.66$ MeV and $(m_n - m_p) = 1.29$ MeV in

(7.46), we get $\dfrac{p}{n} = e^{1.94} \approx 7.0$. Hence, $x \cong \dfrac{1}{12}$. The mass fraction

$$Y = \frac{M(^4He)}{M(^1H_1) + M(^4He)} = \frac{4x}{1+4x} = 0.25.$$

From the observed ^4He abundances and other elements like ^2D etc. a bound on number neutrinos around 5 can be derived. The relative abundance of elements in the universe has been measured by geologists and astronomers. By mass, the universe is about 70% H, 30% ^4He with all other elements amounting 1 to 2%.

Detection of solar/atmospheric neutrinos is a difficult task. Neutrino nucleus interaction cross sections are only of the order of 10^{-42} cm^2 making it very difficult to register these neutrinos on earth despite the very large neutrino flux at the earths surface ($\sim 10^{11}$ cm^{-2} sec^{-1}). Neutrinos can be detected either via neutrino capture nuclear reactions or via neutrino elastic or deep inelastic scattering by electron, proton, neutron etc. In neutrino scattering process the lepton number of each kind of neutrino is conserved.

The pioneering experiment for the solar neutrino detection was started by Raymond Davis Jr et al.[2] at Homestake gold mine in U.S.A. Solar neutrino was detected using the reaction

$$\nu_e + {}^{37}Cl \to {}^{37}Ar + e^-, \tag{7.47a}$$

with the detector threshold at 0.8 MeV.

The neutrino absorption experiment is schematically shown in Fig. 7.8. The interaction between an antineutrino and proton in the target tank according to Eq. (7.47a) produced a neutron and positron. Soon after its formation, the positron is annhilated which results in the emission of two photons each of energy 0.51 MeV. These two photons produce scintillations within the scintillating liquid resulting into two electrical pulses within two photomultiplier tubes the output of which were fed to coincidence circuit.

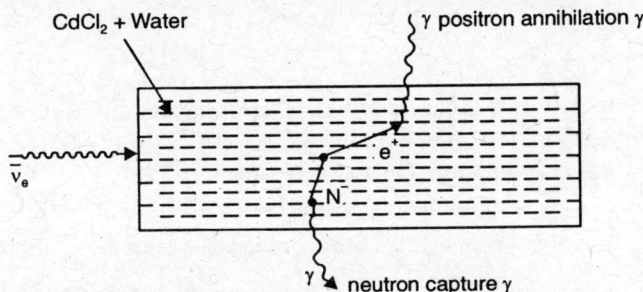

Fig. 7.8: Neutrino Absorption Experiment

The neutrino capture rate is given by:

$$Rate = C \sum_i \int \sigma(E_v) \phi_i(E_v) \, dE_v, \tag{7.47b}$$

where $\sigma(E_v)$ is the capture cross-section on the particular detector material, $\phi_i(E_v)$ are the neutrino fluxes due to the sources i and C is a overall normalization constant. The integration is over the neutrino energy and i sums all neutrino sources.

The detector contains approximately 4×10^5 liters of CCl_4 (tetra chloro ethelene). We can estimate the number of ^{37}Ar atoms produced per day as follows. The threshold energy for the reaction (7.47a) is

$$(M_{Ar} - M_{Cl})c^2 = (0.000874 \times 931.7 \text{ MeV}) = 0.82 \text{ MeV}.$$

The density of $C_2Cl_4 \cong 1.5$ gm ml^{-1}. ^{37}Cl abundance in C_2Cl_4 is around 25%. If we take molecular weight of $C_2Cl_4 = 164$ ($2 \times 12 + 4 \times 35 = 164$) and the total mass of liquid $= 4 \times 10^5 \, \ell \times 1.5 \times 10^3 \, \ell^{-1} = 6 \times 10^8$ gm, then the number Cl nuclei,

$$n = \frac{4 \times 6 \times 10^8 \times 6.02 \times 10^{23}}{4 \times 164} = 2.2 \times 10^{30}.$$

Solar heat flux is 2 Cal cm^{-2} min^{-1} or 8.8×10 MeV cm^{-2} sec^{-1}. Of this, 10% of thermonuclear energy of sun appears in neutrinos of mean energy ~1 MeV. 1% of all neutrinos are energetic enough to induce the above reaction. The solar neutrino flux with $E_n > 0.82$ MeV is available to interact with matter. The interaction cross section of neutrino with matter is a function of energy. The average cross section for $E_n > 0.82$ MeV is $\bar{\sigma} = 10^{-45}$ cm^2/Cl. The calculation of flux of solar neutrino on earth depends on solar model used.

$$f = 8.8 \times 10^8 \text{ cm}^2 \text{ sec}^{-1}.$$

Therefore, the number of the neutrinos detected per day (= number of ^{37}Ar atoms produced per day)

$$N_v = n f \sigma t$$

$$= 2.2 \times 10^{30} \times 8.8 \times 10^8 \text{ cm}^{-2} \text{ sec}^{-1} (10^{-45} \text{ cm}^2) \times 24 \times 3600 \text{ sec}$$

$$= 0.17 \text{ day}^{-1}$$

However, most of the solar neutrinos is not detected this way. We add Ga and Indium to detect low energy neutrinos.

In the Sudbury Neutrino Observatory (SNO)[19], situated 2070 meter underground in the nickel mine in Canada, neutrino detection experiment was carried out utilizing the charged current reaction of neutrino off deuteron: $v_e + D \rightarrow e^- + p + p$. The detector is 1 kton heavy water in 5 cm thick transparent acrylic vessel of diameter 12 m, surrounded by 10,000 photomultiplier tubes mounted on a 7000 tons of light water to absorb γ-rays and neutron which result from the radioactivity in rock.

Ga Experiment (SAGE[20], GNO[21] GALLEX[22]) is a charged current neutrino capture reaction

$$\nu_e + {}^{71}Ga \rightarrow {}^{71}Ge + e^- \tag{7.48}$$

The importance of the gallium (^{71}Ga) detector lies in its low threshold (0.233 MeV), providing it the ability to capture the *pp* neutrinos. Theoretically, predicted neutrino capture rate for Homestake gold mine experiment is (7.9 ± 2.6) SNU, where 1 SNU, solar neutrino unit is 10^{-36} event per target atom/sec. In contrast, for Ga detector it is $\left(132 \pm {}^{20}_{17} \right)$ SNU.

Neutrino deficit problem has been addressed by various group over the past decade. Data from several neutrino detectors around the world, in particular, from the Super-Kamiokande (Super-K)[23] detector in Japan and Sudbury Neutrino Observatory (SNO)[19] in Canada have shown a clear deficit of solar neutrinos from the expected value, which is possible only when neutrinos have mass and they oscillate. This provides the first un-ambiguous evidence for physics beyond the standard model of particle physics.

The experiment of Super-Kamiokanda Collaboration (Koshiba et al.)[23] has been carried in the Kamioka mine of Japan using a 50 kton Water Cerenkov detector. Super-K experiment uses a neutral current elastic scattering of neutrino off electron:

$$\nu + e^- \rightarrow \nu + e^- \tag{7.49}$$

Table 7.4: Results of solar neutrino experiments

	Home stake[9]	*Super-K*[23]	*SNO*[19]	*GALLEX*[22]
R	0.335±0.029	0.459±0.017	0.473±0.074	0.55±0.05
E_{th} (MeV)	0.81	5.0	5.0	0.23

R = Ratio of ratios and E_{th} = Threshold energy for solar neutrino capture.

Kamiokande water tank was lined with 10,000 photomultiplier tubes. The kamiokande detector is direction sensitive and confirmed that neutrinos come from the Sun. In Table 7.4 we give results of four different types of solar neutrino experiments.

The solar neutrino experiments mentioned above, observe a smaller flux than the theoretical prediction. There is a clear deficit of the solar neutrinos from the expected value of $R = 1$. For multi GeV neutrinos, Super-k recently reported $R = 0.66 \pm 0.06$.

The Kamioka liquid scintillator detector (KamLAND) experiment carried by Eguchi et al.[24] represent a major advance in the development of reactor $\bar{v}_e$ measurement. The low energy inverse β-decay antineutrino ($\bar{v}_e$) from reactor, is very susceptible to back ground from cosmic radiation and so must be located deep underground.

The history of $\bar{v}_e$ experiment using reactor as a source dates back to the original experiment performed by Reines and Cowan in 1959 (*see* Fig. 7.8) who used cadmium chloride as target and a 1000 MW reactor giving antineutrino flux of the order of 10^{13} cm^{-2} sec^{-1}. The positron produced in this reaction rapidly comes to rest by ionization loss and forms positronium which annihilates to γ-rays, in turn, producing fast electrons by the Compton effect. The process takes place in the time scale of 10^{-9} sec and electrons are detected. The function of cadmium is to capture the neutron after it has been moderated by successive elastic collisions with protons in water eventually giving γ-rays.

In KamLAND experiment a large liquid scintillator detector is build to study disappearance of $\bar{v}_e$ from nuclear reactors. Seventeen reactors in Japan landscape are used for this experiment. For the first time in the history of $\bar{v}_e$ experiment a substantial deficit event rate was observed (*see* Fig. 7.9). The KamLand detector is a 1000 ton sphere full of mineral oil and an organic substance pseudocummene. When an electron antineutrino ($\bar{v}_e$) strikes a hydrogen nucleus – a proton in the liquid, both particles change identity: $\bar{v}_e + p \rightarrow n + e^+$ in a process known as inverse beta decay. Detector detects e^+ and neutron and confirms that $\bar{v}_e$ has met its demise. In 6 months of observation 54 $\bar{v}_e$ were detected.

Furthermore, antineutrinos are created in the reactor and there is no possibility that they are created in the interior of the Sun due to Sun's magnetic field flipping spins of neutrinos. KamLand results thus correct solar neutrinos in one stroke.

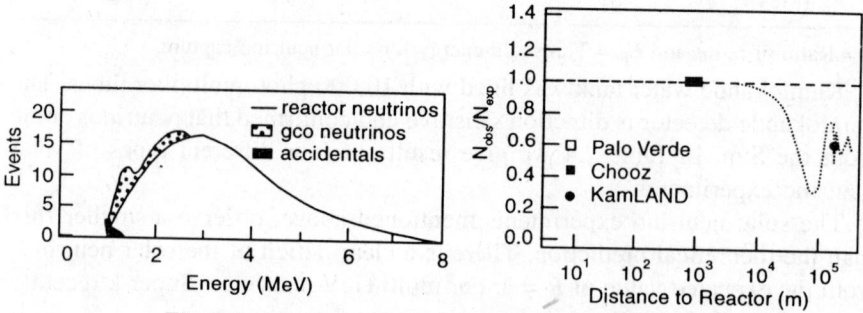

Fig. 7.9: Results of KamLand Experiments (*see* text)

The accelerator based neutrino experiment was carried out in U.S.A. at Los Alamos by Agular et al.[25] using Liquid Scintillator Nuclear Detector (LSND). An intense 800 MeV proton beam from a hadron accelerator were directed to a target for producing pions and kaons, which on stopping downstream produce ν_μ and $\overline{\nu}_\mu$. Due to large cross section for π^- capture by nuclei, only π^+ decay produced $\overline{\nu}_\mu$ and ν_e.

B. Direct Measurement of Neutrino Mass

In a journey to determine neutrino mass, we begin with its direct measurement. Conceptually, the simplest way to explore the neutrino mass is to determine its effect on the momenta and energies of the charged particles emitted in weak decays (Eq. 7.15). In the case of two-body decays e.g., $\pi^+ \rightarrow \mu^+ + \nu_\mu$, the analysis is particularly straightforward at least in principle. In the system where the decaying pion is at rest, the energy and momentum conservation requirements mean that

$E_\pi = E_\mu + E_\nu$ and $p_\mu = -p_\nu$. Hence, $E_\nu^2 = E_\pi^2 + E_\mu^2 - 2E_\pi E_\mu$

or $\qquad p_\nu^2 + m_\nu^2 = m_\pi^2 + m_\mu^2 + p_\mu^2 - 2m_\pi \sqrt{p_\mu^2 + m_\mu^2}$

or $\qquad m_\nu^2 = m_\pi^2 + m_\mu^2 + (p_\mu^2 - p_\nu^2) - 2m_\pi \sqrt{p_\mu^2 + m_\mu^2}$

$$= m_\pi^2 + m_\mu^2 - 2m_\pi \sqrt{p_\mu^2 + m_\mu^2}. \tag{7.50}$$

We see from Eq. (7.50) that the neutrino mass squared appears as a difference of two very large numbers, hence, uncertainties in m_π and p_μ and even m_μ mean that the corresponding mass limit is only < 170 keV. This problem can be avoided by studying the three body decay e.g., $H^3 \rightarrow {}^3He + \beta^- + \overline{\nu}_e$, in which $Q_{\beta^-} = 18.6$ keV. Near the end point of the β-spectrum (Eq. 7.15), a massive neutrino has a little kinetic energy that the effects of neutrino finite mass become visible. Clearly, the effect of finite neutrino mass becomes visible if $(\epsilon_0 - \epsilon) \sim \mu$, i.e. very near the threshold. This is shown in Fig. 7.3b. This nuclear beta decay process has been studied in great detail. Apart from offering a low end point energy E_0 and a moderate half life of 12.3 years, tritium has further advantage that the tritium β-decay is a superallowed nuclear transition. Therefore, no corrections from the nuclear transition matrix elements **M** has to be taken into account. There are some controversies about the various measurements but more recent data seem to be consistent with an upper limit of electron neutrino (ν_e) mass $m_{\nu_e} \approx 2.2$ eV. Kartin collaboration tries to bring it down to 0.2 eV.*

Double Beta Decay

The direct evidence of neutrino mass also comes from double beta decay, which is a rare transition between two nuclei with same mass number A involving change of nuclear charge by two units. There are two modes of the double beta decay. Two neutrino decay $2\nu\beta\beta$,

$$(A, Z) \to (A, Z+2) + e_1^- + e_2^- + \bar{\nu}_{e_1} + \bar{\nu}_{e_2}, \qquad (7.51)$$

which conserves both electric charge and lepton number. On the other hand, neutrinoless double beta decay $0\nu\beta\beta$, $(A,Z) \to (A,Z+2) + e_1^- + e_2^-$, which violates lepton number conservation. The $2\nu bb$ occurs where the mass of a nucleus is such that the regular beta decay cannot occur but a second-order process, called double beta decay, is possible. The decay can proceed only if the initial nucleus is less bound than the final one and both must be more bound than the intermediate nucleus. Typically, the decay can proceed from the ground state (0^+) of the initial nucleus to the ground state of the final nucleus.

In Fig. 7.10 we give a simplified set of level energies for three consecutive nuclei (Z, Z+1, Z+2) where single beta decay from $Z \to Z+1$ is not possible. In this situation a double beta decay process proceeds through a set of intermediate (virtual) levels of the nucleus (Z+1). In other words, double beta decay is a second order process and not a two step process. For example, $^{130}Te \to {}^{130}Xe + 2e^- + 2\bar{\nu}_e$ is allowed but $^{130}Te \to {}^{130}I + e^- + \bar{\nu}_e \to {}^{130}Xe + e^- + \bar{\nu}_e$ is not allowed.

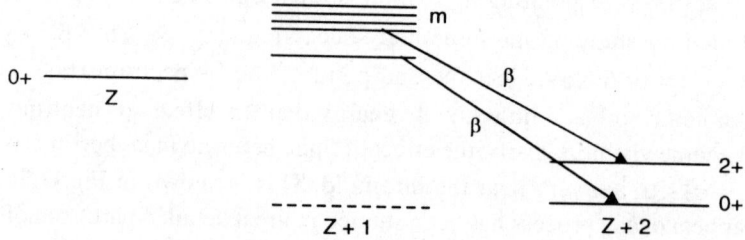

Fig. 7.10: Energy levels for 2ν process

The first laboratory experiment studying the double beta decay process was performed for $^{82}Se \to {}^{82}Kr + 2e^- + 2\bar{\nu}_e$ giving a clear cut evidence for

---

* Katrin Collaboration, Nucl. Phys. **C14**, 752 (2005).

the emission of two electrons in the decay process. In Table 7.5 we give the results of some recent measurements and compare them with the theoretical estimates. The $2\nu\beta\beta$ decay amplitude is described by a sum of products of β^- decay *GT* matrix elements from the initial 0^+ state to a intermediate $1+$ states and those from the intermediate states to the final $0+$ state.

Table 7.5: Results of some $2\nu\beta\beta$ double beta decay ($0^+ \rightarrow 0^+$) Ref. 27

Nucleus	$T_{1/2}$ (Year)		Q (keV)	$M_{GT}^{2\nu}$ (MeV^{-1})
	Experiment	Theory		
^{76}Ge	$(1.3\pm0.1) \times 10^{21}$	2.3×10^{21}	2040	0.15
^{130}Te	$(2.7\pm1.) \times 10^{21}$	2.6×10^{21}	2533	0.017

The nuclear matrix element for 2ν double beta decay connecting 0^+ ground state of two even-even nuclei is given by

$$M_{GT}^{2\nu} = \sum_m \frac{\langle f \mid \vec{\sigma} \, t_+ \mid m \rangle \langle m \mid \vec{\sigma} \, t_+ \mid i \rangle}{E_m - (M_i + M_f)/2},\qquad(7.52)$$

where the sum is taken over nuclear intermediate states m. We use the closure approximation $\sum_m |m\rangle \langle m| = 1$, i.e. to replace intermediate state energy E_m by an average value, $E_m \rightarrow \langle E_m \rangle$. Here M_i and M_f are the masses of the initial and the final nucleus. $2\nu\beta\beta$ is an allowed process with a very long life time $(t_{1/2} \approx 10^{20}$ years) with an energy release of 2.5 MeV. The second mode of double beta decay is the neutrinoless beta decay $0\nu\beta\beta$, which is schematically illustrated in Fig. 7.11 (left).

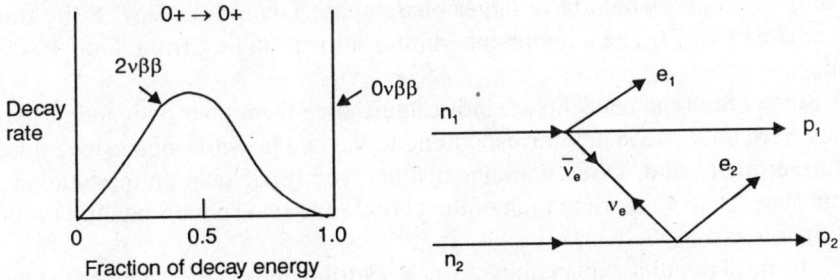

Fig. 7.11: Neutrinoless double beta decay

The $0\nu\beta\beta$ mode differs from the $2\nu\beta\beta$ mode by the fact that the emitted neutrino is reabsorbed. Therefore, $0\nu\beta\beta$ violates lepton number conservation.

It also indicates that neutrino is identical to its own antiparticle, i.e. the charge conjugate state of the original neutrino is identical to the original neutrino itself.

$$C \mid v_{e^-} \rangle \equiv \mid \tilde{v}_{e^-} \rangle = \mid v_{e^-} \rangle.$$

Such a neutrino is called a Majorana neutrino or Majorino (v_M) which is massive and which couples to electrons. Majorana neutrino is virtual and it is emitted between two nucleons in a nucleus. On the other hand for the Dirac neutrino we have

$$C \mid v_{e^-} \rangle \equiv \mid \tilde{v}_{e^-} \rangle \neq \mid v_{e^-} \rangle.$$

This distinction between a Majorana or Dirac nature of neutrino was tested long time ago in the experiment by Davis[2], where a clear cut difference between v_{e^-} and $\overline{v}_{e^-}$ was proven. However, Wu's experiment[10] confirming parity non-conservation showed that right handed $\overline{v}_{e^-}$ could never match left handed v_e:

$$n \rightarrow p + e^- + \overline{v}_e(R) \text{ and } v_e(L) + n \rightarrow p + e^-. \qquad (7.53)$$

Thus, $0\nu\beta\beta$ would be strictly forbidden if v_{e^-} and $\overline{v}_{e^-}$ are massless particles (*see* Exercise 7.5). Therefore, $0\nu\beta\beta$ manifests itself with finite neutrino mass.

The two-electron energy spectrum for $O+ \rightarrow O+$ double beta decay process makes clear distinction between 2ν and 0ν situations as shown schematically in Fig. 7.11. 2ν spectrum is continuous whereas $0n$ shows a sharp peak at E_0.

Therefore, $0\nu\beta\beta$ experiment carried out under strict laboratory conditions, the determination of the character of neutrino is of great importance for testing the standard model. Amongst $0\nu\beta\beta$ candidates the interesting ones are those with higher Q-value, which are less affected by natural radioactive background and which have larger phase space factor (*see* Eq. 7.8-life time decreases by Q^5). The experiments run for a time ranging from 3000-10,000 hours.

Since Majorana neutrino are indistinguishable from their own antiparticle, they have only two states corresponding to the two possible spin orientations. On the other hand, Dirac particles distinct from their own antiparticle have four states: two spin orientation of the particle and two spin orientation for the antiparticle.

The most popular explanation of small neutrino mass is the "see saw mechanism" in which neutrino masses are inversely proportional to some large mass M_H. Four states of each neutrino family have two component light neutrino with mass M_L and two component heavy neutrino (sterile) with mass M_H in such a way that

$$M_L M_H = M_D^2, \tag{7.54}$$

where M_D is the Dirac fermion mass (masses of quarks and charged leptons). Therefore, M_L is small provided M_H is large. If this explanation of neutrino mass is true then the experimentally observed particles are Majorana particles and hence the total lepton number is not conserved. Observation of violation of the total lepton number conservation would be a signal that neutrinos are indeed Majorana particles, that is neutrinos having mass.

The search for neutrinoless double beta decay ($0\nu\beta\beta$) by various international groups through large scale experiments is going on world wide. The $0\nu\beta\beta$ can only proceed via exchange of massive Majorana neutrino and therefore it is extensively studied from the theoretical side. The lower limit of $T_{1/2}^{0\nu}$ obtained for ^{76}Ge and ^{130}Te are shown in Table 7.6.

Table 7.6: Neutrinoless double beta decay

	Experiment		Theory		References
	$T_{1/2}^{0\nu}$ (Year)	$\langle m_\nu \rangle$ (eV)	$T_{1/2}^{0\nu}$ (Year)	$\langle m_\nu \rangle$ (eV)	
^{76}Ge	1.4×10^{25}	(1.8-2.5)	1.4×10^{22}	16	28
^{130}Te	6.6×10^{23}	(1.8-2.5)	2.5×10^{22}	130	28
	1.8×10^{24}	(0.2-1.1)			29

Proposed or suggested future $0\nu\beta\beta$ experiments are separated into two groups based on the magnitude of the proposed isotope mass. They are listed by Elliot and Vogel[27] and briefly summarised in Ref. 43. For example, COBRA[30] uses ^{130}Te source with 10 kg of CdTe and semiconductor detector, GEM[31] uses 1 ton of ^{enr}Ge diodes in liquid nitrogen and MOON[32] uses 34 tons of ^{nat}Mo sheets between plastic scintillators. The next generation of $0\nu\beta\beta$ experiments predict $\langle m_\nu \rangle \approx 10^{-2}$ eV.

Exercise 7.10. 10^8 neutrinos ($N\nu$) traverse earth, which has radius $R = 6000$ km and density $r = 5$ gm/cm^3. How many of them collide if the interaction cross section $\sigma = 7 \times 10^{-41}$ cm^2?

Number of nucleons encountered by a neutrino is then

$$N_N = 2R\sigma \rho N_A$$

where N_A = Avogadros number = 6.02×10^{23}/gm. The total number collisions

$$N = N_N N_\nu = 2 \times 6 \times 10^8 \text{ cm} \times 7 \times 10^{-41} \text{ cm}^2 \times \frac{5 \text{ gm}}{\text{cm}^3} \cdot \frac{6.02 \times 10^{23}}{\text{gm}} \times 10^8$$

$$\cong 25.$$

Exercise 7.11. In a solar neutrino experiment, a large detector contains lithium enriched in the isotope ${}_3^7 Li \left(\text{density} = \dfrac{0.53 \text{ gm}}{\text{cm}^3} \right)$. The detection depends on production, separation and detection of electron capturing isotope ${}^7Be$ in the process: $\nu_e + {}_3^7 Li_4 \rightarrow {}_4^7 Be_3 + e^-$.

To calibrate this detector, a point source emitting 10^{17} monochromatic neutrinos per sec (N_ν) with energy 1.6 MeV is placed at the center of one metric ton sphere of ${}^7Li$. Estimate the total equilibrium disintegration rate of ${}^7Be$.

We consider a shell of ${}^7Li$ container of radius r and thickness dr.

The number lithium nuclei in the shell $= \dfrac{4\pi r^2 dr \, \rho \, N_A}{A}$,

where N_A = Avogadros number, A = mass number of ${}^7Li$. The neutrino flux at $r = N_\nu / 4\pi r^2$. If neutrino $-{}^7Li$ capture cross section $\sigma = 10^{-43}$ cm^2, a is the activity ratio of ${}^7Li$ for forming ${}^7Be$, R = radius of the sphere of ${}^7Li$ then the number ${}^7Be$ nuclei produced per unit time

$$N({}^7Be) = \int \frac{N_\nu}{4\pi r^2} \cdot \rho \, N_A \, \sigma \, a \cdot \frac{4\pi r^2 dr}{A} = N_\nu \rho \, N_A \, \sigma \, a \frac{R}{A}.$$

with $a = 0.93$, $\rho = 0.53$ gm/cm^3, $A = 7$, $N_A = 6.02 \times 10^{23}$ and

$$R = \left(\frac{3 \times 10^6}{4\pi\rho} \right)^{1/3} = 76.6 \text{ cm, we get,}$$

$$N({}^7Be) = 10^{17} \sec^{-1} \, 0.53 \, \frac{\text{gm}}{\text{cm}^3} \cdot \frac{6.02 \times 10^{23}}{\text{gm}} \, 10^{-43} \text{ cm}^2 \, 0.93 \, \frac{76.6 \text{ cm}}{7}$$

$$= 3.2 \times 10^{-2} \sec^{-1}.$$

Exercise 7.12. In the decay $\pi \rightarrow \mu + \nu_\mu$, if the neutrino mass consists of combination of two mass states m_1 and m_2 then show that the difference in energy of the muon in two cases is $\delta E_\mu = (m_1^2 - m_2^2)/2m_\pi$.

If m_ν denotes the mass of the muon neutrino, the muon energy in pion decay at rest (*see* Eq. 7.50) is $E_\mu = (m_\pi^2 + m_\mu^2 - m_\nu^2)/2m_\pi$. If neutrino masses are m_1 and m_2, then $\Delta E_\mu = (m_1^2 - m_2^2)/2m_\pi$.

7.5 NEUTRINO OSCILLATIONS AND MASSES

To explain lower than expected capture rates of solar and atmosphere neutrinos, the most natural framework seems to be a flavor changing oscillation in which neutrinos of a given flavor transform into another flavor. In other words, due to neutrino oscillation some of the ν_e from the Sun appear on the earth as a neutrino of different flavor missed by the detector.

Experimental searches for neutrino oscillation can be classified into two categories:

(1) Disappearance Experiment – in which one looks for the diminution of the neutrino flux due to oscillation to some other flavor to which the detector is not sensitive.

(2) Appearance Experiment – in which one looks for a neutrino flavor not present in the initial beam which can arise from oscillation.

Precise measurements of solar neutrino fluxes coupled with amazing flavour transformation properties of neutrinos provide positive evidence for neutrino mass. Neutrino oscillation is viewed as a quantum mechanical interference phenomena in which neutrinos of a given flavor transform into another. The intrinsic neutrino properties involved are intrinsic mass squared difference Δm^2 and neutrino mixing angle θ in vacuum. To see this we concentrate here on the idea that neutrinos of definite flavor, i.e. the physical states are not states of a definite mass. In other words, flavour eigen states $\neq$ mass eigen states.

A. MODELS FOR NEUTRINO OSCILLATIONS

We build up a model for neutrino oscillation considering only two neutrino flavor states ν_e and ν_μ as the physical states and ν_1 and ν_2 corresponding mass eigen states. As neutrinos travel in the atmosphere (treated as vacuum) only their phases change with time.[34] We write the neutrino flavor eigen states at $t = 0$ as linear superposition of the mass eigen states:

$$|\nu_e(0)\rangle = \cos\theta \, |\nu_1\rangle + \sin\theta \, |\nu_2\rangle$$

$$|\nu_\mu(0)\rangle = -\sin\theta \, |\nu_1\rangle + \cos\theta \, |\nu_2\rangle , \qquad (7.56)$$

where θ is the mixing angle in vacuum. Denoting $\cos\theta = c_\theta$ and $\sin\theta = s_\theta$, we can write Eq. (7.56) in a matrix form:

$$\begin{pmatrix} |\nu_e(0)\rangle \\ |\nu_\mu(0)\rangle \end{pmatrix} = \begin{pmatrix} c_\theta & s_\theta \\ -s_\theta & c_\theta \end{pmatrix} \begin{pmatrix} |\nu_1\rangle \\ |\nu_2\rangle \end{pmatrix},$$

which shows that mixing matrix is real and unitary. We can also express the flavor composition of mass eigen states as

$$\begin{pmatrix} |v_1\rangle \\ |v_2\rangle \end{pmatrix} = \begin{pmatrix} c_\theta & -s_\theta \\ s_\theta & c_\theta \end{pmatrix} \begin{pmatrix} |v_e(0)\rangle \\ |v_\mu(0)\rangle \end{pmatrix}. \tag{7.57}$$

According to Eq. (7.56), the probability of finding electron neutrino flavor, $|v_e(0)\rangle$ in v_1 is $\cos^2\theta$ and that of $|v_\mu(0)\rangle$ in $|v_1\rangle$ is $\sin^2\theta$. We can schematically (Fig. 7.12) show the neutrino flavor states as combination of mass eigen states as follows:

The length of the box gives the probability of finding a mass state in a given flavor. The sum of the length of the boxes is normalized to 1.

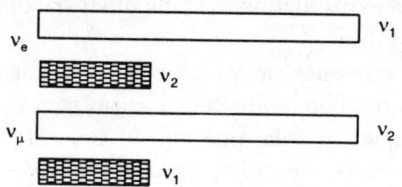

Fig. 7.12: Neutrino flavor states as combination of mass eigen states

In the non-relativistic quantum mechanics, we employ the time dependent S. Eq. to study the time evolution of the state $|v_e(t)\rangle$ as

$$|v_e(t)\rangle = c_\theta \, e^{-iE_1 t} \, |v_1\rangle + s_\theta \, e^{-iE_2 t} \, |v_2\rangle, \tag{7.58}$$

where we use the natural units $\hbar = c = 1$. Assuming ultra-relativistic neutrinos with common definite momentum p, we can write

$$E_i = (p^2 + m_i^2)^{1/2} \quad \text{with } p \gg m_i$$

$$\approx p\left(1 + \frac{m_i^2}{p^2}\right)^{1/2} \quad \cong p + m_i^2/2p \tag{7.59}$$

Therefore, $|v_e(t)\rangle = c_\theta \, e^{-i(p^2 + m_1^2/2p)t} |v_1\rangle + s_\theta \, e^{-i(p^2 + m_2^2/2p)t} |v_2\rangle.$ (7.60)

The dynamics of propagation shown above was the feature, that admixture of mass eigen states in a given flavor state do not change, i.e. there is no $v_1 \rightleftharpoons v_2$ transition. Using (7.57), we can rewrite (7.58) interms of $v_e(0)$ and $v_\mu(0)$ as

$$|v_e(t)\rangle = \Big[c_\theta e^{-i(p^2 + m_1^2/2p)t} \{ c_\theta |v_e(0)\rangle - s_\theta |v_\mu(0)\rangle \}$$

$$+ s_\theta \, e^{-i(p^2 + m_2^2/2p)} \{ s_\theta |v_e(0)\rangle + c_\theta |v_\mu(0)\rangle \} \Big]$$

$$= [(c_\theta^2 \, e^{-im_1^2/2p} + s_\theta^2 \, e^{-im_2^2/2p}) |v_e(0)\rangle$$

$$+ c_\theta \, s_\theta \, e^{-im_2^2 t/2p} - c_\theta \, s_\theta \, e^{-im_1^2/2p}) |v_\mu(0)\rangle. \qquad (7.61)$$

It is now easy to calculate the probability of a muon neutrino which starts evolving at time $t = 0$ and is transformed into an electron neutrino at time t. The required probability is

$$P(v_\mu(0) \rightarrow v_e(t)) = |\langle v_e(t) | v_\mu(0)\rangle|^2$$

$$= s_\theta^2 \, c_\theta^2 \, |e^{-im_2^2 t/2p} - e^{-im_1^2 t/2p}|^2, \quad \text{since} \quad \langle v_\mu(0) | v_e(0)\rangle = 0.$$

In the natural unit $L = t$, where L is the length traveled by neutrino in time t, we get

$$P(v_\mu(0) \rightarrow v_e(t)) = s_\theta^2 \, c_\theta^2 \left| e^{-i(m_1^2 + m_2^2)L/4E} \left\{ e^{-i\Delta m^2 L/4E} - e^{-i\Delta m^2 L/4E} \right\} \right|^2$$

$$= 4 s_\theta^2 \, c_\theta^2 \sin^2 \frac{\Delta m^2 L}{4E} = \sin^2 2\theta \, \sin^2 \frac{\Delta m^2 L}{4E}, \qquad (7.62)$$

where $\Delta m^2 = |m_2^2 - m_1^2|$ and E is the average neutrino kinetic energy. We use the following unit to compare the model predictions with experiments:

$$\left(\frac{\Delta m^2 L}{4E} \right)_{natural\ units} = \frac{\Delta m^2 c^4 L}{4\hbar c E} \frac{10^{-12} \times 10^{18} L \, (km)}{4 \times 197 \times 10^3 E \, (GeV)}$$

$$= 1.27 \frac{L \, (km)}{E \, (GeV)} \Delta m^2 (eV^2), \qquad (7.63)$$

where L is in kilometer and E in GeV (same relation is obtained when L is in meter and E in MeV). The distance travelled by neutrino L (km) and neutrino energy E in (GeV) are called engineering units.

Finally, $\quad P(v_e \rightarrow v_\mu) = \sin^2 2\theta \, \sin^2 \left[\dfrac{1.27 \, \Delta m^2 (eV^2) \, L(km)}{E_v \, (GeV)} \right], \qquad (7.64)$

which shows that $P = 0$ for $\theta = 0$ or $\Delta m^2 = 0$. $\dfrac{\Delta m^2 L}{E_v} \ll 1$ gives the non-oscillation region. The oscillation in this simple case is characterized by a oscillation length

$$L_{osc} (km) = \frac{2.48 \, E_v \, (GeV)}{\Delta m^2 \, (eV^2)} \quad \text{and by} \quad \sin^2 2\theta. \qquad (7.65)$$

At distance $L \gg L_{OSC}$ the oscillation pattern is smeared out and the oscillation probability $P(\nu_\mu \rightarrow \nu_e)$ approaches $P = (1/2) \sin^2 2\theta = 1/2$ for maximal mixing $(\theta = \pi/4)$ since, $\sin^2 aL = (1 - \cos 2aL)/2$ and for L large:

$\cos 2aL \rightarrow 0$. Here a stands for $\dfrac{1.27 \Delta m^2}{E_\nu}$.

The oscillation maxima occurs at $\dfrac{\Delta m^2 L_{OM}}{4 E_\nu} = \dfrac{\pi}{2}(1 + 2n)$ or $L_{OM} = \dfrac{2\pi E_\nu}{\Delta m^2}$,

for $n = 0$.

The variation of flavor transition probability with respect to L for fixed E is shown in Fig. 7.13.

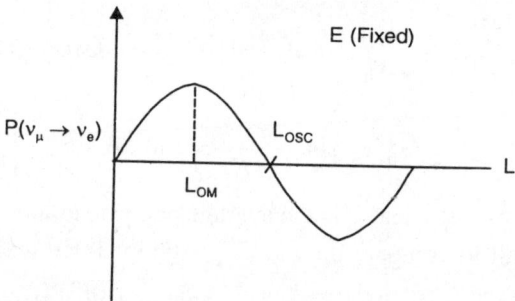

Fig. 7.13: Variation of flavor transition probability with respect to L for fixed E

In a realistic experimental situation, it is quite possible that the wavelength of the neutrino oscillation is much smaller than the uncertainties of the position of the production and detection points of the neutrinos. If it is the case, the relevant quantities are the time average transformation probability or survival probability.

The time average of the transformation probability

$$\bar{P}(\nu_\mu \rightarrow \nu_e) = \frac{1}{2}\sin^2 2\theta \text{ and } \left[\bar{P}(\nu_\mu \rightarrow \nu_e)\right]_{max} = \frac{1}{2}.$$

The survival probability,

$$P(\nu_\mu \rightarrow \nu_\mu) = 1 - P(\nu_\mu \rightarrow \nu_e) = 1 - \sin^2 2\theta \, \sin^2\left(\frac{\Delta m^2 t}{4E}\right). \quad (7.66)$$

The time average survival probability $\bar{P}(\nu_\mu \rightarrow \nu_\mu) = 1 - \dfrac{1}{2}\sin^2 2\theta$.

$$[\bar{P}(\nu_\mu \rightarrow \nu_\mu)]_{min} = 0 \text{ for } \theta = \pi/4.$$

For the obvious reasons the oscillation studies are performed at distances $L \approx L_{OSC}$ from the neutrino source. At shorter distances the oscillation amplitude is reduced and at large distances the neutrino flux is reduced making the experiment more difficult. At distance $L \gg L_{OSC}$ $P(\nu_\mu \rightarrow \nu_e)$ approaches $(1/2)\sin^2 2\theta$ and becomes independent of Δm^2. The survival probability $P(\nu_\mu \rightarrow \nu_\mu)$ data of Super-K experiments for atmospheric neutrinos for maximum mixing angle $\theta = \pi/4$ $(\sin^2 2\theta = 1)$ yields

$$\Delta m^2 = 5 \times 10^{-3} \text{ eV}^2.$$

The probabilities described above satisfy the relation

$$P(\nu_\mu \rightarrow \nu_e) + P(\nu_\mu \rightarrow \nu_\mu) = 1,$$

which reflects the conservation of probability.

In three-neutrino oscillation picture, atmospheric *K2K* data gives

$$\Delta m^2 \Big|_{atom} = (1.6 - 3.9) \times 10^{-3} \text{ eV}^2 \text{ for } \sin^2 2\theta_{atom} \approx 0.91.$$

Solar KamLand[24] data gives

$$\Delta m^2_{sol} = (7.0 - 15.0) \times 10^{-5} \text{ eV}^2 \text{ for } (\sin^2 2\theta_{12})_{sol} \approx 0.85.$$

The size of θ_{13} is $\sin^2 \theta_{13} < 0.03$ and $0.35 \leq \sin^2 \theta_{23} < 0.65$.

At present the observation of nearly maximal atmospheric neutrino mixing and the possibility that also solar mixing is large suggest that all neutrino masses are nearly degenerate (favoured also by data). Moreover, the common value of $|m_\nu|$ could be compatible with a large fraction of hot dark matter in the universe for $|m_\nu| \sim 1\text{-}2$ eV as we shall see later. However, cosmological data do not prefer $0\nu2\beta$ mass range. The neutrino mass spectrum in different 3ν scenarios are: $(m_\nu)_{atmos} \approx 0.06$ eV , $(m_\nu)_{solar} \approx 0.008\text{eV}$. However, there are several open questions: (*i*) Are there more than 3 neutrinos? (*ii*) How precise are measurements? And (*iii*) precision values?

Neutrino oscillation parameters determined from SNO[19] collaboration at 90% CL are:

$$\Delta m^2_{12} = 7.1^{+1.2}_{-0.6} \times 10^{-5} \text{ eV}^2, \qquad\qquad \theta_{12} = 32.50^{+2.4}_{-2.3}.$$

Nishikawa reported (*see* Ref. 45)

$$\Delta m^2_{23} = 2.0^{+0.6}_{-0.4} \times 10^{-3} \text{ eV}^2, \qquad\qquad \sin^2 2\theta_{23} > 0.94 \text{ for } \theta_{13} = 0$$

(note that $\Delta m^2_{12} + \Delta m^2_{23} + \Delta m^2_{31} = 0$).

With the above values of $\Delta m^2 (= m_1^2 - m_2^2)$ it is not possible to form a third value of Δm^2 consistent with the LSND data[25]

$$\Delta m_{LSND}^2 \sim 0.4 - 2 \text{ eV}^2, \ \sin^2 2\theta \sim 0.001 - 0.01,$$

which can not be accommodated in a three generation picture. It has been widely realised that the remedy of this situation might be the introduction of a fourth neutrino usually called sterile neutrino[25].

Neutrino Oscillation in Matter

Mikheyev-Smirnov-Wolfenstein (MSW)[34] made the important suggestion that the solar neutrinos might convey resonant effect on matter induced neutrino missing in the Sun. MSW effect would convert a large fraction of v_e produced in solar core to another type.

When neutrinos propagate in matter, additional contribution to the phase appears, besides the one caused by the non-vanishing mass eigen states $|v_i\rangle$, $i = 1, 2, \ldots$. To see the origin of such phase, we consider the effective Hamiltonian of neutrinos in the presence of matter. The interaction between neutrinos and matter modifies the picture of neutrino oscillations. The effect of the medium is analogous to the appearance of a refractive index.

We can consider that all effective neutrinos interact with quarks and electrons by the neutral current weak interaction via exchange of Z^0 boson but only electron neutrinos and antineutrinos interact with electrons by the exchange of W^- boson referred as charge current interaction. For the charge current process, the current-current interaction can be written as $(V–A) \times (V–A)$ four Fermi contact interaction for $E \ll M_W$ as

$$H_{\text{int}} = j_e^\mu \cdot j_\mu^v = \frac{G_F}{\sqrt{2}} \overline{\psi}_{v_e} \ \gamma^\mu (1 - \gamma_5) \ \psi_e \cdot \overline{\psi}_e \ \gamma_\mu (1 - \gamma_5) \ \psi_{v_e}, \qquad (7.68)$$

which can be rewritten by Fierz transformation as

$$H_{\text{int}} = j_e^\mu \cdot j_\mu^v = \frac{G_F}{\sqrt{2}} \overline{\psi}_e \ \gamma^\mu (1 - \gamma_5) \ \psi_e \cdot \overline{\psi}_{v_e} \ \gamma_\mu (1 - \gamma_5) \ \psi_{v_e}, \qquad (7.69)$$

In the matter rest frame, electrons are non-relativistic and unpolarized so that

$$\overline{\psi}_e \gamma^\mu (1 - \gamma_5) \psi = N_e \delta_{\mu 0}, \qquad (7.70)$$

where N_e is the number density of electrons, assumed to be constant for static unpolarized medium. Hence,

$$H_{\text{int}} = \frac{G_F}{\sqrt{2}} N_e \ \overline{\psi}_{v_e} \ \gamma_0 (1 - \gamma_5) \psi_{v_e}$$

$$= \frac{G_F}{\sqrt{2}} N_e \ \overline{\psi}_{v_e} \ (1 + \gamma_5) \ \gamma_0 \ \psi_{v_e}$$

$$= \sqrt{2}\ G_F\ N_e\ \overline{\psi}_{\nu_e L}\ \gamma_0\ \psi_{\nu_e}\ \text{ since } \overline{\psi}_{\nu_{eL}} = \psi_{\nu_e}\frac{1}{2}(1+\gamma_5). \quad (7.71)$$

In QED a background electric Coulomb potential $V(x)$ interact with electron as $V(x)\ \overline{\psi}_e\ \gamma_0\ \psi_e$. Similarly, above 4 Fermion interaction may be regarded as an interaction of left handed neutrino ν_{eL} with Coulomb type static potential $V_c(x)$ so that

$$H_{\text{int}} = V_c\ \overline{\psi}_{\nu_e L}\ \gamma_0\ \psi_{\nu_e}, \text{ where } V_c = \sqrt{2}\ G_F\ N_e. \quad (7.72)$$

Therefore, the effective potential seen by neutrinos and antineutrinos due to charge current weak interaction in matter are: $V_c(\nu_e) = \sqrt{2}G_F N_e$ and $V_c(\overline{\nu}_e) = -\sqrt{2}G_F N_e$.

Given the effective potential, electron neutrinos traveling distance L in matter of constant density acquire an additional phase

$$\nu_e(L) = \nu_e(0)e^{-i\sqrt{2}G_F N_e L}.$$

Equivalently, one can describe the effect of medium in terms of the refractive index $n_{ref} - 1 = V_c/p$, where p is neutrino momentum. In vacuum, the mass eigen states propagate as a plane wave. Leaving out the common phase, a beam of ultrarelativistic neutrinos $|\nu_i\rangle$ with energy E at a distance L acquires a phase

$$|\nu_i(L)\rangle \sim |\nu_i(L=0)\rangle\ e^{-i\frac{m_i^2 L}{2E}}, \quad (7.73)$$

given that the amplitude of the process $\nu_e \to \nu_{e'}$ is

$$A(\nu_e \to \nu_{e'}) = \sum_i U_{ei}\ e^{-m_i^2 L/2E}\ U_{e'i}^*,$$

where $i(=\nu_e, \nu_\mu, \nu_\tau)$ types of leptons involved. The probability for the flavor change for $i \neq i'$ is the square of the amplitude. It is obvious that due to unitarity of U there is no flavor change if all the masses vanish or exactly degenerate.

Let us now consider the propagation of mass eigen states with time. Again for simplicity we consider two mass eigen states ν_1 and ν_2 that are components of flavor eigen states. The S.Eq. in this case is

$$i\frac{d}{dt}\begin{pmatrix} \nu_1 \\ \nu_2 \end{pmatrix} = \frac{1}{2}\begin{pmatrix} -\dfrac{\Delta m^2}{2E} & 0 \\ 0 & \dfrac{\Delta m^2}{2E} \end{pmatrix}\begin{pmatrix} \nu_1 \\ \nu_2 \end{pmatrix}.$$

We rotate to the flavor basis and get

$$i\frac{d}{dt}\begin{pmatrix} \nu_e \\ \nu_\mu \end{pmatrix} = \frac{1}{2}\begin{pmatrix} c_\theta & s_\theta \\ -s_\theta & c_\theta \end{pmatrix}\begin{pmatrix} -\dfrac{\Delta m^2}{2E} & 0 \\ 0 & \dfrac{\Delta m^2}{2E} \end{pmatrix}\begin{pmatrix} c_\theta & -s_\theta \\ s_\theta & c_\theta \end{pmatrix}\begin{pmatrix} \nu_e \\ \nu_\mu \end{pmatrix}$$

$$= \frac{1}{2}\begin{pmatrix} -\dfrac{\Delta m^2}{2E}c_\theta^2 + \dfrac{\Delta m^2}{2E}s_\theta^2 & \dfrac{\Delta m^2}{2E}2c_\theta\, s_\theta \\ \dfrac{\Delta m^2}{2E}2c_\theta\, s_\theta & -\dfrac{\Delta m^2}{2E}s_\theta^2 + \dfrac{\Delta m^2}{2E}c_\theta^2 \end{pmatrix}\begin{pmatrix} \nu_e \\ \nu_\mu \end{pmatrix}.$$

$$= \frac{1}{2}\begin{pmatrix} -\dfrac{\Delta m^2}{2E}c_{2\theta} & \dfrac{\Delta m^2}{2E}s_{2\theta} \\ \dfrac{\Delta m^2}{2E}s_{2\theta} & -\dfrac{\Delta m^2}{2E}c_{2\theta} \end{pmatrix}\begin{pmatrix} \nu_e \\ \nu_\mu \end{pmatrix}. \tag{7.74}$$

We add an extra term to include matter effect as follows:

$$\sqrt{2}\,G_F N_e\begin{pmatrix} 1 & 0 \\ 0 & 0 \end{pmatrix} = \sqrt{2}\,G_F N_e\frac{1}{2}\left[\begin{pmatrix} 1 & 0 \\ 0 & 1 \end{pmatrix} + \begin{pmatrix} 1 & 0 \\ 0 & -1 \end{pmatrix}\right]$$

$$i\frac{d}{dt}\begin{pmatrix} \nu_e \\ \nu_\mu \end{pmatrix} = \frac{1}{2}\begin{pmatrix} -\dfrac{\Delta m^2}{E}c_{2\theta} + 2\sqrt{2}G_F N_e & \dfrac{\Delta m^2}{2E}\cdot s_{2\theta} \\ \dfrac{\Delta m^2}{2E}s_{2\theta} & \dfrac{\Delta m^2}{2E}c_{2\theta} - 2\sqrt{2}G_F N_e \end{pmatrix}\begin{pmatrix} \nu_e \\ \nu_\mu \end{pmatrix}$$

$$= \frac{1}{2}\begin{pmatrix} -\dfrac{\Delta m_N^2}{E}c_{2\theta_N} & \dfrac{\Delta m^2}{2E}s_{2\theta_N} \\ \dfrac{\Delta m_N^2}{2E}s_{2\theta_N} & \dfrac{\Delta m^2}{2E}c_{2\theta_N} \end{pmatrix}\begin{pmatrix} \nu_e \\ \nu_\mu \end{pmatrix}, \tag{7.75}$$

where

$$\Delta m_N^2 c_{2\theta_N} = \Delta m^2 c_{2\theta} - 2\sqrt{2}\,G_F N_e E \tag{7.76a}$$

$$\Delta m_N^2 s_{2\theta_N} = \Delta m^2 s_{2\theta}. \tag{7.76b}$$

$$\tan 2\theta_N = \frac{\Delta m^2 s_{2\theta}}{\Delta m^2 c_{2\theta} - 2\sqrt{2}\, G_F N_e E}.$$

Now, $\quad \dfrac{\Delta m^2 L_{osc}}{4 E_\nu}\, \pi$ or $L_{osc}\, \dfrac{4\pi E}{\Delta m^2}$

$$\tan 2\theta_N = \frac{\tan 2\theta}{1 - \dfrac{L_{osc}}{L_o \cos 2\theta}}, \text{ where } L_o = \frac{\sqrt{2}\pi}{G_F N_e} \qquad (7.77)$$

is called the matter oscillation length.

Now
$$\sin 2\theta_N = \frac{\tan 2\theta}{\left[\tan^2 2\theta + \left(1 - \dfrac{L_{osc}}{L_o \cos 2\theta}\right)\right]^{1/2}}$$

$$= \sin 2\theta \left[1 + \left(\frac{L_{osc}}{L_o}\right)^2 - \frac{2 L_{osc} \cos 2\theta}{L_o}\right]^{-1/2}. \qquad (7.78)$$

The new mixing angle in matter, θ_N depends on the vacuum mixing angle θ and on the vacuum and matter oscillation lengths L_{osc} and L_o. The effective oscillation in matter is then

$$L_N = L_{osc} \cdot \frac{\sin 2\theta_N}{\sin 2\theta} = L_{osc} \left[1 + \left(\frac{L_{osc}}{L_o}\right)^2 - \frac{2 L_{osc} \cos 2\theta}{L_o}\right]^{-\frac{1}{2}}$$

The probability of detecting ν_e at a distance L from the ν_e source has the usual form with $\theta \to \theta_N$ and $L_{osc} \to L_N$:

$$P(E_\nu, L, \theta, \Delta m^2) = 1 - \sin^2 2\theta_N \, \sin^2 \frac{\pi L}{L_N}.$$

The transition point between the regime of vacuum and matter oscillations is determined by the ratio $\dfrac{L_{osc}}{L_N}$, which is given by

$$\frac{L_{osc}}{L_o} = \frac{2\sqrt{2} G_F N_e E}{\Delta m^2}, \text{ since } L_{osc} = \frac{4\pi E}{\Delta m^2} \text{ and } L_o = \frac{\sqrt{2}\pi}{G_F N_e}.$$

If this fraction is larger than unity, the matter oscillations dominate and when this ratio is less than $\cos 2\theta$, the vacuum oscillations dominate. Generally, there is a smooth transition between two. We can write $N_e = \rho_e Y_e N_o$, where $Y_e = N_e/(N_p + N_n)$, $N_o = 6 \times 10^{23}$ atoms per gm and N_P and N_n are number of protons and neutrons in the matter. For $N_p = N_n = N_e$, we get $Y_e = 1/2$.

Therefore,
$$L_o = \frac{\sqrt{2}\pi}{G_F N_e} = \frac{\sqrt{2}\pi}{G_F \rho_e (gm/cm^3)\, Y_e \dfrac{6 \times 10^{23}}{gm}}. \qquad (7.80)$$

Substituting $G_F = 1.16 \times 10^{-5}\,\mathrm{GeV}^{-2}$ and $10^{-13}\,\mathrm{cm} \equiv 5\,\mathrm{GeV}^{-1}$, we get

$$L_o = \frac{1.414 \times 3.141 \times 25\ \mathrm{cm}}{1.16 \times 10^{-5} \times 10^{-26} Y_e \times 6 \times 10^{23}} \cong \frac{1.7 \times 10^4\ \mathrm{km}}{\rho_e (gm/cm^3)\, Y_e}.$$

In rock, $\rho_e = 4\ \mathrm{gm/cm}^{-3}$ so that $L_o \approx 10^4\ \mathrm{km}$. At the centre of the Sun, $L_o = 200\ \mathrm{km}$. Squaring and adding Eq. (7.76a) and Eq. (7.76b) we get

$$(\Delta m_N^2)^2 = (\Delta m^2 \cos 2\theta - 2\sqrt{2} G_F N_e E)^2 + (\Delta m^2 \sin 2\theta)^2$$

If $\Delta m^2 \cos 2\theta = 2\sqrt{2} G_F N_e E$ then $\Delta m_N^2 = \Delta m^2 \sin 2\theta$.

From (7.75) we get the resonance condition: $2\sqrt{2} G_F N_e E = \Delta m^2 \cos 2\theta$

Substituting $\dfrac{\Delta m^2}{2E} = \dfrac{2\pi}{L_{osc}}$ and $\sqrt{2} G_F N_e = \dfrac{2\pi}{L_o}$ in Eq. (7.75) we get

$$i\frac{d}{dt}\begin{pmatrix} v_e \\ v_\mu \end{pmatrix} = \frac{1}{2}\begin{pmatrix} -2 \cdot \dfrac{2\pi}{L_{osc}} \cos 2\theta + 2 \cdot \dfrac{2\pi}{L_o} & \dfrac{2\pi}{L_{osc}} \sin 2\theta \\[2mm] \dfrac{\sin 2\theta}{L_{osc}} & 0 \end{pmatrix}\begin{pmatrix} v_e \\ v_\mu \end{pmatrix}$$

$$= 2\pi \begin{pmatrix} -\dfrac{\cos 2\theta}{L_{osc}} + \dfrac{1}{L_o} & \dfrac{\sin 2\theta}{2 L_{osc}} \\[2mm] \dfrac{\sin 2\theta}{2 L_{osc}} & 0 \end{pmatrix}\begin{pmatrix} v_e \\ v_\mu \end{pmatrix}. \qquad (7.81)$$

Here one can see clearly that in matter, unlike in vacuum, the oscillation pattern depends on whether the mixing angle θ is larger than or smaller than

$\pi/4$. For antineutrinos the sign in front of L_o is reversed (*see* Eq. 7.80). In matter of a constant density, we can now consider several special cases: In the low density limit $L_{osc} \ll L_o$, matter has a rather small effect on the earth and one is able to observe oscillation provided $L_{osc} <$ earths diameter. Therefore, this limit applies since earth's diameter $\approx 4.8 \times 10^4$ kilometer. In the high density limit $L_{osc} \gg L_o$, the oscillation amplitude is suppressed by a factor $\dfrac{L_o}{|L_{osc}|}$, so that

$$ i\frac{d}{dt}\begin{pmatrix} \nu_e \\ \nu_\mu \end{pmatrix} = \frac{2\pi}{L_{osc}}\begin{pmatrix} -\cos 2\theta + \dfrac{L_{osc}}{L_o} & \dfrac{\sin 2\theta}{2} \\ \dfrac{\sin 2\theta}{2} & 0 \end{pmatrix} \cdot \begin{pmatrix} \nu_e \\ \nu_\mu \end{pmatrix} \psi^- = \psi^\dagger \gamma^0 . $$

For $m_2^2 > m_1^2$ we have, $\tan 2\theta_N = \dfrac{1}{1 - \dfrac{L_{osc}}{L_o \cos 2\theta}} \rightarrow 0$ and $\theta_N \rightarrow \dfrac{\pi}{2}$.

For $L_{osc} \approx L_o$, the matter effect can be enhanced. In particular, for $L_{osc}/L_o = \cos 2\theta$ one has $\sin 2\theta_N = \sin 2\theta \left[1 + \cos^2 2\theta - 2\cos^2 2\theta\right]^{-1/2} = \sin 2\theta\,(\sin^2 2\theta)^{-1/2} = 1$, indicating maximum mixing even for small vacuum mixing angle θ. This is the basis of the Mikheyev-Smirnov-Wolfenstein (MSW) effect[30], resulting from the varying density of the matter.

When considering oscillation in two flavors, the mixing angle θ can be restricted to the interval $(0, \pi/2)$. In vacuum, only $\sin^2 2\theta$ is relevant and, hence, only half of the interval $(0, \pi/4)$ could be used. However, once matter oscillations are present, also the $\cos 2\theta$ becomes relevant and thus the whole $(0, \pi/2)$ interval might be needed. Measured up-down ratios: $\nu_\mu \rightarrow \nu_\tau$ oscillation $\left(N_{up} - N_{down}/N_{up} + N_{down}\right)_{\mu-like} = 0.288 \pm 0.28$.

The neutrino travel distance for 'up' neutrino = 12,800 km. Neutrino travel distance for down neutrino = 15 km. $(\sin^2 2\theta, \Delta m^2) \equiv (1.00, 2.5 \times 10^{-3} \text{ eV}^2)$ for $\chi^2 = 1$.

KamLand Reactor Experiment gives $\dfrac{N(\bar{\nu}_e)_{observed}}{N(\bar{\nu}_e)_{expected}} = 0.611 \pm 0.085$, fitted

by the oscillation formula for $\sin^2 2\theta = 0.83$ and $\Delta m^2 = 5.5 \times 10^{-5} \text{ eV}^2$.

B. Cosmology and Neutrinos

Supernova are very energetic explosions produced when a massive star has reached the end of its nuclear fuel. A supernova arises when the core of the star collapse under its own gravitational attraction releasing energy which causes outer envelope to explode. Thus, the inner part of the star undergoes an implosion while outer part undergoes an explosion. As the gravitational pressure in the remnant (supernova relic) increased, electrons and protons fused through the weak interaction to neutrons and neutrinos $(e^- + p \rightarrow n + \nu_e)$. These neutrinos are called supernova neutrinos. They are temporarily confined to the core and escape to the outer region by diffusion. As neutrinos escape a neutron star with mass $M_{ns} \geq 1.4\, M_0$ and radius $R_{ns} \geq 15\, K_m$ is formed. $1.4\, M_0$ is the Chandrasekhar limit. The neutron pressure then prevents the star to become a black hole, that is, it does not shrink so much that the light from the star cannot get out of its strong inward gravitational pull.

One can make a rough estimate of the energy released and the number of neutrinos produced in the process of neutron star formation in connection with a supernova. The gravitational binding energy as the star radius shrinks is

$$E_b = \frac{3}{5}\left(\frac{GM_{ns}^2}{R_{ns}} \right),$$

where G_N = Newton's gravitational constant $\approx 10^{-38}$ GeV^{-2}, $M_{ns} = 1.4\, M_0 = 1.4 \times 10^{57}$ GeV and $R_{ns} = 15\, Km = 75 \times 10^{18}$ GeV^{-1}. Substituting the above values we get $E_b = 78 \times 10^{58}$ MeV. The star cools through the emission of neutrinos and the total energy carried by neutrinos is approximately E_b. All neutrinos are in thermal equilibrium at the core temperature. If the energy is distributed evenly between $\nu_e, \nu_\mu, \nu_\tau, \bar{\nu}_e, \bar{\nu}_\mu, \bar{\nu}_\tau$, then each particle carries energy $\approx 13 \times 10^{58}$ MeV.

The thermal energy within the new born neutron star is estimated by invoking virial theorem which relates the thermal and potential energy of a self gravitating system. For instance, for a nucleon on the surface of a neutron star, the average kinetic energy must be one half of its gravitational potential energy:

$$\langle E_k \rangle = \frac{1}{2}\frac{G\, M_{ns}\, m_N}{R_{ns}} = \frac{1}{2}\frac{10^{-38}\ \text{GeV}^{-2} \times 1.4 \times 10^{57}\ \text{GeV} \times 1\,\text{GeV}}{15 \times 5 \times 10^{18}\ \text{GeV}^{-1}} = 93\ \text{MeV}.$$

The number of neutrinos produced equals the number of neutrons. Therefore, neutrinos in thermal equilibrium with their environment

$\left(T = \dfrac{2}{3}\langle E_k\rangle \cong 62\ \text{MeV}\right)$ will have similar energies. Thus, it follows that the number of neutrinos produced in supernova explosion,

$$(N_\nu)_{\text{supernova}} = \frac{78}{62}\times 10^{58} \approx 10^{58}.$$

In the early universe, when the temperature was $T > 1$ MeV, the high density of particles allowed weak interaction to occur leading to a substantial density of neutrinos. As the universe expanded and cooled to $T < 1$ MeV, these reactions become much slower than expansion rate and the neutrinos decoupled from the remaining ionized plasma and radiation (photons). After decoupling, the ratio n_ν/n_γ remains constant. Since the present baryon energy density Ω_N is about $\eta\ m_N\ n_\gamma$, neutrinos will dominate the mass density of the universe if $m_\nu \geq \eta\ m_N$, where m_N is the nucleon mass and $\eta = n_B/n_\gamma$, the baryon to photon ratio.

If neutrinos are an important fraction of cosmological density, say $\Omega_\nu \sim 0.1$, then the average neutrino mass must be considerably heavier than the splitting that are indicated by atmospheric and solar neutrinos.

In the present Universe the number density of neutrinos n_ν is comparable to the photon number density n_γ. The theory of BBN and abundances of ^4He, ^7Li and ^2H with 3 neutrino species require $\eta \leq 7\times 10^{-10}$

In the standard hot big bang model, the effective neutrino temperature now is $T_\nu = (4/11)^{1/3}\ T_\gamma \cong 1.9^\circ K$ and the neutrino number density is $n_\nu + n_{\bar\nu} = 110/\text{cc}$ per flavor. Furthermore, if $N (\leq 3)$ families have average mass m_ν then the total neutrino energy density $N\rho_\nu = N m_\nu$ (110) eV/cc, which for 3 generation neutrino gives $\Omega_\nu \approx 0.06$ with $m_\nu < 16$ eV.

If neutrinos were the dominant mass sources in galaxies, one can place a lower limit on neutrino mass by invoking virial theorem, which relates the thermal and potential energy of a self gravitating system. According to this theorem the average kinetic energy must be one half of its gravitational potential energy:

$$\langle \frac{1}{2}m_\nu\ v^2\rangle = \frac{1}{2}\frac{G\ m_\nu M}{R} \quad \text{or}\ v^2 = \frac{GM}{R}.$$

$G(= 10^{-38}\ \text{GeV}^{-2})$ is the Newton's gravitational constant, and R is the size and M the mass of the gravitating system. The mass M_ν due to neutrinos in the system in question is thus

$$M_v \sim m_v \, R^3 \int d^3 p \sim m_v \, R^3 (m_v \, v)^3 = m_v^4 \, R^3 \, v^3.$$

Substituting v^2 from above and the fact $M > M_v$ we get $m_v \geq (G^3 M \, R^3)^{-1/8}$, which shows that the critical neutrino mass decreases with the size and mass of the astrophysical system in question and it works out to be about ~ 20 eV for galactic halo.

Massive neutrinos may affect drastically the picture of galaxy formation because massive neutrinos can start clumping long before photon decoupled since neutrinos interacting so weakly can freely stream through the distribution of particles. Charged particles are prevented from clumping by photons but not neutrinos.

Let M_{Hv} denote the amount of neutrino masses contained within the horizon at a critical time when neutrinos become non-relativistic. This critical time corresponds to a temperature $T \sim m_v$; after this time neutrinos are slow moving and no longer stream freely. Thus, $M_{Hv} = n_v \, m_v \, (\text{horizon size})^3$.

The horizon is, of course, just that distance scale over which photons could have travelled since the big bang. With the application of dimensional analysis we can write, the horizon size $\sim M_{pl} / T^2$ in the radiation era. For relativistic particles such as neutrino we have $n_v = T^3$ and $T = m_v$. Hence,

$$M_{Hv} = m_v^3 \cdot m_v \cdot \frac{M_{Pl}^3}{m_v^6} = \frac{M_{Pl}^3}{m_v^2}, \qquad \text{where } M_{Pl} = \text{Plancks length.}$$

Substituting the value of M_{Pl} (*see* Appendix A) we get

$$M_{Hv} = \frac{10^{57} \text{ GeV}}{m_v^2} = \left(\frac{1 \text{ eV}}{m_v} \right)^2 \times 10^{18} M_{\odot},$$

where $M_{\odot}$ = mass of the Sun = 10^{57} GeV.

For M_{Hv} to be of the order of a typical galactic mass $\sim M_{Galaxy} = 10^{15} M_{\odot}$, one gets the neutrino mass m_v from the following relation:

$$\left(\frac{1 \text{ eV}}{m_v} \right)^2 10^{18} M_{\odot} = 10^{15} M_{\odot}$$

or, $m_v \sim 33$ eV.

The presently observed distribution of matter (through high resolution galaxy survey) and the Cosmic Microwave Background (CMB) would both be affected by the presence of massive neutrinos in the early universe[35]. Thus, comparison of the power spectrum of CMB with the observed distribution of galaxies can provide information on the sum (over all flavors of light neutrino masses:[36] $\sum_f m_{\nu_f} < 0.7$ eV.

In the near-term future, substantial additional information from cosmological observation will be available that will constrain $\sum_f m_{\nu_f}$ compared to the limit above. For example, PLANCK mission[37] gives $\sum_f m_{\nu_f} \sim 0.06$ eV.

The origin of the matter-antimatter asymmetry is one of the most important questions in cosmology. The presently observed baryon asymmetry is

$$Y_B = \frac{n_b - n_{\bar{b}}}{n_b + n_{\bar{b}}} \approx 6.5 \times 10^{-10}.$$

In order to explain baryon asymmetry in terms of microphysical laws, three conditions need to be fulfilled:
1. Baryon number (or Lepton number) violation.
2. C and CP violation. If CP is conserved every reaction which produces a particle will be accompanied by a reaction which gives its antiparticle with no creation of a net baryon number.
3. Departure from thermal equilibrium.

Results from future cosmological studies may answer some questions raised in neutrino physics:
(1) Do neutrinos violate CP symmetry?
(2) Are neutrinos key to the understanding of matter-antimatter asymmetry?

Exercise 7.13. Neutrinos from reactor have energy 1 MeV and the distance between source and detector is 4 meter and an oscillation maxima is observed. Estimate neutrino mass squared difference from the above data.

$$\Delta m^2 (eV^2) = \frac{2\pi E_\nu \text{ (MeV)}}{L \text{ (meter)}} = \frac{2\pi}{4} \cdot 1 \approx 1.5$$

$$\Delta m^2 = 1.5 \text{ eV}^2.$$

Exercise 7.14. Show the variation of $P(\nu_\mu \to \nu_e)$ with respect to E for fixed L and find its minimum value.

The peaks occur at $\frac{\Delta m^2 L}{4 E_{OM}} = \frac{\pi}{2}(1 + 2n)$. For maximum mixing $\sin^2 2\theta = 1$ and $n = 0$, we get

$$\Delta m^2 L = 2\pi \ E_{OM}$$

and $P(v_\mu \rightarrow v_e) = \sin^2\left(\dfrac{\pi \ E_{OM}}{2E}\right)$ for L fixed . For small Δm^2 and $\sin^2 2q = 1$

$$P_{min}(v_\mu \ \rightarrow \ v_e) \ = \ \sin^2 \frac{\Delta m^2 L}{4E} \ = \ \left(\frac{\Delta m^2 L}{4E}\right)^2$$

Exercise 7.15 What is the velocity of a 6°K neutrino in the universe if the neutrino mass is 0.2 eV?

The average kinetic energy of a neutrino gas of temperature T is $\epsilon_k = \dfrac{3kT}{2}$, where k is the Boltzman constant.

$$\beta = \sqrt{2\,\epsilon_k\ /m} = \sqrt{3kT\ /\ m} = \sqrt{3\times 8.62\times 10^{-5}\times \frac{6}{0.2}} = 0.088$$

$$v = 2.64 \times 10^7 \text{ meter/sec .}$$

Exercise 7.16. Estimate the total number of neutrinos reaching the earth from a supernova cloud roughly at a distance of 50 kilometer per sec away from the earth (1 per sec = 3×10^{16} meter).

The given distance = 1.5×10^{21} meters. The integrated flux of neutrino reaching earth

$$= \frac{10^{58}}{4\pi \ (1.50 \times 10^{21})^2 \ \text{meter}^2} = 3.5 \times 10^{14} \ \text{m}^{-2}.$$

Exercise 7.17. In the pre-steller high density and high temperature plasma, complete thermodynamic equilibrium existed between v_e, $\overline{v}_e$, e^+, e^- and γ

(radiation). Show that the number ratio $\dfrac{n_{v_e} + n_{\overline{v}_e}}{n_\gamma} = \dfrac{n_{e^+} + n_{e^-}}{n_\gamma} = 1.5$.

We employ FD distribution for v_e, $\overline{v}_e$, e^+ and e^- and BE distribution for photons (radiation) and obtain

$$\frac{n_{\nu_e} + n_{\bar{\nu}_e}}{n_\gamma} = \frac{n_{e^+} + n_{e^-}}{n_\gamma} = 2 \cdot \frac{\int_0^\infty (e^x + 1)^{-1} x^2 dx}{\int_0^\infty (e^x - 1)^{-1} x^2 dx}.$$

Then (6.83c) gives $\dfrac{n_{\nu_e} + n_{\bar{\nu}_e}}{n_\gamma} = 2 \dfrac{[1 - 2^{(1-3)}] (3-1)! \ \xi(3)}{(3-1)! \ \xi(3)} = 1 \cdot 5.$

However, during the course of cooling to the present time these relations change, since the annihilation of the e^+ e^- increases the number of photons without changing the number of ν_e and $\bar{\nu}_e$.

REFERENCES

1. F. Reines and C.L. Cowan, Phys. Rev. **92**, 830 (1953).
2. R. Davis, Phys. Rev. **97**, 766 (1955).
3. Koshiba et al., Phys. Rev. Lett. **81**, 1774 (1998).
4. E. Fermi, Z. Phys. **88**, 161 (1934).
5. H. Behrens and J. Janecke, Numerical Tables for Beta Decay and Electron Capture. Ed. Landolt and Bornstein, New Series 1/4 Springer Berlin, 1969.
6. John A. Jaros, SLAC Publication # 5958, October (1992).
7. H. Yukawa, Proc. Phys. Math. Soc. (Japan) **17**, 48 (1935).
8. O. Klein, Proceedings of a Symposium held in Warsaw, page 6 (1938).
9. T.D. Lee and C.N. Yang, Phys. Rev. **104**, 254 (1956), Phys. Rev. 105, 1671 (1957).
10. C.S. Wu et al., Phys. Rev. **105**, 1413 (1957).
11. M. Goldhaber et al., Phys. Rev. **109**, 1015 (1958).
12. R. Feynman and M. Gell-Mann, Phys. Rev. **109**, 193 (1958).
13. E.C.G. Sudershan and R.E. Marshak, Phys. Rev. **109**, 1860 (1958).
14. Bob Clare: CERN Courier **39**, 15 (1999). Particle Data Group, Phys. Rev. **D66** (2002).
15. Martin Grunewald, CERN Courier **45**, 25 (2005).
16. N.L. Cabibbo, Phys. Rev. Lett. **10**, 531 (1963).
17. M. Gell-Mann and Y. Ne'eman in "The Eight Fold Way", N.Y., Benjamin, 1964.
18. M. Kobayashi and T. Masakawa, Prog. Theoret. Phys. (Japan) **49**, 652 (1973).
19. SNO Collaboration - S.N. Ahmed et al., Phys. Rev. Lett., **92**, 181301 (2004), ibid **87**, 071301 (2001).
20. SAGE Collaboration - J.N. Abdurashitov et al. Phys. Rev. **C60**, 055801 (1999).
21. GNO Collaboration - M. Altmann et al., Phys. Lett. **B449**, 137 (1999).
22. GALLEX Collaboration - W. Hampel et al., Phys. Lett. **B447**, 127 (1999).
23. Super - Kamiokande Collaboration - Y. Fukuda et al., Phys. Rev. Lett. **82**, 2430; **86**, 5651 (2001) **82**, 2644 (1999); **81**, 1562 (1998).
24. Kam LAND Collaboration - K. Eguchi et al., Phys. Rev. Lett. **90**m 021802 (2003), K2K. Collaboration hep-ex/0106008.

25. LSND Collaboration - c. Athanassoponlos et al. Phys. Rev. Lett. **81**, 1774 (1998); Phys. Rev. **C58**, 2489 (1998). For four neutrino scheme and allowed mass pattern see S. Pakvasa and P. Roy, Phys. Lett. **B535**, 181 (2002).

26. T. Araki et al., Nature **436**, 449 (2005) (KamLand Collaboration).

27. S.R. Elliot, P. Vogel, Ann. Rev. Nucl. Part. Sc. **52**, 481 (2002).

28. Double - Beta Decay and Related Topics Ed. H.V. Klapdor - Kliengrothans World Scientific, 1995.

29. C. Arnaboldi et al. Phys. Rev. Lett. **95**, 142501 (2005).

30. COBRA Collaboration: K. Zuber, Phys. Lett. **B519**, 1 (2001).

31. GEM Collaboration: Yu G. Zdesenko et al. J. Phys. G. **27**, 2129 (2001).

32. MOON Collaboration: H. Ejiri et al. Phys. Rev. Lett. **85**, 2917 (2000).

33. Z. Maki, M. Nakagawa and S. Sakata, Prog. Theoret. Phys. **28**, 870 (1960). J.N. Bahcall, Phys. Lett. **13**, 332 (1964).

34. L. Wolfenstein, Phys. Rev. **D17**, 2369 (1978).
    S.P. Mikheev and A. Yu Smirnov, Sov. J. Nucl. Phys. **42**, 913 (1985) Ibid, Nuvo Cimento **C9**, 17 (1986), Zh. Eksp. Teor. Fiz. **91**, 7 (1986), JETP **64**, 4 (1986).

35. Wayne et al. Phys. Rev. Lett. **80**, 5, 255 (1998).

36. D.N. Spregel et al. Astrophysics J. Suppl. **148**, 175 (2003).

37. http://astro.estic.esa/SA-general/Project/Plausk/.

38. S. Hannestead, Phys. Rev. **D67**, 085017 (2003).

39. W. Marciano in The Standard Model and Higgs Physics, ICTP, Trioste, Italy, SMR1508-11 (2003).

SUGGESTED BOOKS FOR FURTHER READING

40. Quarks and Leptons: An Introductory Course in Modern Particle Physics F. Halzen and A.D. Martin, John Wiley, New York (1984).

41. Neutrino Physics II Ed. K. Winter, Cambridge University Press, Cambridge, 2000. Massive Neutrinos and Physics and Astrophysics, R.N. Mohapatra, IInd Ed. World Scientific, Singapore (1998).

42. Cosmology and Particle Astrophysics, L. Burgstorm and A. Gobar, IInd Ed. John Wiley (2004).

43. Unity of Forces in the Universe, A. Zee, World Scientific (1982).

44. The Physics of the Standard Model and Beyond T. Morii, C.S. Lim and S.N. Mukherjee, World Scientific, Singapore (2004).

45. Neutrino Masses and Oscillations: Triumphs and Challenges, R.D. Mckewon and P. Vogel, Physics Reports, Elsevier **394**, 315-356 (2004) (References therein).

46. Particle Physics in the New Millennium Ed. J. Trampetic and J. Wess Springer (2001).

47. Explosion mechanism, neutrino burst and gravitational wave in core-collapse supernova K. Kotake, K. Sato and K. Takahasi Reports on Prog. Phys. **69**, #4 (2006).

48. Astrophysics in a nutshell Dan Maoz. Princeton University Press, Princeton and Oxford (2007).

49. Neutrino Oscillations, Ed. J.A. Thomas and I. Patricia (*see* theoretical introduction by Stephen Parke) World Scientific, Singapore (2008).

Appendices

A1. NOTATIONS AND SYMBOLS

A nucleus X having mass $\# A$, proton $\# Z$ and neutron $\# N$: $^A_Z X_N$

Centre of mass (c.m.), Laboratory (Lab.), Schrödinger equation (S. Eq.), ground state (g.s.)

$\in$ is an element of, $\supset$ properly includes, $\subset$ is a proper subset of, $\notin$ is not an element of

Symbols above letters		Example	Symbols above letters		Example
Vector quantity	" $\rightarrow$ "	$\vec{A}$	Anti particle	" ~ "	$\tilde{a}$
Time derivative	" . "	$\dot{A}$	Time reversed	" – "	$\overline{v}$
Space derivative		A'	Unit vector	" $\wedge$ "	$\hat{n}$

A2. UNITS AND PHYSICAL CONSTANTS

Sometime for the sake of convenience, we may use the following convention, $\hbar = c = 1$. It follows that $\hbar c = 197.4$ MeV fm $= 1$. In this convention mass and momentum have the dimension of energy and time and length have the dimension of inverse of energy.

$$1 \sec = \frac{1}{6.58 \times 10^{-22}} \text{ MeV}^{-1} = 1.52 \times 10^{24} \text{ GeV}^{-1}.$$

$$1 \text{ fm} = \frac{\text{MeV}^{-1}}{197.4} = 5.068 \text{ GeV}^{-1} \text{ and } e^2 = 1.44 \text{ MeV fm}.$$

Planck units for mass, length and time based on $\hbar$, c and G are:

$$M_{Pl} = \sqrt{\frac{\hbar c}{G}} \approx 10^{19} \text{ GeV}, \quad L_{Pl} = \sqrt{\frac{\hbar G}{c^3}} \approx 10^{-19} \text{ GeV}^{-1}$$

and $\quad T_{Pl} \approx 10^{-19} \text{ GeV}^{-1}$

Table A1.1: Physical constants

Quantity	Value
Velocity of light (c)	3×10^8 ms^{-1}
Placks constant/2π ($h/2\pi$)	6.58×10^{-22} MeV s
Gravitational constant (G)	10^{-38} GeV^{-2}
Boltzman constant (k)	0.861×10^{-10} MeV °K^{-1}
Fine structure constant ($e^2/\hbar c$)	1/137.04
Electronic charge (e^2)	1.44 MeV fm (fm = 10^{-15} m)
Nuclear magneton (μ_0)	0.126 (MeV)$^{1/2}$ fm$^{3/2}$
Proton (Neutron) gyromagnetic ratio: g_S^p (g_S^n)	5.86 (– 3.83)

A3. Four Vectors

The metric in the Minkowski space is defined in accordance to the Bjorken-Drell convention as follows:

The contravariant four-vector denoted by superscript is

$$x^\mu \equiv (x^0, x^1, x^2, x^3) = (t, x, y, z).$$

The corresponding covariant four vector denoted by subscript is

$$x_\mu = g_{\mu\nu} x^\nu,$$

where the summation over repeated index ν is understood. Only the diagonal elements of the matrix tensor $g_{\mu\nu}$ are non-zero:

$$g_{\mu\nu} = g^{\mu\nu} = (1, -1, -1, -1) \equiv \begin{pmatrix} 1 & 0 & 0 & 0 \\ 0 & -1 & 0 & 0 \\ 0 & 0 & -1 & 0 \\ 0 & 0 & 0 & -1 \end{pmatrix}.$$

This gives $x_\mu = (x_0, x_1, x_2, x_3) = (t, -x, -y, -z)$. Space time indices, all Greek letters (μ and ν) take values from 0 to 3. The scalar product of two four vectors is given by

$$a \cdot b = a^\mu b_\mu = g_{\mu\nu} a^\mu b^\nu = a_0 b_0 - \vec{a} \cdot \vec{b}.$$

The covariant four momentum

$$p_\mu = g_{\mu\nu} p^\nu = (p^0, p^1, p^2, p^3) = (E, \vec{p}).$$

The space and time differentiations are defined by

$$\partial_\mu = \frac{\partial}{\partial x^\mu} = \left(\frac{\partial}{\partial t}, \vec{\nabla}\right), \ \partial^\mu = \frac{\partial}{\partial x_\mu} = \left(\frac{\partial}{\partial t}, -\vec{\nabla}\right) \text{ and } \partial_\mu \partial^\mu = \frac{\partial^2}{\partial t^2} - \vec{\nabla} \cdot \vec{\nabla}$$

$$= \frac{\partial^2}{\partial t^2} - \nabla^2 = \square \text{ and } p^\mu = i\partial^\mu = \left(i\frac{\partial}{\partial t}, -i\vec{\nabla} \right); \; p^\mu p_\mu = -\partial^\mu \partial_\mu$$

$$= -\frac{\partial^2}{\partial t^2} + \nabla^2 = -\square.$$

where ∇^2 and $\square$ are called the Laplacian and d' Alembertian operators respectively.

We define the four current $j^\mu \equiv (\rho, \vec{j})$ and using the four gradient $\partial_\mu = \left(\frac{\partial}{\partial t}, \vec{\nabla} \right)$, write the equation of continuity as $\frac{\partial \rho}{\partial t} + \vec{\nabla} \cdot \vec{j} = 0$ or $\partial_\mu j^\mu = 0$, which is Lorentz invariant. Here $\partial_\mu j^\mu (= 0)$ is the four dimensional divergence of j^μ.

The four vector potential $A^\mu = (A^0, \vec{A})$. The auxiliary condition takes the form $\partial_\mu A^\mu = 0$, which is called the Lorentz condition.

A4. Laboratory and Centre-of-Mass Coordinate Systems

Suppose a particle of mass m_1 and velocity $\vec{v}_1$ in a laboratory system collides with a particle of mass m_2 at rest ($\vec{v}_2 = 0$). Let the particle velocities after collision in Lab be $\vec{v}_1'$ and $\vec{v}_2'$ (A). The velocity of centre of mass (c.m) in the Lab. (L) is given by

$$\vec{u} = \frac{m_1 \vec{v}_1 + m_2 \vec{v}_2}{m_1 + m_2} = \frac{m_1}{m_1 + m_2} \vec{v}_1.$$

Fig. A: Laboratory and centre of mass (c.m.) systems for the two-body collisions

The particle velocities before the collision in the c.m. system are:

$$\vec{v}_1 = \vec{v}_1 - \frac{m_1 \vec{v}_1}{m_1 + m_2} = \frac{m_2 \vec{v}_1}{m_1 + m_2} \quad \text{and} \quad \vec{v}_2 = \frac{m_1 \vec{v}_1}{m_1 + m_2}.$$

The relative velocity of the colliding particles $\vec{v}_1 - \vec{v}_2 = \vec{v}_1$. By virtue of energy conservation, c.m. particle velocities maintain the same absolute values after the collision but may change their directions, i.e. $|\vec{v}_1'| = |\vec{w}_1|$ and $|\vec{v}_2'| = |\vec{w}_2|$. Due to the momentum conservation, the final particle velocities have opposite directions. Let the angle between $\vec{v}_1'$ and $\vec{v}_1$ be the c.m. angle $\theta_{c.m.}$. In order to find a relation between the Lab angle θ_L and $\theta_{c.m.}$, we make the projection of the vector equation $\vec{v}_1' = \vec{v}_1' + \vec{u}$ parallel and perpendicular to $\vec{v}_1$:

$$v_1' \sin \theta_L = v_1' \sin \theta_{c.m.},$$

$$v_1' \cos \theta_L = v_1' \cos \theta_{c.m.} + u.$$

Hence,
$$\tan \theta_L = \frac{\sin \theta_{c.m.}}{\dfrac{|\vec{u}|}{|v_1|} + \cos \theta_{c.m.}} = \frac{\sin \theta_{c.m.}}{\gamma + \cos \theta_{c.m.}}.$$

where $\gamma = \dfrac{m_1}{m_2}$. For $m_1 = m_2$ and $\gamma = 1$ we have $\theta_L = \dfrac{1}{2} \theta_{c.m.}$.

By virtue of the Lorentz transformation of total cross section, we have

$$\sigma_L(\theta_L)\, d\Omega_L = \sigma_{c.m.}(\theta_{c.m.})\, d\Omega_{c.m.},$$

where $d\Omega_L = \sin \theta_L\, d\theta_L\, d\phi_L$ and $d\Omega_{c-m} = \sin \theta_{c.m.}\, d\theta_{c.m.}\, d\phi_{c.m.}$.

Therefore, $\sigma_L(\theta_L) \sin \theta_L\, d\theta_L = \sigma_{c.m.}(\theta_{c.m.}) \sin \theta_{c.m.}\, d\theta_{c.m.}$,

taking $\phi_L = \phi_{c.m.}$.

Now
$$\sec^2 \theta_L\, d\theta_L = \left[\frac{\sin^2 \theta_{c.m.}}{(\gamma + \cos \theta_{c.m.})^2} + \frac{\cos \theta_{c.m.}}{(\gamma + \cos \theta_{c.m.})} \right] d\theta_{c.m.},$$

so that
$$\sin \theta_L = \sin \theta_{c.m.} / (1 + 2\gamma \cos \theta_{c.m.} + \gamma^2)^{1/2}$$

and
$$d\theta_L = \frac{(1 + \gamma \cos \theta_{c.m.})}{(1 + 2\gamma \cos \theta_{c.m.} + \gamma^2)} d\theta_{c.m.}.$$

Therefore,
$$\sigma_L(\theta_L) = \frac{[1 + 2\gamma \cos \theta_{c.m.} + \gamma^2]^{3/2}}{(1 + \gamma \cos \theta_{c.m.})} \sigma_{c.m.}(\theta_{c.m.}).$$

Also
$$E_{cm} = \frac{1}{2} \frac{m_1 m_2}{(m_1 + m_2)} (\vec{v}_1 - \vec{v}_2)^2 = \frac{m_2}{m_1 + m_2} E_L.$$

Kinematics of High Energy Scattering

High energy strong interaction data now come from two types of accelerators: (*i*) conventional accelerators, in which a beam of hadrons, leptons, photons or heavy ions interacts with the stationary target and (*ii*) colliding beam accelerators in which two beams of particles of equal but opposite momentum interact. Usually in either type of accelerators we consider the total energy in the c.m. because this is the energy available for the production of additional particles. It has become conventional to describe total c.m. energy by its square usually denoted by s. If two particles a and b of four momentum p_a and p_b respectively collide then $s = (p_a + p_b)^2$. Thus, s does not change under the Lorentz transformation, i.e. s is a Lorentz scalar. In the c.m. system of the scattering process $s = (E_a + E_b, \ 0)^2 = E_{c.m.}^2$ or $\sqrt{s} \equiv E_{c.m.}$.

We now show that in a conventional accelerator, $\sqrt{s}$ increases as $\sqrt{E}$ but in a colliding beam accelerator in which each beam has energy E, $\sqrt{s}$ increases as $2E$.

For the conventional accelerators, let the beam particle denoted by 'a' has the rest mass m_a. For free particles we have (setting $c = 1$):

$$s = E_a^2 - \vec{p}_a^2 + E_b^2 - \vec{p}_b^2 + 2(\vec{p}_a \cdot \vec{p}_b)$$

$$= m_a^2 + m_b^2 + 2(E_a E_b - \vec{p}_a \cdot \vec{p}_b)$$

$$= m_a^2 + m_b^2 + 2E_a m_b, \text{ since } p_b = 0.$$

Since, $s \gg m_a^2$ and $s \gg m_b^2$ then $s \cong 2m_b E_a$ or $\sqrt{s} \propto \sqrt{E}$ (dropping a).
For the colliding beam,

$$s = (p_a + p_b)^2 = p_a^2 + p_b^2 + 2\vec{p}_a \cdot \vec{p}_b$$

$$= E_a^2 - \vec{p}_a^2 + E_b^2 - \vec{p}_b^2 + 2(E_a E_b - \vec{p}_a \cdot \vec{p}_b)$$

$$= E_a^2 + E_b^2 + 2E_a E_b - (\vec{p}_a + \vec{p}_b)^2$$

$$= 4E^2 \text{ for } E_a = E_b = E \text{ and } \vec{p}_a = -\vec{p}_b.$$

Thus, in the colliding beam accelerators $\sqrt{s} = 2E$ and high s can be achieved with beam energies much lower than that required for conventional accelerators.

In the c.m system, the four momentum transfer squared

$$q^2 = (p_a - p_a')^2 = (E_a - \vec{p}_a - E_a' + \vec{p}_a')^2$$

$$= (E_a^2 - \vec{p}_a^2) + (E_a'^2 - \vec{p}_a'^2) - 2E_a E_a' + 2\vec{p}_a \cdot \vec{p}_a'$$

$$= m_a^2 + m_a'^2 - 2E_a E_a' + 2|\vec{p}_a| \cdot |\vec{p}_a'| \cos\theta$$

$$= -2E_a E_a'(1 - \cos \theta) = -\frac{s}{2}(1 - \cos\theta) \quad \text{for } E_a = E_a' = E,$$

since at high energies masses are negligible and the cross term vanishes for $E_a = E_a'$.

The energy component E is said to be time like and 3-momentum $\vec{p}$ components space like. For elastic scattering in the c.m. system

$$q^2 = -\vec{q}^2 \text{ or } -q^2 = |\vec{q}|^2 = Q^2,$$

that is, momentum transfer squared is negative and in order to work with +ve quantity we use $Q^2 = -q^2 > 0$.

APPENDIX–B

B1. MATHEMATICAL PRELIMINARIES

We shall describe a method of handling the Time Independent Schrödinger Equation (TISE) for a system of two nucleons moving under the influence of a force derived from a potential between two nucleons. The TISE has the following form:

$$\left[-\frac{\hbar^2}{2m_1} \cdot \nabla_1^2 - \frac{\hbar^2}{2m_2} \nabla_2^2 + V(\vec{r}_1, \vec{r}_2) \right] \Psi = E\Psi, \tag{1}$$

where m_1 and m_2 are the masses of nucleon '1' and '2', and r_1 and r_2 their respective position coordinates. We define a set of new coordinates namely the relative and 'centre of mass' (c.m.) as follows:

$$\vec{r} = \vec{r}_1 - \vec{r}_2 \text{ and } \vec{R} = (\vec{r}_1 + \vec{r}_2)/2,$$

where $\vec{r}$ and $\vec{R}$ denote respectively the relative and c.m. coordinates. The S. Eq. (1) resolves into two equations; one for the relative motion and another for the centre of mass motion:

$$\frac{-\hbar^2}{2\mu} \nabla_r^2 \, \psi(\vec{r}) + V(\vec{r}) \, \psi(\vec{r}) = E_{\text{Re}l} \, \psi(\vec{r}), \tag{2a}$$

$$\frac{-\hbar^2}{2m^*} \nabla_R^2 \, \phi(\vec{R}) = E_{c \cdot m} \, \phi(\vec{R}) , \tag{2b}$$

where $\mu = (m_1 m_2)/(m_1 + m_2)$, $m^* = m_1 + m_2$ and $E_{c \cdot m} = E - E_{\text{Re}l}$.

In the above we have used the factorization $\Psi = \phi(\vec{R}) \psi(\vec{r})$. In the absence of external force, the c.m. moves as a free particle and we will not be concerned with this motion. Hence, we investigate only the relative motion. In the c.m. system, $E - E_{\text{Re}l} = 0$.

We assume a central interaction, i.e. $V(\vec{r}) = V(r)$ and use the spherical coordinates:

$$x = r\cos\theta \, \sin\phi, \quad y = r\sin\theta \, \sin\phi, \quad z = r\cos\theta \text{ and get}$$

$$\nabla^2 \equiv \frac{1}{r^2} \frac{\partial}{\partial r}\left(r^2 \frac{\partial}{\partial r} \right) + \frac{1}{r^2 \sin\theta} \frac{\partial}{\partial\theta}\left(\sin\theta \, \frac{\partial}{\partial\theta} \right) + \frac{1}{r^2\sin\theta} \frac{\partial^2}{\partial\varphi^2} .$$

Further, we can separate Eq. (2a) in the radial and angular variables as

$$\frac{1}{R(r)}\left[\frac{\partial}{\partial r}\left(r^2 \frac{\partial}{\partial r} \right) + \frac{2\mu r^2}{\hbar^2} \, [E - V(r)] \right] R(r)$$

$$= -\frac{1}{S(\theta,\phi)} \left[\frac{1}{\sin\theta} \frac{\partial}{\partial\theta} \left(\sin\theta \frac{\partial}{\partial\theta} \right) + \frac{1}{\sin^2\theta} \frac{\partial}{\partial\phi^2} \right] S(\theta,\phi)$$

by using the factorization $\psi(r) \equiv R(r)\, S(\theta, \phi)$.

Since the left hand side depends only r and the right hand side on θ and ϕ, we could equate each side to a separation constant λ, so that

$$\frac{1}{\sin\theta} \frac{\partial}{\partial\theta} \left(\sin\theta \frac{\partial S}{\partial\theta} \right) + \frac{1}{\sin^2\theta} \frac{\partial^2 S}{\partial\phi^2} + \lambda S = 0. \tag{3}$$

Again by writing $S(\theta, \phi) = W(\theta)\, Z(\phi)$ we get

$$\frac{\partial^2 Z(\phi)}{\partial\phi^2} + m^2 Z(\phi) = 0, \qquad \text{where } m^2 \text{ is a separation constant.} \tag{4a}$$

The solution of (4a) is $Z(\phi) = e^{im\phi} = \cos\, m\phi + i\sin\, m\phi$. The single valuedness of $Z(\phi)$ demands $Z(\phi) = Z(\phi + 2\pi)$.

The above condition requires $m = 0, \pm 1, \pm 2, \ldots$ The normalized $Z(\phi)$ is then

$$Z(\phi) = \frac{1}{\sqrt{2\pi}} e^{\pm im\phi}. \tag{4b}$$

The q-equation is:

$$\sin\theta \frac{\partial}{\partial\theta} \left(\sin\theta \frac{\partial W(\theta)}{\partial\theta} \right) + \lambda\, \sin^2\theta\, W(\theta) = m^2 W(\theta). \tag{5a}$$

On substitution $\cos\theta = z, \lambda = \ell(\ell+1)$ and $W(\theta) \equiv P_\ell^m(z)$; (5a) reduces to

$$(1-z^2)\frac{\partial^2}{\partial z^2} P_\ell^m(z) - 2z\frac{\partial}{\partial z} P_\ell^m(z) + \left[\ell(\ell+1) - \frac{m^2}{1-z^2} \right] P_\ell^m(z) = 0, \tag{5b}$$

which is an eigen value equation. Solutions of (5b) that are compatible with the boundary conditions, exist only when ℓ is a non-negative whole number, i.e. $\ell = 0,1,2\ldots$ and $|m| < \ell$. For $m = 0$, the solution functions are called the Legendre polynomials, and for $m \neq 0$, they are called the associated Legendre polynomials. The Legendre polynomials have the property

$$\int_{-1}^{+1} P_\ell^m(z)\, P_{\ell'}^m(z)\, dz = \begin{cases} 0 & \text{for } \ell' \neq \ell \\[2mm] \dfrac{2(\ell+|m|)!}{(2\ell+1)\,(\ell-|m|)!} & \text{for } \ell' = \ell \end{cases}$$

The solution of (5b) then becomes $Y_\ell^m(\theta, \phi) = N_{\ell,m} P_\ell^m(\cos\theta)e^{im\phi}$,

where $N_{\ell,m} = (-)^m \sqrt{\dfrac{2\ell+1}{4\pi}\dfrac{(\ell-|m|)!}{(\ell+|m|)!}}.$ (6)

The functions $Y_\ell^m(\theta, \phi)$ are called the spherical harmonics:

$$Y_0^0 = \frac{1}{\sqrt{4\pi}} \qquad Y_1^{\pm1} = \pm\sqrt{\frac{3}{8\pi}}\sin\theta\, e^{\pm i\phi} \qquad Y_2^0 = \sqrt{\frac{5}{4\pi}}\left(\frac{3}{2}\cos^2\theta - 1\right)$$

$$Y_2^{\pm2} = \sqrt{\frac{15}{32\pi}}\sin^2\theta\, e^{\pm 2i\phi} \qquad\qquad Y_2^{\pm1} = \mp\sqrt{\frac{15}{8\pi}}\sin\theta\cos\phi\, e^{\pm i\phi}$$

Let us now return to the radial equation

$$\frac{1}{R}\left\{r^2\frac{d^2R}{dr^2} + 2r\frac{dR}{dr}\right\} + \frac{2\mu r^2}{\hbar^2}\{E - V(r)\}R = \ell(\ell+1).$$

or $\quad \dfrac{d^2R}{dr^2} + \dfrac{2}{r}\dfrac{dR(r)}{dr} + \dfrac{2\mu}{\hbar^2}\{E - V(r)\}R = \dfrac{\ell(\ell+1)}{r^2}R.$

The field free equation for $E > 0$ is then

$$\frac{d^2R(r)}{dr^2} + \frac{2}{r}\frac{dR(r)}{dr} + \left\{k^2 - \frac{\ell(\ell+1)}{r^2}\right\}R(r) = 0, \text{ where } k^2 = \frac{2\mu E}{\hbar^2}. \quad (7a)$$

Eliminate the first order differential by substitution $R(r) = \dfrac{u(r)}{r}$:

$$\frac{d^2u(r)}{dr^2} - \frac{\ell(\ell+1)}{r^2}u(r) + k^2u(r) = 0.$$

We substitute $u(r) = r^{1/2}v(r)$ in the above equation and get

$$r^2\frac{d^2v}{dr^2} + r\frac{dv}{dr} + \left[k^2r^2 - \left(\ell+\frac{1}{2}\right)^2\right]v(r) = 0, \quad (7b)$$

which should be treated separately for $k^2 > 0$ and $k^2 < 0$.

For $k^2 > 0$, i.e. positive E, Eq. (7b) on substitution $kr = x$ can be brought to the form:

$$x^2\frac{d^2v(x)}{dx^2} + x\frac{dv(x)}{dx} + \left[x^2 - \left(\ell+\frac{1}{2}\right)^2\right]v(x) = 0,$$

which is a standard Bessel equation.

The solution of (7b) is then the Bessel functions of half integral order:

$$v(r) = A J_{\ell+\frac{1}{2}}(kr) + B J_{-\left(\ell+\frac{1}{2}\right)}(kr),$$

where both $J_{\ell+\frac{1}{2}}(kr)$ and $J_{-\left(\ell+\frac{1}{2}\right)}(kr)$ are given in terms of certain power series expansion of (kr) for any arbitrary order (*see* CRC Standard Mathematical Tables, Ed. Robert C. Weast et. al. The Chemical Rubber Co. Ohio).

It is useful to define the spherical Bessel functions as

$$j_\ell(kr) = \sqrt{\frac{\pi}{2kr}}\, J_{\ell+\frac{1}{2}}(kr) \text{ and the spherical Neumann functions as}$$

$$\eta_\ell(kr) = (-)^{\ell+1} \sqrt{\frac{\pi}{2kr}}\, J_{-\left(\ell+\frac{1}{2}\right)}(kr).$$

The advantage of using the spherical Bessel functions and the Neumann functions is that both can be expressed in terms of trigonometric functions:

$$j_0(kr) = \frac{\sin kr}{kr} \qquad\qquad j_1(kr) = \frac{\sin kr}{(kr)^2} - \frac{\cos kr}{kr}$$

$$j_2(kr) = \left[\frac{3}{(kr)^3} - \frac{1}{kr}\right] \sin kr - \frac{3 \cos kr}{(kr)^3} \text{ etc.}$$

The general recursion formula is

$$j_{\ell+1}(kr) = \frac{2\ell+1}{kr} j_\ell(kr) - j_{-\ell-1}(kr). \tag{8}$$

Further, $\quad j'_\ell(kr) = j_{\ell-1}(kr) - \dfrac{\ell+1}{kr} j_{\ell-1}(kr). \tag{9}$

By analogous method we get:

$$\eta_0 = -\frac{\cos kr}{kr}, \; \eta_1 = -\frac{\cos kr}{(kr)^2} - \frac{\sin kr}{kr}$$

$$\eta_2 = -\left[\frac{3}{(kr)^3} - \frac{1}{kr}\right] \cos kr - \frac{3 \sin kr}{(kr)^2} \text{ etc.}$$

The functions $j_\ell(kr)$ and $\eta_\ell(kr)$ have the following asymptotic expansions $(kr \to \infty)$:

$$j_\ell(kr) \to \frac{1}{kr} \cos(kr - \ell\pi/2) \text{ and } \eta_\ell(kr) \to \frac{1}{kr} \sin(kr - \ell\pi/2).$$

Finally, the radial function $u(r)$ can be written as

$$u(r) = r\sqrt{\frac{2k}{\pi}} \left\{ A\, j_\ell(kr) + B(-)^{\ell+1} \eta_\ell(kr) \right\}.$$

The S-wave radial wave function asymptotically then behaves as

$$u(r) \xrightarrow[r\to\infty]{} \sqrt{\frac{2}{\pi k}} [A \sin kr + B \cos kr].$$

B2. Some Results of Scattering Theory

In a typical scattering experiment a well collimated beam emerging from an accelerator is allowed to strike a target, kept in a scattering chamber, and after scattering particles are detected by a suitable detector at a great distance from the scatterer. The detector subtends a cone of solid angle $d\Omega$ at the scattering centre and particles scattered into this cone are counted [B]. If N_i is the number of particles incident from left per unit area per second and $N\text{-}_{sc}$ is the number of these scattered per unit solid angle per scatterer, then the differential

scattering cross section $\dfrac{d\sigma}{d\Omega} = \dfrac{N_{SC}}{N_i}$.

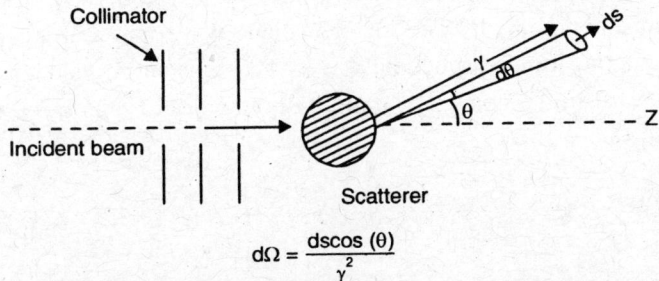

$$d\Omega = \frac{ds\cos(\theta)}{\gamma^2}$$

Fig. B: Schematic diagram for scattering experiment

This is the quantity the experimentalist gives and theoretician interprets it in terms of probability calculated from the wave function. Once the beam has left the source, its propagation is guided by the laws of quantum mechanics and experimentalist has no control over the particles unless they reach his detector. Let us first consider the scattering of spinless particles neglecting the Coulomb force.

When a nucleon is scattered by another nucleon the energy of the relative motion of the particles is positive. The wave function $\psi(\vec{r})$ describing the scattering is given by the solution of Eq. (2a) in Appendix B1 for $M/2 = \mu$ and positive E. In this case, the wave function $\psi(\vec{r})$ at a large distance must have the form of a sum of incident plane wave and scattered outgoing spherical waves: $\psi(\vec{r}) = \psi_{inc} + \psi_{sc}$. When the potential is spherically symmetric the angular momentum of the scattered particle is a constant of motion. Then we develop solutions of TISE interms of angular momentum eigen functions. The field free equation is

$$(\nabla^2 + k^2)\psi_{inc} = 0. \tag{10}$$

The corresponding radial equation can be transformed to a standard Bessel equation so that the solution $R(r) = \dfrac{u(r)}{r}$ can be written in terms of a spherical Bessel functions. A plane wave, which means physically, a particle beam parallel to the Z-axis, has no angular momentum component along the Z-axis. Hence, the solution of the angular equation corresponding to the field free TISE is $P_\ell(\cos\theta)$. Thus, any bounded solution of Eq. (10) having azimuthal symmetry can be written as a linear combination of $j_\ell(kr)$ and $P_\ell(\cos\theta)$. Therefore,

$$e^{ikz} = e^{ikr\cos\theta} = \sum_{\ell=0}^{\infty} A_\ell \, j_\ell(kr) P_\ell(\cos\theta), \tag{11}$$

where A_ℓ are arbitrary constants, determined by the orthogonal properties of the Legendre polynomials. The incident plane wave may be considered as coherent superposition of spherical waves. The constant A_ℓ is determined as follows: We substitute $\cos\theta = t$, multiply both sides of (11) by $P_\ell(t)$, then integrate so that

$$\int_{-1}^{+1} e^{ikrt} P_\ell(t)\, dt = \sum_{\ell'=0}^{\infty} \int_{-1}^{+1} A_{\ell'} \, j_{\ell'}(kr)\, P_{\ell'}(t)\, P_\ell(t)\, dt = A_\ell \, j_\ell(kr) \frac{2}{2\ell+1}.$$

By repeated integration of LHS, asymptotically we get,

$$\int_{-1}^{+1} e^{ikrt} P_\ell(t)\, dt \to \frac{1}{ikr}\left[e^{ikr} - (-)^\ell e^{-ikr} \right] = \frac{2i^\ell}{kr} \sin(kr - \ell\pi/2),$$

where we have used $P_\ell(\pm 1) = (\pm)^\ell$ and $(-1)^\ell = e^{i\ell\pi}$. Asymptotically

$$j_\ell(kr) \to \frac{1}{kr}\sin(kr - \ell\pi/2), \text{ hence } A_\ell = i^\ell(2\ell+1).$$

If $P_\ell(\cos\theta)$ in (11) is replaced by $\sqrt{\dfrac{4\pi}{2\ell+1}}\, Y_\ell^0$ then $A_\ell = i^\ell \sqrt{4\pi(2\ell+1)}$.

Finally, $\Psi_{inc} = e^{ikz} = \sum_{\ell=0}^{\infty} (2\ell+1)(1/2ikr)\left[e^{ikr} - (-)^{\ell} e^{-ikr}\right] P_{\ell}(\cos\theta).$

The first term inside bracket represents an outgoing spherical wave, because the time dependent wave function $\sim e^{i(kr-\omega t)}$ keeps a given value if r increases. The second term represents an ingoing wave shrinking towards the origin. Ψ_{inc} is normalized to have one particle per unit volume. If the plane wave propagate not in the Z-direction but in any other direction $\hat{k}$, then

$$e^{i\vec{k}\cdot\vec{r}} = 4\pi \sum_{\ell=0}^{\infty} \sum_{m=-\ell}^{+\ell} i^{\ell} j_{\ell}(kr) Y_{\ell}^{m*}(\hat{k}) Y_{\ell}^{m}(\hat{r}).$$

The incident flux along the Z-direction

$$N_{inc} = j_{inc} = -\frac{i\hbar}{2\mu}\left[\Psi_{inc}^{*}\frac{\partial}{\partial z}\Psi_{inc} - \Psi_{inc}\frac{\partial}{\partial z}\Psi_{inc}^{*}\right]\frac{k\hbar}{\mu} = v,$$

where v is the velocity of the incident particle with respect to the target.

In the presence of the target, the incident particle sense a potential and for a central field we must solve S. Eq. for a known potential. It is likely that the solution of this equation will be similar to the solution of the field free equation (10) but it will only differ in phase because the particle sensed the potential before moving to the far out distance. Therefore, in nuclear reaction or scattering, the amplitude of the outgoing spherical part of the plane wave is modified. The wave function $\psi(\vec{r})$ of the complete system asymptotically is

$$\psi(\vec{r}) = \sum_{\ell=0}^{\infty} (2\ell+1)\frac{1}{2ikr}\left[\eta_{\ell}\, e^{ikr} - (-)^{\ell} e^{-ikr}\right] P_{\ell}(\cos\theta), \qquad (12)$$

where, η_{ℓ} is a complex amplitude for the ℓth partial wave and is related to the phase shift δ_{ℓ} as: $\eta_{\ell} = |\eta_{\ell}|\, e^{2i\delta_{\ell}}$. If $|\eta_{\ell}| = 1$, there is no change in the number of particles in the channel 'ℓ' and only elastic scattering will take place. If, however, $|\eta_{\ell}| < 1$ then both elastic scattering and nuclear reaction will take place. In the case of charged particle the exponential term must be replaced by an appropriate Coulomb wave function. When the spin degree of freedom is included, the factor $(2\ell+1)$ must be replaced by a weighted sum over magnetic quantum number. The scattering amplitude might depend upon the spin states of the two nucleons and their charges. Hence it is likely to be a matrix in spin and isospin space. The scattered wave function $\psi_{sc} = \psi(\vec{r}) - \Psi_{inc}$. Therefore,

$$\Psi_{sc} = \sum_{\ell=0} \frac{(2\ell+1)}{-2ik}(1-\eta_{\ell}) P_{\ell}(\cos\theta)\frac{e^{ikr}}{r} = f(\theta)\frac{e^{ikr}}{r},$$

where
$$f(\theta) = \sum_{\ell=0}^{\infty} \frac{(2\ell+1)}{-2ik} (1 - \eta_\ell) P_\ell(\cos\theta).$$

$$= \sum_{\ell=0}^{\infty} \frac{(2\ell+1)}{k} e^{i\delta_\ell} \sin\delta_\ell \, P_\ell(\cos\theta). \tag{13}$$

The scattered wave has only outgoing component. The phase shifts δ_ℓ describe the phase difference between the scattered and unscattered waves. They contain the information about the shape and strength of the potential and the energy dependence of the cross section. The fact that δ_ℓ appear not only as a phase factor but also in the amplitude (13) follows from the conservation of particle current in elastic scattering. The advantage of using phase shifts is that they are quite closely related to the dynamics of *N-N* interactions as represented by the potential $V(r)$. The number of scattered particles which cross an element of area ds perpendicular to the radius vector in unit time, is $n_{sc} = j_{sc}(ds)$, where

$$j_{sc} = -\frac{i\hbar}{2\mu} \left[\psi_{sc}^+ \frac{\partial \psi_{sc}}{\partial r} - \psi_{sc} \frac{\partial \psi_{sc}^*}{\partial r} \right] = \frac{k\hbar}{\mu r^2} \, |f(\theta)|^2 = v \, |f(\theta)|^2 / r^2.$$

Hence, $n_{sc} = \left[v \, |f(\theta)|^2 \right] \dfrac{ds}{r^2} = v \, |f(\theta)|^2 \, d\Omega$. The number of scattered particles per unit solid angle $N_{sc} = v \, |f(\theta)|^2$ so that $\dfrac{d\sigma}{d\Omega} = |f(\theta)|^2$.

The total scattering cross section

$$\sigma_{sc} = \int |f(\theta)|^2 \, d\Omega$$

$$= \int (1/4k^2 \sum_{\ell'} (2\ell+1)(2\ell'+1)(1-\eta_\ell)^*(1-\eta_{\ell'}) \, P_\ell(\cos\theta) P_{\ell'}(\cos\theta) \, d\Omega$$

$$= \frac{\pi}{k^2} \sum_{\ell=0}^{\infty} (2\ell+1)|1-\eta_\ell|^2 = \pi\lambda^2 \sum_{\ell=0}^{\infty} (2\ell+1)|1-\eta_\ell|^2 = \sum_{\ell=0}^{\infty} \sigma_{sc}^\ell,$$

where $\sigma_{sc}^\ell = \pi\lambda^2(2\ell+1) \, |1-\eta_\ell|^2$

and $\displaystyle\int P_\ell(\cos\theta) P_{\ell'}(\cos\theta) \, d\Omega = \frac{4\pi}{(2\ell+1)} \delta_{\ell\ell'}.$

Finally, $\sigma_{sc} = \dfrac{4\pi}{k^2} \displaystyle\sum_{\ell=0}^{\infty} (2\ell+1) \sin^2\delta_\ell. \tag{14}$

Scattering of Particles with Spin

A spin dependent interaction causes the polarization of the scattered particle along direction normal to the scattered plane even for an unpolarized projectile with spin. It may happen that spin flip occurs, namely the projection of the projectile spin along Z-axis changes from up to down or vice versa. When a particle is scattered elastically only two things change in the c.m., its direction and its spin state. The asymptotic wave function for spin 1/2 particle is

$$\phi(\vec{r},\vec{\sigma}) \rightarrow \chi_{1/2}^{\mu} \, e^{ikz} + \sum_{\nu} \chi_{1/2}^{\nu} \, f_{\nu\mu}(\theta,\varphi) \frac{e^{ikr}}{r}, \tag{15}$$

where $f_{\nu\mu}(\theta,\varphi)$ is the scattering amplitude for scattering from spin state $\frac{1}{2}\mu$ to $\frac{1}{2}\nu$.

The partial cross section, $\dfrac{d\sigma_{\nu\mu}}{d\Omega} = |f_{\nu\mu}(\theta,\varphi)|^2$. If the final spin state is not determined and incident beam with spin degeneracy g has no preferred direction then

$$\frac{d\sigma}{d\Omega} = \frac{1}{g} \sum_{\nu\mu} |f_{\nu\mu}(\theta,\varphi)|^2$$

We assume that the interaction involved in scattering has a spin-orbit term. We further assume that the incident particle is uncharged so that one needs not include the Coulomb potential. The interaction conserves the total angular momentum $\vec{j}(=\vec{\ell}+\vec{s})$ and also the parity. $\phi(\vec{r},\vec{\sigma})$ satisfies a S. Eq. with corresponding spin-orbit term and the radial wave function $u_{n\ell j}(j=\ell \pm 1/2)$, satisfies radial equation (3.14b).

Now, $\quad e^{ikz} \chi_{1/2}^{\mu} = \sum\limits_{\ell=0} \sqrt{4\pi(2\ell+1)} \, j_{\ell}(kr) \, Y_{\ell}^{0} \chi_{1/2}^{\mu}$

$$= \sum_{j} \sum_{\ell=|j-\frac{1}{2}|} i^{\ell} \sqrt{4\pi(2\ell+1)} \, \langle \ell \, \tfrac{1}{2} \, 0 \, \mu | m_j \rangle \, j_{\ell}(kr) \, |jm_j\rangle.$$

The scattered wave corresponding to this incident wave is obviously obtained by multiplying by an appropriate factor. Asymptotically $u_{n\ell j}(kr)$ behaves as

$$u_{n\ell j}(kr) \rightarrow \frac{1}{2ik} \left[i^{-\ell} e^{2i\delta_{\ell j}} \frac{e^{ikr}}{r} - i^{\ell} \frac{e^{-ikr}}{r} \right].$$

Equating the coefficients of $\dfrac{e^{ikr}}{r}$ in the asymptotic expansion of

$(\phi\,(\vec{r},\vec{\sigma}) - e^{ikz}\,\chi_{1/2}^{\mu})$, we get,

$$f_{\nu\mu}(\theta,\,\phi)\,\chi_{1/2}^{\nu} = \frac{1}{2ik}\sum_{j\ell m} e^{2i\delta_{\ell j}}\,[4\pi(2\ell+1)]^{1/2}\,\langle\ell\,\tfrac{1}{2}\,0\,\mu|jm_j\rangle$$

$$\times\langle\ell\,\tfrac{1}{2}\,m\,\nu|jm_j\rangle\left\{\frac{(2\ell+1)\,(\ell-m)!}{4\pi(\ell+m)!}\right\}^{1/2}(-)^m\,e^{im\varphi}\,P_{\ell}^{m}(\cos\theta)\,\chi_{1/2}^{\nu}$$

$$-\sum_{\ell}(2\ell+1)\,P_{\ell}^{0}(\cos\theta)\,\chi_{1/2}^{\mu}. \tag{16}$$

Let us consider here the scattering of spin $\dfrac{1}{2}$ particle with spin zero target. For example, a polarized neutron beam scattered on a ^{12}C target (spinless). Therefore, $\mu = \pm\dfrac{1}{2}$, $\nu = \pm\dfrac{1}{2}$, $g = 2$ and $j = \ell \pm\dfrac{1}{2}$.

For $\mu = \nu = \dfrac{1}{2}$, $m = 0$, $m_j = \dfrac{1}{2}$, we get

$$f_{1/2\,1/2}(\theta,\,\varphi) = \frac{1}{2ik}\left[\sum_{\ell}(2\ell+1)\,\langle\ell\,\tfrac{1}{2}\,0\,\tfrac{1}{2}\,|\,\ell+\tfrac{1}{2}\,\tfrac{1}{2}\rangle^2\,e^{2i\delta_+}\right.$$

$$\left.+\,(2\ell+1)\,\langle\ell\,\tfrac{1}{2}\,0\,\tfrac{1}{2}\,|\,\ell-\tfrac{1}{2}\,\tfrac{1}{2}\rangle^2\,e^{2i\delta_-} - (2\ell+1)\right]P_{\ell}^{0}(\cos\theta),$$

where $\delta_{\pm}$ are phase shifts $\delta_{\ell j}$ for $j = \ell \pm 1/2$. Substituting the values of C-coefficients $\langle\ell\,\tfrac{1}{2}\,0\,\tfrac{1}{2}\,|\,\ell+\tfrac{1}{2}\,\tfrac{1}{2}\rangle = \left\{\dfrac{\ell+1}{(2\ell+1)}\right\}^{1/2}$ and

$\langle\ell\,\tfrac{1}{2}\,0\,\tfrac{1}{2}\,|\,\ell-\tfrac{1}{2}\,\tfrac{1}{2}\rangle = \left\{\dfrac{\ell+1}{(2\ell+1)}\right\}^{1/2}$, we get

$$f_{1/2\,1/2}(\theta,\,\varphi) = \frac{1}{2ik}\left[\sum_{\ell}\left\{(\ell+1)\,e^{2i\delta_+} + \ell\,e^{2i\delta_-} - (2\ell+1)\right\}\right]$$

$$P_{\ell}^{0}(\cos\theta) = g(\theta),$$

which is the non-spin flip amplitude. We also have $f_{1/2\ 1/2} = f_{-1/2-1/2}$. Again for

$$\mu = \frac{1}{2}, \nu = -\frac{1}{2}, m = 1, m_j = \frac{1}{2}, j = \ell \pm \frac{1}{2}, \text{ we have}$$

$$f_{1/2\ -1/2}(\theta,\varphi) = \frac{1}{2ik}\left[\sum_{\ell}\left\{(2\ell+1)\ \langle\ell\ \frac{1}{2}\ 0\ \frac{1}{2}\ |\ \ell+\frac{1}{2}\ \frac{1}{2}\rangle e^{2i\delta_+}\right.\right.$$

$$\left.+ (2\ell+1)\ \langle\ell\ \frac{1}{2}\ 0\ \frac{1}{2}\ |\ \ell-\frac{1}{2}\ \frac{1}{2}\rangle\ \langle\ell\ \frac{1}{2}\ 1-\frac{1}{2}\ |\ \ell-\frac{1}{2}\ \frac{1}{2}\rangle\ e^{2i\delta_-}\right\}$$

$$\left.\times\left[\frac{(\ell-1)!}{(\ell+1)!}\right]^{1/2}\ (-)\ e^{i\varphi}\ P_\ell^1(\cos\theta)\right].$$

Substituting the values of *C*-coefficients

$$\langle\ell\ \frac{1}{2}\ 1-\frac{1}{2}\ |\ \ell+\frac{1}{2}\ \frac{1}{2}\rangle = \left(\frac{\ell}{2\ell+1}\right)^{1/2} \text{ and}$$

$$\langle\ell\ \frac{1}{2}\ 1-\frac{1}{2}\ |\ \ell-\frac{1}{2}\ \frac{1}{2}\rangle = \left\{\frac{\ell+1}{2\ell+1}\right\}^{1/2},$$

we get

$$f_{1/2\ -1/2}(\theta,\varphi) = \frac{i}{2k}\left[\sum_{\ell}\left\{e^{2i\delta_+} - e^{2i\delta_-}\right\}\right]P_\ell^1(\cos\theta)\ e^{i\varphi} = ih(\theta)\ e^{i\varphi}, \quad (17)$$

where $h(\theta) = \frac{1}{2k}\sum_{\ell}\left\{e^{2i\delta_+} - e^{2i\delta_-}\right\}P_\ell^1(\cos\theta)$ is the spin flip amplitude.

We have also $f_{1/2\ -1/2} = f_{-1/2\ 1/2}$. When the projectile is charged

$$g(\theta) = f_c(\theta) + \frac{1}{2ik}\left[\sum_{\ell}\left\{(\ell+1)\ e^{2i\delta_+} + \ell\ e^{2i\delta_-}\right\} - (2\ell+1)\ e^{2i\sigma_\ell}\right]P_\ell^0(\cos\theta),$$

$$(18)$$

$$h(\theta) = \frac{i}{2k}\sum_{\ell}\left[\left\{e^{2i\delta_+} - e^{2i\delta_-}\right\}e^{2i\sigma_\ell}\right]P_\ell^1(\cos\theta), \quad (19)$$

where σ_c is the Coulomb phase shift. If there are no spin dependent term in the potential $\delta_+ = \delta_-$, then the formalism of spin $\frac{1}{2}$ projectiles gives the same results as spin less projectiles. Finally,

$$\frac{d\sigma}{d\Omega} = |g(\theta)|^2 + |h(\theta)|^2 \tag{20}$$

Scattering Matrix: Some time it is more convenient to express scattering and reaction cross sections in terms of scattering matrix. If the amplitude of incoming wave in the channel α is A_α then amplitude of the outgoing wave in the channel β,

$$B_\beta = \sum_\alpha S_{\alpha\beta} A_\alpha,$$

where $S_{\alpha\beta}$ is the scattering matrix. Then the total wave function for $\ell = 0$ can be written as

$$\psi_0(r) = \frac{1}{2ikr}\left[\sum_\beta S_{\alpha\beta} e^{ik_\beta r} - e^{-ik_\alpha r}\right], \tag{21}$$

where the summation runs over all energetically possible reaction channels β for an ingoing channel α. The element of the scattering matrix $S_{\alpha\beta}$, thus, is the amplitude of the outgoing wave in β when there is an incoming wave of unit amplitude in α.

Conservation of flux: It requires that the outgoing intensity summed over all channels is equal to the incoming intensity summed over all channels:

$$\sum_\beta |B_\beta|^2 = \sum_\alpha |A_\alpha|^2$$

Now $\displaystyle\sum_\beta B_\beta^* B_\beta = \sum_{\alpha\beta} B_\beta^* S_{\alpha\beta} A_\alpha = \sum_{\alpha\beta\gamma} S_{\gamma\beta}^* A_\gamma^* S_{\alpha\beta} A_\alpha = \sum_{\alpha\gamma}\sum_\beta S_{\gamma\beta}^* S_{\alpha\beta} A_\gamma^* A_\alpha.$

Therefore, $\displaystyle\sum_\beta S_{\gamma\beta}^* S_{\alpha\beta} = \delta_{\alpha\gamma}$. The conservation flux implies the unitarity of the scattering matrix: $\displaystyle\sum_\beta S_\beta^* S_\beta = 1$. It follows from (14) that:

$\sigma_{\alpha\alpha} = \sigma_{sc} = \pi\,\lambda_\alpha^2\,|1 - S_{\alpha\alpha}|^2$, where the scattering matrix element $S_{\alpha\alpha} = \eta_\ell$ and $\sigma_R = \pi\lambda_\alpha^2(1 - |S_{\alpha\alpha}|^2)$ is the reaction cross section for the non-elastic channels.

Again, according (13) the scattering amplitude in the forward direction $(\theta = 0)$,

$$f(0) = \sum_{\ell=0}^\infty \frac{(2\ell+1)}{k} e^{i\delta_\ell}\frac{(e^{i\delta_\ell} - e^{-i\delta_\ell})}{2i}$$

$$= \sum_{\ell=0}^\infty \frac{(2\ell+1)}{k}(\cos\delta_\ell + i\sin\delta_\ell)\sin\delta_\ell.$$

Hence,
$$\operatorname{Im} f(0) = \sum_{\ell=0}^{\infty} \frac{(2\ell+1)}{k} \sin^2 \delta_\ell = \frac{k\sigma_{sc}}{4\pi}.$$

or
$$\sigma_{sc} = \frac{4\pi}{k} \operatorname{Im} f(0), \tag{22}$$

a relation called the "optical theorem". It states that the total (here elastic) cross section is related to the amount of particles removed from the original beam. Indeed σ_{sc} can be deduced from the measurement of transmission loss through a target in the forward direction ($\theta = 0$).

If time reversal invariance is satisfied, as in the case of strong interactions, then S is also symmetric, i.e. $S_{\alpha\beta} = S_{-\beta-\alpha}$, where the negative signs indicate that all spins in the channel concerned are reversed. The above relation is called the "reciprocity theorem", which demands that the direct and inverse transition probabilities referred to the same final states are equal, i.e. $S_{\beta\alpha} = S_{\alpha\beta}$, where $\alpha = 1...N$ and $\beta = 1...N$; N being the number of channels.

The unitarity (conservation of probability) demands $\sum_{\beta} S_{\alpha\beta} S_{\alpha\beta}^{\dagger} = 1$.

Lippmann-Schwinger Equation: The formal quantum scattering theory has been developed both in terms of the Schrödinger differential equation in the position space and the Lippmann-Schwinger (LS) integral equation in the momentum space. LS equation for state vector/transition matrices provides the basis for deriving the Born series in wave mechanics, and reaction amplitudes in the rearrangement collisions. We consider here two-body collision and its method of solution. The time independent S. Eq. for potential scattering is

$$(E - H_0) \, \psi(k,r) = V(r) \, \psi(k,r), \tag{23}$$

where $H_0 \equiv$ kinetic energy operator and $V(r)$ is interaction potential (operator).
The plane wave solution satisfies

$$(E - H_0)\phi(k,r) = 0 \text{ also } (E - H_0) \, \phi(k',r) = (E - E') \, \phi(k',r).$$

The orthogonality relations are given by

$$\int \phi^*(k',r) \, \phi^*(k,r) \, dr = (2\pi)^3 \, \delta(k-k')$$

$$\int \phi^*(k,r') \, \phi(k,r) \, dk = (2\pi)^3 \, \delta(r-r'). \tag{24}$$

We expand the scattering solution interms of plane wave states:

$$\psi(k,r') = \int a(k'') \, \phi \, (k'',r') \, dk'' \tag{25}$$

Substituting (25) in (23), we get

$$\int (E - H_0) a(k'') \, \phi(k'', r') \, dk'' = V(r') \, \psi(k, r').$$

We then multiply on the left by $\phi^*(k', r')$ integrate and use (24) to get:

$$\int \phi^*(k', r') \, (E - H_0) \, a(k'') \, \phi(k'', r') \, dk'' dr' = \int \phi^*(k', r') \, V(r') \psi(k, r') \, dr'$$

$$(2\pi)^3 \int (E - E') \, a(k'') \delta(k' - k'') \, dk'' = \int \phi^*(k', r') \, V(r') \, \psi(k, r') \, dr'. \quad (26)$$

Employing the properties of the delta function we get

$$(2\pi)^3 \, (E - E') \, a(k') = \int \phi^*(k', r') \, V(r') \, \psi(k, r') \, dr'$$

Substituting the above relation back into (26) we get

$$\psi(k, r) = \iint \frac{1}{(2\pi)^3 \, (E - E')} \, \phi^*(k', r') \, V(r') \phi(k', r') \times \psi(k, r') \, dr' dk'$$

$$= \int G_0^k(r, r') V(r') \psi(k, r') \, dr'$$

where $\quad G_0^k(r, r') = (2\pi)^{-3} \, (E - E')^{-1} \int \phi^*(k', r') \, \phi(k', r) \, dk' \quad$ (27)

The general solution $\quad \psi(k, r) = \phi(k, r) + \int G_0^{k'}(r, r') V(r') \, \psi(k, r') \, dr'$, where

$$G_0^{k'}(r, r') = (2\pi)^{-3} \frac{2\mu}{\hbar^2} \int \frac{\phi^*(k', r') \, \phi(k', r') \, dk'}{(k^2 - k'^2)}, \quad (28)$$

which is the Lippmann-Schwinger (LS) integral equation and $\langle k' | V(r, r') | k \rangle$ is the two-body t-matrix. We must now choose the form of $G_0^{k'}(r, r')$ so that the wave function $\psi(k, r)$ satisfies the required boundary condition. Eq. (27) is unsatisfactory as it stands because of the presence of poles at $k' = \pm k$. To eliminate this we redefine

$$G_0^{k'}(\vec{r}, \vec{r}') = \lim_{\epsilon \to 0} (2\pi)^{-3} \frac{2\mu}{\hbar^2} \int \frac{\phi(\vec{k}, \vec{r}) \, \phi^*(\vec{k}', \vec{r}') \, d\vec{k}'}{k^2 - k'^2 + i \epsilon},$$

where ϵ is small +ve quantity which is allowed to go to zero after the integration has been performed. We can now perform the integration over the angular coordinates of k' followed by the integration in the complex k' plane taking the residue at the poles $k' = \sqrt{k^2 + i \epsilon}$ and take the limit $\epsilon \to 0$. We rewrite S. Eq. (23) in the c.m. system:

$$(H_0 + V - E)\, \psi(r) = 0, \text{ where } V\,\psi(r) = \int \langle r \mid V \mid r' \rangle\, \psi(r)\, dr', \quad (29)$$

for a non-local potential; which means that the interaction is not confined to a single radius r, and the final integration must include the integration over r'. We carry out calculation in the momentum representation. Therefore, we introduce Fourier transform for the potential:

$$\langle k \mid V \mid k' \rangle = \int e^{-ikr} \langle r \mid V \mid r' \rangle\, e^{ik'r'}\, dr\, dr'. \text{ For a local potential}$$

$$\langle r \mid V \mid r' \rangle = V(r)\, \delta(r - r') \text{ and } \langle k \mid V \mid k' \rangle = \int V(r)\, e^{i(k'-k)r}\, dr. \quad (30)$$

C1. THE DIRAC EQUATION

The relativistic energy for the field free particle of rest mass m is

$$E^2 = p^2 + m^2$$

or
$$E = \omega = \pm(p^2 + m^2)^{1/2} = (k^2 + m^2)^{1/2}. \tag{1}$$

We have used the standard unit $\hbar = c = 1$. The quantum mechanical wave equation for a relativistic particle may be obtained by replacing $E \rightarrow i\dfrac{\partial}{\partial t}$ and $p \rightarrow -i\nabla$, i.e. $-\dfrac{\partial^2}{\partial t^2} = -\nabla^2 + m^2$, yielding the Klien Gordon equation :

$$[\Box + m^2]\,\psi = 0, \quad \text{where } \Box = -\nabla^2 + \frac{\partial^2}{\partial t^2}. \tag{2}$$

is Lorentz invariant and so also Eq. (2), as m is a constant quantity.

Dirac approached the problem of finding relativistic wave equation by starting from the time dependent S. Eq.

$$i\,\frac{\partial\psi(r,t)}{\partial t} = H\,\psi(r,t), \tag{3}$$

where $H = \dfrac{p^2}{2m}$ is the Hamiltonian. Now, if p is replaced by $-i\nabla$, the resulting wave equation is unsymmetrical with respect space and time derivatives and, hence, it is not relativistic. Dirac therefore modified the Hamiltonian in such a way as to make it linear in space derivative (or linear in momentum). With the Dirac linearization, we get

$$H = \hat{\alpha} \cdot \hat{p} + \hat{\beta}\,m = -i\,\hat{\alpha} \cdot \vec{\nabla} + \hat{\beta}m. \tag{4}$$

Substitution of (4) in (3) leads to $\left[i\dfrac{\partial}{\partial t} + i\,\hat{\alpha} \cdot \vec{\nabla} + \hat{\beta}\,m\right]\psi = 0,$ (5)

where $\hat{\alpha}$ and $\hat{\beta}$ are the Dirac matrices (hermitian and square) and are given by

$$\alpha_x = \begin{bmatrix} 0 & \hat{\sigma}_x \\ \hat{\sigma}_x & 0 \end{bmatrix}, \alpha_y = \begin{bmatrix} 0 & \hat{\sigma}_y \\ \hat{\sigma}_y & 0 \end{bmatrix}, \alpha_z = \begin{bmatrix} 0 & \hat{\sigma}_z \\ \hat{\sigma}_z & 0 \end{bmatrix}, \beta = \begin{bmatrix} I & 0 \\ 0 & -I \end{bmatrix}, \tag{6}$$

where $\hat{\sigma}_x$, $\hat{\sigma}_y$ and $\hat{\sigma}_z$ are the 2×2 Pauli-matrices and I is a 2×2 unit matrix. The physical results do not depend on special choice of Dirac matrices $\hat{\alpha}_i$ and $\hat{\beta}$.

Free Particle Solutions: Since α and β are represented by matrices, Eq. (4) or Eq. (5) will have meaning if we represent ψ by a matrix with four rows and one column. That is, these matrices operate on a four dimensional "spinor space", which is spanned by a set of four basis functions:

$$\psi(r,t) = \begin{bmatrix} \psi_1(r,t) \\ \psi_2(r,t) \\ \psi_3(r,t) \\ \psi_4(r,t) \end{bmatrix}. \tag{7}$$

By simple matrix multiplication, we get from (5), (6) and (7):

$$i\frac{\partial}{\partial t}\begin{bmatrix} \psi_1 \\ \psi_2 \\ \psi_3 \\ \psi_4 \end{bmatrix} + i\nabla_x \begin{bmatrix} \psi_4 \\ \psi_3 \\ \psi_2 \\ \psi_1 \end{bmatrix} + i\nabla_y \begin{bmatrix} -\psi_4 \\ \psi_3 \\ -\psi_2 \\ \psi_1 \end{bmatrix} + i\nabla_z \begin{bmatrix} \psi_3 \\ -\psi_4 \\ \psi_1 \\ -\psi_2 \end{bmatrix} - m \begin{bmatrix} \psi_1 \\ \psi_2 \\ -\psi_3 \\ -\psi_4 \end{bmatrix} = 0. \tag{8}$$

The plane wave solutions of (8) of the form

$$\psi_j(r,t) = u_j \, e^{i(\vec{p}.\vec{r} - \omega t)}, \quad j = 1, 2, 3, 4 \tag{9}$$

can be found with u_j as numbers. Substituting (9) in (8) we get

$$(E-m)u_1 - p_z u_3 - (p_x - ip_y)u_4 = 0,$$
$$(E-m)u_2 + p_z u_4 - (p_x + ip_y)u_3 = 0,$$
$$(E+m)u_3 - p_z u_1 - (p_x - ip_y)u_2 = 0,$$

and
$$(E+m)u_4 + p_z u_2 - (p_x + ip_y)u_1 = 0, \tag{10}$$

where $E(=\omega)$ and $\vec{p}(=\vec{k})$ are now numbers.

These equations (10) are homogeneous in the u_j and hence have solutions only if the determinant of the coefficient is zero. This determinant is $(E^2 - m^2 - p^2)$, so that the relation between E and p is in agreement with (1). Explicit solution can be obtained for any momentum p by choosing a sign for the energy say $E_+ = +(p^2 + m^2)^{1/2}$. Then there are two linearly independent solutions which are conveniently written as

$$u_1 = 1, u_2 = 0, u_3 = p_z/E_+ + m, u_4 = \frac{p_x + ip_y}{E_+ + m} \tag{11a}$$

$$u_1 = 0, u_2 = 1, u_3 = \frac{(p_x - ip_y)}{E_+ + m}, u_4 = \frac{p_z}{E_+ + m}, \tag{11b}$$

Similarly, if we choose the negative square root $E_- = -(p^2 + m^2)^{1/2}$, we obtain two new solutions which are conveniently written as

$$u_3 = 1, u_4 = 0, u_1 = \frac{p_z}{E_- - m}, u_2 = \frac{p_x + ip_y}{E_- - m} \tag{12a}$$

$$u_3 = 0, u_4 = 1, u_1 = \frac{-(p_x - ip_y)}{E_- - m}, u_2 = -\frac{p_z}{E_- - m} \tag{12b}$$

Each of these four solutions is multiplied by normalization constant

$$N = \left[1 + \left\{ \frac{p^2}{(E+m)} \right\} \right]^{-\frac{1}{2}}. \tag{13}$$

The Dirac spinor can therefore be written as

$$E > 0, \alpha \qquad \psi_+^\alpha = N\, e^{i(\vec{k}\cdot\vec{r} - \omega t)} \begin{bmatrix} 1 \\ 0 \\ p_z/(E_+ + m) \\ p_+/(E_+ + m) \end{bmatrix} \tag{14a}$$

$$E > 0, \beta \qquad \psi_+^\beta = N\, e^{i(\vec{k}\cdot\vec{r} - \omega t)} \begin{bmatrix} 1 \\ 0 \\ p_-/(E_+ + m) \\ -p_z/(E_+ + m) \end{bmatrix} \tag{14b}$$

$$E < 0, \alpha \qquad \psi_-^\alpha = N e^{i(\vec{k}\cdot\vec{r} - \omega t)} \begin{bmatrix} p_z/(E_- - m) \\ p_+/(E_- - m) \\ 1 \\ 0 \end{bmatrix} \tag{15a}$$

$$E < 0, \beta \qquad \psi_-^\alpha = N e^{i(\vec{k}\cdot\vec{r} - \omega t)} \begin{bmatrix} p_z/(E_- - m) \\ p_+/(E_- - m) \\ 1 \\ 0 \end{bmatrix} \tag{15b}$$

Just as the two components of the spin wave function $\alpha = \begin{pmatrix} 1 \\ 0 \end{pmatrix}$ and $\beta = \begin{pmatrix} 0 \\ 1 \end{pmatrix}$ are connected by two spin states in non-relativistic mechanics, the four components of ψ have also physical significance. In order to see this let us split the four component spinor ψ into two component spinor ϕ and χ, i.e.

$$\psi = \begin{bmatrix} \psi_1 \\ \psi_2 \\ \psi_3 \\ \psi_4 \end{bmatrix} = \begin{pmatrix} \phi \\ \chi \end{pmatrix} \text{ with } \phi = \begin{pmatrix} \psi_1 \\ \psi_2 \end{pmatrix} \text{ and } \chi = \begin{pmatrix} \psi_3 \\ \psi_4 \end{pmatrix}. \tag{16}$$

Using explicit form for $\hat{\alpha}$ and $\hat{\beta}$ matrices, Eq. (4) can be written as

$$\begin{pmatrix} 0 & \hat{\sigma} \\ \hat{\sigma} & 0 \end{pmatrix} \cdot \hat{p} \begin{pmatrix} \phi \\ \chi \end{pmatrix} + m \begin{pmatrix} I & 0 \\ 0 & -I \end{pmatrix} \begin{pmatrix} \phi \\ \chi \end{pmatrix} = E \begin{pmatrix} \phi \\ \chi \end{pmatrix},$$

$$\hat{\alpha} \cdot \hat{p}\chi + m\phi = E\phi \quad \text{and} \quad \hat{\alpha} \cdot \hat{p}\phi - m\chi = E\chi, \tag{17}$$

where $E = E_\pm = \pm(p^2 + m^2)^{1/2}$. From (17), we see that

$$\chi = \frac{\hat{\sigma} \cdot \hat{p}}{E + m} \phi \quad \text{and} \quad \phi = \frac{\hat{\sigma} \cdot \hat{p}}{E - m} \chi. \tag{18}$$

From these equations, it is clear that, if E is positive, i.e. $E = E_+$ then χ is very small compared to ϕ in the non-relativistic limit, whereas for negative energies $E = E_-$ the opposite is true. Thus, for the positive and negative energies

the Dirac spinors are: $\Psi_+ = \begin{pmatrix} \phi \\ \dfrac{\hat{\sigma} \cdot \hat{p}}{E_+ + m} \phi \end{pmatrix}$ and $\Psi_- = \begin{pmatrix} \chi \\ -\dfrac{\hat{\sigma} \cdot \hat{p}}{E_- - m} \chi \end{pmatrix}.$ (19)

Covariant Form: According to the principle of relativity, the basic equations of physics are the same in every inertial system. Therefore, the Dirac equation should be written in a relativistically covariant form and there must be a explicit rule to calculate $\psi'(x')$ in the moving system, if $\psi(x)$ is given for the fixed system. For many purposes, it is convenient to write the Dirac equation in a more symmetric form, treating space and time coordinates on a more equal footing. The Minkowaski space is defined in accordance to Bjorken-Drell.

If we multiply the time dependent Dirac equation (5) by β and use $\beta^2 = 1$, then

$$\left[i\beta \, \frac{\partial}{\partial x^0} + i \, \hat{\beta} \, \hat{\alpha} \cdot \hat{\nabla} - m \right] \psi = 0. \tag{20}$$

We now introduce γ-matrices following the Bjorken-Drell convention:

$\gamma^0 = \hat{\beta}$, $\gamma^k = \hat{\beta} \, \hat{\alpha}_k$, $k = 1, 2, 3$ and write (20) as

$$i\left(\gamma^0 \frac{\partial}{\partial x^0} + \gamma^1 \frac{\partial}{\partial x^1} + \gamma^2 \frac{\partial}{\partial x^2} + \gamma^3 \frac{\partial}{\partial x^3} \right) \psi - m\psi = 0. \tag{21}$$

Hence, the Dirac equation in covariant form is

$$\left[i \, \gamma^\mu \, \partial_\mu - m \right] \psi = 0, \tag{22}$$

where $\partial_\mu = \dfrac{\partial}{\partial x^\mu}$ and summation over Lorentz index μ is implied.

The Dirac 4×4 γ-matrices satisfy the following relations:

$$\gamma_5 = \gamma^5 = i \, \gamma^0 \gamma^1 \gamma^2 \gamma^3 \text{ and } \gamma_5^1 = 1 \tag{23}$$

$\{\gamma^\mu, \gamma^\nu\} = 2g^{\mu\nu} I$, where I is a 4×4 unit matrix and $\{\gamma_5, \gamma^\mu\} = 0$. (24)

With our definition of γ-matrices γ_0 and γ_5 are hermition, while the three matrices (γ^k, $k = 1, 2, 3$) are anti-harmitian. The scalar product of a γ-matrix and a 4 vector is

$$\gamma_\mu a^\mu \equiv \phi = \gamma^0 a^0 - \vec{\gamma} \cdot \vec{a} \tag{25}$$

and similarly $\gamma_\mu \, p^\mu \equiv p = i\partial = i\gamma^0 \partial_t + i\vec{\gamma} \cdot \vec{\nabla}.$ (26)

Trace Theorems: In calculating transition matrix elements, the following trace theorems are useful

$Tr \, I = 4$, $Tr \, \gamma_5 = 0$, Tr (odd number of γ's) $= 0$ and $Tr(\gamma_\mu \, \gamma_\nu) = g_{\mu\nu}$.   (27)

$$Tr(\gamma^\mu \, p_1 \, \gamma^\nu \, \gamma_5 \, p_2) \, Tr(\gamma_\mu \, p_3 \, \gamma_\nu \, \gamma_5 \, p_4)$$

$$= 32 \left[(p_1 \cdot p_3)(p_2 \cdot p_4) - (p_1 \cdot p_4)(p_2 \cdot p_3) \right]. \tag{28}$$

$$Tr \left[\gamma^\mu (1 - \gamma_5) \, p_1 \gamma^\nu (1 - \gamma_5) \, p_2 \right] Tr \left[\gamma_\mu (1 - \gamma_5) \, p_3 \, \gamma_\nu (1 - \gamma_5) p_4 \right]$$

$$= 256 \, (p_1 \cdot p_3)(p_2 \cdot p_4). \tag{29}$$

Bilinear Covariants: By the Lorentz invariance, we mean that for every solution ψ (x) to a Dirac equation and every Lorentz transformation

$$x'^\mu = a_\nu^\mu \, x^\nu \quad (\mu, \nu = 0, 3),$$

there exists a transformed solution in new coordinate x'. This transformed solution is linearly related to the original solution, i.e.

$$\psi'(x') = S(a)\,\psi(x)\,.$$

The transformation law for γ^μ is $S^{-1}(a)\,\gamma^\mu\,S(a) = a^\mu_\nu\,\gamma^\nu$. The Lorentz transformation may be regarded as a rotation in four dimensional space. If ψ obeys the Dirac equation, then $\psi'(x')$ also obeys the Dirac equation in the new coordinate x' as follows:

$$S^{-1}(a)\,(i\,\gamma^\mu\,\partial'_\mu - m)\,\psi'(x') = (S^{-1}(a)\,i\,\gamma^\mu\,\partial'_\mu - m)\,S(a)\,\psi(x)$$

$$= (S^{-1}(a)\,i\,\gamma^\mu S(a)\,\partial'_\mu - m)\,\psi(x) = (i\,a^\mu_\nu\,\gamma^\nu\,a^\lambda_\mu\,\partial_\lambda - m)\,\psi(x)$$

$$= (i\,\delta^\lambda_\nu\,\gamma^\nu\,\partial_\lambda - m)\,\psi(x) = (i\,\gamma^\nu\,\partial_\nu - m)\,\psi(x),$$

where $\partial'_\mu = \dfrac{\partial}{\partial x'_\mu}$. Thus, if $\psi(x)$ obeys the Dirac equation so also $\psi'(x')$.

It is well known that the products of the Dirac matrices give sixteen linearly independent quantities which can be classified into five groups in accordance to their transformation properties under the Lorentz transformation. They are called the bilinear covariants under the Lorentz transformation. By inserting appropriate γ's between $\bar\psi$ and ψ, we can construct these.

Let us consider the quantity of the form $\bar\psi\,\gamma^\mu\,\psi$ and verify its transformation:

$$\psi'(x')\,\gamma^\mu\,\psi'(x') = \bar\psi(x)\,S^{-1}(a)\gamma^\mu\,S(a)\psi(a)$$

$$= a^\mu_\nu\,\bar\psi(x)\gamma^\nu\,\psi(x). \tag{30}$$

The transformation laws of other bilinear expressions may be found as well. Let us consider space inversion (parity operation), which is a discrete Lorentz transformation:

$$\psi'^\mu = (x^0, -\bar x) = a^\mu_\nu\,x^\nu, \text{ where } a^\mu_\nu = \begin{bmatrix} 1 & 0 & 0 & 0 \\ 0 & -1 & 0 & 0 \\ 0 & 0 & -1 & 0 \\ 0 & 0 & 0 & -1 \end{bmatrix}. \tag{31}$$

The corresponding $S(a) = \gamma^0$, satisfying $S^{-1}\gamma^\mu S = (\gamma^0, -\gamma^k)$. Under the parity operation, the Dirac equation takes the form:

$$\left(i\,\gamma^0\,\frac{\partial}{\partial x_0} - i\,\gamma^k\,\frac{\partial}{\partial x^k} - m\right)\psi(x_0, -x) = 0,$$

which is not a Dirac equation. However, on multiplying from the left by γ^0 we get

$$\left(i \, \gamma^{0^2} \frac{\partial}{\partial x_0} - i \, \gamma^0 \gamma^k \frac{\partial}{\partial x^k} - m \gamma_0 \right) \psi(x_0, -x) = 0$$

or

$$\left(i \gamma^\mu \frac{\partial}{\partial x^\mu} - m \right) \gamma^0 \, \psi(x_0, -x) = 0.$$

Therefore, $\gamma^0 \, \psi(x_0, -x)$ satisfies the Dirac equation. Hence, the transformed Dirac wave function under space inversion is $\overline{\psi}'(x') = \gamma^0 \, \psi(x)$. Therefore, under the parity operation

$$\overline{\psi}'(x') \, \gamma^\mu \, \psi'(x') = \begin{cases} \overline{\psi}(x) \, \gamma^0 \, \psi(x) \\ -\overline{\psi}(x) \, \gamma^k \, \psi(x) \end{cases}, \tag{32}$$

which behaves like four vectors. Thus, we can show $\overline{\psi} \psi$ transforms as a scalar (invariant) under the Lorentz transformations and parity transformation whereas $\overline{\psi} \, \gamma_5 \, \psi$ transform as a scalar under the Lorentz transformation but changes sign under parity transformation and therefore, is known as pseudoscalar. On the other hand, $\overline{\psi} \, \gamma^\mu \, \psi$ transforms as a vector under the Lorentz transformation and changes sign under the parity transformation. We can show, $\overline{\psi} \, \gamma_\mu \, \gamma_5 \, \psi$ transforms as a vector under the Lorentz transformations but is invariant under the parity transformation and therefore, known as pseudovector or axial vector:

$$\overline{\psi}'(x') \, \gamma_\mu \, \gamma_5 \, \psi'(x') = \overline{\psi}(x) \gamma_0 \, \gamma_\mu \, \gamma_5 \, \gamma_0 \, \psi(x) = \overline{\psi}(x) \begin{pmatrix} -\gamma_0 \\ \gamma_k \end{pmatrix} \psi(x). \tag{33}$$

Since, $S^{-1} \gamma^5 S = -\gamma^5$, the presence of γ^5 gives rise to the pseudo nature of the axial vector. Bilinear quantities $\overline{\psi}(x) \, O_i \, \psi(x)$, $i = 1, 2, 3, 4$ and 5 for S, V, T, A and P respectively transform like scalar, vector, tensor, axial vector and pseudoscalar under the proper Lorentz transformation and space inversion.

C2. SECOND QUANTIZATION NOTATIONS

Let us introduce an operator $a_m^\dagger$ which, operating on any state $|\phi\rangle$ creates an extra particles in the state m, in addition to what is already contained in $|\phi\rangle$. If $|\phi\rangle$ represents "Vacuum" (i.e., a state with no particle), then $a_m^\dagger |\phi\rangle$ will represent a state $|m\rangle$ of the single particle. On the other hand if $|\psi\rangle$ denotes a determinantal wave function for A particles, then $a^\dagger |\psi\rangle$ is a state of A+1 particle. $a_m^\dagger$ is called the 'creation operator' for single particle state $|m\rangle$.

Similarly, the Hermitean conjugate operator a_m removes a single particle in a state m operating on any state $|\phi\rangle$. Thus, if $|\psi\rangle$ is an A particle state, then $a_m |\psi\rangle$ is the state of $(A-1)$ particles which is obtained by eliminating the state m contained in $|\psi\rangle$. It therefore follows that, if the state m is not originally present in $|\psi\rangle$ then $a_m |\psi\rangle = 0$. The operator a_m is called the destruction operator for the single particle in state m. Thus, in this language, the Pauli exclusion principle is stated as $a_m^\dagger |\psi\rangle = 0$ if m is an occupied state, because it is not possible to create another particle in the same state 'm'. Similarly, $a_m |\psi\rangle = 0$ if m is not an occupied state. Let us denote the vacuum state, i.e. a state with no particle by $|0\rangle$. Then $a_m^\dagger |0\rangle = |m\rangle$, $a_m |0\rangle = 0$,

$$a_n a_m^\dagger |0\rangle = |0\rangle \text{ if } m = n \quad \text{and} \quad a_n a_m^\dagger |0\rangle = 0 \text{ if } m \neq n. \tag{34}$$

In other words, $a_n a_m^\dagger |0\rangle = \delta_{mn} |0\rangle$ and $a_m^\dagger a_n |0\rangle = |0\rangle$. Therefore,

$$(a_n a_m^\dagger + a_m^\dagger a_n) = \delta_{mn} |0\rangle, [a_n, a_m^\dagger] = \delta_{mn} \text{ and } [a_n^\dagger, a_m^\dagger] = [a_n, a_m] = 0. \tag{35}$$

Now, $\quad a_n^\dagger a_m^\dagger = -a_m^\dagger a_n^\dagger$ and making $m = n$; $a_n^\dagger a_n^\dagger |\psi\rangle = -a_n^\dagger a_n^\dagger |\psi\rangle$.

So, a quantity can be equal to negative of itself only if it is zero. We have established the desired result that two particles can not occupy the same state (the Pauli exclusion principle).

In the language of second quantization, a single-particle operator destroys a particle in state r and creates it in state s. Symbolically, $V_i\, \psi_r \rightarrow \psi_s$, where r and s are suitable quantum levels. Therefore, a single-particle $(s \cdot p)$ operator in second quantized form is $\sum_{rs} \langle s|V|r\rangle a_s^\dagger a_r$. The quantity $\langle s|V|r\rangle$ is called the amplitude. The expectation value of a $s.p.$ operator in single particle state 'i' is given by

$$\langle 0| a_i \left\{ \sum_{r,s} \langle s|V|r\rangle a_s^\dagger a_r \right\} |a_i^\dagger |0\rangle$$

$$= \sum_{r,s} \langle s|V|r\rangle \langle 0| a_i\, a_s^\dagger\, a_r\, a_i^\dagger |0\rangle = \sum_{r,s} \langle s|V|r\rangle \langle 0| \left\{ a_i\, a_s^\dagger (\delta_{ir} - a_i^\dagger a_r) \right\} |0\rangle$$

$$= \sum_{r,s} \langle s|V|r\rangle \langle 0| a_i\, a_s^\dagger\, \delta_{ir} |0\rangle = \sum_{r,s} \langle s|V|r\rangle \delta_{is}\, \delta_{ir} = \langle i|V|i\rangle, \tag{36}$$

which is similar to (3.4).

The two-particle operator destroys particle in states m, n and creates them in r, s. Hence, it is given by $V_{ij} = \sum_{\substack{r,s \\ m,n}} \langle rs|V|mn\rangle a_r^\dagger a_s^\dagger a_n a_m$.

D1. GROUP THEORY AND SYMMETRY TRANSFORMATIONS

In order to understand the symmetry properties of a physical system it is necessary to know some terminology and concepts of group theory, which is a branch of mathematical physics dealing with symmetry. In non-relativistic quantum mechanics we examine the Hamiltonian (or Lagrangian in the relativistic quantum mechanics) describing a physical system for its invariance properties under various transformations. If the Hamiltonian is the same after transformation as before, then we say that it has a symmetry under transformation. A set of transformations U_θ forms a mathematical group G having the following group properties:

(1) The product of any two elements U_θ and $U_{\theta'}$ is equal to $U_{\theta+\theta'}$ and so is itself a transformation which leaves H invariant. In group theoretical terminology, if $U_\theta U_{\theta'} \in G$ then $U_{\theta+\theta'} \in G$.

(2) The transformations are associative, that is,

$$(U_\theta \ U_{\theta'})U_{\theta''} \ = \ U_\theta(U_{\theta'} \ U_{\theta''}). \tag{1}$$

(3) An identity element exists ($\theta = 0$).

(4) Inverse of any element exists ($\theta \to -\theta$).

If all members of the group commute then the group is called *Abelian* and if they don't, the group is non-Abelian. Further if $U_\theta^\dagger = U_\theta^{-1}$ then U_θ is *unitary*. Here dagger stands for Hermitian conjugate.

Example: Symmetry transformations like parity operation, translation and rotation of a coordinate frame form a group. Translations in space and time are *Abelian*, whereas successive rotations about different axes are non-*Abelian*. The set of rotations of a system forms a group, each rotation being an element of the group. Two successive rotations R_1 followed by R_2 (written as product $R_2 R_1$) are equivalent to a single rotation (that is to another element). There is an identity element (no rotation) and every rotation has an inverse. The product is not necessarily commutative: $R_1 R_2 \neq R_2 R_1$, i.e. the rotation forms non-Abelian group. The associative law always holds: $R_3(R_2 R_1) = (R_3 R_2)R_{-1}$. The rotation group is continuous.

If the Hamiltonian is invariant under translation and rotations in space, then measurements can not distinguish an absolute position or orientation in space. In other words, we do not want an experimental result to depend upon the specific laboratory orientation of the system we are measuring. Rotation therefore must form a symmetry group of a system. For example, if under a rotation R, the states of a system transform as

$$|\psi\rangle \to |\psi'\rangle = U_\theta|\psi\rangle,$$

then the probability that a system described by $|\psi\rangle$ will be found in state $|\phi\rangle$ unchanged by rotation R requires

$$|\langle\phi|\psi\rangle|^2 = |\langle\phi'|\psi'\rangle|^2 = |\langle\phi|U_\theta^\dagger \, U_\theta|\psi\rangle|^2,$$

so that $U_\theta^\dagger \, U_\theta$ must be unitary. Under unitary transformation, the expectation value of a dynamical quantity remains unchanged. For example, the Hamiltonian is unchanged by a symmetry operation R of the system and the matrix elements are preserved. Therefore,

$$\langle\phi'|H|\psi'\rangle = \langle\phi|U_\theta^\dagger \, U_\theta|\psi\rangle = \langle\phi|H|\psi\rangle,$$

so that, $\qquad\qquad H = U_\theta^\dagger \, HU_\theta \text{ or } [U_\theta, H] = 0.$ \hfill (2)

The transformation U_θ has no explicit time dependence and the equation of motion

$$i\frac{d}{dt}|\psi(t)\rangle = H|\psi(t)\rangle \text{ is unchanged by the symmetry operation.}$$

As a consequence, the expectation value of U_θ is a constant of motion, i.e.

$$i\frac{d}{dt}\langle\psi(t)|U_\theta|\psi(t)\rangle = \langle\psi(t)|U_\theta \, H - HU_\theta|\psi(t)\rangle = 0.$$

If any unitary operator U which does not explicitly depend on time and commutes with the Hamiltonian H, i.e. the Hamiltonian is invariant under the transformation generated by U, then U is a conserved quantity. All the group properties follow on considering infinitesimal rotation through an infinitesimal angle $\in$ about the 3-axis (or Z-axis). We assume the Cartesian axes. The coordinate transformation resulting from the said operation is

$$x \to x' = x + \in y, \quad y \to y' = y - \in x \quad \text{and} \quad z \to z'$$

The unitary operator $U_\in$ transforming the state function is

$$U_\in|\psi(x, y, z)\rangle = |\psi(x + \in y, \ y - \in x, z)\rangle$$

$$= |\psi(x, y, z)\rangle + \in\left(y\frac{\partial}{\partial x} - x\frac{\partial}{\partial y}\right)|\psi(x, y, z)\rangle$$

$$= (1 - i \in J_z)|\psi(x, y, z)\rangle,$$

where J_z is the Z-component of angular momentum in units of $\hbar$. To the first order in $\in$

$$U_\in = 1 - i \in J_3.$$ \hfill (3)

The operator J_3 is called generator of rotations about the 3-axis. Now

$$U_\epsilon^\dagger U_\epsilon = (1 + i\epsilon\, J_3^\dagger)\,(1 - i\epsilon\, J_3) \quad \text{or} \quad J_3^\dagger = J_3 \text{ since } U_\epsilon^\dagger U_\epsilon = 1.$$

Therefore, J_3 is Hermitian and hence a quantum mechanical observable. The i was introduced in Eq. (3) to make it Hermitian. We may introduce similar Hermitian generators of rotations about the '1' and '2' axes respectively J_1 and J_2. Generally, a rotation θ about an axis specified by a unit vector $\bar{n}$ is generated by the operator $U_\theta = e^{-i\bar{n}\cdot\bar{J}}$, where the angular momentum $\bar{J}$ is the generator of the transformation. Since the laws of physics must be independent of the orientation of the coordinate axes in space, the frames of reference linked by the rotation operators are equivalent. So rotation in three dimension forms a symmetry group. The conservation of angular momentum is thus due to the invariance of the Hamiltonian under rotation (isotropy of space). Three generators of rotation group (for the rotation in 3-dimensional space) satisfy the commutation rule:

$$[J_i, J_j] = i\,\epsilon_{ijk}\, J_k, \tag{4}$$

where $\epsilon_{ijk} = +1\,(-1)$ if ijk are a cyclic (anticyclic) permutation of 1, 2, 3 and $\epsilon_{ijk} = 0$ otherwise. The quantity ϵ_{ijk} is called structure constant of the group. Eq. (4) completely defines the group properties. The rotation operator U_ϵ is a Lie group because it is a continuous function of its parameter ϵ. The generators (satisfying algebra (4)) of the rotation group in three dimensions $O(3)$, act on three dimensional ordinary vector space and they represent angular momentum.

The angular momentum algebra is an essential part of nuclear theory. Defining J^2 and $J_\pm$ operator as $J^2 = J_1^2 + J_2^2 + J_3^2$, $J_+ = J_1 + iJ_2$ and $J_- = J_1 - iJ_2$, it is easily seen that J^2 commutes with $J_\pm$ and J_3:

$$[J^2 J_\pm] = 0,\ [J^2 J_3] = 0 \text{ and } J^2 = J_+ J_- + J_3^2 \tag{5}$$

Therefore, we can seek a basis $|jm\rangle$ in which J^2 and J_3 operators are simultaneously diagonal with eigen value $j\,(j+1)$ and m, such that

$$J^2|jm> = j(j+1)\ |jm> \text{ and } J_3|jm> = m\ |jm>,$$

also
$$J_\pm|jm> = \sqrt{(j \mp m)\,(j \pm m + 1)}\ \ |jm \pm 1>;\ -j \le m \le +j$$

An operator that commutes with all the generators of the group is known as the Casimir operator, e.g., in the $O(3)$ group, three generators J_1, J_2 and J_3 commutes with J^2, i.e. $[J^2 J_1] = [J^2 J_2] = [J^2 J_3]$, and, therefore, J^2 is the quadratic Casimir operator of the group $O(3)$.

The generators of $O(3)$ group transform $|jm\rangle$ into each other such that j remains unchanged, i.e. generators of $O(3)$ group $J_\pm$ and J_3 operating on state $|jm\rangle$ do not change the quantum number j, the set of states $|jm=j\rangle, |jm=j-1\rangle \cdots |jm_j=-j\rangle$ form an irreducible representation of $O(3)$. The basis states are distinguished by quantum number 'm', which forms an irreducible representation of group $O(2)$, which is a subgroup of $O(3)$. Hence, Casimir operators of $O(3)$ and $O(2)$ are respectively J^2 and J_3 and representation levels are j and m, so that the basis states are $|jm\rangle$. Now any Hamiltonian that can be written using the invariant operators J^2 and J_3 only can be diagonalized within the basis states that characterize the $O(3) \supset O(2)$ group reduction. As an example we consider the Hamiltonian

$$H = aJ^2 + b J_3 \text{ or } H|jm> = aJ^2 |jm\rangle + bJ_3 |jm\rangle = \{aj(j+1)+bm\}|jm\rangle$$

Therefore, the eigen value $E = aj(j+1)+bm$. \hfill (6)

Vector Spherical Harmonics: We describe the electromagnetic field mathematically by a vector field $\vec{A}(\vec{r})$. We shall now consider its transformation properties under the finite rotations. The rotational invariance implies that the field components A_x, A_y and A_z transform under finite rotations of coordinate frame in the same manner as the components x, y, z of the coordinate vector $\vec{r}$ itself. The rotation operator

$$\hat{R} = \begin{pmatrix} \cos\theta & \sin\theta & 0 \\ -\sin\theta & \cos\theta & 0 \\ 0 & 0 & 1 \end{pmatrix}.$$

An alternate point of view is possible in which the coordinate frame is kept fixed but the vector $r' \to r$ so that $\vec{r} = \hat{R}\, \vec{r}'$. Similarly for a vector function $\vec{A}(\vec{r})$, we get $\vec{A}(\vec{r}) = R\vec{A}'(\vec{r}')$, or $\vec{A}'(\vec{r}') = R^\dagger A(Rr')$. Since R is unitary i.e. $R\,R^\dagger = I$, we can drop the prime on $\vec{r}'$ and obtain $\vec{A}'(\vec{r}) = R^\dagger A(R\vec{r})$. For an infinitesimal rotation $\theta = \epsilon$, we have

$$\begin{pmatrix} A'_x \\ A'_y \\ A'_z \end{pmatrix} = \left[1 + \epsilon\left(y\frac{\partial}{\partial x} - x\frac{\partial}{\partial y} \right) + \epsilon \begin{pmatrix} 0 & -1 & 0 \\ 1 & 0 & 0 \\ 0 & 0 & 0 \end{pmatrix} \right] \begin{pmatrix} A_x \\ A_y \\ A_z \end{pmatrix} = \tilde{R} \begin{pmatrix} A_x \\ A_y \\ A_z \end{pmatrix}$$

where $\qquad \vec{R} = 1 - \dfrac{i}{\hbar}\epsilon\, \vec{n}.\vec{J}.$

In the above $\vec{n}$ defines the direction of the rotation axis. In the present case $\vec{n}\cdot\vec{J} = J_z$,

where
$$J_z = \frac{\hbar}{i}\left(x\frac{\partial}{\partial y} - y\frac{\partial}{\partial x}\right) + \hbar\begin{pmatrix} 0 & -i & 0 \\ i & 0 & 0 \\ 0 & 0 & 0 \end{pmatrix} = L_z + S_z. \tag{7}$$

The first term in (7) is immediately recognized to be the operator L_z for the orbital angular momentum. It is natural to associate the second term with the intrinsic spin of the vector field:

$S_z = \hbar\begin{pmatrix} 0 & -i & 0 \\ i & 0 & 0 \\ 0 & 0 & 0 \end{pmatrix}$. Similarly for the infinitesimal rotation about X and

Y axes

we get
$$S_x = \hbar\begin{pmatrix} 0 & 0 & 0 \\ 0 & 0 & -i \\ 0 & i & 0 \end{pmatrix} \text{ and } S_y = \hbar\begin{pmatrix} 0 & 0 & i \\ 0 & 0 & 0 \\ -i & 0 & 0 \end{pmatrix}. \tag{8}$$

In the present form none of the spin component matrices is diagonal. We immediately see that $S^2 = S_x^2 + S_y^2 + S_z^2 = 2\hbar^2 I_3$, where I_3 is a 3×3 unit matrix and hence, $S(S+1) = 2$. The fact that the electromagnetic field can be described by a vector field leads to the assignment of spin $S = 1$ to the electromagnetic field due to its transformation properties as vector. The spin components of photons are $S_z = 0$ and $S_z = \pm 1$. A photon with $S_z = 0$ is called longitudinal and a photon with both $S_z = \pm 1$ taken together is called transverse.

Thus, spin operator S is a consequence of vector nature of A. As a result of the unit spin of photon field, the total angular momentum J of a photon emitted in a 2^L pole mode is the vector sum of orbital angular momentum L and spin S, and eigen functions of radiation field are given by the vector spherical harmonics y_{L1J}^M defined by

$$y_{L1J}^M(\theta, \phi) = \sum_{\mu=-1}^{+1} \langle L1M - \mu\mu \mid JM \rangle Y_L^{M-\mu} \chi_1^\mu, \tag{9}$$

where $\langle L1M - \mu\mu \mid JM \rangle$ are the Clebs Gordon Coefficients. The y_{L1J}^M's satisfy

$$\int [y_{L1J}^{M*} \, y_{L'1J'}^{M'}(\theta, \phi)] \, d\Omega = \delta_{LL'} \, \delta_{JJ'} \, \delta_{MM'}.$$

For a given value of J there are three values of L ($J \pm 1$, J), the vector spherical harmonics, which we shall use most are the ones $L = J$.

SU(2) **Group:** Because of the importance of nucleon spin and isospin in nuclear physics we discuss here the properties of special unitary group in two dimensions $SU(2)$. In the lowest dimension, nontrivial representation of the rotation group (fundamental representation), the generators may be written

$$J_i = \frac{1}{2}\sigma_i \text{ with } i = 1, 2, 3, \text{ where } \sigma_i \text{ are:}$$

$$\sigma_1 = \begin{pmatrix} 0 & 1 \\ 1 & 0 \end{pmatrix}, \sigma_2 = \begin{pmatrix} 0 & -i \\ i & 0 \end{pmatrix}, \sigma_3 = \begin{pmatrix} 1 & 0 \\ 0 & -1 \end{pmatrix} \text{ which satisfy}$$

$$[\sigma_i, \sigma_j] = 2i \in_{ijk} \sigma_k. \tag{10}$$

The above set of unitary 2×2 Pauli matrices with determinant 1 form the group SU(2). The rank of the group is 1, which is also the number of diagonal traceless matrix. The basis for the above representation is conveniently chosen to be the eigen vector of σ_3 that is, the column vectors $\begin{pmatrix} 1 \\ 0 \end{pmatrix}$ and $\begin{pmatrix} 0 \\ 1 \end{pmatrix}$ describing a spin half particle of spin projection up $\left(m = +\frac{1}{2} \right)$ and a spin projection down $\left(m = -\frac{1}{2} \right)$ along the third axis respectively. σ_i are Hermitian, and the transformation matrices $U_{\theta_i} = e^{-\theta_i \sigma_i / 2}$ are unitary. The set of all unitary 2×2 matrices is known as the group $U(2)$. However, $U(2)$ is larger than the group of matrices U_{θ_i}, since generator all have zero trace. Since the unit determinant is preserved in matrix multiplication, the traceless unitary 2×2 matrices form a subgroup $SU(2)$ of $U(2)$. $SU(2)$ denotes a special unitary group in two dimensions. The term "special" signifies that $Det\, U = 1$. Quite simply, in matrix representation, $SU(n)$ is the group of n dimension having all $(n \times n)$ independent unitary matrices with determinant +1. It has $(n^2 - 1)$ generators, that obey closed algebra specified by structure constants.

As already discussed in the beginning of Chapter 1 (Sec. 1.1B), the basis states on which the isospin acts can be identified with particle multiplets, which can be characterized by the eigen values of T^2; their individual states can be labeled by T_3. There can exist several isospin multiplets with same T but distinguished by other quantum number S as well as Y ($Y = S+B$). The new conserved quantum number, the strangeness S or equivalently hyper charge Y leads to the so called $SU(3)$ flavor symmetry, which will be discussed next.

$SU(3)$ Group: We introduce two new operators U and V which have commutation properties similar to those of T and those of ordinary angular momentum. In the case of ordinary angular momentum [$SU(2)$ group] the simplest representation is obtained by the Pauli matrices and by their commutation rules. Similarly, for three operators T, U and V there is a fundamental representation by nine 3×3 matrices (including unit matrix). The group properties of these operators are called 'special unitary group in three complex dimensions, i.e. $SU(3)$. The generators are the eight $(3^2 - 1 = 8)$ linearly independent traceless and Hermitian matrices with det = 1. The SU(3) symmetry was first noted in flavor space consisting of u, d and s light quarks.

We represent the three quarks q_1, q_2 and q_3 in matrix form:

$$q_1 = \begin{pmatrix} 1 \\ 0 \\ 0 \end{pmatrix}, q_2 = \begin{pmatrix} 0 \\ 1 \\ 0 \end{pmatrix}, q_3 = \begin{pmatrix} 0 \\ 0 \\ 1 \end{pmatrix}$$

Employing the matrix representations of T_3 and Y, i.e.

$$T_3 = \begin{pmatrix} \frac{1}{2} & 0 & 0 \\ 0 & -\frac{1}{2} & 0 \\ 0 & 0 & 0 \end{pmatrix} \text{ and } Y = \begin{pmatrix} \frac{1}{3} & 0 & 0 \\ 0 & \frac{1}{3} & 0 \\ 0 & 0 & -\frac{2}{3} \end{pmatrix},$$

we can show that the eigenvalues of the charge operator $\hat{Q}$ acting on the fundamental triplet are $\frac{2}{3}, -\frac{1}{3}$ and $-\frac{1}{3}$:

$$\hat{Q} = \left(T_3 + \frac{1}{2} Y \right) = \begin{pmatrix} \frac{2}{3} & 0 & 0 \\ 0 & -\frac{1}{3} & 0 \\ 0 & 0 & -\frac{1}{3} \end{pmatrix}.$$

Hence, $\hat{Q} \begin{pmatrix} 1 \\ 0 \\ 0 \end{pmatrix} = \begin{pmatrix} \frac{2}{3} & 0 & 0 \\ 0 & -\frac{1}{3} & 0 \\ 0 & 0 & -\frac{1}{3} \end{pmatrix} \begin{pmatrix} 1 \\ 0 \\ 0 \end{pmatrix} = \frac{2}{3} \begin{pmatrix} 1 \\ 0 \\ 0 \end{pmatrix},$

similarly $\hat{Q} \begin{pmatrix} 0 \\ 1 \\ 0 \end{pmatrix} = -\frac{1}{3} \begin{pmatrix} 0 \\ 1 \\ 0 \end{pmatrix}$ and $\hat{Q} \begin{pmatrix} 0 \\ 0 \\ 1 \end{pmatrix} = -\frac{1}{3} \begin{pmatrix} 0 \\ 0 \\ 1 \end{pmatrix}$.

In this $SU(3)$ model we consider unitary unimodular transformations on three quarks q_1, q_2 and q_3 as follows: $q' = U(\theta) q$ where U is a unimodular, unitary transformation matrix given by $U = \exp [i \; \theta_j \; G_j]$, where θ_j are real parameters.

The eight generators G_j play an analogous role of three Pauli matrices in $SU(2)$, which act on the fundamental doublets $\begin{pmatrix} 1 \\ 0 \end{pmatrix}, \begin{pmatrix} 0 \\ 1 \end{pmatrix}$. The generators G_j of the $SU(3)$ group can have one representation

$$G_j = \frac{1}{2} \lambda_j, \tag{11a}$$

where $\lambda_1 = \begin{pmatrix} 0 & 1 & 0 \\ 1 & 0 & 0 \\ 0 & 0 & 0 \end{pmatrix}, \lambda_2 = \begin{pmatrix} 0 & -i & 0 \\ i & 0 & 0 \\ 0 & 0 & 0 \end{pmatrix}, \lambda_3 = \begin{pmatrix} 1 & 0 & 0 \\ 0 & -1 & 0 \\ 0 & 0 & 0 \end{pmatrix}.$

$\lambda_4 = \begin{pmatrix} 0 & 0 & 1 \\ 0 & 0 & 0 \\ 1 & 0 & 0 \end{pmatrix}, \lambda_5 = \begin{pmatrix} 0 & 0 & -i \\ 0 & 0 & 0 \\ i & 0 & 0 \end{pmatrix}, \lambda_6 = \begin{pmatrix} 0 & 0 & 0 \\ 0 & 0 & 1 \\ 0 & 1 & 0 \end{pmatrix}.$

$$\lambda_7 = \begin{pmatrix} 0 & 0 & 0 \\ 0 & 0 & -i \\ 0 & i & 0 \end{pmatrix}, \lambda_8 = \frac{1}{\sqrt{3}} \begin{pmatrix} 1 & 0 & 0 \\ 0 & 1 & 0 \\ 0 & 0 & -2 \end{pmatrix}. \tag{11b}$$

λ_i $(j = 1...8)$ are called the Gell-Mann matrices, which like the Pauli matrices are traceless and Hermitian. Only λ_3 and λ_8 are diagonal. The first three matrices are just like the Pauli matrices with just one extra zero row and zero column.

Since the number of traceless diagonal matrices is 2, the rank of SU(3) is 2. The number of Casimir operators is also 2. λ_1, λ_2 and λ_3 form a subgroup of $SU(3)$. λ_8 commutes with λ_3 and thus has no analogue in $SU(2)$. The matrices λ_k fulfill the following commutation relations:

$$\left[\frac{\lambda_i}{2}, \frac{\lambda_j}{2} \right] = i f_{ijk} \frac{\lambda_k}{2}, i, j, k = 1, 2, \cdots 8, \tag{12}$$

where the "structure constants" f_{ijk} are completely antisymmetric under the exchange of any two indices. The relations (12) define the algebra of SU(3). Non-zero values of f_{ijk} are obtained by the rule of matrix multiplication as follows:

$$f_{123} = +1, f_{147} = f_{516} = f_{246} = f_{257} = f_{345} = f_{637} = \frac{1}{2} \text{ and } f_{458} = f_{678} = \frac{\sqrt{3}}{2}.$$

$$\text{trace}(\lambda_k \lambda_\ell) = 2\delta_{k\ell}.$$

$SU(2)$ being a group of rank 1, has only one Casimir operator T^2. $SU(3)$, on the other hand is of rank 2 and has two Casimir operators. These, however, have no direct physical significance, so that generally in the case of $SU(3)$ a pair of numbers (p, q) is used to label the $SU(3)$ multiplets. Status within a multiplet can be labeled by the eigen values of T_3 and Y which are simultaneously diagonal in the representation (11b).

The operators U, V and T have the following physical meaning:

$$T_3 = Q - \frac{1}{2}Y; \ U_3 = -\frac{1}{2}Q + Y; \ V_3 = -\frac{1}{2}(Q + Y) \text{ and } T_3 + V_3 + U_3 = 0. \quad (13)$$

The eight quantum numbers are 3 components of isospin, hyper charge Y, 2 components of U-spin and 2 components of V-spin. It is convenient to have a short hand notation which will describe the character of a particular $SU(3)$ representation.

A general unitary irreducible representation of $SU(3)$ is designated by two parameters p and q mentioned earlier, each takes the non-negative values 0, 1, 2, 3.... UIR (p, q) means 3-representation p times and 3 conjugate representation q-times symmetrically and combining them to get largest representation of $SU(3)$. The number 'n' of substrates of a multiplet contained in $UIR(p,q)$ is $n = \frac{1}{2}(p+1)(q+1)(p+q+2)$.

$UIR(p, q)$ is a conjugate representation, i.e. $UIR(p, q) \equiv UIR(q, p)$. the conjugate representation $\overline{3}$ refer to the multiplets of anti-quark. Thus, we can obtain the weight diagrams for $UIR(1,0)$, $UIR(0,1)$, $UIR(1,1)$ and $UIR(3,0)$, which give triplet, octet and decuplet representation respectively. The fundamental unitary triplet of $SU(3)$ namely $UIR(1,0)$ or the antitriplet $UIR(0,1)$ has not been found. However, other representation resulting from $SU(3)$ theory are identified as baryon and meson unitary octets and baryon unitary decuplet.

D2. COUPLING RULES FOR ANGULAR MOMENTUM

Let $\vec{j}_1$ and $\vec{j}_2$ represent different angular momenta, either the angular momenta of different particles or the spin and orbital angular momenta of the same particle. Our aim is to find simultaneous eigen vectors of j_1 and j_2. They are linear combination of the products $|j_1m_1\rangle|j_2m_2\rangle$. We consider the coupled basis denoted by the eigen vectors $|j_1\, j_2\, j\, m\rangle$ with $\vec{j} = \vec{j}_1 + \vec{j}_2$,

$$j^2|j_1\, j_2\, j\, m\rangle = j(j+1)|j_1\, j_2\, jm\rangle \text{ and } j_z|j_1\, j_2\, j_m\rangle = m|j_1\, j_2\, m\rangle$$

We express

$$|j_1 j_2 jm\rangle = \sum_{m_1 m_2} \langle j_1 j_2 m_1 m_2 \,|\, jm\rangle |j_1 m_1\rangle\, j_2 m_2\rangle. \tag{15a}$$

The coefficient $\langle j_1 j_2 m_1 m_2 \,|\, jm\rangle$ is the Clebsch-Gordon coefficients (C-coefficient) or vector coupling coefficient. From the definition, it is clear that for $\langle j_1 j_2 m_1 m_2 \,|\, jm\rangle$ we have the triangle rule:

$$|j_1 + j_2| \le j \ge |j_1 - j_2|.$$

The projection quantum numbers satisfy the sum rule: $m_1 + m_2 = m$.

We note that both the basis are diagonal and orthonomal. They have the same dimensionality $(2j_1 + 1)(2j_2 + 1)$. The transformation from the product basis to the coupled basis is unitary.

The transformation is reversible:

$$|j_1\, m_1\rangle|j_2\, m_2\rangle = \sum_{jm}\langle jm|j_1\, j_2\, m_1\, m_2\rangle|j\, m\rangle \tag{15b}$$

Wigner-Eckart Theorem: Quite often we were required to compute matrix elements of the multipole operator of the form $\langle f\,|\,\hat{Q}_{\lambda,\mu}\,|\,i\rangle$ where $\langle f\,\|\,i\rangle$ and $Q_{\lambda\mu}$ are functions of the same coordinates, r, θ, ϕ, σ. The multipole operator $Q_{\lambda\mu}$ is expanded in spherical harmonics. If $|f\rangle$ and $|i\rangle$ do not depend on spin, then the matrix element reduces to a linear combination of matrix element of the following form:

$$M = \int d\Omega\, Y_{\ell_1}^{m_1^*}(\Omega)\, Y_{\lambda}^{\mu}(\Omega)\, Y_{\ell_2}^{m_2}(\Omega). \tag{16}$$

To compute this matrix element, we use the expansion of the product of two spherical harmonics of the same coordinates, as a linear combination of a single spherical harmonics:

$$Y_{\lambda}^{\mu}(\Omega)Y_{\ell_2}^{m_2}(\Omega) = \sum_{L,M}\left[\frac{(2\lambda+1)\,(2\ell_2+1)}{4\pi(2L+1)}\right]^{1/2}$$

$$\times \langle \ell_2\lambda m_2\mu\,|\,LM\rangle\, \langle \ell_2\lambda 00\,|\,L0\rangle\, Y_L^M(\Omega).$$

Multiplying by $Y_{\ell_1}^{m_1^*}(\Omega)$ and using the orthogonality of the spherical harmonics, we have

$$M = \left[\frac{(2\lambda + 1)\,(2\ell_2 + 1)}{4\pi(2\ell_1 + 1)}\right]^{1/2} \langle \ell_2 \lambda m_2 \mu \mid \ell_1 m_1 \rangle \langle \ell_2 \lambda 00 \mid \ell_1 0 \rangle$$

for $\qquad\qquad\qquad\qquad \ell_1 + \ell_2 + \lambda = \text{even}$

$\qquad = 0 \qquad\qquad\qquad \text{for } \ell_1 + \ell_2 + \lambda = \text{odd}$

This is a special case of the Wigner-Eckart theorem which states that a matrix element of the form $\langle f \mid \hat{Q}_{\lambda,\mu} \mid i \rangle$ can always be reduced to the product of a C-coefficient depending on the projection quantum numbers and a reduced matrix element that is independent of the projection quantum numbers:

$$\langle \ell_1 m_1 \mid Y_\lambda^\mu \mid \ell_2 m_2 \rangle = \langle \ell_2 \lambda m_2 \mu \mid \ell_1 m_1 \rangle \langle \ell_1 \| Y_\lambda \| \ell_2 \rangle,$$

where $\qquad\qquad \langle \ell_1 \| Y_\lambda \| \ell_2 \rangle = \left[\frac{(2\lambda + 1)\,(2\ell_2 + 1)}{4\pi(2\ell_1 + 1)}\right]^{1/2} \langle \ell_2 \lambda 00 \mid \ell_1 0 \rangle.$

Table D1: C-coefficients $\langle j_1 j_2 m_1 m_2 \mid jm \rangle$ for positive values of m_2 and for (i) $j_2 = \frac{1}{2}$ (ii) $j_2 = 1$ (iii) $j_2 = \frac{3}{2}$.

(i)	$j = \frac{1}{2}$	$m_2 = \frac{1}{2}$
	$j = j_1 + \frac{1}{2}$	$\left[\dfrac{j_1 + m + \frac{1}{2}}{2j_1 + 1}\right]^{1/2}$
	$j = j_1 - \frac{1}{2}$	$\left[\dfrac{j_1 - m + \frac{1}{2}}{2j_1 + 1}\right]^{1/2}$

(ii)	$j_2 = 1$	$m_2 = 1$	$m_2 = 0$
	$j = j_1 + 1$	$\left[\dfrac{(j_1 + m)(j_1 + m + 1)}{(2j_1 + 1)(2j_1 + 2)}\right]^{1/2}$	$\left[\dfrac{(j_1 - m + 1)(j_1 + m + 1)}{(2j_1 + 1)(j_1 + 1)}\right]^{1/2}$
	$j = j_1$	$-\left[\dfrac{(j_1 + m)(j_1 - m + 1)}{2j_1(j_1 + 1)}\right]^{1/2}$	$-\left[\dfrac{m^2}{j_1(j_1 + 1)}\right]^{1/2}$
	$j = j_1 - 1$	$-\left[\dfrac{(j_1 - m)(j_1 - m + 1)}{2j_1(2j_1 + 1)}\right]^{1/2}$	$-\left[\dfrac{(j_1 - m)(j_1 + m)}{j_1(2j_1 + 1)}\right]^{1/2}$

Contd...

Contd...

Contd....

(iii)	$j_2 = \dfrac{3}{2}$	$m_2 = \dfrac{3}{2}$	$m_2 = \dfrac{1}{2}$
	$j = j_1 + \dfrac{3}{2}$	$\left[\dfrac{(j_1+m-\frac{1}{2})(j_1+m+\frac{1}{2})(j_1+m+\frac{3}{2})}{(2j_1+1)(2j_1+2)(2j_1+3)}\right]^{\frac{1}{2}}$	$\left[\dfrac{3(j_1+m+\frac{1}{2})(j_1+m+\frac{1}{2})(j_1-m+\frac{3}{2})}{(2j_1+1)(2j_1+2)(2j_1+3)}\right]^{\frac{1}{2}}$
	$j = j_1 + \dfrac{1}{2}$	$-\left[\dfrac{3(j_1+m-\frac{1}{2})(j_1+m+\frac{1}{2})(j_1-m+\frac{3}{2})}{2j_1(2j_1+1)(2j_1+3)}\right]^{\frac{1}{2}}$	$-\left[\dfrac{(j_1-3m+\frac{3}{2})^2(j_1+m+\frac{1}{2})}{2j_1(2j_1+1)(2j_1+3)}\right]^{\frac{1}{2}}$
	$j = j_1 - \dfrac{1}{2}$	$\left[\dfrac{3(j_1+m-\frac{1}{2})(j_1-m+\frac{1}{2})(j_1-m+\frac{3}{2})}{(2j_1-1)(2j_1+1)(2j_1+2)}\right]^{\frac{1}{2}}$	$-\left[\dfrac{(j_1+2m-\frac{1}{2})^2(j_1-m+\frac{1}{2})}{(2j_1-1)(2j_1+1)(2j_1+2)}\right]^{\frac{1}{2}}$
	$j = j_1 - \dfrac{3}{2}$	$-\left[\dfrac{(j_1-m-\frac{1}{2})(j_1-m+\frac{1}{2})(j_1-m+\frac{3}{2})}{2j_1(2j_1-1)(2j_1+1)}\right]^{\frac{1}{2}}$	$\left[\dfrac{3(j_1+m-\frac{1}{2})(j_1-m-\frac{1}{2})(j_1-m+\frac{1}{2})}{2j_1(2j_1-1)(2j_1+1)}\right]^{\frac{1}{2}}$

Contd...

(iv)	$j_2 = 2$	$m_2 = 2$
	$j = j_1 + 2$	$\left[\dfrac{(j_1-m+1)(j_1+m)(j_1+m+1)(j_1+m+2)}{(2j_1+1)(2j_1+2)(2j_1+3)(2j_1+4)}\right]^{\frac{1}{2}}$
	$j = j_1 + 1$	$-\left[\dfrac{(j_1+m-1)(j_1+m)(j_1+m+1)(j_1-m+2)}{2j_1(j_1+1)(j_1+2)(2j_1+1)}\right]^{\frac{1}{2}}$
	$j = j_1$	$-\left[\dfrac{3(j_1+m-1)(j_1+m)(j_1-m+1)(j_1-m+2)}{(2j_1-1)2j_1(j_1+1)(2j_1+3)}\right]^{\frac{1}{2}}$
	$j = j_1 - 1$	$-\left[\dfrac{(j_1+m-1)(j_1-m)(j_1-m+1)(j_1-m+2)}{2j_1(j_1-1)(j_1+1)(2j_1+1)}\right]^{\frac{1}{2}}$
	$j = j_1 - 2$	$\left[\dfrac{(j_1-m+1)(j_1-m)(j_1-m+1)(j_1-m+2)}{(2j_1-2)(2j_1-1)2j_1(2j_1+1)}\right]^{\frac{1}{2}}$

Contd...

Contd...

Contd...

(v)		$m_2 = 1$
	$j_2 = 2$	
$j = j_1 + 2$		$\left[\dfrac{(j_1-m+2)(j_1+m+2)(j_1+m+1)(j_1+m)}{(2j_1+1)(j_1+1)(2j_1+3)(j_1+2)}\right]^{\frac{1}{2}}$
$j = j_1 + 1$		$-\left[\dfrac{(j_1-2m+2)^2(j_1+m)(j_1+m+1)}{2j_1(2j_1+1)(j_1+1)(j_1+2)}\right]^{\frac{1}{2}}$
$j = j_1$		$\left[\dfrac{3(j_1-2m)^2(j_1+1)(j_1+m)}{(2j_1-1)(2j_1+2)(2j_1+3)}\right]^{\frac{1}{2}}$
$j = j_1 - 1$		$\left[\dfrac{(j_1+2m-1)^2(j_1-m+1)(j_1-m)}{(j_1-1)j_1(2j_1+1)(2j_1+2)}\right]^{\frac{1}{2}}$
$j = j_1 - 2$		$-\left[\dfrac{(j_1-m+1)(j_1-m)(j_1-m-1)(j_1+m-1)}{(j_1-1)(2j_1-1)(2j_1+1)j_1}\right]^{\frac{1}{2}}$

Contd...

(vi)	$j_2 = 2$	$m_2 = 0$
	$j = j_1 + 2$	$\left[\dfrac{3(j_1 - m + 2)(j_1 - m + 1)(j_1 + m + 2)}{(2j_1 + 1)(2j_1 + 2)(2j_1 + 3)(j_1 + 2)}\right]^{\frac{1}{2}}$
	$j = j_1 + 1$	$\left[\dfrac{3m^2(j_1 - m + 1)(j_1 + m + 1)}{j_1(2j_1 + 1)(j_1 + 1)(j_1 + 2)}\right]^{\frac{1}{2}}$
	$j = j_1$	$\left[\dfrac{\{3m^2 - j_1(j_1 + 1)\}^2}{(2j_1 - 1)(2j_1 + 1)j_1(2j_1 + 3)}\right]^{\frac{1}{2}}$
	$j = j_1 - 1$	$-\left[\dfrac{2m^2(j_1 - m)(j_1 + m)}{(j_1 - 1)(2j_1 + 1)j_1(j_1 + 1)}\right]^{\frac{1}{2}}$
	$j = j_1 - 2$	$\left[\dfrac{3(j_1 - m)(j_1 - m - 1)(j_1 + m)(j_1 + m - 1)}{(2j_1 - 2)(2j_1 - 1)(2j_1 + 1)j_1}\right]^{\frac{1}{2}}$

Subject Index

Reader's Notes

Reader's Notes